Mycorrhiza - Function, Diversity, State of the Art

Ajit Varma • Ram Prasad • Narendra Tuteja
Editors

Mycorrhiza - Function, Diversity, State of the Art

Fourth Edition

Springer

Editors
Ajit Varma
Amity Institute of Microbial Technology
Amity University Uttar Pradesh
Noida, Uttar Pradesh
India

Ram Prasad
Amity Institute of Microbial Technology
Amity University Uttar Pradesh
Noida, Uttar Pradesh
India

Narendra Tuteja
Amity Institute of Microbial Technology
Amity University Uttar Pradesh
Noida, Uttar Pradesh
India

ISBN 978-3-319-53063-5   ISBN 978-3-319-53064-2 (eBook)
DOI 10.1007/978-3-319-53064-2

Library of Congress Control Number: 2017940190

Printed on acid-free paper

This Springer imprint is published by Springer Nature
The registered company is Springer International Publishing AG
The registered company address is: Gewerbestrasse 11, 6330 Cham, Switzerland

Foreword

In the first half of the nineteenth century, Justus von Liebig popularized the "Law of the Minimum", stating that plant growth is controlled by the scarcest resource—being the "limiting factor"—rather than the total amount of resources available. Although sunlight and temperature may be such limiting factors, the term mostly refers to mineral nutrients. In view of the recognition of this principle, Liebig highlighted that trace minerals—next to major nutrients—are essential for plant growth and may often be a limiting factor for plant development and even survival. This is particularly critical under conditions where micronutrients are in low supply or even immobilized and, hence, cannot be taken up by the plant roots alone.

Fungi generally inhabit the rhizosphere of vascular plants, where they live off organic material. They access their carbon and nutrient source *via* an extensive network of hyphae with a very large surface area, aiding the decomposition of the organic material in the root zone. Mycorrhizal fungi form a symbiotic association within the plants' rhizosphere. This may occur—mostly as a mutualistic association—either intracellular (arbuscular mycorrhizal fungi) or extracellular (ectomycorrhizal fungi). Due to their large reactive surface area, mycorrhizal fungi are attributed with a significant function in biogeochemical cycles. They play an important role in agricultural and natural ecosystems, where they increase production in the former and sustainability of the ecosystem in the latter case, particularly under low nutrient conditions (Pate and Beard 1984). The extent to which plants depend on mycorrhizae varies, but most plants studies so far show augmented development with mycorrhizal associations. Some orchid species are even facultatively myco-heterotrophic and form a parasitic relationship with mycorrhizal fungi for part of their life cycle. Other myco-heterotrophic plants are totally dependent on mycorrhizal associations, in which case the relationship becomes entirely parasitic in favour of the plant. On example is the ghost plant (*Monotropa uniflora*), which is an herbaceous perennial plant devoid of chlorophyll (Yang and Pfister 2006). Due to its lack of photosynthetic capacity, it is myco-heterotroph on mycorrhizal fungi associated with certain trees (mostly beech trees), providing the energy for the fungus and ultimately also for its parasite.

In symbiotic plant associations with mycorrhizae, the benefit for the fungus is generally attributed to the access to photosynthetically produced carbohydrates, translocated from the plant's leaves to the roots and its fungal associate. The plant—on the other hand—benefits from the high absorptive capacity of the large surface area of the hyphal system (the mycelium) for mineral elements and even water. The fungal hyphae are much finer and longer than the plants roots, allowing access and direct contact to a larger volume of soil. This improves access to mineral nutrients, particularly also micronutrients, which often can be a limiting factor for plant growth, as the "Law of the Minimum" suggests. The mycelium of micorrhizal fungi can even mobilize nutrients that are physically or chemically immobilized and, hence, cannot be taken up by the plant roots alone. Heavy clay soils are prone to immobilization of certain micronutrients and even phosphate. Under such conditions, mycorrhizal–plant associations may be essential for the survival of the plant. Consequently, one may argue that the plant–mycorrhizal symbiosis is next to the plant–rhizobial symbiosis (performing symbiotic nitrogen fixation) the most important symbiotic system for sustainability and productivity of terrestrial plant systems.

Plant–mycorrhizal associations are found under most environmental conditions on the planet and in a vast number of combinations with varying associated partners—multiple or singular. Consequently, we are still far from understanding the details of the various combinations of relationships of plants with micorrhizae formed in different habitats.

This volume is the fourth in a series of books on mycorrhizae, which on the one hand acknowledges the vastness of the research area and on the other hand provides a condensed insight into the most recent discoveries in the field. The book picks up the points outlined above and exemplarily elucidates them in various settings and environmental conditions. It mainly focuses on natural ecosystems while spanning the bridge to agricultural systems where it is called for. Keeping to a holistic approach, the various chapters explain how recent research results on plant–mycorrhizal associations in combination with new information on mycorrhizal and rhizobial symbionts can help refining existing and new concepts on how such symbiotic systems can augment ecosystem sustainability and vigour. In order to outline the basis for progress in the field, the first chapter of this volume puts mycorrizal research in a historical perspective. The following four chapters employ various angles to focus on ways how the plant–micorrhizal system can generate enhanced nutrient uptake and how bacteria can provide additional benefits. Chapters 6, 7, 9 and 10 cover the role of the symbiosis in essentially undisturbed and disturbed ecosystems, including the role in early succession. The signalling processes in the establishment of rhizobial and mycorrhizal symbiotic endophytes are discussed in Chap. 8, while several specific adaptations of fungi, e.g. Truffle (Chap. 11), and plants, e.g. Grapevine (Chap. 13), or specific environmental conditions, e.g. wetlands (Chap. 14), or hypoxic conditions (Chap. 16) are covered in the second half of the book. Climate change and the question how arbuscular mycorrhizal fungi are affected by this recent, global phenomenon is the topic of

Chap. 15. The final two chapters introduce a new symbiont *Piriformospora indica* (*Serendipita indica*) and outline its large-scale cultivation in a bioreactor.

With the intensive research in the area continuing, this volume presents the current state of the art. As such, it is a valuable reference and basis for future investigations. With more scientific progress in the field, we can look forward to enhanced insights into this important and exciting symbiotic system.

Alexander P. Hansen

References

Pate JS, Beard JS (1984) Kwongan: plant life in the sandplain: biology of a south-west Australian shrubland. University of Western Australia Press, p 284.

Yang S, Pfister DH (2006) *Monotropa uniflora* plants of eastern Massachusetts form mycorrhizae with a diversity of rassulacean fungi. Mycologia 98(4):535–540.

Preface

We have already published mycorrhiza volumes in the years 1991, 1992, 1995, 1999, and 2008. Since then, a lot of new research findings and approaches have been published in literature. The present volume will emphasize the current perspectives of mycorrhiza around the globe. Springer-verlag, Heidelberg, Germany, has invited us to compile three new volumes in quick succession (2017) highlighting the work on mycorrhiza after 2008. Mycorrhizas have been an essential stabilizing factor in the terrestrial ecosystems for centuries. These fungi aid in the productivity of plants via the formation of dynamic associations with plant roots. The symbiotic associations formed are an important subject to evaluate numerous opportunities using modern tools of biotechnology. The possibilities of genetically manipulating these associations have led to optimization of plant productivity in ecosystems with minimal risk of environmental damage. The resourceful management of mycorrhizal associations has the potential to favor the sustainable production of quality foods while ensuring environmental quality for future generations.

The unique associations formed by these fungi have sparked a vast array of interest in mycorrhizal studies. Recent developments in the study of mycorrhizas have encouraged us to present a new book on progress in this field. The fourth edition of the mycorrhiza book gives exemplary insight into the advancements in mycorrhizal studies. It is hoped that this new edition will interest readers in the latest results of mycorrhiza research and also encourage young researchers to prove the challenging field of mycorrhizal studies.

This volume consists of 18 chapters covering the diverse mycorrhizal associations by 46 subject specialists.

We are grateful to the many people who helped us to bring this volume to light. We wish to thank Drs. Jutta Lindenborn, Isabel Ullmann, Man-Thi Tran, and Hanna Hensler-Fritton Springer Heidelberg for generous assistance and patience in continuing the volume. Finally, special thanks go to our families, immediate and extended, not forgetting those who have passed away, for their support or their incentives in putting everything together. Editors in particular are very thankful to Dr. Ashok K. Chauhan, Founder President of the Ritnand Balved Education

Foundation (an umbrella organization of Amity Institutions), New Delhi, for the kind support and constant encouragement received. Special thanks are due to my esteemed faculty colleagues and dear student Ms Diksha Bhola and other technical staff.

Ajit Varma
Ram Prasad
Narendra Tuteja

Contents

1 **Introduction to Mycorrhiza: Historical Development** 1
Ram Prasad, Diksha Bhola, Khalid Akdi, Cristina Cruz, Sairam KVSS, Narendra Tuteja, and Ajit Varma

2 **Mobilization of Micronutrients by Mycorrhizal Fungi** 9
Priyanku Teotia, Manoj Kumar, Ram Prasad, Vivek Kumar, Narendra Tuteja, and Ajit Varma

3 **Soil: Do Not Disturb, Mycorrhiza in Action** 27
Cristina Cruz, Alessandro Ramos, Olubukola Oluranti Babalola, Hessini Kamel, Teresa Dias, and Ajit Varma

4 **Mycorrhiza: Creating Good Spaces for Interactions** 39
Geetanjali Manchanda, Raghvendra Pratap Singh, Zhi Feng Li, and Jun Jie Zhang

5 **Mycorrhizal Helper Bacteria: Sustainable Approach** 61
Devendra K. Choudhary, Ajit Varma, and Narendra Tuteja

6 **Mycorrhization of Fagaceae Forests Within Mediterranean Ecosystems** 75
Francisca Reis, Rui M. Tavares, Paula Baptista, and Teresa Lino-Neto

7 **Ectomycorrhizal Mushrooms: Their Diversity, Ecology and Practical Applications** 99
Rohit Sharma

8 **Plant Flavonoids: Key Players in Signaling, Establishment, and Regulation of Rhizobial and Mycorrhizal Endosymbioses** 133
Priyanka Singla and Neera Garg

9 **Mycorrhizas in Forest Tree Health** 177
Vivek Kumar, Manoj Kumar, Ram Prasad, Narendra Tuteja, and Ajit Varma

10 Ectomycorrhizal Fungi: A Major Player in Early Succession 187
Izabela L. Kałucka and Andrzej M. Jagodziński

11 Truffle Ecology: Genetic Diversity, Soil Interactions and Functioning 231
Antonietta Mello, Elisa Zampieri, and Alessandra Zambonelli

12 Inter- and Intraspecific Fungal Diversity in the Arbuscular Mycorrhizal Symbiosis 253
Brandon Monier, Vincent Peta, Jerry Mensah, and Heike Bücking

13 Arbuscular Mycorrhizal Fungi and Dark Septate Endophytes in Grapevine: The Potential for Sustainable Viticulture? 275
M. Likar and M. Regvar

14 What Have We Learnt from Studying Mycorrhizal Colonisation of Wetland Plant Species? 291
Alenka Gaberščik, Nataša Dolinar, Nina Šraj, and Marjana Regvar

15 Response of Arbuscular Mycorrhizal Fungi to Global Climate Change and Their Role in Terrestrial Ecosystem C and N Cycling 305
Bhoopander Giri and Bhawna Saxena

16 Arbuscular Mycorrhizal Fungi in Hypoxic Environments 329
Irena Maček

17 *Piriformospora indica* (*Serendipita indica*): The Novel Symbiont . . . 349
Uma Singhal, Ram Prasad, and Ajit Varma

18 Mass Cultivation of Mycorrhiza-Like Fungus *Piriformospora indica* (*Serendipita indica*) by Batch in Bioreactor 365
Uma Singhal, Manpreet Kaur Attri, and Ajit Varma

Index 385

Contributors

Khalid Akdi AMC Chemical, Poligono Industrial La Isla, Dos Hermanas, Sevilla, Spain

Olubukola Oluranti Babalola Food Security and Safety Niche, Faculty of Agriculture, Science and Technology, North-West University, Mmabatho, South Africa

Paula Baptista REQUIMTE - School of Agriculture, Polytechnic Institute of Bragança, Bragança, Portugal

Diksha Bhola Amity Institute of Microbial Technology (AIMT), Amity University Uttar Pradesh, Noida, Uttar Pradesh, India

Heike Bücking Biology and Microbiology Department, South Dakota State University, Brookings, SD, USA

Devendra K. Choudhary Amity Institute of Microbial Technology (AIMT), Amity University Uttar Pradesh, Noida, Uttar Pradesh, India

Cristina Cruz CE3C—Centre for Ecology, Evolution and Environmental Changes, Faculdade de Ciências da Universidade de Lisboa, Lisbon, Portugal

Teresa Dias Center for Ecology, Evolution and Environmental Changes (Ce3C), Faculty of Sciences, Universidade de Lisboa, Lisbon, Portugal

Nataša Dolinar Biotechnical Faculty, Department of Biology, University of Ljubljana, Ljubljana, Slovenia

Alenka Gaberščik Biotechnical Faculty, Department of Biology, University of Ljubljana, Ljubljana, Slovenia

Neera Garg Department of Botany, Panjab University, Chandigarh, India

Bhoopander Giri Department of Botany, Swami Shraddhanand College, University of Delhi, Delhi, India

Andrzej M. Jagodziński Polish Academy of Sciences, Institute of Dendrology, Kórnik, Poland

Faculty of Forestry, Department of Game Management and Forest Protection, Poznań University of Life Sciences, Poznań, Poland

Izabela L. Kałucka Faculty of Biology and Environmental Protection, Department of Algology and Mycology, University of Łódź, Łódź, Poland

Hessini Kamel Faculty of Science, Biology Department, Taif University, Taif, Saudi Arabia

Manpreet Kaur Attri Amity Institute of Microbial Technology (AIMT), Amity University Uttar Pradesh, Noida, Uttar Pradesh, India

Manoj Kumar Amity Institute of Microbial Technology (AIMT), Amity University Uttar Pradesh, Noida, Uttar Pradesh, India

Vivek Kumar Amity Institute of Microbial Technology (AIMT), Amity University Uttar Pradesh, Noida, Uttar Pradesh, India

Zhi Feng Li State Key Laboratory of Microbial Technology, College of Life Science, Shandong University, Jinan, Shandong Province, People's Republic of China

M. Likar Biotechnical Faculty, Department of Biology, University of Ljubljana, Ljubljana, Slovenia

Teresa Lino-Neto BioSystems and Integrative Sciences Institute (BioISI), Plant Functional Biology Centre, University of Minho, Braga, Portugal

Irena Maček Biotechnical Faculty, Department of Agronomy, University of Ljubljana, Ljubljana, Slovenia

Faculty of Mathematics, Natural Science and Information Technologies, University of Primorska, Koper, Slovenia

Geetanjali Manchanda Department of Botany, D.A.V. University, Jalandhar, Punjab, India

Antonietta Mello Istituto per la Protezione Sostenibile delle Piante, SS Torino - CNR, Torino, Italy

Jerry Mensah Biology and Microbiology Department, South Dakota State University, Brookings, SD, USA

Brandon Monier Biology and Microbiology Department, South Dakota State University, Brookings, SD, USA

Rekha Pandey Department of Botany, Panjab University, Chandigarh, India

Vincent Peta Biology and Microbiology Department, South Dakota State University, Brookings, SD, USA

Ram Prasad Amity Institute of Microbial Technology (AIMT), Amity University Uttar Pradesh, Noida, Uttar Pradesh, India

Alessandro Ramos Laboratory of Biochemistry and Physiology of Microorganisms, Universidade Estadual do Norte Fluminense (UENF), Campos dos Goitacazes, Brazil

Marjana Regvar Biotechnical Faculty, Department of Biology, University of Ljubljana, Ljubljana, Slovenia

Francisca Reis BioSystems and Integrative Sciences Institute (BioISI), Plant Functional Biology Centre, University of Minho, Braga, Portugal

Sairam KVSS Prathista Biotech Limited, Secundrabad, Andhara Pradesh, India

Bhawna Saxena Department of Botany, Swami Shraddhanand College, University of Delhi, Delhi, India

Rohit Sharma Microbial Culture Collection (MCC), National Centre for Cell Science, S.P. Pune University Campus, Pune, Maharashtra, India

Raghvendra Pratap Singh State Key Laboratory of Microbial Technology, College of Life Science, Shandong University, Jinan, Shandong Province, People's Republic of China

Uma Singhal Amity Institute of Microbial Technology (AIMT), Amity University Uttar Pradesh, Noida, Uttar Pradesh, India

Nina Šraj Biotechnical Faculty, Department of Biology, University of Ljubljana, Ljubljana, Slovenia

Rui M. Tavares BioSystems and Integrative Sciences Institute (BioISI), Plant Functional Biology Centre, University of Minho, Braga, Portugal

Priyanku Teotia Laboratory of Microbiology, Department of Botany, CCS University, Meerut, India

Narendra Tuteja Amity Institute of Microbial Technology (AIMT), Amity University Uttar Pradesh, Noida, Uttar Pradesh, India

Ajit Varma Amity Institute of Microbial Technology (AIMT), Amity University Uttar Pradesh, Noida, Uttar Pradesh, India

Alessandra Zambonelli Dipartimento di Scienze Agrarie (DipSA), Alma Mater StudiorumUniversità di Bologna, Bologna, Italy

Elisa Zampieri Dipartimento di Scienze Agrarie, Forestali e Alimentari (DISAFA), Università di Torino, Grugliasco, TO, Italy

Present Address: Rice Research Unit, CREA, Council for Agricultural Research and Economics, Vercelli, Italy

Jun Jie Zhang College of Food and Bioengineering, Zhengzhou University of Light Industry, Zhengzhou, Henan Province, China

Chapter 1
Introduction to Mycorrhiza: Historical Development

Ram Prasad, Diksha Bhola, Khalid Akdi, Cristina Cruz, Sairam KVSS, Narendra Tuteja, and Ajit Varma

Abstract Arbuscular mycorrhizal fungi (AMF) are vital component of natural ecosystem, being gifted to form symbiont with plant roots. AM fungi have mutualistic relationships with more than 80% of terrestrial plant species. Because of wide range of relationships with host plants, it becomes difficult to identify the species on the morphological bases as the spores are to be extracted from the soil. In spite of their abundance and wide range of relationship with plant species, AMF have shown low species diversity. AMF have high functional diversity because different combinations of host plants and AMF have different effects on the numerous aspects of symbiosis. Recent fossil evidence has dated the appearance of arbuscular mycorrhizae back to 460 million years, preexisting vascular plants. These studies benefit the paleoecological importance of mycorrhizae and enhancement to our understanding of the evolution of mutualisms.

1.1 Introduction

Arbuscular mycorrhiza (AM) association is the most widespread plant–fungal symbiosis on earth, in the majority of extant land plants, and can be traced back to the Early Devonian, more than 400 million years ago, which corresponds to the

R. Prasad (✉) • D. Bhola • N. Tuteja • A. Varma
Amity Institute of Microbial Technology (AIMT), Amity University Uttar Pradesh, Block 'E-3', 4th Floor, Sector-125, Noida, Gautam Buddha Nagar, Uttar Pradesh 201313, India
e-mail: rprasad@amity.edu

K. Akdi
AMC Chemical, Poligono Industrial La Isla, Rio Viejo 40, 41703 Dos Hermanas, Sevilla, Spain

C. Cruz
Faculty of Sciences, Center for Ecology, Evolution and Environmental Changes (Ce3C), Universidade de Lisboa, Lisboa, Portugal

S. KVSS
Prathista Biotech Limited, Vaishnavi Bhawan 1-5-1015/80&81, Manzeera Colony, Old Alwal, Secundrabad, Andhra Pradesh, India

A. Varma et al. (eds.), *Mycorrhiza - Function, Diversity, State of the Art*,
DOI 10.1007/978-3-319-53064-2_1

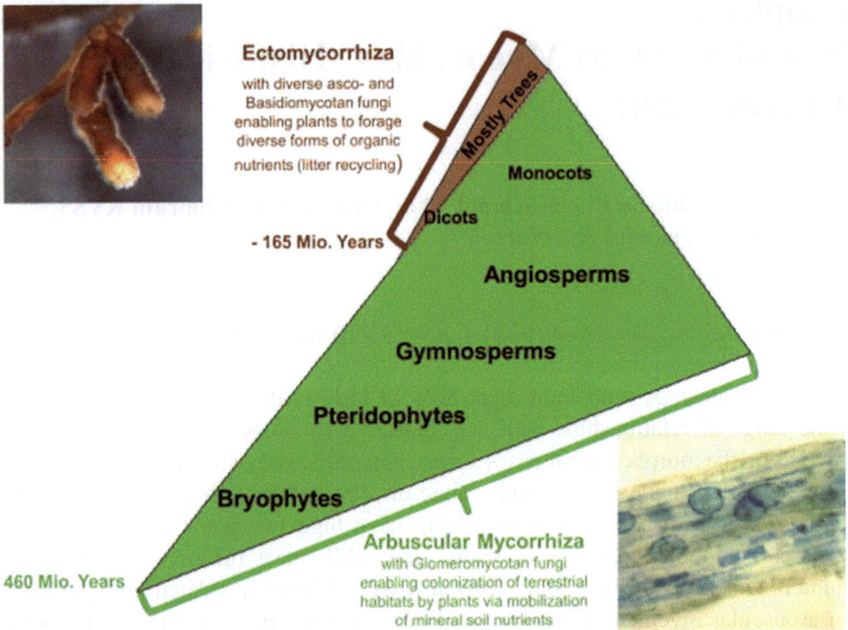

Fig. 1.1 Summary of the natural historical development of arbuscular mycorrhizal (AM) and ectomycorrhizal (EM) symbioses in relation to the ecological radiation of terrestrial plants (Reprinted with permission from Buscot 2015)

onset of the colonization of terrestrial habitats by plants (Fig. 1.1) (Remy et al. 1994; Schwendemann et al. 2011; Buscot 2015). Presently, the majority of land plants have mutually beneficial symbiotic relationship with a fungal partner in which the plant normally receives soil nutrients in exchange for photosynthates. However, this understanding is based almost exclusively on investigations of the AM associations of higher plants, whereas virtually nothing is known about their functionality in the "lower," more evolutionarily ancient land plant clades (liverworts, clubmosses, and ferns). There is evidence that AM is an evolutionary ancient partnership that commenced with the earliest land plants, but the absence of information on the functioning of the symbiosis in lower plants is a major gap in our understanding of its significance in driving colonization of land plant, and the costs–benefits of plant–fungal co-evolution in lower plants are mysterious.

1.2 Structure of Mycorrhizal Fungi

Frank (1885) was probably the first to identify the extensive nature of associations between plant roots and mycorrhizal fungi (Frank and Trappe 2005). Four major mycorrhizal types have been described based on their structure and function,

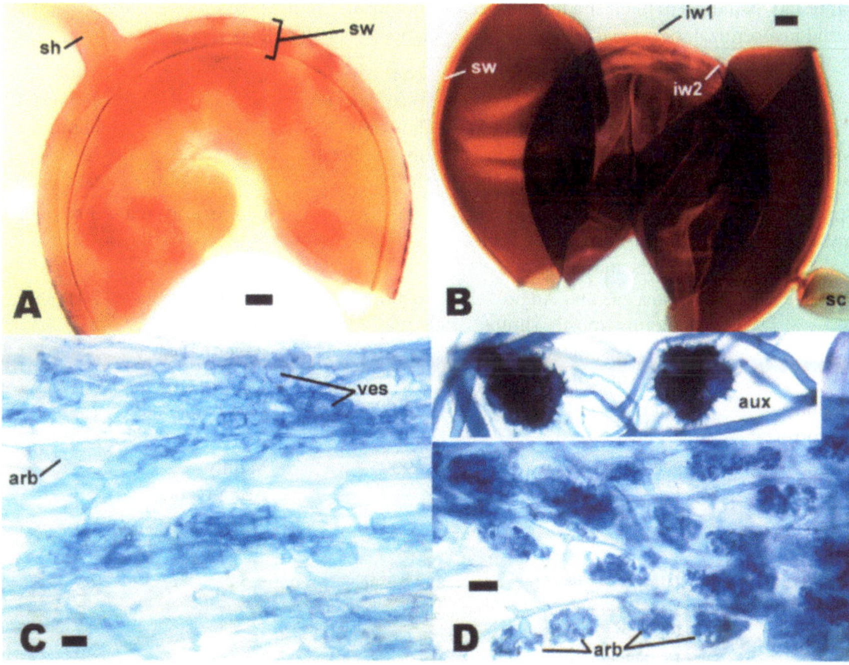

Fig. 1.2 Examples of spores and colonization by arbuscular mycorrhizal fungi: (**a**) Subcellular structure of a *Glomus clarum* spore broken and mounted in Melzer's reagent. (**b**) Subcellular structure of a *Scutellospora pellucida* spore broken and mounted in Melzer's reagent. (**c**) Typical mycorrhizae of *Acaulospora morrowiae* stained in 0.05% trypan blue. (**d**) Typical mycorrhizae of *Gigaspora rosea* stained in 0.05% trypan blue; *inset* shows auxiliary cells. Abbreviations: *sw* spore wall; *sh* subtending hyphae; *sc* sporogenous cell; *iw1* first inner wall; *iw2* second flexible inner wall; *arb* arbuscule; *ves* vesicle; *aux* auxiliary cells. *Scale bar* = 20 μm (Reprinted with permission from Bever et al. 2001)

namely, arbuscular mycorrhiza (AM), ectomycorrhiza (EM), orchid mycorrhiza, and ericoid mycorrhiza (Fig. 1.2 and Table 1.1).

Today, despite the large number of plant species forming AM associations worldwide, two types of arbuscular mycorrhizal (AM) associations, the Arum type and the Paris type, have been recognized based on morphological characteristics of the colonization process. In the Arum type, the fungal symbiont spread in the root cortex via intercellular hyphae. Short side branches penetrate the cortex cells and produce arbuscules. The Arum type (linear) is commonly described in fast-growing root systems of crop plants. In the Paris type (coiled), the hyphae grow intracellular coils and spread directly from cell to cell inside the cortex. Co-occurrence of Arum- and Paris-type morphology of AM is found in cucumber and tomato (Kubota et al. 2005).

Arbuscules are relatively short-lived, at least in the Arum-type mycorrhiza, and the hyphae are comparatively long-lived (Holley and Peterson 1979; Smith and Dickson 1991). The arbuscules gradually degenerate, while the plant cell remains

Table 1.1 Numbers of plant and fungal species forming arbuscular mycorrhizal, ectomycorrhizal, orchid mycorrhizal, or ericoid mycorrhizal associations (van der Heijden et al. 2015)

Mycorrhizal type	Major groups of plants	Number of plant species hosting mycorrhizal fungi	Fungal identity	Total estimated number of fungal taxa
Arbuscular mycorrhiza	Most herbs, grasses and many trees, many hornworts and liverworts	200,000	Glomeromycota	300–1600
Ectomycorrhiza	Pinaceae and Angiosperms (mostly shrubs and trees, mostly temperate), some liverworts	6000	Basidiomycota and Ascomycota	20,000
Orchid mycorrhiza	Orchids	20,000–35,000	Basidiomycota	25,000
Ericoid mycorrhiza	Members of the Ericaceae, some liverworts	3900	Mainly Ascomycota, some Basidiomycota	>150
Nonmycorrhizal plant species	Brassicaceae, Crassulaceae, Orobanchaceae, Proteaceae, etc.	51,500	–	0

alive, which is a difference related to many plant pathogenic fungi which cause plant cell death. For rapidly growing crop species, the formation of arbuscules may take 2–3 days and the whole arbuscular cycle may take around 1 week.

Mycorrhizal fungi live inside the cortex of plant roots, on the surface of the root, or around the epidermal cells of the root. The hyphae of these fungi also grow out from the roots into the soil where they forage for nutrients that are limiting to plant growth, especially nitrates and phosphates, but organically bound nutrients are also acquired by some mycorrhizal types (e.g., EM and ericoid mycorrhizal fungi) (Read and Perez-Moreno 2003). These nutrients as well as other benefits are then delivered to their host plants in return for carbohydrates (Smith and Read 2008). Consequently, the mycorrhizal symbiosis exerts a strong influence on plant growth and fitness.

1.3 Taxonomy of Arbuscular Mycorrhizal Fungi

Taxonomy identifies and describes names, generating tools for taxonomic identification of fungi (species lists, descriptions of formally delimited species, and identification keys, among others). Arbuscular mycorrhizal fungi (AMF) are characterized by the development of branching structures called arbuscules inside the cortical cells of roots (Fig. 1.2). Arbuscules increase the contact area between plant root and fungus and are thought to be the primary sites of exchange of the plant's carbon for the fungus's phosphorus. One suborder of these fungi, Glomineae, also forms vesicles, or sack-like reservoirs, within plant cortical cells (Fig. 1.2c). The presence of the fungi (members of the Glomeromycota) within the root assists the uptake of nutrients and water and may also have a protective effect against soil pathogens.

AMF are believed to propagate via infective hyphae, hyphal fragments, or asexual spores (Fig. 1.3). A general life history begins with colonization of a root and the development of arbuscules from branch hyphae within the root. Hyphae may extend from one infected root to another or from an infected root to the root of another plant. Spores form in the root cortex or in the soil. These spores may be dormant for a period, but they will eventually germinate and colonize another root. Spores may be dispersed away from the site in which they were formed. Viable spores are generally short-lived (exceptions: some spores of *Acaulospora* species), and viability is limited by dormancy, susceptibility to pathogens, and some

Fig. 1.3 Spores of arbuscular mycorrhizal fungi: the central picture is a collective spore from nine species of AM fungi. Around this central photo, we have arranged pictures of individual species. Starting in the *upper left* corner and moving clockwise around the central photo, these species are *Scutellospora calospora, S. pellucida, S. heterogama, Archaeospora trappei, Gigaspora gigantea, Gi. rosea, Acaulospora colossica*, and *Ac. morrowiae*. *Scale bar* = 200 μm (Reprinted with permission from Bever et al. 2001)

supplementary factors. Although the morphology and architecture of external hyphae and internal mycorrhizal structures can differ between families of AM fungi (e.g., Fig. 1.2c, d), there are few differences between species to species within the same genus. Therefore, taxonomy of these fungi is based on the discrete characters of the spore subcellular structure, which can vary from simple to very complex for a single multinucleate cell (e.g., Fig. 1.2a, b; Morton 1988; Morton and Bentivenga 1994). On the basis of characteristic features of spore wall and spore ontogeny, AMF are grouped into genera that include approximately 145 species described to date. Undoubtedly, the majority of AM fungal species remain undefined. The International Culture Collection of Arbuscular and Vesicular Arbuscular Mycorrhizal Fungi (INVAM) at West Virginia University, for example, currently maintain approximately 40 isolates that do not belong to currently described species (Morton et al. 1993). For more information on this collection, see the INVAM Web site at http://invam.caf.wvu.edu.

1.4 Conclusion and Future Prospects

A major development have been through in the arena of mycorrhizal research after the discovery that mycorrhizal associations are abundant and important for plant nutrition. Currently, more than 100 years later, the ecological function of symbiosis is much better implicit, the biodiversity and evolution of this symbiosis is no longer a black box, genomes of a widespread array of mycorrhizal fungi have been sequenced, and molecular and nanoscale level interactions establishing the symbiosis are starting to be revealed.

Acknowledgment Ajit Varma is thankful to Department of Science and Technology and Department of Biotechnology for partial financial funding and to DST-FIST for providing confocal microscope facility.

References

Bever JD, Schultz PA, Pringle A, Mortan JB (2001) Arbuscular mycorrhizal fungi: more diverse than meets the eye, and the ecological tale of why. BioScience 51:923–931

Buscot F (2015) Implication of evolution and diversity in arbuscular and ectomycorrhizal symbioses. J Plant Physiol 172:55–61

Frank AB (1885) Ueber die auf Wurzelsymbiose beruhende Ernaehrung gewisser Baeume durch unterirdische Pilze. Ber Dtsch Bot Ges 3:128–145

Frank AB, Trappe JM (2005) On the nutritional dependence of certain trees on root symbiosis with belowground fungi (an English translation of A.B. Frank's classic paper of 1885). Mycorrhiza 15:267–275

Holley JD, Peterson RL (1979) Development of a vesicular-arbuscular mycorrhiza in bean roots. Can J Bot 57:1960–1978

Kubota M, McGonigle TP, Hyakumachi M (2005) Co-occurrence of Arum- and Paris-type morphologies of arbuscular mycorrhizae in cucumber and tomato. Mycorrhiza 15:73–77

Morton JB (1988) Taxonomy of VA mycorrhizal fungi: classification, nomenclature, and identification. Mycotaxon 32:267–324

Morton JB, Bentivenga SP (1994) Levels of diversity in endomycorrhizal fungi (Glomales, Zygomycetes) and their role in defining taxonomic and non-taxonomic groups. Plant Soil 159:47–59

Morton JB, Bentivenga SP, Wheeler WW (1993) Germ plasm in the International Collection of Arbuscular Mycorrhizal Fungi (INVAM) and procedures for culture development, documentation and storage. Mycotaxon 48:491–528

Read DJ, Perez-Moreno J (2003) Mycorrhizas and nutrient cycling in ecosystems—a journey towards relevance? New Phytol 157:475–492

Remy W, Taylor TN, Hass H, Kerp H (1994) Four hundred-million-year-old vesicular arbuscular mycorrhizae. Proc Natl Acad Sci USA 91:11841–11843

Schwendemann AB, Decombeix A-L, Taylor TN, Taylor EL, Krings M (2011) Morphological and functional stasis in mycorrhizal root nodules as exhibited by a Triassic conifer. Proc Natl Acad Sci USA 108:13630–13634

Smith SE, Dickson S (1991) Quantification of active vesicular-arbuscular mycorrhizal infection using image analysis and other techniques. Aust J Plant Physiol 18:637–648

Smith SE, Read DJ (2008) Mycorrhizal symbiosis, 3rd edn. Academic Press, Amsterdam

van der Heijden MG, Martin FM, Selosse MA, Sanders IR (2015) Mycorrhizal ecology and evolution: the past, the present, and the future. New Phytol 205:1406–1423

Chapter 2
Mobilization of Micronutrients by Mycorrhizal Fungi

Priyanku Teotia, Manoj Kumar, Ram Prasad, Vivek Kumar, Narendra Tuteja, and Ajit Varma

Abstract Mycorrhizal fungus constitutes heterogeneous fungal taxa embracing an array of plant species. This group is found allied with the roots of beyond 90% of the plant species in this world. There is a range of mycorrhizal associations, among which arbuscular and ectotrophic mycorrhizal interactions are of high biological and economic significance. This chapter gives details about habitation, host range, and structural components of these mycorrhizal groups, along with a meticulous discussion on the mineral absorption, mechanisms involved in different absorption pathways. In addition to enhancement of mineral nutrient uptake by plants in soil, several mycorrhizal fungi execute an important task in mobilizing mineral nutrients from inaccessible organic substrate, mineral particles, and rock surfaces. Mycorrhizal fungi adopt various methods to achieve the purpose effectively, like greater area of absorption for the roots of plant, liberation of biochemical compounds, and consortium with different microbes. Furthermore, mycorrhizal fungi also provide an imperative C sink in soil other than mobilizing nutrients, consequently playing an important role in the cycling of these mineral elements. The role of every partner in a mycorrhizal association is to be exposed by the application of molecular and genetic tools, coupled with high-throughput sequencing and advanced microscopy. The signaling pathways between plants and fungi have recently been elucidated, and recognition of a range of novel nutrient transporters has unveiled a number of cellular processes which are fundamental to the mycorrhizal symbiosis. Various transporters, particularly proton-coupled phosphate transporters, have been documented on both the fungal and plant membranes which contribute to transmission of phosphate from fungi to plants. Even though much work has been formerly done on several aspects, such as symbioses, the extent to which these are functionally essential in agriculture remains uncertain. It is a vital need to spotlight on the questions, whose answers will offer novel perspectives on mycorrhizal utility.

Priyanku Teotia and Manoj Kumar: equal contributor.

P. Teotia
Laboratory of Microbiology, Department of Botany, CCS University, Meerut, India

M. Kumar • R. Prasad • V. Kumar (✉) • N. Tuteja • A. Varma
Amity Institute of Microbial Technology (AIMT), Amity University Uttar Pradesh, Block 'E-3', 4th Floor, Sector-125, Noida, Gautam Buddha Nagar, Uttar Pradesh 201313, India
e-mail: vkumar51@amity.edu; vivekbps@gmail.com

A. Varma et al. (eds.), *Mycorrhiza - Function, Diversity, State of the Art*,
DOI 10.1007/978-3-319-53064-2_2

2.1 Introduction

Even the modest things are vital to the world particularly in relation to getting plants established. Under natural environmental condition, plants remain in close association with soil microorganisms called mycorrhizal fungi. The mycorrhizal fungi inhabit plant roots and extend the root system into the adjoining soil. Unexpected quantities of mycorrhizal filaments are found available in healthy soil. An extremely small section of soil associated with dynamically growing plants may be full of numerous fungal filaments. The affiliation is favorable for the reason that the plants have the benefit of improved uptake of water and mineral nutrient, resistance against diseases, greater survival, and enhanced growth.

The term "mycorrhiza" was coined by the German scientist, A. B. Frank, about a century ago. Factually, word "Mycorrhiza" stands for fungus root; however, it is a symbiotic association existing among a group of soil fungi and the roots of higher plant (Habte 2000). This is a mutualistic organization that depicts the bidirectional interaction and exchange of resources across the mycorrhizal interface. In this association, the mycorrhizal fungus provides the host plant with mineral nutrients, like phosphate and nitrogen, and amplifies the abiotic stress tolerance against conditions of drought, salinity, and heavy metal and biotic stress resistance from various root pathogens, and in return, the host plant transports about 4–20% of its photosynthetic product, i.e., carbon compound to the mycorrhizal fungus (Wright et al. 1998). Records of fossil study indicate that mycorrhizal association commenced about 400–450 million years back, and these mycorrhizal interactions played a significant role in colonization of land by the plants (Smith and Read 2008). Even though mycorrhiza came to light nearly 100 years back, their significance in enhancing plant productivity did not get appropriate credit until past 50 years, until molecular biology got highly developed and gave an insight into the mode of action of mycorrhizal fungi. Presently, numerous scientists around the globe are engaged in study of the mycorrhizal interactions, and any research on plant productivity can hardly be considered as complete, without inclusion of mycorrhizal associations (Habte 2000). About 90% of the identified land plant species formulate mycorrhizal relationship with the ubiquitous fungi in soil (Bonfante and Genre 2010). In dissimilarity with the reciprocal beneficial mycorrhizal association, several mycoheterotrophic plants, nearly 400 species from diverse plant families of bryophytes, pteridophytes, and angiosperms, rely on mycorrhizal fungi to fulfill their carbon need. Such plants lose their photosynthetic efficacy and become parasitic on mycorrhizal fungi coupled with adjacent autotrophic plant (Bücking et al. 2012).

In the following chapter, the main prominence is given to mutually beneficial ectotrophic and arbuscular mycorrhizal associations, since they have great ecological and economic implications (Marschner and Dell 1994). In environmentally sustainable agriculture, arbuscular mycorrhizal fungi can be regarded as "biofertilizer and bioprotector" owing to their capability to colonize and facilitate ample variety of food and cash crops. In contrast, ectomycorrhizal fungi colonize a smaller number of plant species and operate as symbiotic cohort of trees and shrubs; these play a foremost role in the forest ecosystem (Finlay 2008) and could be a fundamental element in phytoremediation as well as revegetation purposes (Bücking 2011; Giri et al. 2005).

2.2 Occurrence and Host Specificity of Mycorrhizal Fungi

AM fungi in general belong to six genera from the class azygosporous zygomycetes. While ectomycorrhizal fungi largely belong to the class basidiomycetes, a few belong to the class zygosporic zygomycetes and ascomycetes. AM are remarkably proficient in mobilizing the inorganic phosphorus (P) and thus prevail well in temperate and arid climates where P is frequently a limiting factor. The AM associations subsist in a large variety of tropical and temperate tree species, as they are not much specific in forming association with the host plant species (Bücking et al. 2002). These associations are known to rarely exist in members of the plants belonging to families Amaranthaceae, Pinaceae, Betulaceae, Brassicaceae, Chenopodiaceae, Cyperaceae, Juncaceae, Proteaceae, and Polygonaceae. As compared to AM, ectomycorrhizal fungi are more efficient in captivating N, and these are more frequent in boreal zone as well as in the temperate zone with high humidity, as the occurrence of low temperature with high humidity promotes the accretion of organic matter, reduced pH, and less N availability (Kilpeläinena et al. 2016). Ectomycorrhizal fungi interact with reasonably lesser section of all plant species, probably just about 3%; however, this 3% embodies almost all of trees of the temperate and boreal forests (especially the plant species of family Fagaceae and Pinaceae); therefore, it can be said that the majority of the forests (in terms of land area) on the surface of earth are dependent on ectomycorrhizal fungi (Habte 2000; Smith and Read 2008; Bonfante and Genre 2010).

2.3 Variation in Structure Among Mycorrhizal Fungi

Mycorrhizal fungi form a variety of associations with the plants; among these, endomycorrhizal association of the arbuscular (AM) type and ectomycorrhizal (ECM) associations have a greater economic and ecological importance. Arbuscular mycorrhizal (AM) and ECM associations differ in their structural aspects as well as the plant and fungal species these embrace (Fig. 2.1 and Table 2.1).

In the ECM, the fungal hyphae infringe the cortex section in the root of the host plant but do not penetrate the cortical cells. The ECM forms hyphal network around cortical cells of the root; this hyphal network is known as the "Hartig Net." In addition to Hartig net, the ECM also forms a thick layer of hyphal mat on the surface of roots, known as sheath or mantle; this sheath covers feeder roots. Thus,

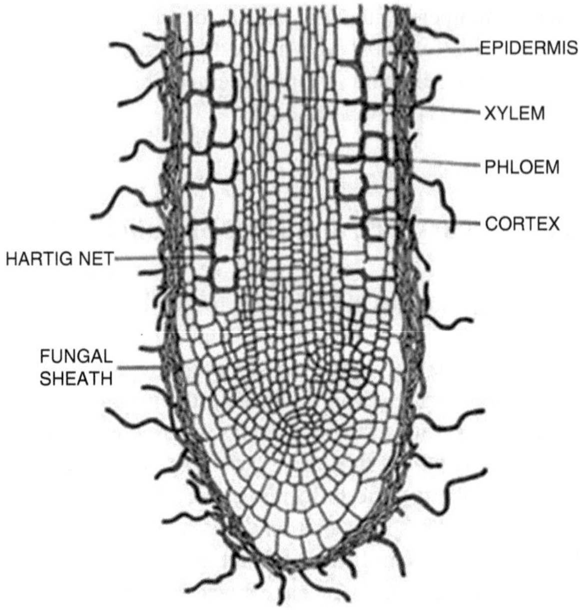

Fig. 2.1 Ectomycorrhizal fungi showing its structure in an infecting root (Source: http://www.biologydiscussion.com/plants/absorption-of-mineral/absorption-of-mineral-salts-by-higher-plant-with-diagram/22764)

Table 2.1 Comparison between AM and ECM associations (Modified after Bücking et al. 2012)

Features	AM fungi association	Ectomycorrhizal fungi association
Transport of nutrients to plant	Specifically important for phosphorus transport, also contribute to nitrogen transport	Specifically important for nitrogen Transport but also have significant Contribution in P transport
Occurrence of fungi	Mainly in warm and dry climates where phosphorus availability is low	Climates with low temperature and high humidity, where nitrogen availability is low
Plant host range	Associates with a very wide range of hosts	Associates with comparatively lower portion of plant species
Type of fungal nutrition	Obligate biotrophophic fungi	Facultative saprotrophytic fungi
Structural elements in fungi	Arbuscules, ERM, and vesicles in some types	Mantle, Hartig net, and ERM
Fungal mode of penetration in host plant	Both inter- and extracellular penetration	Only intercellular penetration
Pathway of nutrient uptake	Both plant and mycorrhizal pathway	Mainly mycorrhizal pathway

they are also known as "sheathing mycorrhiza." Infectivity of the host plants by ectomycorrhizal fungi generally leads to modification in the feeder roots that can be observed by naked eyes (Genre 2010). The feeder roots inhabited by fungi are thicker, show more branching, and are differently colored as compared to uncolonized roots. Usually, the ectomycorrhizas initiate in between the fine roots and dikaryotic mycelia, which are formed by the union of two different monokaryotic hyphae that germinate from spores. The distinctive fungal sheath or mantle which is composed of aggregated hyphae appends to the surface of roots. This mycelium is correlated to the extramatrical hyphae to facilitate exploration of the substrate; moreover, these are accountable for mineral nutrition mobilization and uptake of water in the symbiotic tissues (Fig. 2.2). "Hartig Net" in the inner zone of the mantle forms an interface where exchange of metabolites takes place. The root cells bounded by fungal hyphae are living; the fungal hyphae are apoplastic but can colonize the epidermal cells as in angiosperms or cortical cell as in gymnosperms (Barker et al. 1998).

In AM or in the fungi, hyphae enter into the cortical cells of the roots and either may produce balloon-like, membrane-bound organelles of diverse shapes, outside or inside the cortical cells, called vesicles, or may constitute finely divided dichotomously branched hyphal invaginations called arbuscules. These structures are supposed to be the site for the exchange of materials among the host plant and fungi. Vesicles, on the other hand, have twin function, they commonly act as storage structure, and lately after they are aged, they function as reproductive structures. The characteristic features of the VA mycorrhizas are vesicles and arbuscules together with large spores. Vesicles are mostly invisible in these types of mycorrhizal associations; therefore, several scientists recommend the use of the term AM, more favorable over the designation vesicular–arbuscular (VA) mycorrhiza. Both AM fungi and ECM fungi expand their hyphae from the root into soil (extraradical hyphae), which are responsible for mobilization of nutrients from soil into the roots (Fig. 2.2).

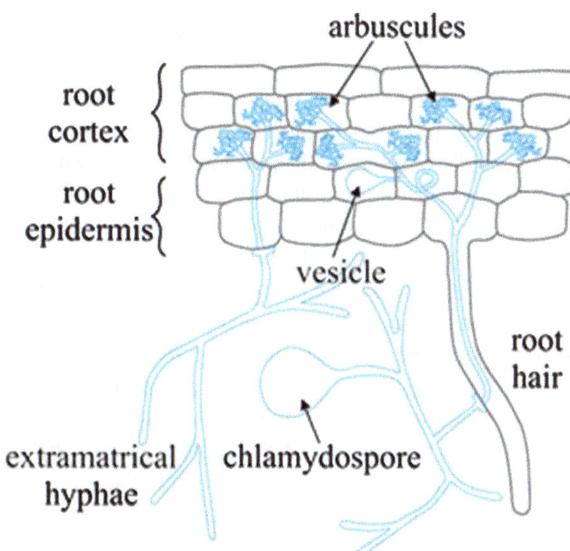

Fig. 2.2 AM fungi showing its structures in infecting root (Adopted from: © http://www.davidmoore.org.uk/assets/mostly_mycology/diane_howarth/am.htm)

2.4 Nutrient Uptake Pathways in Mycorrhizal Roots

There are two pathways *via* which the plants take up nutrients from the soil (Smith et al. 2011). The intake could be either by "plant pathway" that comprises of unmediated uptake of the nutrients by the epidermal cells of the root hairs from the soil or the nutrient uptake by plants can take place through the "mycorrhizal pathway," which consists of nutrient uptake by the extraradical mycelium of its fungal associate which further transfers the nutrients to "Hartig net" in the ECM association, or else to the intraradical mycelium in the AM association, and ultimately to the plant from the interfacial apoplast (Harrison et al. 2002). The nutrient uptake from the soil by means of plant pathway, on the other hand, is consistently constrained by the reduced mobility of nutrients in the soil (Bücking and Kafle 2015). AM and ECM roots diverge in their structural aspects, and this dissimilarity has correlation with their slightly diverse method of nutrient uptake in the AM and ECM plants (Fig. 2.3 and Table 2.2).

The AM roots do not create the fungal sheath and therefore can apparently exploit both the pathways for uptake of nutrient from soil (Bücking et al. 2012). The AM symbionts show a collective mode of the nutrient uptake, which has also been suggested previously (Bücking and Kafle 2015). This led to the supposition that the nutrient uptake through the mycorrhizal pathway can be evaded while availability of nutrient in the soil is excess. Moreover, the plants associated with mycorrhiza do not always show a affirmative growth response. However, such notion has become contentious now (Smith and Read 1997; Smith et al. 2009, 2011), and it has been established that the mycorrhizal pathway can direct the entire P uptake as well as that the factual role of mycorrhizal pathway toward complete P uptake can be "veiled" (Smith et al. 2003; Nagy et al. 2009). The transporters in plants, which are concerned with P uptake by means of plant pathway, are downregulated in reaction to AM symbiosis (Harley and Smith 1983; Chiou et al. 2001; Grunwald et al. 2009); at the same time, mycorrhizal transporters that are particularly involved in the P uptake from mycorrhizal interface are upregulated (Xu et al. 2007; Paszkowski et al. 2002). The total amount of P uptake by common investment of the pathways also depends on plant and the fungal species. Zhang et al. (2015) confirmed that *Rhizophagus irregularis* was more proficient in P absorption as compared to *Acaulospora longula* and *Gigaspora margarita* in *Lotus japonicus*. Grunwald et al. (2009) have established that the *Glomus intraradices* species has the utmost capability to repress the expression of P transporters in plants within the plant pathway; at the same time, *G. mosseae* showed the slightest outcome. This evidence also supports the concept that the positive input of mycorrhizal pathway to the nutrient accessibility is reliant on the effectiveness with which the AM associates interact along with exchange of nutrient across the mycorrhizal interface (Bücking et al. 2012). The suppression of the plant pathway by AM fungi can result in growth reductions in mycorrhizal plants, once the mycorrhizal pathway is unable to reimburse the reduced uptake by the plant pathway (Smith and Smith 2011). The assumption is that the AM fungus can induce the downregulation of the transporters

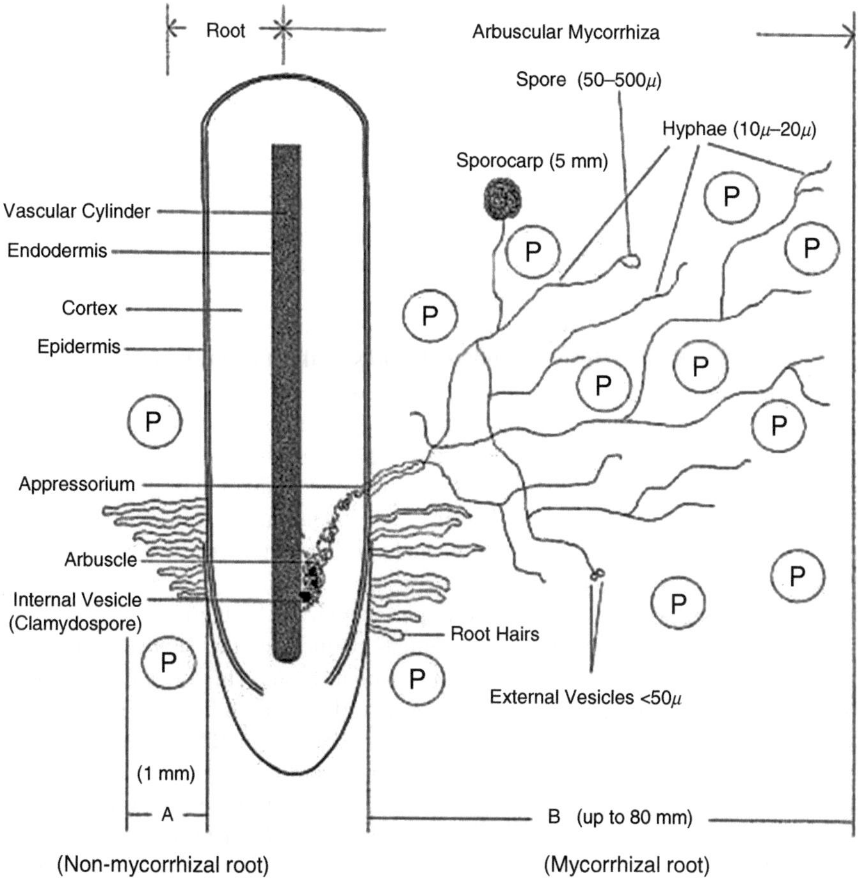

Fig. 2.3 Nutrient uptake pathways in non-mycorrhizal and mycorrhizal roots (Adapted from: Available from: http://dx.doi.org/10.5772/52570)

Table 2.2 Transportation of various nutrients by AM fungi and ectomycorrhizal fungi

Nutrient transported	AM fungi	Ectomycorrhizal fungi
P	+	+
NH_4^+	+	+
Co	+	+
NO_3	–	+
K	+	+
Ca	+	–
SO_4^+	+	–
Cu	+	–
Zn	+	–
Fe	+	+
Mn	+	+
Mg	+	+

of plant pathway to augment its C accessibility. The higher reliance on mycorrhizal pathway for the nutrient acquirement has shown to stimulate the C circulation to root system of the plant (Nielsen et al. 1998; Postma and Lynch 2011).

Of the tree species associated with ectomycorrhiza, most parts of their root surface consist of the region that is nonfunctional in nutrient uptake and the regions that are actively responsible for nutrient acquisition like non-mycorrhizal white or the ECM roots that stand for merely 2 or 16% of the entire root length, correspondingly (Taylor and Peterson 2002). In this condition, the role of the fungal mantle or the sheath that surrounds the root tips is mainly essential (Taylor and Peterson 2005). In a situation, when the fungal layer restricts the infusion of nutrient ions through it, the root tissue lying beneath the fungal mantle would be separated from the nutrient solution in the soil, and such roots will be exclusively reliant upon mycorrhizal pathway for the nutrient acquirement. The ECM fungal species and its structural and functional properties of the mantle are responsible for determining whether the fungal sheath would present an apoplastic barrier or not. Taylor and Peterson (2005) carried out research in relation to the evaluation of permeability of the *Pinus banksiana/Hebeloma cylindrosporum* fungal mantle to the berberine and the radioactive sulfate ions. They established that the fungal mantle was absolutely impervious to the tracer dye. Above an exposure period of 24 h to sulfate ions, the fungal mantle still demonstrated to be impermeable. Such outcomes revealed that plant can exceedingly depend on the fungal partner for the supply of mineral nutrients, because there is modest amount of plant tissue which has the ability of nutrient absorption from exterior of fungal mantle. Some other fungi have been shown to release hydrophobins during ECM development (Coelho et al. 2010). Hydrophobin is a diminutive hydrophobic protein that is accountable for clasping of the fungal hyphae to a surface; moreover, it may add to the impermeability of water in the fungal sheath (Unestam 1991; Unestam and Sun 1995). Therefore, only 2% of root surface of the pines is non-mycorrhizal, as well as ERM of the ECM fungus can signify nearly 99% of the nutrient-exchange interface along the length of the roots in pine (Rousseau et al. 1992), ECM-associated tree species like pine are thought to be greatly dependent on their fungal cohort (Ouahmane et al. 2009; Brundrett 2002), and it may be established that the mycorrhizal pathway plays a more significant role for nutrient acquisition in the ECM root systems as compared to the AM root systems (Bücking et al. 2012).

2.5 Possible Mechanisms of Nutrient Acquisition by Mycorrhizal Fungi

Mycorrhizal fungi are capable of absorbing and transporting almost all the 15 essential macro- and micronutrients vital for growth of the plant. Mycorrhizal fungi ooze out strong chemical compounds into the soil that mobilize firm or rock-bound nutrients such as phosphorous, iron, and other "tightly arrested" mineral nutrients in the soil. The entire process of dissolution and transportation of nutrients is of great importance in providing nutrition to the plant, and this requires the consideration of high levels of fertility by the non-mycorrhizal plants for maintaining their health. Mycorrhizal fungi create an elaborate web of hyphae that confines and absorbs nutrients restoring the nutritional assets in soils. In the non-mycorrhizal situation, much of this fertility is exhausted or mislaid from the soil system. Mycorrhizal interactions may directly influence the growth of the host plant through the improvement in nutritional attainment by the fungal associate or obliquely by altering the transpiration rates and constitution of the rhizospheric microflora (Marschner and Dell 1994), mobilization of nutrient from the organic substrates (Finlay 2008), by improving the fertilizer use efficacy (Jeff et al. 2005), or by advantageous alliance with other soil microbes (Finlay 2008).

The two key steps in nutrient absorption from the soil and release of the nutrients through mycorrhizal association involve:

(1) Mobilization and acquisition by the fungal mycelia
(2) Transportation of absorbed nutrients across the fungal–root interface

2.5.1 Mobilization and Absorption of Nutrients

In addition to the hyphae that are in the direct touch with the surface of the root, every mycorrhizal fungi also builds up extramatrical mycelium that extends from surface of infected root into the adjacent soil. Both the fungi, arbuscular mycorrhizal (AM) and ECM, manufacture huge quantity of the extramatrical mycelium. Among these, arbuscular mycorrhizal mycelium extends up to many centimeters from the surface of the infected root while ECM mycelium most likely spreads up to some meters (Goltapeh et al. 2008). In both cases, the mycelium stretches adequately afar from the nutrient depletion zone for inaccessible and bound mineral nutrients around each root; moreover, it also exhibits an intricate structure that provides it with an efficient nutrient gathering network (Schachtman et al. 1998; Bücking and Heyser 2001; Goltapeh et al. 2008). One of the components of mycorrhiza is the extramatrical mycelium that competently exhumes bulk soil for sparse nutrients plus transports obtained nutrients to the fungal–root interface where nutrients are transferred to the host plant (Bücking and Kafle 2015). Many ectomycorrhizal fungi spread extramatrical mycelium in the form of a dispersed mat of the individual hyphae forming intricate linear multi-hyphal arrangement recognized as rhizomorphs. The hyphae that are at the center of rhizomorphs are devoid of cell wall and measure about 35 μm in diameter; these play an important role in the transport of photosynthetic assimilates and inorganic mineral nutrients (Table 2.2). Conversely, diffused hyphae in the disperse mat that grow in front of the arbuscular mycorrhizas that measure nearly diameter 1–5 μm in diameter make available a widespread surface area for the absorption of nutrient from the soil. At the same time, hyphae with larger diameter of up to 10 μm are liable for an exceptional translocatory infrastructure for efficient transfer of solutes from the bulk soil from the rhizospheric soil to the surface of root (Ravnskov and Jakobsen 1995). Other than increasing uptake of mineral nutrients by the plant, which are previously there in soil, numerous mycorrhizal fungi could perhaps perform a major function in mobilization of the mineral nutrients from the organic substrate (Hodge and Fitter 2010), mineral element, or else cover rock surfaces (Finlay and Rosling 2006).

Several mycorrhizal fungi can possibly perform an important task in the mobilization of nutrients, for example, nitrogen and phosphorus from the structural or any other polymers that are, however, inaccessible to the plant roots. Withdrawal of nutrients like N and P by means of mycorrhizal fungi from a variety of organic substrates like saprotrophic mycelia (Lindahl et al. 1999), dead and decaying nematodes (Perez-Moreno and Read 2001b), pollen grains (Perez-Moreno and Read 2001a; Finlay 2008), and *Collembola* (Klironomos and Hart 2001) has been verified by many researchers. The association of mycorrhizal fungus in the microbe-based mobilization and immobilization cycle leads to mobilization of the N and the P from plant litter, microfaunal, mesofaunal, and microbe base, permitting the unique plant commune to flourish alongside the altitudinal or the latitudinal ascent (Smith et al. 2003, 2009).

The ectomycorrhizal fungi that inhabit boreal forest ecosystems are the appropriate example of events like mobilization and transportation of nutrients. In such ecosystems, N and P are present in the organic form that is not easily available for utilization by the autotrophs. In these forests, the foremost plant species are significantly dependent on the mycorrhizal symbionts to gratify their nutritional requirements. Ectomycorrhizal symbiont has the capability to immediately act on the structural polymers that might be a cause for nutrient unavailability and in the mobilization of N as well as P from organic polymers (Read and Perez-Moreno 2003). Lindahl et al. (2007) observed that the saprotrophic microbes and fungi produce a harmonized assembly of debris-degrading enzymes that are essential for early phase of decomposition process; in addition, the N mobilized by such saprotrophic fungi is reserved in their mycelia. With the fall of C:N ratio in the decomposed organic matter, the saprotrophs contemplate to be less vigorous with respect to mycorrhizal species that directly receive the host assimilates (Hodge et al. 2000). The occurrence of ectomycorrhizal fungi in the finely degraded litter along with humus presents signal that the mycorrhizal hyphae play a substantial role in the mobilization of N from well-degraded organic waste in the boreal forest soils. Moreover, the unsteady carbon librated in the soil *via* roots and allied mycorrhizal fungi could play a crucial role to mobilize N. The production of extracellular enzymes such as proteinases and peptidases by ectomycorrhizal fungi enables them to competently hydrolyze the organic nitrogen resource to liberate amino acids; these can be taken up by fungi. Also, the secretion of extracellular phosphomonoesterase and phosphodiesterase enzymes by Ectomycorrhizal fungi enables to mobilize mineral nutrients in the soil. The enzyme phosphodiesterase is capable of mobilizing phosphorus, which is confiscated inside nucleic acids. Several ectomycorrhizal fungi also produce enzymes which are hydrolytic in action; these fall within the family of cellulase, hemicellulase, or lignase. The enzymes support the entry of hyphae into the dead and decaying organic matter in the soil and get in touch with the mineral nutrients seized within. In this way, the ectomycorrhizal fungi condense the typical mineral nutrient cycles, releasing nutrients seized within the organic matter of the soil. There are reports that suggest that the Ectomycorrhizal fungi are capable of siderophores production, which bind and form complexes with iron and oxalate that amplify potassium uptake by the symbiont. Production of reducing agents by the ectomycorrhizal fungi magnifies the acquisition of ions from then stable oxides like MnO_2, consequently serving in enhanced plant nutrition (Lindahl et al. 2001, 2007).

The strict biotrophic character of AM fungi suggests that such fungi are unable to utilize organic nitrogen sources (Bücking and Kafle 2015); nevertheless, a number of studies reveal that the hyphae of AM fungi develop on the organic matter and relocate nitrogen to its host plant (Leigh et al. 2009; Hodge and Fitter 2010), which results in elevated plant nitrogen content in the mycorrhizal plants (Thirkell et al. 2015). Reynolds et al. (2005) established that there is no confirmation about the promotion of plant N acquisition by AM fungi and the better growth of old field perennial trees under low N supply situation; however, AM fungi could be associated with the decomposing organic material in several ecosystems. While Hodge et al. (2001) verified the improved decomposition and N mobilization from dead and decaying grass foliage in the existence of AM fungi, Leigh et al. (2009) established that AM fungi did not exhibit saprophytic competence and the fungus captures N from organic matter almost certainly as the product of decomposition. Though AM fungus speeds up the N absorptions from organic substance (Atul-Nayyar et al. 2009) and manipulates the C exchange within the soil microbe population during the decomposition process (Herman et al. 2012), advanced research is still required to discriminate between direct competence of AM fungi to mobilize organic material along with their probable, indirect consequence on putrefaction and nutrient uptake by plant, which occurs by stimulation by decomposers and followed by uptake of the decomposed products by the mycorrhizal fungus (Li et al. 2006; Finlay 2008).

In addition to the organic matter, mycorrhizal fungi are also accountable for dynamically mobilizing nutrients from the mineral particles and the rock surfaces by means of weathering; this may occur either by mycorrhizal fungi alone or in alliance with other microbes like bacteria or any other fungi (Wallander et al. 1997; Landeweert et al. 2001; Finlay and Rosling 2006; Finlay 2008). The role of arbuscular mycorrhizal (AM) fungi in mineral/rock weathering is contradictory; moreover, there are barely few verifications that suggest improved consumption of comparatively insoluble type of inorganic phosphorus like rock phosphate by the AM fungi. Such effects may perhaps depend on synergistic association among AM fungi and the P solubilizing microbial community. Wallander (2006) confirmed the vital contribution of mycorrhizal fungi in mineral weathering of forest soils. Reports suggest that Ectomycorrhizal fungi produce certain low-molecular-weight (LMW) organic acids, which are utilized in weathering of the minerals rocks

(Ahonen-Jonnarth et al. 2000). Breemen et al. (2000) reported that several open tubular apertures of about 3–10 μm in size were existing in the weatherable minerals in every podzol surface soil and the shallow granitic rock under European coniferous forests, and they suggested that these pores were created by complexing LMW organic acids, which leach out in association with mycorrhizal fungi. The hyphae of ectomycorrhizal fungi penetrate and, perhaps, generate microsites that are otherwise far from the contact of the plant roots and inaccessible from the bulk soil solution. The mobilized and dissolved nutrients can be further carried to roots of the host plant, shunning soil solution with the frequently toxic concentrations of the Al^{3+} ions from the acid rain (Clark 1997), and also avoiding antagonism for uptake of nutrient with other microorganisms.

2.5.2 Movement of Carbon and Nutrients Across the Fungus–Root Interface

Whatsoever the form of mycorrhizal fungi is or the approach they take up for mobilization of the nutrients, these ultimately reach at the fungal–root interface within symplasm of fungus. The transfer of nutrients to the host plant engrosses the efflux of mineral ions through plasma membrane of fungus followed by the inclusion from apoplasm interface across the plasma membrane of the cells of host root (Cairney and Burke 1996). Escape of the nutrient substances across the interface is lowered by complex fungal arrangements.

It has been suggested that the local physiochemical conditions are directed by the series of events occurring in either of the symbionts in an association. This includes accumulation of impermeable extracellular resources between mycelium within the mantle in several ectomycorrhiza and on the tip of the hyphal ingress into root cells in the arbuscular mycorrhiza. Ectomycorrhizas produce an explicit apoplasmic compartment, which averts surfeit of the nutrients from interface apoplasm. The carbon required for growth and metabolism of the mycorrhizal fungi is largely acquired as photoassimilate from the roots of the host plant (Smith and Read 2008; Bonfante and Genre 2010). In distinction to ericoid mycorrhizal fungi and phytopathogenic fungi, arbuscular mycorrhizal (AM) and ectomycorrhizal fungi (ECM) are unable to utilize sucrose as a source of carbon, and therefore they seize on simple sugars like glucose and fructose, from mycorrhizal interface. The fungal genome contains invertase gene, which is associated with its mode of nutrition, and in divergence to various plant-allied fungi, like pathogenic fungi and endophytic fungi, there does not exist any proposal that confirms the presence of invertase genes in AM or ECM fungi (Parrent et al. 2009; Bonfante and Genre 2010; Wahl et al. 2010) or else holding invertase activity (Salzer and Hager 1996). As a result, mycorrhizal fungus depends on host cell for its invertase activity, specifically in the region of interfacial apoplast, in favor of sucrose hydrolysis. This hydrolysis of sucrose makes simple sugars such as hexoses, glucose, or fructose, available for fungal utilization. Further, the reports suggest that the glucose is mostly engrossed by the hyphae of "Hartig net" while fructose is largely taken up by the hyphae present in the inner layers of mantle (Nehls et al. 2002). Numerous carriers have been recognized that are present uniformly on plant and the fungal membranes, which are mainly responsible for transporting nutrients from the fungal to plant cell. In perspective of the ECM relationship, the transporter AmAMT2 of *Amanita muscaria* has high affinity for NH4+ ions. This importer is induced in extraradical mycelium, while it is downregulated in fungal sheath and "Hartig net" (Willmann et al. 2007, Martin and Nehls 2009). The good expression of AmAMT2 transporter in ERM implies a high competence of ERM for uptake of NH4+ ions. On the other hand, the reduced level of expression in "Hartig net" specifies that NH4+ ions serve as imminent nitrogen source, which is transported by mycorrhizal fungus to the roots of host plant. The reduced level of expression of the NH4+ importer in "Hartig net" reduces the reabsorption of the NH4+ in the fungual hyphae from interfacial apoplastic zone along with rise of the total NH4+ transport to the host. The existence of upregulated high-affinity NH4+ importers of plant in ECM roots also supports the transport of NH4+ across the ECM interface (Selle et al. 2005; Couturier et al. 2007). A study conducted on the rice and *Medicago truncatula* by Wang and Qiu (2006) suggested that the transporter enzymes in plasma membrane, proton-ATPases (H+-ATPase), are specifically upregulated in arbuscule enclosing cells, and these are requisite for the improved proton pumping action in the membrane vesicles of AM colonized roots (Harrison et al. 2002). Any alteration in the regulation and function of H+-ATPase decreases the arbuscule size and diminutive uptake of nutrients by the host plant via mycorrhizal symbiosis. Overexpression of the gene regulating H+-ATPase Os-HA1 improved the phosphate uptake and the plasma membrane efficacy, indicating that the H+-ATPase performs a significant role stimulating peri-arbuscular membrane, so as to facilitate the nutrient exchange in plant cells having arbuscule. A high-affinity phosphate (P) transporter, Pt4, is entirely manifested in the mycorrhizal roots; besides, it is concerned with the getting hold of the P delivered *via* the fungus (Xu et al. 2007). Another transporter, AMT2;2, is a high-affinity ammonium transporter, which is positioned in peri-arbuscular membrane (Guether et al. 2009). Furthermore, the occurrence of mycorrhizal induced sulfate transporters in the AM roots proposes that the sulfate too is transferred across the mycorrhizal interface from AM fungus to host plant (Casieri et al. 2012; Allen and Shachar-Hill 2009; Helber et al. 2011).

2.6 Conclusions

A greater fraction of the higher plants are found to be allied with the Mycorrhizal fungi. These symbiotic interactions differ extensively in their structures as well as functions. Of the numerous kinds of mycorrhizal fungi, Arbuscular mycorrhizal (AM) and ECM fungi play an important role in nature. Both the categories of mycorrhizal fungi not merely assist in uptake of the major plant nutrients such as P and N but also help in captivating other micronutrients like Fe, Cu, Zn, etc. Mycorrhizal fungi implement various means to achieve the task effectively: measuring the greater absorbing surface area of the plants, releasing biochemical compounds along with alliance with other microbes in its ambience. Other than mobilizing the mineral nutrients, mycorrhizal fungi also provide significant C sink in soil; hence, these have a critical impact on cycling of the elements within soil. Consequently, mycorrhiza is established as a significant association for nutrient management in the ecosystem.

Acknowledgment Ajit Varma is thankful to Department of Science and Technology and Department of Biotechnology for partial financial funding and to DST-FIST for providing confocal microscope facility.

References

Ahonen-Jonnarth U, Van Hees PAW, Lundström US, Finlay RD (2000) Production of organic acids by mycorrhizal and non-mycorrhizal *Pinus sylvestris* L. seedlings exposed to elevated concentrations of aluminium and heavy metals. New Physician 146:557–567

Allen JW, Shachar-Hill Y (2009) Sulfur transfer through an arbuscular mycorrhiza. Plant Physiol 149:549–560

Atul-Nayyar A, Hamel C, Hanson K, Germida J (2009) The arbuscularmycorrhizal symbiosis links N mineralization to plant demand. Mycorrhiza 19:239–246

Barker SJ, Stummer B, Gao L, Dispain I, O'Connor PJ, Smith SE (1998) A mutant in *Lycopersicon esculentum* Mill, with highly reduced VA mycorrhizal colonization: isolation and preliminary characterization. Plant J 15:791–797

Bonfante P, Genre A (2010) Mechanisms underlying beneficial plant–fungus interactions in mycorrhizal symbiosis. Nat Commun 1:48. doi:10.1038/ncomms1046

Breemen NV, Finlay R, Lundstrom U, Jongmans AG, Giesler R, Olsson M (2000) Mycorrhizal weathering: a true case of mineral plant nutrition? Biogeochemistry 49:53–67

Brundrett MC (2002) Tansley review no. 134: coevolution of roots and mycorrhizas of land plants. New Physician 154:275–304

Bücking H (2011) Ectomycoremediation: an eco-friendly technique for the remediation of polluted sites. In: Rai M, Varma A (eds) Diversity and biotechnology of ectomycorrhizae. Soil biology. Springer, Berlin, Heidelberg, pp 209–229

Bücking H, Heyser W (2001) Microautoradiographic localization of phosphate and carbohydrates in mycorrhizal roots of *Populustremula* × *Populus alba* and the implications for transfer processes in ectomycorrhizal associations. Tree Physiol 21:101–107

Bücking H, Kafle A (2015) Role of arbuscular mycorrhizal fungi in the nitrogen uptake of plants: current knowledge and research gaps. Agronomy 5:587–612

Bücking H, Kuhn AJ, Schröder WH, Heyser W (2002) The fungal sheath of ectomycorrhizal pine roots: an apoplastic barrier for the entry of calcium, magnesium, and potassium into the root cortex? J Exp Bot 53:1659–1669

Bücking H, Liepold E, Ambilwade P (2012) The role of the mycorrhizal symbiosis in nutrient uptake of plants and the regulatory mechanisms underlying these transport processes. In: Dhal NK, Sahu SC (eds) Plant science. ISBN:978-953-51-0905-1

Cairney JWG, Burke RM (1996) Physiological heterogeneity within fungal mycelia: an important concept for a functional understanding of the ectomycorrhizal symbiosis. New Physician 134:685–695

Casieri L, Gallardo K, Wipf D (2012) Transcriptional response of *Medicagotruncatula* sulphate transporters to arbuscular mycorrhizal symbiosis with and without sulphur stress. Planta. doi:10.1007/s00425-012-1645-7

Chiou TJ, Liu H, Harrison MJ (2001) The spatial expression patterns of a phosphate transporter (MtPT1) from *Medicago truncatula* indicate a role in phosphate transport at the root/soil interface. Plant J 25:281–293

Clark RB (1997) Arbuscular mycorrhizal adaptation, spore germination, root colonization, and host plant growth and mineral acquisition at low pH. Plant Soil 192:15–22

Coelho ID, de Queiroz MV, Costa MD, Kasuya MCM, de Araujo EF (2010) Identification of differentially expressed genes of the fungus *Hydnangium* sp during the pre-symbiotic phase of the ectomycorrhizal association with *Eucalyptus grandis*. Mycorrhiza 20:531–540

Couturier J, Montanini B, Martin F, Brun A, Blaudez D, Chalot M (2007) The expanded family of ammonium transporters in the perennial poplar plant. New Phytol 174:137–150

Finlay RD (2008) Ecological aspects of mycorrhizal symbiosis: with special emphasis on the functional diversity of interactions involving the extraradial mycelium. J Exp Bot 59:1115–1126

Finlay RD, Rosling A (2006) Integrated nutrient cycles in forest ecosystems, the role of ectomycorrhizal fungi. In: Gadd GM (ed) Fungi in biogeochemical cycles. Cambridge University Press, Cambridge, pp 28–50

Genre A (2010) Mechanisms underlying beneficial plant-fungus interactions in mycorrhizal symbiosis. Nat Commun 1:48–52

Giri B, Giang PH, Kumari R, Prasad R, Sachdev M, Garg AP, Oelmüller R, Varma A (2005) Mycorrhizosphere: strategies and functions. Soil Biol 3:213–252

Goltapeh EM, Danesh YR, Prasad R, Varma A (2008) Mycorrhizal fungi: what we know and what should we know. In: Varma A (ed) Mycorrhiza, 3rd edn. Springer, Heidelberg, pp 3–28

Grunwald U, Guo WB, Fischer K, Isayenkov S, Ludwig-Müller J, Hause B, Yan XL, Kuster H, Franken P (2009) Overlapping expression patterns and differential transcript levels of phosphate transporter genes in arbuscularmycorrhizal, pi-fertilised and phytohormone-treated *Medicagotruncatula* roots. Planta 229:1023–1034

Guether M, Neuhauser B, Balestrini R, Dynowski M, Ludewig U, Bonfante P (2009) A mycorrhizal-specific ammonium transporter from *Lotus japonicus* acquires nitrogen released by arbuscular mycorrhizal fungi. Plant Physiol 150:73–83

Habte M (2000) Mycorrhizal fungi and plant nutrition. In: Silva JA, Uchida R (eds) Plant nutrient management in Hawaii's soils, approaches for tropical and subtropical agriculture. College of Tropical Agriculture and Human Resources, University of Hawaii at Manoa

Harley JL, Smith SE (1983) Mycorrhizal symbiosis. Academic Press, Toronto, pp 112–115

Harrison MJ, Dewbre GR, Liu J (2002) A phosphate transporter from *Medicagotruncatula* involved in the acquisition of phosphate released by arbuscular mycorrhizal fungi. Plant Cell 14:2413–2429

Helber N, Wippel K, Sauer N, Schaarschmidt S, Hause B, Requena N (2011) A versatile monosaccharide transporter that operates in the arbuscular mycorrhizal fungus *Glomus*sp. is crucial for the symbiotic relationship with plants. Plant Cell 23:3812–3823

Herman DJ, Firestone MK, Nuccio E, Hodge A (2012) Interactions between an arbuscular mycorrhizal fungus and a soil microbial community mediating litter decomposition. FEMS Microbiolol Ecol 80:236–247

Hodge A, Fitter AH (2010) Substantial nitrogen acquisition by arbuscular mycorrhizal fungi from organic material has implications for N cycling. Proc Natl Acad Sci USA 107:13754–13759

Hodge A, Robinson D, Fitter AH (2000) An arbuscular mycorrhizal inoculum enhances root proliferation in, but not nitrogen capture from, nutrient-rich patches in soil. New Phytol 145:575–584

Hodge A, Campbell CD, Fitter AH (2001) An arbuscular mycorrhizal fungus accelerates decomposition and acquires nitrogen directly from organic material. Nature 413:297–299. doi:10.1038/35095041

Jeff H, Taylor, Peterson CA (2005) Ectomycorrhizal impacts on nutrient uptake pathways in woody roots. New For 30:203–214

Kilpeläinena J, Vestbergb M, Repoc T, Lehtoa T (2016) Arbuscular and ectomycorrhizal root colonisation and plant nutrition in soils exposed to freezing temperatures. Soil Biol Biochem 99:85–93

Klironomos JN, Hart MM (2001) Animal nitrogen swap for plant carbon. Nature 41:651–652

Landeweert R, Hoffland E, Finlay RD, Kuyper TW, van Breemen N (2001) Linking plants to rocks: ectomycorrhizal fungi mobilize nutrients from minerals. Trends Ecol Evol 16:248–254

Leigh J, Hodge A, Fitter AH (2009) Arbuscular mycorrhizal fungi can transfer substantial amounts of nitrogen to their host plant from organic material. New Phytol 181:199–207

Li HY, Smith SE, Holloway RE, Zhu YG, Smith FA (2006) Arbuscular mycorrhizal fungi contribute to phosphorus uptake by wheat grown in a phosphorus-fixing soil even in the absence of positive growth responses. New Phytol 172:536–5343

Lindahl B, Stenlid J, Olsson S, Finlay RD (1999) Translocation of 32P between interacting mycelia of a wood decomposing fungus and ectomycorrhizal fungi in microcosm systems. New Phytol 44:183–193

Lindahl B, Olsson S, Stenlid J, Finlay RD (2001) Effects of resource availability on mycelial interactions and 32P-transfer between a saprotrophic and an ectomycorrhizal fungus in soil microcosms. FEMS Microbiol Ecol 38:43–52

Lindahl BD, Ihrmark K, Boberg J, Trumbore SE, Hogberg P, Stenlid J, Finlay RD (2007) Spatial separation of litter decomposition and mycorrhizal nitrogen uptake in a boreal forest. New Phytol 173:611–620

Marschner H, Dell B (1994) Nutrient uptake in mycorrhizal symbiosis. Plant Soil 159:89–102

Martin F, Nehls U (2009) Harnessing ectomycorrhizal genomics for ecological insights. Curr Opin Biotechnol 12:509–515

Nagy R, Drissner D, Amrhein N, Jakobsen I, Bucher M (2009) Mycorrhizal phosphate uptake pathway in tomato is phosphorus-repressible and transcriptionally regulated. New Phytol 181:950–959

Nehls S, Mikolajewski, Magel E, Hampp R (2002) Carbohydrate metabolism in ectomycorrhizas: gene expression, monosaccharide transport and metabolic control. New Phytol 150:533–541

Nielsen KL, Bouma TJ, Lynch JP, Eissenstat DM (1998) Effects of phosphorus availability and vesicular–arbuscular mycorrhizas on the carbon budget of common bean (*Phaseolus vulgaris*). New Phytol 139:647–656

Ouahmane L, Revel JC, Hafidi M, Thioulouse J, Prin Y, Galiana A, Dreyfus B, Duponnois R (2009) Responses of *Pinus halepensis* growth, soil microbial catabolic functions and phosphate-solubilizing bacteria after rock phosphate amendment and ectomycorrhizal inoculation. Plant Soil 320:169–179

Parrent JL, James TY, Vasaitis R, Taylor AFS (2009) Friend or foe? Evolutionary history of glycoside hydrolase family 32 genes encoding for sucrolytic activity in fungi and its implications for plant-fungal symbioses. BMC Evol Biol 9:148–154

Paszkowski U, Kroken U, Roux C, Briggs SP (2002) Rice phosphate transporters include an evolutionarily divergent gene specifically activated in arbuscular mycorrhizal symbiosis. Proc Natl Acad Sci USA 99:13324–13329

Perez-Moreno J, Read DJ (2001a) Exploitation of pollen by mycorrhizalmycelial systems with special reference to nutrient cycling in boreal forests. Proc R Soc B 268:1329–1335

Perez-Moreno J, Read DJ (2001b) Nutrient transfer from soil nematodes to plants: a direct pathway provided by the mycorrhizalmycelial network. Plant Cell Environ 24:1219–1226

Postma JA, Lynch JP (2011) Root cortical aerenchyma enhances the growth of maize on soils with suboptimal availability of nitrogen, phosphorus, and potassium. Plant Physiol 156:1190–2001

Ravnskov S, Jakobsen I (1995) Functional compatibility in arbuscular mycorrhizas measured as hyphal p transport to the plant. New Phytol 129:611–618

Read DJ, Perez-Moreno J (2003) Mycorrhizas and nutrient cycling in ecosystems: a journey towards relevance? New Phytol 157:475–492

Reynolds HL, Hartley AE, Vogelsang KM, Bever JD, Schultz PA (2005) Arbuscular mycorrhizal fungi do not enhance nitrogen acquisition and growth of old-field perennials under low nitrogen supply in glasshouse culture. New Phytol 167:869–880

Rousseau JVD, Reid CPP, English RJ (1992) Relationship between biomass of the mycorrhizal fungus *Pisolithustinctorius* and phosphorus uptake in loblolly pine seedlings. Soil Biol Biochem 24:183–184

Salzer P, Hager A (1996) Sucrose utilization of the ectomycorrhizal fungi *Amanita muscaria* and *Hebelomacrus tuliniforme* depends on the cell wall-bound invertase activity of their host *Piceaabies*. Bot Acta 104:439–445

Schachtman DP, Reid RJ, Ayling SM (1998) Phosphorus uptake by plants: from soil to cell. Plant Physiol 116:44–53

Selle A, Willmann M, Grunze N, Gessler A, Weiss M, Nehls U (2005) The high-affinity poplar ammonium importer PttAMT1.2 and its role in ectomycorrhizal symbiosis. New Phytol 168:697–706

Smith SE, Read DJ (1997) Mycorrhizal symbiosis. Academic Press, London

Smith SE, Read DJ (2008) Mycorrhizal symbiosis, 3rd edn. Academic Press, London

Smith SE, Smith FA (2011) Roles of arbuscular mycorrhizas in plant nutrition and growth: new paradigms from cellular to ecosystem scales. Annu Rev Plant Biol 62:227–250

Smith SE, Smith FA, Jakobsen I (2003) Mycorrhizal fungi can dominate phosphate supply to plants irrespective of growth responses. Plant Physiol 133:16–20

Smith FA, Grace EJ, Smith SE (2009) More than a carbon economy: nutrient trade and ecological sustainability in facultative arbuscular mycorrhizal symbioses. New Phytol 182:347–358

Smith SE, Jakobsen I, Grønlund M, Smith FA (2011) Roles of arbuscular mycorrhizas in plant phosphorus nutrition: interactions between pathways of phosphorus uptake in arbuscular mycorrhizal roots have important implications for understanding and manipulating plant phosphorus acquisition. Plant Physiol 156:1050–1057

Taylor JH, Peterson CA (2002) Morphometric analysis of *Pinus banksiana* Lamb. root anatomy during a 3-month field study. Trees 14:239–247

Taylor JH, Peterson CA (2005) Ectomycorrhizal impacts on nutrient uptake pathways in woody roots. New For 30:203–214

Thirkell JD, Cameron DD, Hodge A (2015) Resolving the "nitrogen paradox" of arbuscular mycorrhizas: fertilization with organic matter brings considerable benefits for plant nutrition and growth. Plant Cell Environ. doi:10.1111/pce.12667

Unestam T (1991) Water repellency, mat formation, and leaf-stimulated growth of some ectomycorrhizal fungi. Mycorrhiza 1:13–20

Unestam T, Sun YP (1995) Extramatrical structures of hydrophobic and hydrophilic ectomycorrhizal fungi. Mycorrhiza 5:301–311

Wahl R, Wippel K, Goos S, Kämper J, Sauer N (2010) A novel high-affinity sucrose transporter is required for virulence of the plant pathogen *Ustilago maydis*. PLoS Biol 8:1000303. doi:10.1371/journal.pbio.1000303

Wallander H (2006) Uptake of P from apatite by *Pinus sylvestris* seedlings colonized by different ectomycorrhizal fungi. Plant Soil 218:249–256

Wallander H, Bonfante P, Wickman T, Jacks G (1997) Apatite as a source of mycorrhizal and non-mycorrhizal *Pinus sylvestris*. Plant Soil 196:123–131

Wang B, Qiu YL (2006) Phylogenetic distribution and evolution of mycorrhizae in land plants. Mycorrhiza 16:299–363

Willmann A, Weiss M, Nehls U (2007) Ectomycorrhiza-mediated repression of the high affinity ammonium importer gene AmAMT2 in *Amanita muscaria*. Curr Genet 51:71–78

Wright DP, Read DJ, Scholes JD (1998) Mycorrhizal sink strength influences whole plant carbon balance of *Trifoliumrepens* L. Plant Cell Environ 21:881–891

Xu GH, Chague V, Melamed-Bessudo C, Kapulnik Y, Jain A, Raghothama KG, Levy AA, Silber A (2007) Functional characterization of LePT4: a phosphate transporter in tomato with mycorrhiza-enhanced expression. J Exp Bot 258:2491–2501

Zhang X, Chen B, Ohtomo R (2015) Mycorrhizal effects on growth, P uptake and Cd tolerance of the host plant vary among different AM fungal species. Soil Sci Plant Nutr 61:359–368

Chapter 3
Soil: Do Not Disturb, Mycorrhiza in Action

Cristina Cruz, Alessandro Ramos, Olubukola Oluranti Babalola, Hessini Kamel, Teresa Dias, and Ajit Varma

Abstract Fungi are key actors in controlling primary productivity. Depending on the fungal functional group, they may promote or decrease plant productivity. It is consensual that arbuscular mycorrhizal and some endophytic fungi contribute to promote plant productivity and defense against phytopathogenic organisms, including fungi. However, there is not much information about the relation between the distinct functional groups of fungi. In this chapter, we aim at understanding the importance of arbuscular mycorrhizal fungi (*Glomus intraradices*) and *Piriformospora indica* (*Serendipita indica*) inoculation in the tolerance of Tomato (*Solanum lycopersicum*) plants to fusarium wilt.

Tomato plants grown at two nutritional levels were inoculated with *Glomus intraradices* and/or *Piriformospora indica* (*Serendipita indica*) and then infected with *Fusarium oxysporum*. Plant biomass accumulation showed that plants inoculated with *Glomus intraradices* and *Piriformospora indica* (*Serendipita indica*) accumulated more biomass and were more tolerant to Fusarium wilt. The analysis of the root exudates showed that fungal infection changed the composition of the root exudates and pointed out the importance of antioxidant compounds.

C. Cruz (✉) • T. Dias
Faculty of Sciences, Center for Ecology, Evolution and Environmental Changes (Ce3C), Universidade de Lisboa, Lisbon, Portugal
e-mail: ccruz@fc.ul.pt

A. Ramos
Laboratory of Biochemistry and Physiology of Microorganisms, Universidade Estadual do Norte Fluminense (UENF), Campos dos Goytacazes, Brazil

O.O. Babalola
Food Security and Safety Niche, Faculty of Agriculture, Science and Technology, North-West University, Private Bag X2046, Mmabatho 2735, South Africa

H. Kamel
Biology Department, Faculty of Science, Taif University, Al-Haweiah, Taif 888, Saudi Arabia

A. Varma
Amity Institute of Microbial Technology (AIMT), Amity University Uttar Pradesh, Block 'E-3', 4th Floor, Sector-125, Noida, Gautam Buddha Nagar, Uttar Pradesh 201313, India
e-mail: ajitvarma@amity.edu

A. Varma et al. (eds.), *Mycorrhiza - Function, Diversity, State of the Art*,
DOI 10.1007/978-3-319-53064-2_3

3.1 Introduction

Fungi cannot fix energy and nutrients directly, so they use the energy stored in plant and animal biomass to grow. They are a key group of organisms that interact with other organisms and the abiotic environment to regulate ecosystem processes. Fungi are able to share resources and span a range of spatial and temporal scales to mediate flows of energy and materials.

The role of fungi in primary production goes behind making nutrients available to plants. They form symbiotic associations with bacteria, algae, and plants and work as a supportive network for photobionts (Dighton 2016). These symbioses include mycorrhiza and endophytes, which can occur in multiple plant parts. Such interactions improve nutrient availability for primary production, confer plant tolerance to drought, salt, and heavy metals, and provide a degree of protection from pathogenic fungi or bacteria and herbivory (Smith and Read 1997).

However, fungi may also have negative effects on the primary production. Some of them are phytopathogenic and may attack plants below and above ground. The effects of these phytopathogens are particularly evident in agrosystems, especially in intensive monocrop agriculture (Termoshuizen 2014). Both plant pathogens and mycorrhiza may play an important role in regulating plant productivity and plant community composition. The degree of impact of the phytopathogen on plant host depends on the fungal species and on the environmental conditions. In many cases, the effect of the phytopathogen increases when plants are grown under suboptimal conditions and are already under stress. The role of mycorrhiza is probably to improve nutrient uptake by the host plant or to alter its physiology so the plant is better able to defend itself against the pathogen (Volpin et al. 1994; Sharma and Mukerji 2007) rather than direct competition between the symbiotic fungi.

In a meta-analysis of interactions between arbuscular mycorrhizal fungi (AMF) and phytopathogen fungi, Borowicz (2001) concluded that about 50% of the studies showed that AMF gave a degree of protection to their host plant against the phytopathogens. The effect of the pathogenic fungus was usually to reduce the growth of the AMF. The interactions between the functional groups of fungi resulted in the reduction of the growth of each one in about 16%.

The aim of this work was to understand if the presence of the endophyte *Piriformospora indica* (*Serendipita indica*) affected the relation between AMF and plant phytopathogenic fungi. Tomato (*Solanum lycopersicum*) plants were inoculated with AMF (*Glomus intraradices, Gi*) and or *P. indica* (*Serendipita indica*) (Pi) and then infected or not with *Fusarium oxysporum* under low and high nutrient availability. Results point out to a synergistic effect between *P. indica* (*Serendipita indica*) and *Glomus intraradices* toward plant defense against fusariosis.

3.2 Protocol Applied

F. oxysporum was isolated from the host plant. The stem of a diseased plant was cut lengthwise to reveal the xylem and trimmed of all the leaves and secondary roots leaving only the main stem and root. The stem was surface sterilized by soaking in 10% bleach solution for 5 min and then cut into 2–4 mm pieces. Sterilized stem pieces were incubated on Potato Dextrose Agar (PDA) plates under light conditions. Once the fungus was grown sufficiently from the pieces, mycelium isolates were transferred to new PDA plates and incubated for 10–14 days. Colonies of *F. oxysporum* were recognized by the characteristic reddish-purple centre surrounded by a pinkish white aerial mycelium. The morphological identification was confirmed using molecular methods (Amaral et al. 2013).

Tomato seeds (Wt and AVP1OX), in the proportion of 1:10 (seeds: solution), were surface sterilized first with sodium hypochlorite 10% (v/v) for 5 min and then with 70% alcohol for 3 min. Seeds were then rinsed five times with sterile water. After sterilization, seeds were plated in polypropylene trays filled with 20 cm^3 autoclaved substrate composed of sand and vegetable soil in the proportion 1:2. Seedlings were allowed to grow for 15 days in a growth chamber (16/8 h light/dark; 350 $\mu mol\ m^{-2}\ s^{-1}$ at the leaf surface level during the light period, 25 ± 1 °C, 70% humidity).

Rhizoglomus intraradices, formerly *Glomus intraradices* BEG 24 (Gi), was obtained from the Bank Exchange of Glomales (BEG, Dijon—France). Spores were multiplied, in association with *Sorghum tricolor* roots grown in culture pots (5 L) filled with sterilized sand:vermiculite (1:1) mixture, and allowed to grow for three months under greenhouse conditions. The mixture of soil, fungal spores, hyphae fragments, and roots of colonized *S. tricolor* was used as AMF inoculant.

Inoculant quality was assessed by fungal spore counting. Fungal spores were extracted from the inoculant mixture by wet sieving (Vierheilig et al. 1998), followed by centrifugation in 20 and 60% sucrose gradient. Fifty grams of each were suspended in 2 L of water, and the suspension passed through overlapping sieves of 710 and 45 μm. The material collected in the 45 μm sieve was transferred to 50 mL tubes containing the sucrose gradient and centrifuged at 2000 g for 1 min. Spores were collected and transferred to crosshatch Petri dishes for counting.

P. indica (*Serendipita indica*) was multiplied in PDA plates. 0.5 cm diameter discs collected from the periphery of 9-day-old fungal colonies were used as inoculants.

After 15 days of growth, tomato seedlings were transplanted into pots (500 mL) filled with autoclaved sand:clay (2:1) mixture. AMF inoculation was performed by applying, at a depth of 2–3 cm below surface level, 30 mL per pot of the AMF inoculants containing Gi spores and hyphae, and Pi inoculation was performed by introducing a disc of culture in contact with the root. Non-inoculated treatment (control) received the same volume of substrate (30 mL). Seedlings were divided into two groups: one received just water and the other received 100 mL of full-strength Hoagland solution. Seedlings were irrigated with sterile water in order to maintain substrate water content between 40 and 60% of the substrate water capacity. Fifteen days after the inoculation, seedlings were (or not) infected with 1 mL of 10^4 mL^{-1} spores of *F. oxysporum* per plant. Plants were allowed to grow for more 30 days. Each treatment had 15 replicates.

Disease incidence was calculated as the ratio between the number of infected plants and the total number of plants multiplied by 100. Disease severity was determined according to Wellman (1939), where numerical values were assigned to five groups ($g_1 = 0.5, 1$; $g_2 = 2$; $g_3 = 3, 4$; $g_4 = 5, 6$; $g_5 = \geq 7$). Disease severity was calculated by the following formula: Disease severity $= (ng_1 + 2\,ng_2 + 5\,ng_3 + 10\,ng_4 + 20\,ng_5)/(n$ diseased plants). For confirmation of *F. Oxysporum* f. sp. *Lycopersici* infection, segments of 2-cm length starting upwards the shoot basis were dipped in 70% ethanol, flamed, and put into Petri dishes containing PDA amended with antibiotics to prevent bacterial growth (Steinkellner et al. 2011). The identification of *F. oxysporum f. sp. Lycopersici* was done according to Nelson (1990) by visual and microscopic analyses.

At the end of the experiment, five plants per treatment receiving low nutrient availability were gently removed from the substrate and washed thoroughly with water. Roots were submerged in acetate buffer (25 mM, pH 5.5) for 6 h, in a proportion of 10 mL:g root fresh weight. At the end, root exudates were stored at −20 °C until further processing. After that plant root and shoot fresh weight were determined. One root exudate per plant and five per treatment were prepared.

Root exudates were fractioned on an Amberlite XAD-1180 (Fluka Chemie GmbH) to hydrophilic and lipophilic compounds using MQ water and absolute ethanol as eluants. For GC-MS analyses, the hydrophilic fraction was hydrolyzed with 10% HCl for 3 h at room temperature, followed by derivatization (Kanani et al. 2008). Lyophilized samples were dissolved in 50 μl of a solution of methoxyamine hydrochloride in pyridine (20 mg/mL). After incubation at room temperature for 18 h, 50 μl *N*-methyl-*N*-TMS-trifluoroacetamide was added for derivatization into trimethyl-silyl ethers and esters. 0.5 μl of the obtained solution was injected into an AutoSystem XL gas chromatograph (Perkin Elmer Inc., Waltham, MA, USA) in the split-less mode following the procedure described by Hage-Ahmed et al. (2013). The obtained chromatograms were integrated with Turbomass 4.1.1 (Perkin Elmer Inc.), and the peak areas of measured compounds were converted to relative amounts (% of the total peak area of the chromatogram). Mass spectra were tentatively identified by comparison with commercial (NIST 08, National Institute of Standards and Technology, Gaithersburg, MD) and noncommercial databases (Kopka et al. 2005) and grouped into classes of compounds. For HPLC-UV analyses of lipophilic analytes, dried exudates were diluted in methanol (10 mg/ml). The analyses were carried out using a Dionex Summit HPLC with a photodiode array detector following the procedure described by Hage-Ahmed et al. (2013).

Germination of *F. oxysporum* spores (100 μl 10^7 microconidia/mL) was assayed in 500 μl root exudates or acetate buffer (25 mM, pH 5.5) in 24-well culture plates and incubated at 24 °C in the dark for 20 h, under shaking 200 rpm. After 20 h, plates were observed under the microscope and 200 spores in each well checked for germination.

The degree of colonization of the tomato roots was assessed, based on 100 root fragments from five plants of each treatment (Giovanetti and Mosse 1980) and stained with trypan blue (Phillips and Hayman 1970). Briefly, thin roots were rinsed in running water, cut into pieces of 1 cm, and treated with KOH 10% for 1 h. Then, roots were rinsed five times with sterile water, treated with HCl 1% for 3 min, and stained with trypan blue 0.05% diluted in lactoglycerol 0.02%. Roots were observed with a stereoscope at 10–40× magnification.

3.3 Salient Features

The *Fusarium* wilt symptoms include loss of turgor and leaf chlorosis, starting from the lower leaves, sometimes followed by leaf abscission and plant death. It may be manifested on only one side of the host plant (Schwartz et al. 2005). Symptoms of *F. oxysporum* infection were visible on control and Gi-inoculated plants. Symptomatic plants were much smaller, chlorotic, and wilted.

The observation of root segments showed that all plants inoculated with Gi or Pi were colonized independently of the presence of *F. oxysporum*. Colonization rates ranged between 40 and 60% without presenting any relation with treatments.

Nutrient availability in the root medium determined plant biomass accumulation only in plants not inoculated with Gi or Pi. As it would be expected, higher nutrient availability allowed higher plant biomass accumulation and higher resistance to *F. oxysporum* infection. Plant inoculation with Gi increased biomass accumulation to the levels of the non-inoculated plants grown under high nutrient availability and substantially increased plant tolerance to *F. oxysporum*. Plant inoculation with Pi increased biomass accumulation by 10–20% in relation to control plants, but although small, the effect of the infection with *F. oxysporum* was evident. However, plants inoculated with both Gi and Pi produced high biomass accumulation without any effect of the *F. oxysporum* (Fig. 3.1). The simultaneous inoculation of the plants with both fungi (Gi and Pi) reduced disease incidence and severity to less than 25% of that observed for the control plants.

Disease incidence and severity were lower for control plants grown with higher than with lower nutrient availability. The same tendency was observed for Gi- and Pi- inoculated plants (Table 3.1). There was no correlation between disease incidence or severity and the degree of Gi or Pi root colonization. Inoculation with Gi and/or Pi was effective in decreasing the incidence and the severity of *Fusarium* wilt. In particular, the co-inoculation of the plants with Gi and Pi reduced disease incidence in 50% and severity in more than 75%, even at low nutrient availability (Table 3.1).

In order to understand the mechanisms involved in the increased tolerance of tomato plants to *Fusarium* wilt after inoculation with Gi and Pi, *F. oxysporum* spores were germinated in root exudates obtained from the distinct treatments. Only low nutrient availability treatments were considered. In comparison with buffer, root exudates promoted *F. Oxysporum* germination. However, the germination of

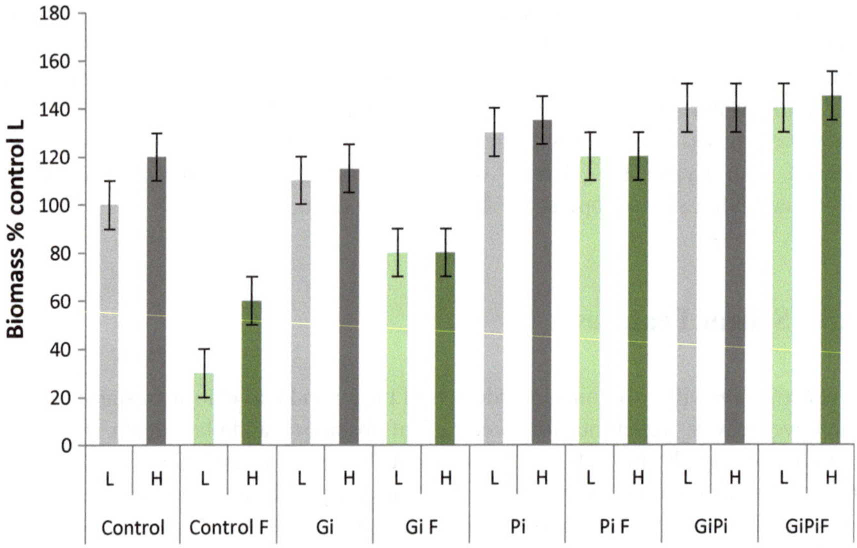

Fig. 3.1 Biomass accumulation of plants grown with the distinct treatments. Control plants were not inoculated. L—low nutritional level. H—high nutritional level. F—*Fusarium oxysporum*. Gi—*Glomus intraradices*. Pi—*Piriformospora indica* (*Serendipita indica*). GiPi—Co-inoculation of *Glomus intraradices* and *Piriformospora indica* (*Serendipita indica*). *Values* represent the mean percentage of the control plants under low nutritional level and the bars the standard deviation. $N = 10$

Table 3.1 *F. Oxysporum* f. sp. Lycopersici disease incidence (%) and severity in the different treatments (mean ± sd)

	Disease incidence (%)		Disease severity (%)	
Inoculants	Low nutrient	High nutrient	Low nutrient	High nutrient
Control	86	57	2.0 ± 0.2	1.3 ± 0.3
G. intraradices	71	43	1.4 ± 0.3	1.0 ± 0.3
P. indica (*Serendipita indica*)	57	40	1.1 ± 0.3	0.8 ± 0.2
Both	41	30	0.3 ± 0.2	0.3 ± 0.8

Plants not infected with *F. Oxysporum* f. sp. Lycopersici did not present disease symptoms

F. Oxysporum spores was much lower (10–30%) when roots were inoculated with Gi and/or Pi. It is interesting that only in the presence of Pi root infection with *F. oxysporum* affected the rate of spore germination in the root exudates (Fig. 3.2).

3.3.1 *Chemical Composition of the Root Exudates*

In the polar fraction of the root exudates, sugars and organic acids had the highest peak areas ranging between 30–50% and 28–48%, respectively (Fig. 3.3); amino

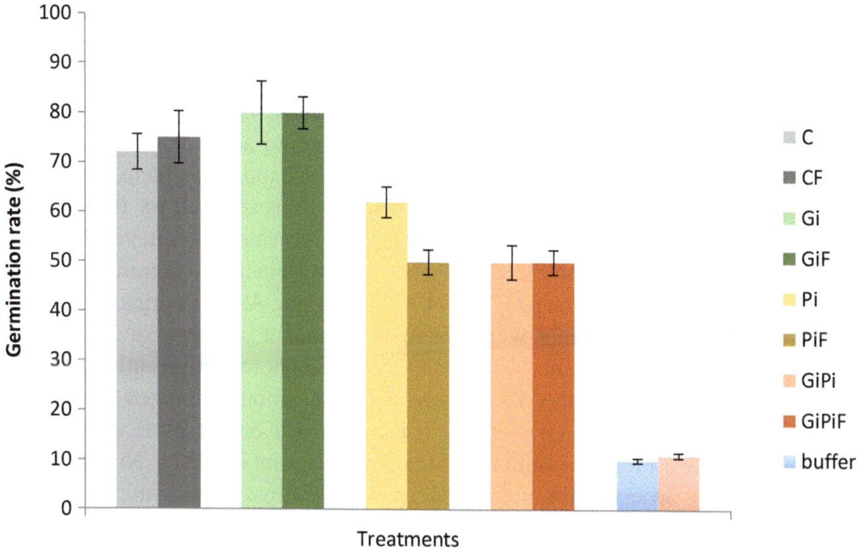

Fig. 3.2 Germination rate of the *Fusarium oxysporum* spores in buffer or in the root exudates of the low nutritional level treatments. F—*Fusarium oxysporum*. Gi—*Glomus intraradices*. Pi—*Piriformospora indica* (*Serendipita indica*). GiPi—Co-inoculation of *Glomus intraradices* and *Piriformospora indica* (*Serendipita indica*). *Values* represent the mean percentage of the control plants under low nutritional level and the bars the standard deviation. N = 5

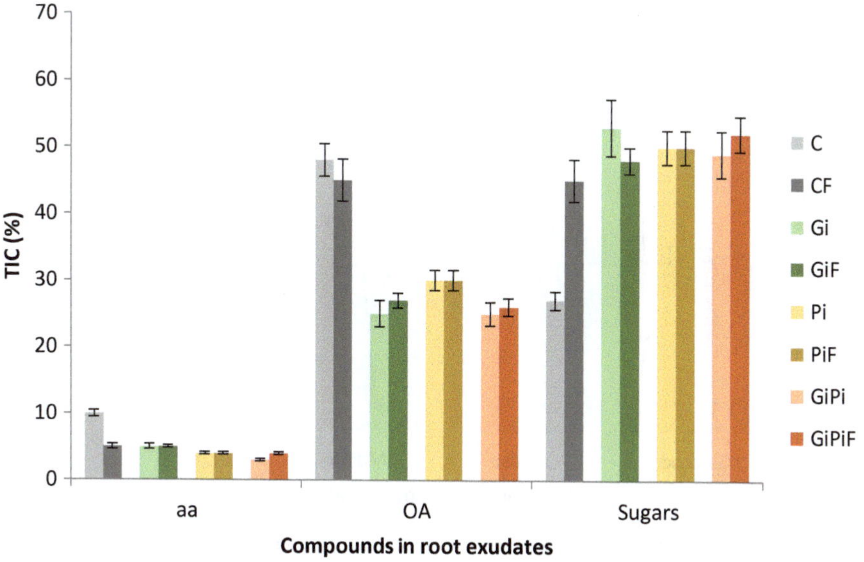

Fig. 3.3 Relative peak area (TIC) of selected compounds present in the root exudates from GC–MS analyses (mean ± Sd, $n = 5$). The low level nutritional treatments comprised: F—*Fusarium oxysporum*. Gi—*Glomus intraradices*. Pi—*Piriformospora indica* (*Serendipita indica*). GiPi—Co-inoculation of *Glomus intraradices* and *Piriformospora indica* (*Serendipita indica*). *Values* represent the mean percentage of the control plants under low nutritional level and the bars the standard deviation. $N = 5$

acids represented a much lower percentage of the pic area of the polar fraction of the root exudates. Taking into consideration the general composition of the root exudates, those of the control plants, especially when not infected with *F. oxysporum*, differed from the other root exudates. Control plants had root exudates less enriched in sugars and enriched in amino acids and organic acids. According to their prevalence in the analyzed root exudates, glucose, fructose, malate, citrate, and succinate were the polar metabolites selected for further analysis (Fig. 3.4). Glucose and fructose were more prevalent whenever root was colonized by fungi. With the exception of the control plants, no significant differences were detected due to the presence of *F. oxysporum*. As far as organic acids are concerned, malate and citrate were the more abundant across treatments. And, contrary to sugars, the presence of fungi in the rhizosphere significantly decreases the presence of malate in the rhizosphere. From the roots colonized with fungi, those colonized with both Gi and Pi (in the presence or not of *F. oxysporum*) presented the lowest relative amounts of malate in the root exudates. No significant differences were observed for citrate concentrations across treatments.

In the nonpolar fraction of the root exudates, five main substances were identified (Fig. 3.5), namely, tryptophan, protocatechuic acid, chlorogenic acid, salicylic acid, and caffeic acid. The relative peak area ranged from 0 to 8%. Chlorogenic and caffeic acids were the only compounds where differences among treatments could

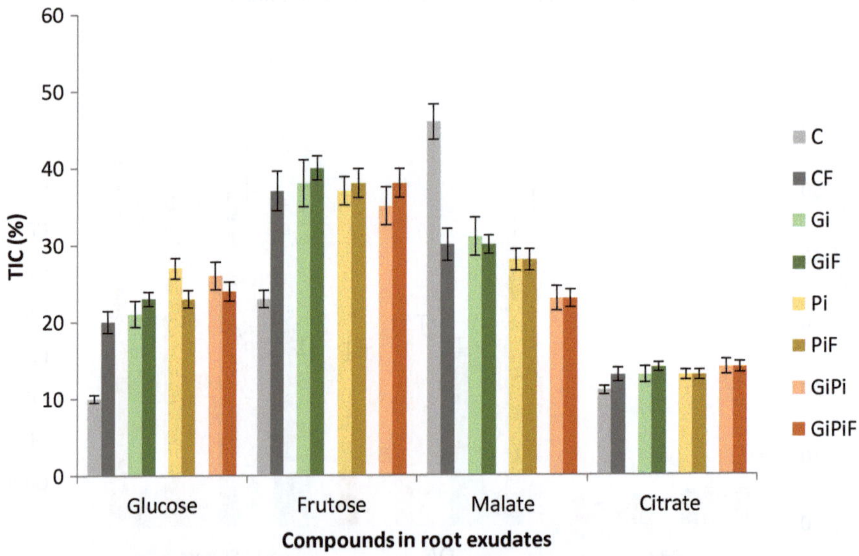

Fig. 3.4 Relative peak area (TIC) of selected compounds present in the root exudates from GC–MS analyses grouped into substance classes and treatments (mean ± Sd, $n = 5$). The low level nutritional treatments comprised: F—*Fusarium oxysporum*. Gi—*Glomus intraradices*. Pi—*Piriformospora indica* (*Serendipita indica*). GiPi—Co-inoculation of *Glomus intraradices* and *Piriformospora indica* (*Serendipita indica*). *Values* represent the mean percentage of the control plants under low nutritional level and the bars the standard deviation. $N = 5$

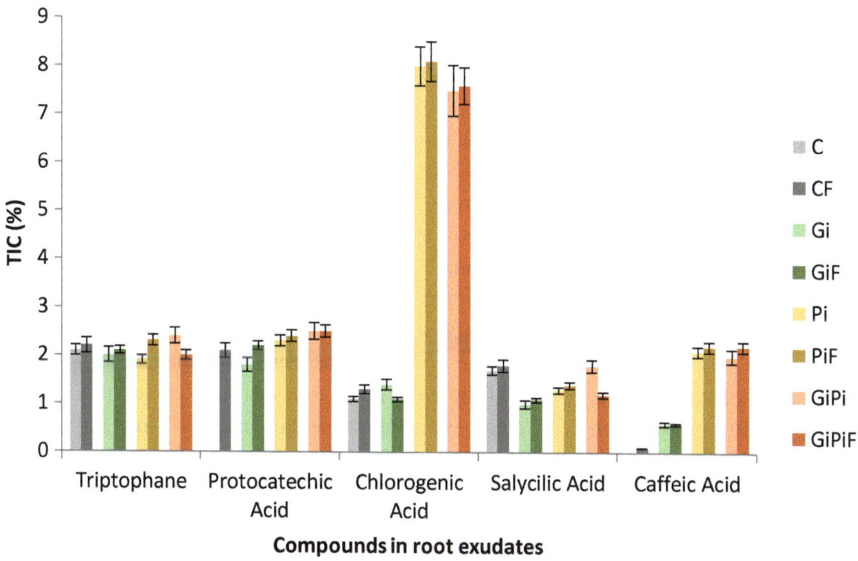

Fig. 3.5 Relative peak area (TIC) of selected compounds present in the root exudates from relative peak area (TIC) of HPLC–UV analyses (mean ± Sd, $n = 5$). The low level nutritional treatments comprised: F—*Fusarium oxysporum*. Gi—*Glomus intraradices*. Pi—*Piriformospora indica* (*Serendipita indica*). GiPi—Co-inoculation of *Glomus intraradices* and *Piriformospora indica* (*Serendipita indica*). *Values* represent the mean percentage of the control plants under low nutritional level and the bars the standard deviation. $N = 5$

be detected. Roots colonized with Pi presented higher levels of both acids in relation to the other ones.

3.4 Interpretation

In this study, we attempt to highlight the potential importance of fungi as regulators of ecosystem functions, namely of primary productivity. Depending on the functional group considered, fungi may contribute to increase or decrease plant biomass productivity. It is consensual that AMF and some endophytic fungi promote while phytopathogenic fungi decrease plant productivity. However, the rhizosphere is a very heterogeneous place and plants are simultaneously colonized by many fungi belonging to distinct functional groups. The output of this interaction is not well known and the results obtained so far are contradictory. However, soil microbes belonging to *Trichoderma* (Melo 1998) and *Bacillus* genera demonstrated to be efficient in controlling *Fusarium* wilt. Our dataset showed that Gi and Pi may be very efficient in controlling *F. oxysporum* disease (Fig. 3.1 and Table 3.1). Although these effects are very dependent on all biotic and abiotic variables, they show the potential of AMF and endophyte fungi to control *Fusarium* wilt.

The mechanisms behind these interactions are complex and result from multifactorial responses. However, root exudates are certainly involved. The effect may be a direct one resulting from the production of compounds by the AMF or endophytic fungi, competition between the AMF and endophytic fungi with the phytopathogen, or induction of plant defense pathways due to the priming effect of AMF and endophytic fungi. Independently of the mechanism involved, a change in the composition of the root exudates must be observed (Fig. 3.2–5). Although our results must be taken with caution due to technical limitations in the analysis of the composition of the root exudates, some changes were consistent with the treatments. The fungal (beneficial or pathogenic) colonization increased the sugar component of the root exudates, which makes sense since the fungus increases the root sink for carbon (Fig. 3.1). The opposite was observed for malate.

Priming of the immune system of the plant leads to a systemic protection against pathogens in combination with changes in the secondary metabolism (Jung et al. 2012), which may explain the relative increase in the amount of chlorogenic and caffeic acids when roots were colonized with Pi (Fig. 3.5). Chlorogenic acids are a group of phenolic secondary metabolites produced by certain plant species and an important component of coffee (*Coffea* spp.). Chlorogenic acid has been implicated in biotic and abiotic stress responses, while the related shikimate esters are key intermediates for lignin biosynthesis (Lallemand et al. 2012). Elevated levels of chlorogenic acid in transgenic tomato plants increased protection from UV light (Clé et al. 2008) and enhanced microbial resistance (Niggeweg et al. 2004). More recently, it has been shown that chlorogenic acid can act as a pest resistance factor in ornamental plants (Leiss et al. 2009). Meanwhile, the closely related shikimate esters are known to be key intermediates in the synthesis of lignin (Hoffmann et al. 2004; Chen and Dixon 2007).

3.5 Conclusion

Our results point out the relevance of mycorrhizal and endophytic fungi in the regulation of phytopathogenic fungi and show that the effect may be mediated by changes in the composition of the root exudates calling attention to chlorogenic acid, which is an important group of antioxidants, soluble esters formed between phenolic hydroxycinnamates and quinic acid. However, root exudates consist of many different compounds with different effects on microorganisms in the rhizosphere. Therefore, other compounds may be responsible or involved in the biological control of *Fusarium* wilt mediated by AMF Gi and Pi. Taking into consideration that fruits and vegetables are major sources of antioxidants, and high levels of these compounds in the diet are believed to contribute to improved health condition (Bazzano et al. 2002; Astley 2003), it would also be important to investigate the levels of these compounds in comestible parts of the plants.

These results apart from highlighting the importance of AMF and AMF-like fungi such as *P. indica* (*Serendipita indica*) in biocontrol also raise the question of crop management, since intensive soil management destroys the mycelium networks and may substantially minimize the work of fungi in controlling primary productivity.

Acknowledgment Ajit Varma is thankful to Department of Science and Technology and Department of Biotechnology for partial financial funding and to DST-FIST for providing confocal microscope facility.

References

Amaral DOJ, Almeida CMA, Malafaia CB, Silva MLRB, Correia MTS, Lima VLM, Silva MV (2013) Identification of races 1, 2 and 3 of *Fusarium oxysporum* f. sp. lycopersici by molecular markers. Afr J Microbiol Res 7:2324–2331. doi:10.5897/AJMR12.2234

Astley S (2003) Dietary antioxidants: past, present and future? Trends Food Sci Technol 14:93–98

Bazzano LA, He J, Ogden LG, Loria CM, Vupputuri S, Myers L, Whelton PK (2002) Fruit and vegetable intake and risk of cardiovascular disease in US adults: the first National Health and Nutrition Examination Survey Epidemiologic Follow-Up Study. Am J Clin Nutr 76:93–99

Borowicz VA (2001) Do arbuscular mycorrhizal fungi alter plant–pathogen relations? Ecology 82:3057–3068

Chen F, Dixon RA (2007) Lignin modification improves fermentable sugar yields for biofuel production. Nat Biotechnol 25:759–761

Clé C, Hill LM, Niggeweg R, Martin CR, Guisez Y, Prinsen E, Jansen MA (2008) Modulation of chlorogenic acid biosynthesis in Solanum lycopersicum: consequences for phenolic accumulation and UV-tolerance. Phytochemistry 69:2149–2156

Dighton J (2016) Fungi in ecosystem processes, 2nd edn. CRC Press, New York, pp 155–189

Giovanetti M, Mosse B (1980) An evaluation of techniques for measuring vesicular arbuscular mycorrhizal infection in roots. New Phytol 84:489–500

Hage-Ahmed K, Moyses A, Voglgruber A, Hadacek F, Steinkellner S (2013) Alterations in root exudation of intercropped tomato mediated by the arbuscular mycorrhizal fungus *Glomus mosseae* and the soilborne pathogen *Fusarium oxysporum* f.sp. lycopersici. J Phytopathol 161:763–773

Hoffmann L, Besseau S, Geoffroy P, Ritzenthaler C, Meyer D, Lapierre C, Pollet B, Legrand M (2004) Silencing of hydroxycinnamoyl-coenzyme A shikimate/quinate hydroxycinnamoyl transferase affects phenylpropanoid biosynthesis. Plant Cell 16:1446–1465

Jung SC, Martinez-Medina A, Lopez-Raez JA, Pozo MJ (2012) Mycorrhiza-induced resistance and priming of plant defenses. J Chem Ecol 38:651–664

Kanani H, Chrysanthopoulos PK, Klapa MI (2008) Standardizing GC–MS metabolomics. J Chromatogr B 871:191–201

Kopka J, Schauer N, Krueger S (2005) GMD.CSB.DB: the Golm Metabolome Database. Bioinformatics 21:1635–1638

Lallemand LA, Zubieta C, Lee SG, Wang Y, Acajjaoui S, Timmins J, McSweeney S, Jez JM, McCarthy JG, McCarthy AA (2012) A structural basis for the biosynthesis of the major chlorogenic acids found in coffee. Plant Physiol 160:249–260

Leiss KA, Maltese F, Choi YH, Verpoorte R, Klinkhamer PG (2009) Identification of chlorogenic acid as a resistance factor for thrips in Chrysanthemum. Plant Physiol 150:1567–1575

Melo IS (1998) Agentes microbianos de controle de fungos fitopatogênicos. In: Melo IS, Azevedo JL (eds) Controle Biológico, vol 1. Embrapa Meio Ambiente, Jaguariúna, pp 17–67

Nelson EB (1990) Exudate molecules initiating fungal responses to seeds and roots. Plant Soil 129:61–73

Niggeweg R, Michael AJ, Martin C (2004) Engineering plants with increased levels of the antioxidant chlorogenic acid. Nat Biotechnol 22:746–754

Phillips JM, Hayman DS (1970) Improved procedures for clearing roots and staining parasitic and vesicular-arbuscular mycorrhizal fungi for rapid assessment of infection. Trans Br Mycol Soc 55:158–160

Schwartz HF, Steadman JR, Hall R, Forster RL (eds) (2005) Compendium of bean diseases, 2nd edn. American Phytopathological Society, St. Paul, p 120

Sharma MPAG, Mukerji KG (2007) Arbuscular mycorrhiza mediated plant pathogen interactions and the mechanisms involved. In: Chincholkar SB, Mukerji KJ (eds) Biochemical control of plant diseases. Haworth Press, New York, pp 47–74

Smith SE, Read DJ (1997) Mycorrhizal symbiosis, 2nd edn. Academic Press, San Diego, pp 110–130

Steinkellner S, Hage-Ahmed K, Garcıa-Garrido JM, Illana A, Ocampo JA, Vierheilig H (2011) A comparison of wild-type, old and modern tomato cultivars in the interaction with the arbuscular mycorrhizal fungus *Glomus mosseae* and the tomato pathogen *Fusarium oxysporum* f.sp. lycopersici. Mycorrhiza 22:189–194

Termoshuizen AJ (2014) Root pathogens. In: Dighton J, Krumins JA (eds) Interactions in soil: promoting plant growth. Springer Science and Baseness Media, Dordrecht, pp 119–137

Vierheilig H, Coughlan AP, Wyss U, Piche Y (1998) Ink and vinegar, a simple staining technique for arbuscular-mycorrhizal fungi. Appl Environ Microbiol 64:5004–5007

Volpin HE, Okon Y, Kapulnik Y (1994) A vesicular-arbuscular mycorrhiza (*Glomus intraradices*) induces a defense response in alfalfa roots. Plant Physiol 104:683–689

Wellman FL (1939) A technique for studying host resistance and pathogenicity in tomato *Fusarium* wilt. Phytopathology 29:945–956

Chapter 4
Mycorrhiza: Creating Good Spaces for Interactions

Geetanjali Manchanda, Raghvendra Pratap Singh, Zhi Feng Li, and Jun Jie Zhang

Abstract Soil is a complicate environment, where complex systems of multiple interactions between the organisms take place. Plant health is majorly determined by these vital interactions in the soil. The ubiquitous arbuscular mycorrhizal (AM) fungi and a number of microbes interact synergistically to enhance the fitness of each other as well as plants they are associated with. Both the interacting partners are cross facilitators, where AM fungi provide suitable specialized ecological niches as well as nutrients for bacteria, and in turn bacteria improves the mycorrhization, provides pool of available P and N, and helps in management of biotic and abiotic stresses. Given the importance of AM and the interacting microbes in low-input sustainable agriculture, it is important to understand their interactions.

4.1 Introduction

Web of interactions between organisms spans all ecosystems and strongly influences the structure of natural populations and communities. As an example, over 90% of all land plants depend on symbiotic mycorrhizal associations for their survival. Estimates suggest that 74% of all plant species form arbuscular mycorrhiza (AM) with fungi of the Glomeromycota clade (Brundrett 2009; Smith and Read 2008), 2% of plants form ectomycorrhiza (EM) associations, 9% of plants form orchid mycorrhizas, and 1% of plants form ericoid mycorrhizas (Brundrett 2009; van der Heijden et al. 2015). Frank, in the year 1885, for the first time, used

G. Manchanda (✉)
Department of Botany, D.A.V. University, Jalandhar, Punjab, India
e-mail: g_manchanda@rediffmail.com; geetanjali10194@davuniversity.org

R.P. Singh • Z.F. Li
State Key Laboratory of Microbial Technology, College of Life Science, Shandong University, Jinan 250100, Shandong, People's Republic of China

J.J. Zhang
College of Food and Bioengineering, Zhengzhou University of Light Industry, Zhengzhou, Henan 450000, China

A. Varma et al. (eds.), *Mycorrhiza - Function, Diversity, State of the Art*,
DOI 10.1007/978-3-319-53064-2_4

the term “mykorrhiza” to describe the modified root structures of forest trees. The term has since been extended to cover a range of symbiotic associations between fungi and plant roots (Smith and Read 2008). About 6000 species in the Glomeromycotina, Ascomycotina, and Basidiomycotina have been recorded as mycorrhiza. Arbuscular mycorrhiza is the oldest and most widespread symbiosis present in all taxa of extant flora, occurring in the roots of most angiosperms and pteridophytes, along with some gymnosperms and the gametophytes of some lower plants like mosses and lycopods (Smith and Read 1997). Paleobotanical and molecular sequence data suggest an arbuscular mycorrhiza originate before the Devonian period when the first land plants formed associations with Glomalean fungi from the Glomeromycota about 460 million years ago (Heckman et al. 2001; Kistner and Parniske 2002; Redecker et al. 2000). This is estimated to be some 300–400 million years before the appearance of root nodule symbioses with nitrogen-fixing bacteria. Arbuscular mycorrhizal symbioses can be formed with a very wide range of plant species, as many as 250,000.

The mycobionts (fungal partners) facilitate the phytobionts (plant partners) to acquire soil resources (nutrients and water), while the photobionts deliver photoassimilates to the mycobionts (Buscot et al. 2000). Mycorrhizal fungi connect their plant hosts to the heterogeneously distributed nutrients in soil areas, not accessible by roots via a vast network of extraradical mycelium. Thus, the plant’s access to nutrients and water is improved by mycorrhizal fungi. It has been suggested that this symbiosis was fundamental for the evolution of early land plant and invade the harsh terrestrial environment, where depletion zones rapidly develop in the soil after element absorption by roots (Cairney 2000; Corradi and Bonfante 2012; Jansa et al. 2013; Simon et al. 1993; Taylor and Krings 2005). Therefore, it seems that the AM association could have evolved as a mutualistic symbiosis, facilitating the adaptation of plants to the terrestrial environment (Schüßler 2000).

The extraradical mycelium of mycorrhizal fungi not only increases the nutrient absorptive surface area of their host plant root systems but also provides a direct conduit for translocation of photosynthetically derived carbon to microsites in the soil and a large surface area for interaction with other microorganisms. The bacteria also occupy certain specific fungal niches, i.e., spores, extraradical hyphae, and intraradical mycelia. Furthermore, AM spores and hyphae are also a valuable source of food for many soil microorganisms (i.e., bacteria, other fungi, and nematodes) (Corradi and Bonfante 2012). Microscopic and molecular analysis showing bacterial colonization on the surface of AMF hyphae and spores demonstrates that an intimate relationship between AMF and microbes exists (Agnolucci et al. 2015; Scheublin et al. 2010; Toljander et al. 2006). Additionally, many of them harbor endobacteria in their cytoplasm (Bonfante and Anca 2009). These bacteria can also influence AMF fitness (Frey-Klett et al. 2007) and ecological function (Cheng et al. 2012; Feng et al. 2003; Hodge et al. 2001; Zhang et al. 2014). Consequently, microbes are recognized as a third part of the AM symbiosis, not just soilborne “free riders” (Jansa et al. 2013). The mycorrhizal symbiosis results in changes in the rhizosphere microbes and beneficial effects are the result of synergistic interactions among all rhizosphere microbes and AM (Linderman 1992). Therefore, a better understanding of interactions between AM and microbes in the rhizosphere is required, owing to their crucial ecological consequences and implications in sustainable agriculture. This review is an attempt to present a comprehensive account of AMF–bacterial interactions prevalent in the soil.

4.2 The Interactive Space

The rhizosphere (Hiltner 1904), defined as the narrow zone of soil subject to the influence of living roots, is characterized by intense bacterial activity as a result of a leakage or exudation of substances from the root (Curl and Truelove 1986; Pinton et al. 2001) further redefined rhizosphere as the zone that includes the soil influenced by the root along with the root tissues colonized by microorganisms (Morgan et al. 2005; Pinton et al. 2001). The rhizosphere concept was further expanded to include the ubiquitous mycorrhizal fungal associates of roots (Rawlings 1958). Arbuscular mycorrhizal fungi are naturally abundant in agricultural soils and may account for some 25% of the soil microbial biomass (Hamel 2007; Hamel et al. 1991; Olsson et al. 1999) and up to 80% of the fungal biomass (Baath et al. 2004; Kabir et al. 1997). A gram of soil can contain up to 30 m of AM fungal extraradical hyphae (Smith and Read 1997). The term mycorrhizosphere (Oswald and Ferchau 1968) refers to the zone of influence of the mycorrhiza (fungus root) in the soil. In the mycorrhizosphere, two different zones are discernible: first is the rhizosphere, a thin layer of soil that surrounds the root and is under the joint direct influence of the root, root hairs, and AM hyphae adjacent to the root, and second is the hyphosphere, a zone of AM hypha–soil interactions, not directly influenced by the root (Marschner 1995). The hyphosphere may be more or less densely permeated by the AM soil mycelium (8–20 km hyphae L^{-1} soil) (Schreiner et al. 1997).

Mycorrhizosphere soil is influenced by exudates from both the root tissue and the fungal hyphae. Root exudates play an important role in communication with rhizosphere-inhabiting microorganisms, and, for this communication, a broad range of substrates and signaling molecules are produced by plants. Plant roots release 5–21% of their photosynthetically fixed carbon as soluble sugars, amino acids, or secondary metabolites (Badri et al. 2013; Badri and Vivanco 2009; Chaparro et al. 2013), and these are used by the microbial communities in the rhizosphere. It is reported that plants produce a diverse range of more than 100000 different low-molecular mass natural products, known as secondary metabolites (Bais et al. 2004). Root exudates are important in establishment of AM symbiosis (Vierheilig 2004). The establishment of AMF symbiosis and infection structure can occur only in the presence of signals released by host roots (Czarnota et al. 2003; Smith and Read 1997). Mycorrhizal establishment is also known to change mineral nutrient composition, C allocation patterns, and hormonal balance in addition to other aspects of plant physiology (Barea 2000; Marschner and Timonen 2006) such as root exudate composition (Bansal and Mukerji 1994), which further influences the rhizosphere environment. These factors in turn affect the chemo tactic response of soil bacteria and composition of bacterial populations in the mycorrhizosphere (Buee et al. 2000; Soderberg et al. 2002; Sood 2003). Thus, the mycorrhizal symbiotic status changes the chemical composition of root exudates while the development of the fungal soil mycelium serves as a carbon source to rhizosphere microbial communities and introduces physical modifications into the environment surrounding the roots.

In addition to mycorrhiza-induced changes in the rhizosphere, there are specific modifications in the microenvironments surrounding the mycorrhizal mycelium itself (Gryndler et al. 2000). The extraradical mycelium of AM fungi not only constitutes a large surface area that interacts with the surrounding soil environment but also represents an important source of C for the soil microflora (George et al. 1995). A number of compounds are exuded into the soil by the AM extraradical mycelium (Wright et al. 1996; Wright and Upadhyaya 1996) that influence the chemical composition and pH of the soil environment (Bago and Azcón-Aguillar 1997). The analysis of exudates from a *Glomus* revealed the presence of low-molecular-weight sugars and organic acids and also unidentified high-molecular-weight compounds (Toljander et al. 2007). The release of carbonaceous exudates by AM hyphae (Toljander et al. 2007) and the rapid (5–6 days) turnover of these hyphae (Staddon et al. 2003) result in hyphosphere effect. The C compounds released have different bioactivity on different soil microorganisms of the hyphosphere, and as a result, mycelial exudates can not only increase microbial growth and vitality but can also change the composition of the microbial community (Duponnois et al. 2008; Toljander et al. 2007). The hyphosphere (Marschner 1995) supports biotic activities distinct from those in soils under joint root and fungal influence (Linderman 1988). Microbial activity may thus differ in mycorrhizosphere, hyphosphere, rhizosphere, and bulk soils (Andrade et al. 1997).

4.3 AM: A Specialized Ecological Niche

The interactions between AM fungi and bacteria associated with their extraradical mycelium as well as the presence of endosymbiotic intracellular bacteria living inside certain fungal isolates have been reported by a number of studies (Bianciotto et al. 1996, 2003; Bianciotto and Bonfante 2002; Roesti et al. 2005; Toljander et al. 2006). Bianciotto et al. (1996) investigated interaction between germinated spores of *Gigaspora margarita* with strains of either *Rhizobium leguminosarum* or *Pseudomonas fluorescens*. They observed that these rhizobacteria interacted with spore and hyphae (from germ tubes) of *Gi. Margarita* under sterile conditions. They also suggested that AMF are a vehicle for the colonization of plant roots by soil rhizobacteria. Thus, AM fungal spores and hyphae provide specific ecological niches for various populations of bacteria. The fungal partner often selects and hosts soil microorganisms, which could be of beneficial influence for fungal physiology and development (Artursson et al. 2006; Lumini et al. 2006; Xavier and Germida 2003). The microorganisms may either be derived from the surrounding soil or could be carried by fungal spores. These interactions are ecologically very significant (Buzzini et al. 2005). Bacteria such as *Paenibacillus brasiliensis* and *Pseudomonas fluorescens* have been found to be in direct physical association with extraradical hyphae of *Glomus intraradices* (Toljander et al. 2006). Bacteria have also been reported to be associated with AMF spores, and these bacteria colonize mainly the outer wall layer (Bonfante-Fasolo and Schubert 1987; Filippi et al. 1998; Maia and Kimbrough 1998; Walley and Germida 1996). *Alcaligenes*, *Bacillus*, *Corynebacterium*, and *Pseudomonas* have been reported to be associated with the spores of *Glomus versiforme* (Mayo et al. 1986) and spore walls of *Glomus clarum* NT4 (Xavier and Germida 2003). *Cellvibrio*, *Chondromyces*, *Flexibacter*, *Lysobacter*, and *Pseudomonas* spp. were found to be associated with spores of *Glomus geosporum* and *Glomus constrictum* (Roesti et al. 2005). Some bacteria have also been found in the cytoplasm of AMF spores (Bianciotto et al. 2000; MacDonald and Chandler 1981). A *Burkholderia* sp. bacterium was found in the cytoplasm of spores of *Gigaspora margarita*, which has recently been assigned the new taxonomical identity *Candidatus* Glomeribacter gigasporarum (Bianciotto et al. 1996, 2003).

The mycorrhizal hyphae and spores present a stable microhabitat for the establishment of bacterial biofilms in soil. A biofilm is a physical structure formed by aggregation of microorganisms, in which cells adhere to each other and/or to a surface. Bacterial attachment generally proceeds through two consecutive steps (Broek and Vanderleyden 1995). In the first step, the bacteria adhere loosely as single cells, whereas in the second step, the bacteria become more firmly attached to the plant root, and additional free bacteria are entrapped, resulting in the formation of large bacterial clusters at the attachment site to form biofilms. In natural ecosystems, it has been shown that up to 99% of all bacterial activities are associated with biofilms attached to solid surfaces (Costerton et al. 1987; Potera 1996). Biofilm represents a dominant organization of bacteria in nature, in which population of bacteria are embedded in an exopolysaccharide matrix secreted on a surface (Fujishige et al. 2006). This organization has several advantages for bacteria because it promotes higher resistance to environmental and biological stresses than planktonic cells (Burmolle et al. 2007). At the same time, a biofilm can facilitate hyphal penetration through the soil, and when hyphae colonize plant tissues, the bacteria can continue their functions. The bacteria located in hyphae can be released to the intercellular spaces of roots after hyphae are digested by endogenous enzymes (Fig. 4.1). In the intercellular spaces, these bacteria can interact with other endophytic microorganisms. Bacteria able to form biofilms on the surface of AMF mycelia might play an important role in some of the functions associated with AMF such as nutrient mobilization and protection against pathogens. Also, bacteria may have physical effects through biofilm formation, as well as chemical effects through the release of compounds in the exudates (Cruz and Ishii 2012).

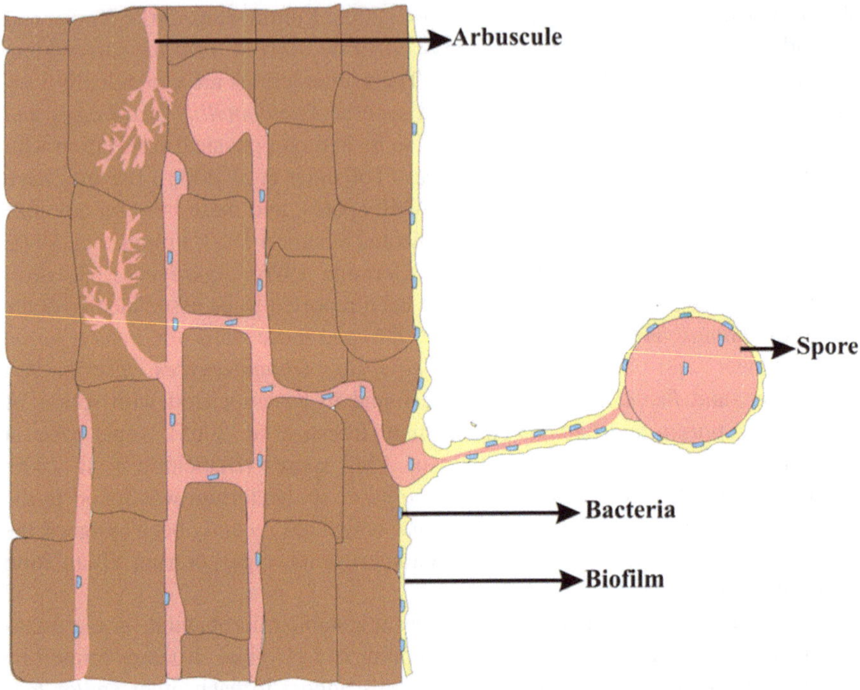

Fig. 4.1 Association between the AM and interacting bacteria

When associated with plant roots, mycorrhizal fungi receive up to 30% of the total carbon fixed and frequently transform it into trehalose, a disaccharide that has been proposed to behave as a carbon sink (Lopez et al. 2007; Wiemken 2007). Several studies have highlighted the possible role of trehalose in the interactions between bacteria and mycorrhizal fungi. It has been hypothesized that trehalose, secreted from fungal cells, can facilitate the colonization of the hyphae and the formation of biofilms on them. Trehalose was reported to be responsible for the selection of specific bacterial communities in the mycorrhizospheres of tree roots in forest nurseries and plantations (Frey et al. 1997; Izumi et al. 2006; Uroz et al. 2007).

The rhizobacteria with biocontrol abilities, *P. fluorescens* CHAO, formed sparse spots while two mucoid mutants of this strain (with increased production of acidic extracellular polysaccharides (EPS), essential for biofilm formation) formed a large number of clusters on non-mycorrhizal carrot roots, roots colonized with *Gi. margarita,* and extraradical hyphae of this AM fungus, demonstrating that EPS are involved in the in vitro association of *P. fluorescens* CHAO to these biological surfaces (Bianciotto et al. 2001a). Moreover, mutants of *Azospirillum brasilense* and *Rhizobium leguminosarum* affected in EPS production were strongly impaired in the capacity to attach to mycorrhizal root, AM, and inert structures (Bianciotto et al. 2001b). Various strains of *Burkholderia* inoculated on the germinating spores of *Gi. decipiens* were able to colonize the interior of the spores, demonstrating that AM colonization does not occur on AM surfaces only through biofilm formation (Levy et al. 2003; Lioussanne 2010). Further, Toljander et al. (2006) examined the attachment of five different strains of gfp-tagged soil bacteria [*Paenibacillus brasilensis* PB177 (pnf8), *Bacillus cereus* VA1 (pnf8), *Pseudomonas fluorescens* SBW25::gfp/lux, *Arthrobacter chlorophenolicus* A6G, and *Paenibacillus peoriae* BD62 (pnf8)] to vital and nonvital extraradical hyphae of the arbuscular mycorrhizal fungi *Glomus* sp. MUCL 43205 and *Glomus intraradices* MUCL 43194. The study revealed that *Arthrobacter chlorophenolicus* did not attach to hyphae, whereas the other bacterial strains did to a varying degree, indicating that that soil bacteria differed in their ability to colonize vital and nonvital hyphae and that this can also be influenced by the arbuscular mycorrhizal fungal species involved. Different bacterial populations may establish themselves under the influence of different AM plant–fungus combinations. The size and composition of bacterial populations in the rhizosphere has been shown to depend on the nature and quantity of root exudates that are controlled by AM fungi (Azaizeh et al. 1995). The subsequent competition between bacteria and AM fungi for this C source (Christensen and Jakobsen 1993) may determine the size and composition of bacterial populations in the rhizosphere.

It has been hypothesized that AM fungi stimulate the growth of rhizobacteria by serving of nutritional source through the liberation of exudates. Growth of *Pseudomonas chlororaphis* was shown to be stimulated in presence of crude extracts (containing not only AM exudates but also mycelial compounds) from the extraradical network of in vitro grown *G. intraradices* (Filion et al. Filion et al. 1999). Roesti et al. (2005) have suggested bacterial saprophytic activity in *G. geosporum* spores by scanning electron microscopy observations. The direct consumption of AM fungi was evidenced by the erosion of the spore's outer layer and its covering by mucilaginous products. AMF exudates were also shown to influence the vitality of soil bacteria. Toljander et al. (2007) investigated the effects of exudates produced by AM extraradical mycelia on the growth and development of an extracted soil bacterial community in vitro. The bacterial community extracted from soil was shown to be significantly affected after 48 h when inoculated with exudates produced by AM mycelia in comparison to a control composed of culture medium. Their study demonstrated the direct effects of mycelial exudates on a soil bacterial community. Sood (2003) investigated the chemotactic responses of the plant-growth-promoting rhizobacteria *Azotobacter chroococcum* and *Pseudomonas fluorescens* to roots of arbuscular mycorrhizal (*Glomus fasciculatum*) tomato plants. They reported a stronger attraction of *Azotobacter chrococcum* and *P. fluorescens* by exudates collected from tomato roots colonized by *G. fasciculatum* than by exudates collected from non-colonized roots. Scheublin et al. (2010) found specificity in bacterial attachment to AM hyphae and suggested that hyphal exudates shape hyphosphere bacterial communities through specific signaling that makes AM hyphae attractive to certain types of bacteria and not to others.

Molecular and morphological analyses demonstrate that some microbes select mycorrhizal fungi as a special niche for accomplishing their life cycle. AM fungi, as biotrophic and obligate plant symbionts, themselves host additional endosymbionts in their cytoplasm, biotrophic endobacteria (Bianciotto et al. 2003; Naumann et al. 2010; Torres-Cortes et al. 2015). AM fungi host intracellular structures very similar to bacteria, called bacteria-like organisms (BLOs) and first described in 1970s on the basis of electron microscope observations (Mosse 1970; Scannerini and Bonfante 1991). These bacteria have been found in several members of the Gigasporaceae (Bianciotto et al. 1996, 2000; Jargeat et al. 2004). Ultrastructural observations revealed rod-shaped BLOs in the vacuoles of germinating spores, often associated with large protein bodies (Bonfante and Balestrini 1994). Amplification of bacterial 16S RNA gene from total spore DNA followed by direct sequencing indicated that these bacteria are closely related to the genus *Burkholderia*, a group belonging to the β subdivision of the Proteobacteria and that they are present throughout the fungal life cycle (Bianciotto et al. 1996). On the basis of 16S ribosomal DNA (rDNA) sequence analysis, these bacteria had been initially assigned to the genus *Burkholderia* (Bianciotto et al. 1996) but were recently reassigned to a new taxon named "*Candidatus* Glomeribacter gigasporarum" (Bianciotto et al. 2003). They have been detected in all fungal compartments (spores, germ tube, and extra- and intraradical hyphae) except arbuscules (Bianciotto et al. 1996). Microbes, such as the endosymbiont *Burkholderia*, possess genes which may have an impact on bacterial, fungal, and plant metabolisms (Minerdi et al. 2002). Two types of endosymbionts are known in AMF: (1) a rod-shaped, Gram-negative beta-proteobacterium, related to *Burkholderia* (Bonfante and Balestrini 1994), *Candidatus* Glomeribacter gigasporarum (CaGg), common in several species of the family Gigasporaceae (Bianciotto et al. 2003; Desiro et al. 2014; Mondo et al. 2012), and (2) a much more widespread type, coccoid bacterium, displaying a homogeneous Gram-positive-like wall structure (Desiro et al. 2014; Scannerini and Bonfante 1991), which represents a currently undescribed taxon of Mollicutes related endobacteria (MRE) with a wide distribution across Glomeromycota (Naumann et al. 2010). The CaGg genome sequence (Ghignone et al. 2012) revealed that Glomeribacter endobacteria are nutritionally dependent on the fungal host and have a possible role in providing the fungus with essential factors like vitamin B12 (Ghignone et al. 2012). Phenotypic consequences of CaGg removal from the host include important morphological changes as well as reduced proliferation of host presymbiotic hyphae. Yet, the host is not obligately dependent on the bacteria (Lumini et al. 2007; Mondo et al. 2012). These features suggest that Glomeribacter endobacteria are mutualistic associates of AMF (Lumini et al. 2007). Comparisons of host and symbiont phylogenies indicate that while CaGg is a heritable endosymbiont (Bianciotto et al. 2004), it also engages in recombination and host switching, which play an important role in stabilizing this 400-million-year-old association (Mondo et al. 2012). MRE are associated with all major phylogenetic lineages of AMF studied so far and, thus, indirectly also with more than 80% of all land plants. They are coccoid, located in the cytoplasm without a surrounding host–membrane, and appear to possess a Gram-positive cell wall (Naumann et al. 2010). MRE were frequently detected in the intraradical and extraradical mycelium and in spores of AMF; however, they could never be detected free living (Naumann et al. 2010; Torres-Cortes et al. 2015). MRE have recently been demonstrated to also occur in several non-AMF species from the genus *Endogone* (Mucoromycotina), where some members are also plant symbionts (Desiro et al. 2015). Based on the 16S rRNA gene sequences, this novel lineage is sister to a clade encompassing the Mycoplasmatales and Entomoplasmatales (Naumann et al. 2010). During their long-lasting co-evolution, MRE have formed distinct, monophyletic evolutionary lineages within their fungal hosts, with a 16S rRNA gene (16S) sequence divergence of up to 20% (Desiro et al. 2014). The MRE have been detected in 17 out of 28 investigated AMF samples from culture collections, including members of Archaeosporales, Diversisporales, Glomerales (Naumann et al. 2010), as well as in mycorrhizal thalli of liverworts (Desiro et al. 2013). In most of the AMF hosts and irrespectively of the AMF identity, these endobacteria displayed a conspicuous variability in their 16S rRNA gene sequence. Torres-Cortes et al. (2015) have hypothesized that MRE play an important biological role in AM, consistent with the observation that they have been maintained as ancient endosymbionts in major evolutionary AMF lineages that separated hundreds of million years ago. The occurrence of these bacteria in AM fungi is intriguing, and their physiological role in fungal fitness as well as their potential role in mycorrhizal symbiosis are completely unknown. Collectively, these observations indicate that CaGg is a stable associate of Gigasporaceae, whereas the lifestyle of the MRE and the nature of their association with Glomeromycota are uncertain. Furthermore, Desiro et al. (2013) investigated the patterns of distribution and coexistence of the two endosymbionts, CaGg and MRE, in spore samples of several strains of *Gigaspora margarita*. They found that a single AM host can harbor both types of endobacteria, with MRE population being more abundant, variable, and prone to recombination than the CaGg one. Both endosymbionts seem to retain their genetic and lifestyle peculiarities regardless of whether they colonize the host alone or together. These findings showed for the first time that fungi support an intracellular bacterial microbiome, in which distinct types of endobacteria coexist in a single cell.

Salvioli et al. (2016) used next-generation sequencing to analyze the transcriptional profile of *Gigaspora margarita* in the presence and in the absence of its endobacterium in order to understand the effect of bacteria on fungal fitness. Transcriptome analysis revealed that the endobacterium had a stronger effect on the pre-symbiotic phase of the fungus in terms of increased fungal sporulation success, elevation of the fungal bioenergetic capacity, increased ATP production, and elicitation of mechanisms to detoxify reactive oxygen species. They showed that the bacteria seemed to enhance the fungal responsiveness to strigolactones. According to them, although the endobacterium exacts a nutritional cost on the AMF, endobacterial symbiosis improves the fungal ecological fitness by priming mitochondrial metabolic pathways and giving the AMF more tools to face environmental stresses. Thus, they hypothesized that as described for the human microbiota, endobacteria may increase AMF innate immunity.

4.4 The Functional Aspects of Interactions

4.4.1 Cross Facilitation

Mycorrhiza and bacteria act as cross facilitators, where each of them increases the fitness of the other interactive partner. Specific bacterial populations that colonize the spores and external mycelia of mycorrhizal fungi may actively influence the spore germination, growth of external fungal mycelia, and mycorrhizal root colonization. Mycorrhization helper bacteria (MHB) (Garbaye 1994) are known to stimulate the growth of mycelia and enhance the formation of mycorrhizas (Barea et al. 2002; Garbaye 1994; Gryndler et al. 2000). Mosse (1962) showed that some MHB as well as their culture filtrates were able to stimulate the spore germination of *G. mosseae*. Mayo et al. (1986) observed that the germination of *G. versiforme* spores was greatly reduced by surface sterilization as compared to the spores with naturally associated microbial communities. Further, the addition of bacteria (including *Pseudomonas* and *Corynebacterium* strains) isolated from nonsurface-disinfected spores also increased spore germination compared to disinfected spores. Tylka et al. (1991) proposed that volatile compounds secreted by streptomycetes positively impacted AM fungal germination.

Xavier and Germida (2003) reported that direct contact between the spores and bacteria was necessary for the induction of spore germination in *Glomus clarum*, indicating a ligand–receptor interaction between the two microbes. Root exudates stimulate the growth of the biopolymer-degrading populations that would in turn accelerate the decay of the outer spore walls. The presence of active biopolymer-degrading bacterial populations on the spore surface could support spore germination by releasing nutrients or degrading toxic compounds that inhibit germination. Thus, the process of maturation and eventual germination of AMF spores might benefit from the activity of the surface microorganisms degrading the outer hyaline layer (Roesti et al. 2005). Bacteria can stimulate the mycorrhizal spore germination by mechanisms such as erosion of spore walls (Filippi et al. 1998; Maia and Kimbrough 1998), production of stimulatory compounds such as CO_2 and other volatiles (Carpenter-Boggs et al. 1995), or by influencing AMF phosphorus acquisition (Ruiz-Lozano and Bonfante 2000).

Hildebrandt et al. (2002), while working with the *Paenibacillus validus–Glomus intraradices* interactions, showed that the otherwise obligately symbiotic *G. intraradices* could grow and sporulate in fungus–bacterium cocultures. During the initiation of mycorrhizal symbiosis, signal molecules such as phytohormones, enzymes, polysaccharides, phenolic compounds, adhesins, and volatiles are produced by the plants (Akiyama et al. 2005). MHB can influence through the synthesis of many of these chemicals. MHB may also detoxify metabolites produced by the fungus that inhibit mycelia growth (Duponnois and Garbaye 1990). Examples of MHB strains are predominantly *Bacillus* and *Pseudomonas*, but examples have also been found in the genera *Bradyrhizobium*, *Burkholderia*,

Paenibacillus, *Rhodococcus*, and *Streptomyces*. Lumini et al. (2007) proved that the presence of endosymbiotic bacteria, *Candidatus* Glomeribacter gigasporarum in *Gigaspora margarita*, strongly improves the presymbiotic growth of the fungus, as shown by increased hyphal elongation and branching following treatment with root exudates. Soil bacteria may stimulate the root exudation by production of certain compounds. This further stimulates the growth of mycorrhizal fungal mycelia in the rhizosphere or facilitates root penetration by the fungus (Barea 2000; Marulanda et al. 2006). The arbuscular mycorrhizal colonization of roots with *Glomus fistulosum* and the growth rate of the hyphae in the soil substrate were significantly higher when the fungus was co-inoculated with *Pseudomonas putida* or with the low-molecular-weight fraction of the bacterial culture supernatant (Vósatka and Gryndler 1999), indicating that the effective substances were in this fraction. Two isolates of *Paenibacillus validus* (DSM ID617 and ID618) were shown to stimulate growth of the arbuscular mycorrhizal fungus *Glomus intraradices* Sy167 up to the formation of fertile spores by Hildebrandt et al. (2006). They reported that a specific carbon source, raffinose, was present in bacterial cultures and mycelial growth was supported by this sugar. Salvioli et al. (2016) observed that the endobacterial symbiosis improved the fungal ecological fitness by priming mitochondrial metabolic pathways and giving the AMF more tools to face environmental stresses. The presence of the endobacterium tunes a huge number of metabolic pathways, including spore production, fungal wall remodeling, and mineral nutrient uptake and transport, also leading to a positive impact on the phosphate content of plant roots and shoots. Thus, although the endobacterium has a nutritional cost on the AMF, the symbiosis has proven beneficial to the AMF, from an evolutionary point of view.

In addition to providing suitable niches as well as nutrition for bacteria, the fungal partner also might influence coinhabiting species via the secretion of chemical substances. The nature of fungal exudates may decide the type of interaction with the bacteria. Mycorrhiza selects the prokaryotes with which it wants to cooperate. The exudation of antibiotics is an important fungal contribution to the mutualism. Therefore, the mycorrhiza specifically favors the growth of symbiosis-promoting bacteria and prevents infection with antagonistic ones. Andrade et al. (1998) showed that root and fungal components of mycorrhiza affect soil bacteria indirectly by influencing water–stable soil aggregates and providing a favorable, protective environment for the organisms. The endocellular mode of living may be advantageous for many bacteria as it provides physical protection of these bacteria from competition of space and nutrition, as well as grazing by the predators in the soil.

4.4.2 Nutrition

It has been previously reported that AM fungi have no known saprotrophic capability, which makes them incapable of breaking down the organic nutrients

(Gryndler 2003; Joner et al. 2000; Leigh et al. 2009; Tisserant et al. 2013). However, microbes play important and varied roles in elemental [e.g., C, nitrogen (N) and phosphorus (P)] biogeochemical cycles (Nannipieri et al. 2003; Torsvik and Øvreas 2002). Since microbes are able to release various enzymes to decompose organic matter, it has been hypothesized that in doing so, they provide the AMF hyphae with inorganic nutrients (Hodge 2014; Zhang et al. 2014, 2016). Mycorrhizal hyphae exude labile C and thus increase local nutrient availability in the hyphosphere, which in turn results in stimulation of surrounding soil microbes (Cheng et al. 2012; Jansa et al. 2013). The microbes utilize AM-released C and pay back the benefits in terms of released mineral nutrients. The transcriptome of *Glomus intraradices* (DAOM 197198) has revealed that this AM fungus may have a low capability of utilizing phytate because it lacks phytase protein (Tisserant et al. 2012). Also, although arbuscular mycorrhizal fungi have crucial role in the phosphorus cycle that is fundamental to sustainable crop plant productivity, they lack the ability to secrete phosphatases (Tisserant et al. 2013). Therefore, the AM fungi are unable to utilize organic P from the soil and then provide to the plant. However, >40% of culturable bacteria are able to mineralize organic P (the so-called phosphate-solubilizing bacteria (PSB) by releasing numerous phosphatases into the surrounding soil (Jorquera et al. 2008). Phosphate solubilizing bacteria are free-living soil microorganisms that are present in most soils (Rodriguez and Fraga 1999) and have the capacity to solubilize organic P, improving its ability for uptake. PSB solubilize organic and inorganic P, through the action of synthesized phosphatases, by lowering the pH of the soil and/or chelating P from Al^{3+}, Fe^{3+}, Fe^{2+}, and Ca^{2+} with the help of organic acids (Browne et al. 2009; Rodriguez and Fraga 1999). Ordonez et al. (2016) reported that P solubilizing-capable *Pseudomonas* bacteria and AMF can have positive effects on each other's growth that could potentially lead to synergism between some combinations. Studies by Zhang et al. (2016) revealed that beneficial interactions between an AMF and a PSB occur, with each providing a key resource for the other. PSB were shown to be responsible for organic P hydrolysis by releasing phosphatases, while AMF could acquire the inorganic P subsequently released, and AMF hyphal growth was enhanced. AMF release C compounds into the hyphosphere which the PSB were demonstrated to utilize.

Studies by Wang et al. (2016) provide evidence that a PSB strain colonizing the hyphosphere assimilates photosynthates of the host plant from hyphal exudates of an AM fungus. In addition, the hyphal exudates prime the activity of the PSB, which further accelerates phytate-P turnover in the hyphosphere. Their results provide the first in situ demonstration of the pathway underlying the carbon flux from plants to the AM mycelium-associated PSB, and the PSB assimilated the photosynthates exuded by the fungus and promoted mineralization and turnover of organic P in the soil. Further, Minerdi et al. (2001) demonstrated in germinating spores of *G. margarita,* the expression of *nifHDK* (for nitrogen fixation) genes could indicate that the *Burkholderia* endobacteria supply the fungus with nitrogen during its pre-infection growth.

4.4.3 Biotic and Abiotic Stress Management

The bacteria associated with the mycorrhiza perform important functions such as protection against plant pathogens. The possible role of AM fungi in reduction of the deleterious effects of soilborne pathogens has been previously highlighted (Gerdemann 1974; Whipps 2004). The potential of AMF to control various plant pathogenic fungi has been clearly demonstrated (Boyetchko and Tewari 1996; Kapoor et al. 1998; Kasiamdari et al. 2002; Krishna and Bagyaraj 1983). Meyer and Linderman (1986) found a lower number of sporangia and zoospores formed by cultures of *Phytophthora cinnamomi* in leachates of rhizosphere soil from AM plants than from non-AM plants, suggesting that sporangium-inducing microorganisms had declined or sporangium-inhibitors had increased. Secilia and Bagyaraj (1987) found that there were more pathogen-antagonistic actinomycetes in the rhizosphere of mycorrhizal plants than in that of non-mycorrhizal controls. It has been suggested that higher level interactions between the host plants, AMF, and AMF-associated microbes may also be responsible for some of the reported bioprotection effects conferred to the AMF as well as to the associated plants by the AMF. Inhibition of different plant fungal pathogens, such as *Rhizoctonia solani* by a *Paenibacillus* strain isolated from surface-sterilized *Glomus mosseae* spores was reported (Budi et al. 1999). The occurrence of antagonistic isolates depended on AM fungal species, but not plant host, and originated from *G. intraradices* spores. Out of 18 cultivable isolates from surface-disinfected spores of *G. mosseae*, 14 (especially isolates identified as *Bacillus simplex* but also as *B. niacini*, *B. drententis*, *Paenibacillus* spp., and *Methylobacterium* spp.) showed antagonism against various soilborne pathogens (*P. nicotianae* particularly, but also *F. solani* and three stains of *F. oxysporum*) (Lioussanne 2007). Bharadwaj et al. (2008) also studied the effect of AM fungal spore-associated bacteria on plant pathogens and evaluated the formation of siderophores. They found that found that species assemblages of cultivable bacteria from surface-disinfected spores of *G. mosseae* and *G. intraradices* were influenced both by fungal and plant species, with spore-type being the most prominent factor. This specificity of interaction AM species dependent was hypothesized to be related to spore size and surface roughness. Their study revealed that a high number of bacteria inhibit the growth of the plant pathogen *Rhizoctonia solani*, although the active compounds have not been identified in this study. In addition, it was shown that 16 of 57 antagonistic isolates (fluorescent pseudomonads) produce siderophores (Bharadwaj et al. 2008). Prokaryotes such as fluorescent pseudomonads and actinomycetes, capable of producing a broad range of antibiotics and chitinolytic enzymes, have been found to be closely associated with the AMF. This is likely one of the mechanisms of suppression of certain saprophytic or parasitic soil fungi by the AMF (Harrier and Watson 2004; Pozo et al. 2002; Toussaint et al. 2008), which may be beneficial for both the AMF (suppression of mycoparasitic *Trichoderma*) and for the mycorrhizal plants (biocontrol of *Fusarium* and other pathogens) (Barea et al. 2005; Jaderlund et al. 2008; Toussaint et al. 2008).

Citernesi et al. (1994) studied the influence of the biocontrol compound iturin A2, secreted by *Bacillus subtilis* strain M51, on AM fungi. The saprophytic growth of the fungus *G. mosseae* was inhibited by iturin A2, and no retardation in growth or establishment of symbiosis was noticed in the presence of the tomato host plant (*Lycopersicon esculentum*), whereas infection with competing species was hindered. The ability of AM fungi to specifically harbor and then to stimulate rhizobacteria with biocontrol properties suggests that these bacteria would directly reduce pathogen development within the mycorrhizosphere and would consequently strongly contribute to the biocontrol mediated by AM fungi on soilborne diseases.

Marulanda et al. (2006) investigated the interactions between *Bacillus thuringiensis*, a drought-adapted bacterium, and two isolates of *Glomus intraradices* on *Retama sphaerocarpa*, a drought-adapted legume. Increased root growth and relative water uptake was observed by co-inoculation of *B. thuringiensis* and the *G. intraradices*. *G. intraradices*-colonized roots showed the highest intensity and arbuscule richness when associated with *B. thuringiensis*. Co-inoculation of autochthonous microorganisms reduced by 42% the water required to produce 1 mg of shoot biomass. Thus, their study provided evidence of the effectiveness of rhizosphere bacterium, singly or associated with AM fungus, in increasing plant water uptake, which represents a positive microbial effect on plants grown under drought environments. Vivas et al. (2003) studied the effects of bacterial inoculation (*Bacillus* sp.) on symbiosis between lettuce and the *Glomus mosseae* and *Glomus intraradices*. They reported that the effects of each fungus on plant physiology were modulated by the bacterium. *Bacillus* sp. inoculation improved all plant and fungal parameters under stress. The highest amount of live and active AM mycelium for both fungi was obtained after co-inoculation with *Bacillus* sp. suggesting that selected free-living bacteria and AM fungi can be co-inoculated to optimize the formation and functioning of the AM symbiosis in both normal and adverse environments. Sarand et al. (1998) suggested that mycorrhizal hyphae were able to support microbial biofilms of catabolic plasmid (Tol+)-harboring bacteria which could be active in bioremediation of petroleum-contaminated soil. In further experiments, these authors (Sarand et al. 2000) demonstrated that the number of Tol+ bacteria was higher in mycorrhizospheric soil compared with bulk soil and inoculation with bacteria had a positive effect on plant and fungal development. The presence of easily available plant-derived carbon sources did not impede the degradation of the m-toluate by the bacteria (Sarand et al. 1999).

Mycorrhizal symbioses are ubiquitous in terrestrial ecosystems and therefore have important roles in improving the fitness of plants as well as other interacting microbes. Given the importance of mycorrhizal fungi and the associated microbes, it is vital to understand their interactions. Analysis of the interacting microbes will not only highlight their respective ecological roles but also will point toward the evolutionary significance of these interactions. Further research should focus on the functional mechanisms of such cross-facilitative interactions and identification of specific efficient microbes that can interact positively with the mycorrhizal fungi.

The results from such investigations can be used for designing specialized mixed inocula for specific plants present in specific environments so as to benefit low-input sustainable cropping systems.

References

Agnolucci M, Battini F, Cristani C, Giovannetti M (2015) Diverse bacterial communities are recruited on spores of different arbuscular mycorrhizal fungal isolates. Biol Fertil Soils 51:379–389

Akiyama K, Matsuzaki K, Hayashi H (2005) Plant sesquiterpenes induce hyphal branching in arbuscular mycorrhizal fungi. Nature 435:824–827

Andrade G, Mihara KL, Linderman RG, Bethlenfalvay GJ (1997) Bacteria from rhizosphere and hyphosphere soils of different arbuscular-mycorrhizal fungi. Plant Soil 192:71–79

Andrade G, Linderman RG, Bethlenfalvay GJ (1998) Bacterial associations with the mycorrhizosphere and hyphosphere of the arbuscular mycorrhizal fungus *Glomus mosseae.* Plant Soil 202:79–87

Artursson V, Finlay RD, Jansson JK (2006) Interactions between arbuscular mycorrhizal fungi and bacteria and their potential for stimulating plant growth. Environ Microbiol 8:1–10

Azaizeh H, Marschner H, Romheld V, Wittenmayer L (1995) Effects of a vesicular-arbuscular mycorrhizal fungus and other soil microorganisms on growth, mineral nutrient acquisition and root exudation of soil grown maize plants. Mycorrhiza 5:321–327

Baath E, Nilsson L, Goransson H, Wallander H (2004) Can theextent of degradation of soil fungal mycelium during soil incubation be used to estimate ectomycorrhizal biomass in soil. Soil Biol Biochem 36:2105–2109

Badri DV, Vivanco JM (2009) Regulation and function of root exudates. Plant Cell Environ 32:666–681

Badri DV, Chaparro JM, Zhang R, Shen Q, Vivanco JM (2013) Application of natural blends of phytochemicals derived from the root exudates of *Arabidopsis* to the soil reveal that phenolic-related compounds predominantly modulate the soil microbiome. J Biol Chem 288:4502–4512

Bago B, Azcón-Aguillar C (1997) Changes in the rhizosphere pH induced by arbuscular mycorrhiza formation in onion (*Allium cepa* L.) Zt Pflanzenern Bodenkde 160:333–339

Bais HP, Park SW, Weir TL, Callaway RM, Vivanco JM (2004) How plants communicate using the underground information superhighway. Trends Plant Sci 9:26–32

Bansal M, Mukerji K (1994) Positive correlation between VAM induced changes in root exudation and mycorrhizosphere mycoflora. Mycorrhiza 5:39–44

Barea J (2000) Rhizosphere and mycorrhiza of field crops. In: Balázs E, Galante E, Lynch J, Schepers J, Toutant J, Werner D, Werry P (eds) Biological resource management: connecting science and policy. Springer/INRA Editions, Berlin/New York, pp 110–125

Barea JM, Rosario A, Aguilar CA (2002) Mycorrhizosphere interactions to improve plant fitness and soil quality. Antonie Van Leeuwenhoek 81:343–351

Barea JM, Pozo MJ, Azcon R, Azcon-Aguilar C (2005) Microbial co-operation in the rhizosphere. J Exp Bot 56:1761–1778

Bharadwaj DP, Lundquist PO, Persson P, Alstrom S (2008) Evidence for specificity of cultivable bacteria associated with arbuscular mycorrhizal fungal spores. FEMS Microbiol Ecol 65:310–322

Bianciotto V, Bonfante P (2002) Arbuscular mycorrhizal fungi: a specialised niche for rhizospheric and endocellular bacteria. Antonie Van Leeuwenhoek 81:365–371

Bianciotto V, Bandi C, Minerdi D, Sironi M, Tichy HV, Bonfante P (1996) An obligately endosymbiotic mycorrhizal fungus itself harbors obligately intracellular bacteria. Appl Environ Microbiol 62:3005–3010

Bianciotto V, Lumini E, Lanfranco L, Minerdi D, Bonfante P, Perotto S (2000) Detection and identification of bacterial endosymbionts in arbuscular mycorrhizal fungi belonging to the family Gigasporaceae. Appl Environ Microbiol 66:4503–4509

Bianciotto V, Andreotti S, Balestrini R, Bonfante P, Perotto S (2001a) Mucoid mutants of the biocontrol strain *Pseudomonas fluorescens* CHA0 show increased ability in biofilm formation on mycorrhizal and nonmycorrhizal carrot roots. Mol Plant-Microbe Interact 14:255–260

Bianciotto V, Andreotti S, Balestrini R, Bonfante P, Perotto S (2001b) Extracellular polysaccharides are involved in the attachment of *Azospirillum brasilense* and *Rhizobium leguminosarum* to arbuscular mycorrhizal structures. Eur J Histochem 45:39–49

Bianciotto V, Lumini E, Bonfante P, Vandamme P (2003) 'Candidatus glomeribacter gigasporarum' gen. nov., sp. nov., an endosymbiont of arbuscular mycorrhizal fungi. Int J Syst Evol Microbiol 53:121–124

Bianciotto V, Genre A, Jargeat P, Lumini E, Becard G, Bonfante P (2004) Vertical transmission of endobacteria in the arbuscular mycorrhizal fungus *Gigaspora margarita* through generation of vegetative spores. Appl Environ Microbiol 70:3600–3608

Bonfante P, Anca IA (2009) Plants, mycorrhizal fungi, and bacteria: a network of interactions. Annu Rev Microbiol 63:363–383

Bonfante P, Balestrini A (1994) Storage and secretion processes in the spore of *Gigaspora margarita* Becker & Hall as revealed by high-pressure freezing and freeze substitution. New Phytol 128:93–101

Bonfante-Fasolo P, Schubert A (1987) Spore wall architecture of *Glomus* spp. Can J Bot 65:539–546

Boyetchko S, Tewari J (1996) Use of VA mycorrhizal fungi in soil-borne disease management. In: Utkhede R, Gupta V (eds) Management of soil borne diseases. Kalyani Publishers, New Delhi, pp 146–163

Broek A, Vanderleyden J (1995) The role of bacterial motility, chemotaxis and attachment in bacteria–plant interaction. Mol Plant-Microbe Interact 8:800–810

Browne P, Rice O, Miller S, Burke J, Dowling D, Morrissey J (2009) Superior inorganic phosphate solubilization is linked to phylogeny within the *Pseudomonas fluorescens* complex. Appl Soil Ecol 43:131–138

Brundrett M (2009) Mycorrhizal associations and other means of nutrition of vascular plants: understanding the global diversity of host plants by resolving conflicting information and developing reliable means of diagnosis. Plant Soil 320:37–77

Budi SW, van Tuinen D, Martinotti G, Gianinazzi S (1999) Isolation from the *Sorghum bicolor* mycorrhizosphere of a bacterium compatible with arbuscular mycorrhiza development and antagonistic towards soilborne fungal pathogens. Appl Environ Microbiol 65:5148–5150

Buee M, Rossignol M, Jauneau A, Ranjeva R, Becard G (2000) The pre-symbiotic growth of arbuscular mycorrhizal fungi is induced by a branching factor partially purified from plant root exudates. Mol Plant-Microbe Interact 13:693–698

Burmolle M, Hansen LH, Sorensen SJ (2007) Establishment and early succession of a multispecies biofilm composed of soil bacteria. Microb Ecol 54:352–362

Buscot F, Munch JC, Charcosset JY, Gardes M, Nehls U, Hampp R (2000) Recent advances in exploring physiology and biodiversity of ectomycorrhizas highlight the functioning of these symbioses in ecosystems. FEMS Microbiol Rev 24:601–614

Buzzini P, Gasparetti C, Turchetti B, Cramarossa MR, Vaughan-Martini A, Martini A, Pagnoni UM, Forti L (2005) Production of volatile organic compounds (VOCs) by yeasts isolated from the ascocarps of black (*Tuber melanosporum* Vitt.) and white (*Tuber magnatum* Pico) truffles. Arch Microbiol 184:187–193

Cairney JWG (2000) Evolution of mycorrhiza systems. Naturwissenschaften 87:467–475

Carpenter-Boggs L, Loynachan T, Stahl P (1995) Spore germination of *Gigaspora margarita* stimulated by volatiles of soil-isolated Actinomycetes. Soil Biol Biochem 27:1445–1451

Chaparro JM, Badri DV, Bakker MG, Sugiyama A, Manter DK, Vivanco JM (2013) Root exudation of phytochemicals in *Arabidopsis* follows specific patterns that are developmentally programmed and correlate with soil microbial functions. PLoS One 8:e55731
Cheng L, Booker FL, Tu C, Burkey KO, Zhou L, Shew HD, Rufty TW, Hu S (2012) Arbuscular mycorrhizal fungi increase organic carbon decomposition under elevated CO_2. Science 337:1084–1087
Christensen H, Jakobsen I (1993) Reduction of bacterial growth by a vesicular-arbuscular mycorrhizal fungus in the rhizosphere of cucumber (*Cucumis sativus* L.) Biol Fertil Soils 15:253–258
Citernesi AS, Filippi C, Bagnoli G, Giovannetti M (1994) Effects of the antimycotic molecule Iturin A2, secreted by *Bacillus subtilis* strain M51, on arbuscular mycorrhizal fungi. Microbiol Res 149:241–246
Corradi N, Bonfante P (2012) The arbuscular mycorrhizal symbiosis: origin and evolution of a beneficial plant infection. PLoS Pathog 8:e1002600
Costerton JW, Cheng KJ, Geesey GG, Ladd TI, Nickel JC, Dasgupta M, Marrie TJ (1987) Bacterial biofilms in nature and disease. Annu Rev Microbiol 41:435–464
Cruz AF, Ishii T (2012) Arbuscular mycorrhizal fungal spores host bacteria that affect nutrient biodynamics and biocontrol of soil-borne plant pathogens. Biol Open 1:52–57
Curl E, Truelove B (1986) The rhizosphere. Advanced series in agricultural sciences. Springer, Berlin, vol 15
Czarnota MA, Rimando AM, Weston LA (2003) Evaluation of root exudates of seven sorghum accessions. J Chem Ecol 29:2073–2083
Desiro A, Naumann M, Epis S, Novero M, Bandi C, Genre A, Bonfante P (2013) Mollicutes-related endobacteria thrive inside liverwort-associated arbuscular mycorrhizal fungi. Environ Microbiol 15:822–836
Desiro A, Salvioli A, Ngonkeu EL, Mondo SJ, Epis S, Faccio A, Kaech A, Pawlowska TE, Bonfante P (2014) Detection of a novel intracellular microbiome hosted in arbuscular mycorrhizal fungi. ISME J 8:257–270
Desiro A, Faccio A, Kaech A, Bidartondo MI, Bonfante P (2015) Endogone, one of the oldest plant-associated fungi, host unique Mollicutes-related endobacteria. New Phytol 205:1464–1472
Duponnois R, Garbaye J (1990) Some mechanisms involved in growth stimulation of ectomycorrhizal fungi by bacteria. Can J Bot 68:2148–2152
Duponnois R, Kisa M, Prin Y, Ducousso M, Plenchette C, Lepage M, Galiana A (2008) Soil factors influencing the growth response of *Acacia holosericea* A. Cunn. ex G. Don to ectomycorrhizal inoculation. New For 2:105–117
Feng G, Song YC, Li XL, Christie P (2003) Contribution of arbuscular mycorrhizal fungi to utilization of organic sources of phosphorus by red clover in a calcareous soil. Appl Soil Ecol 22:139–148
Filion M, Starnaud M, Fortin JA (1999) Direct interaction between the arbuscular mycorrhizal fungus *Glomus intraradices* and different rhizosphere microorganisms. New Phytol 141:525–533
Filippi C, Bagnoli G, Citernesi AS, Giovannetti M (1998) Ultrastructural spatial distribution of bacteria associated with sporocarps of *Glomus mosseae*. Symbiosis 24:1–12
Frey P, Frey-Klett P, Garbaye J, Berge O, Heulin T (1997) Metabolic and genotypic fingerprinting of fluorescent Pseudomonads associated with the Douglas Fir-laccaria bicolor mycorrhizosphere. Appl Environ Microbiol 63:1852–1860
Frey-Klett P, Garbaye J, Tarkka M (2007) The mycorrhiza helper bacteria revisited. New Phytol 176:22–36
Fujishige NA, Kapadia NN, De Hoff PL, Hirsch AM (2006) Investigations of *Rhizobium* biofilm formation. FEMS Microbiol Ecol 56:195–206
Garbaye (1994) Helper bacteria: a new dimension to the mycorrhizal symbiosis. New Phytol 128:197–210

George E, Marschner H, Jakobsen I (1995) Role of arbuscular mycorrhizal fungi in uptake of phosphorus and nitrogen from soil. Crit Rev Biotechnol 15:257–270

Gerdemann JW (1974) Vesicula-arbuscular mycorrhiza. Academic, New York

Ghignone S, Salvioli A, Anca I, Lumini E, Ortu G, Petiti L, Cruveiller S, Bianciotto V, Piffanelli P, Lanfranco L, Bonfante P (2012) The genome of the obligate endobacterium of an AM fungus reveals an interphylum network of nutritional interactions. ISME J 6:136–145

Gryndler MJJ (2003) Chitin stimulates development and sporulation of arbuscular mycorrhizal fungi. Appl Soil Ecol 22:283–287

Gryndler M, Hrselova H, Striteska D (2000) Effect of soil bacteria on hyphal growth of the arbuscular mycorrhizal fungus *Glomus claroideum*. Folia Microbiol (Praha) 45:545–551

Hamel C (2007) Extraradical arbuscular mycorrhizal mycelia: shadowy figures in the soil. In: Hamel C, Plenchette C (eds) Mycorrhizae in crop production: applying knowledge. Haworth, Binghampton, pp 1–36

Hamel C, Barrantes-Cartin U, Furlan V, Smith D (1991) Endomycorrhizal fungi in nitrogen transfer from soybean to maize. Plant Soil 138:33–40

Harrier LA, Watson CA (2004) The potential role of arbuscular mycorrhizal (AM) fungi in the bioprotection of plants against soil-borne pathogens in organic and/or other sustainable farming systems. Pest Manag Sci 60:149–157

Heckman DS, Geiser DM, Eidell BR, Stauffer RL, Kardos NL, Hedges SB (2001) Molecular evidence for the early colonization of land by fungi and plants. Science 293:1129–1133

Hildebrandt U, Janetta K, Bothe H (2002) Towards growth of arbuscular mycorrhizal fungi independent of a plant host. Appl Environ Microbiol 68:1919–1924

Hildebrandt U, Ouziad F, Marner FJ, Bothe H (2006) The bacterium *Paenibacillus validus* stimulates growth of the arbuscular mycorrhizal fungus *Glomus intraradices* up to the formation of fertile spores. FEMS Microbiol Lett 254:258–267

Hiltner L (1904) Uber neuere Erfahrungen und Probleme auf dem Gebiete der Bodenbakteriologie unter besonderden berucksichtigung und Brache. Arb Dtsch Landwirtsch Gesellschaft 98:59–78

Hodge A (2014) Interactions between arbuscular mycorrhizal fungi and organic material substrates. Adv Appl Microbiol 89:47–99

Hodge A, Campbell CD, Fitter AH (2001) An arbuscular mycorrhizal fungus accelerates decomposition and acquires nitrogen directly from organic material. Nature 413:297–299

Izumi H, Anderson IC, Alexander IJ, Killham K, Moore ER (2006) Endobacteria in some ectomycorrhiza of Scots pine (*Pinus sylvestris*). FEMS Microbiol Ecol 56:34–43

Jaderlund L, Arthurson V, Granhall U, Jansson JK (2008) Specific interactions between arbuscular mycorrhizal fungi and plant growth-promoting bacteria: as revealed by different combinations. FEMS Microbiol Lett 287:174–180

Jansa J, Bukovska P, Gryndler M (2013) Mycorrhizal hyphae as ecological niche for highly specialized hypersymbionts – or just soil free-riders? Front Plant Sci 4:134. doi:10.3389/fpls.2013.00134

Jargeat P, Cosseau C, Ola'H B, Jauneau A, Bonfante P, Batut J, Becard G (2004) Isolation, free-living capacities, and genome structure of "*Candidatus* glomeribacter gigasporarum," the endocellular bacterium of the mycorrhizal fungus *Gigaspora margarita*. J Bacteriol 186:6876–6884

Joner EJ, Van Aarle IM, Vosátka M (2000) Phosphatase activity of extra-radical arbuscular mycorrhizal hyphae: a review. Plant Soil 226:199–210

Jorquera M, Hernandez M, Rengel Z, Marschner P, Mora M (2008) Isolation of culturable phosphor bacteria with both phytate-mineralization and phosphate-solubilization activity from the rhizosphere of plants grown in a volcanic soil. Biol Fertil Soils 44:1025–1034

Kabir Z, O'Halloran I, Fyles J, Hamel C (1997) Seasonal changes of arbuscular mycorrhizal fungi as affected by tillage practices and fertilization: hyphal density and mycorrhizal root colonization. Plant Soil 192:285–293

Kapoor A, Mukherji K, Kapoor R (1998) Microbial interactions in mycorrhizosphere of *Anethum graveolens* L. Phytomorphology 48:383–389

Kasiamdari R, Smith S, Smith F, Scott E (2002) Influence of the mycorrhizal fungus, *Glomus coronatum*, and soil phosphorus on infection and disease caused by binucleate *Rhizoctonia* and *Rhizoctonia solani* on mung bean (*Vigna radiata*). Plant Soil 238:235–244

Kistner C, Parniske M (2002) Evolution of signal transduction in intracellular symbiosis. Trends Plant Sci 7:511–518

Krishna K, Bagyaraj D (1983) Interaction between *Glomus fasciculatum* and *Sclerotium rolfsii* in peanut. Can J Bot 61:2349–2351

Leigh J, Hodge A, Fitter AH (2009) Arbuscular mycorrhizal fungi can transfer substantial amounts of nitrogen to their host plant from organic material. New Phytol 181:199–207

Levy A, Chang BJ, Abbott LK, Kuo J, Harnett G, Inglis TJ (2003) Invasion of spores of the arbuscular mycorrhizal fungus *Gigaspora decipiens* by *Burkholderia* spp. Appl Environ Microbiol 69:6250–6256

Linderman R (1988) Mycorrhizal interactions with the rhizosphere microflora – the mycorrhizosphere effect. Phytopathology 78:366–371

Linderman RG (1992) Vesicular-arbuscular mycorrhizae and soil microbial interactions. In: Bethlenfalvay G, Linderman R (eds) Mycorrhizae in sustainable agriculture. American Society of Agronomy, Special Publication No. 54. Madison, WI, pp 45–70

Lioussanne L (2007) Rôles des modifications de la microflore bactérienne et de l'exudation racinaire de la tomate par la symbiose mycorhizienne dans le biocontrôlesur le Phytophthora nicotianae. Doctoral thesis, University of Montreal, Montreal [In French]

Lioussanne L (2010) The role of the arbuscular mycorrhiza-associated rhizobacteria in the biocontrol of soilborne phytopathogens. Span J Agric Res 8:51–61

Lopez MF, Manner P, Willmann A, Hampp R, Nehls U (2007) Increased trehalose biosynthesis in hartig net hyphae of ectomycorrhizas. New Phytol 174:389–398

Lumini E, Ghignone S, Bianciotto V, Bonfante P (2006) Endobacteria or bacterial endosymbionts? To be or not to be. New Phytol 170:205–208

Lumini E, Bianciotto V, Jargeat P, Novero M, Salvioli A, Faccio A, Becard G, Bonfante P (2007) Presymbiotic growth and sporal morphology are affected in the arbuscular mycorrhizal fungus *Gigaspora margarita* cured of its endobacteria. Cell Microbiol 9:1716–1729

MacDonald RM, Chandler MR (1981) Bacterium-like organelles in the vesicular-arbuscular mycorrhizal fungus *Glomus caledonius*. New Phytol 89:241–246

Maia LC, Kimbrough JW (1998) Ultrastructural studies of spores and hypha of a *Glomus* species. Int J Plant Sci 159:581–589

Marschner H (1995) Mineral nutrition of higher plants, 2nd edn. Academic, London

Marschner P, Timonen S (2006) Bacterial community composition and activity in rhizosphere of roots colonized by arbuscular mycorrhizal fungi. In: Mukerji KG, Manoharachary C, Singh J (eds) Microbial activity in the rhizoshere. Springer, Berlin, pp 139–154

Marulanda A, Barea JM, Azcon R (2006) An indigenous drought-tolerant strain of *Glomus intraradices* associated with a native bacterium improves water transport and root development in *Retama sphaerocarpa*. Microb Ecol 52:670–678

Mayo K, Davis R, Motta J (1986) Stimulation of germination of spores of *Glomus versiforme* by spore-associated bacteria. Mycologia 3:426–431

Meyer J, Linderman R (1986) Response of subterranean clover to dual inoculation with vesicular-arbuscular mycorrhizal fungi and a plant growth promoting bacterium *Pseudomonas putida*. Soil Biol Biochem 18:185–190

Minerdi D, Fani R, Gallo R, Boarino A, Bonfante P (2001) Nitrogen fixation genes in an endosymbiotic *Burkholderia* strain. Appl Environ Microbiol 67:725–732

Minerdi D, Banciotto V, Bonfante P (2002) Endosymbiotic bacteria in mycorrhizal fungi: from their morphology to genomic sequences. Plant Soil 244:211–219

Mondo SJ, Toomer KH, Morton JB, Lekberg Y, Pawlowska TE (2012) Evolutionary stability in a 400-million-year-old heritable facultative mutualism. Evolution 66:2564–2576

Morgan JA, Bending GD, White PJ (2005) Biological costs and benefits to plant-microbe interactions in the rhizosphere. J Exp Bot 56:1729–1739

Mosse B (1962) The establishment of vesicular-arbuscular mycorrhiza under aseptic conditions. J Gen Microbiol 27:509–520

Mosse B (1970) Honey-coloured sessile Endogone spores: II Changes in fine structure during spore development. Arch Mykrobiol 74:146–159

Nannipieri P, Ascher J, Ceccherini MT, Landi L, Pietramellara G, Renella G (2003) Microbial diversity and soil functions. Eur J Soil Sci 54:655–670

Naumann M, Schussler A, Bonfante P (2010) The obligate endobacteria of arbuscular mycorrhizal fungi are ancient heritable components related to the Mollicutes. ISME J 4:862–871

Olsson PA, Thingstrup I, Jakobsen I, Baath E (1999) Estimation of the biomass of arbuscular mycorrhizal fungi in a linseed field. Soil Biol Biochem 31:1879–1887

Ordonez YM, Fernandez BR, Lara LS, Rodriguez A, Uribe-Velez D, Sanders IR (2016) Bacteria with phosphate solubilizing capacity alter mycorrhizal fungal growth both inside and outside the root and in the presence of native microbial communities. PLoS One 11:e0154438

Oswald ET, Ferchau HA (1968) Bacterial associations of coniferous mycorrhizae. Plant Soil 28:187–192

Pinton R, Varanini Z, Nannipieri P (2001) The rhizosphere: biochemistry and organic substances at the soil-plant interface. Marcel Dekker, New York

Potera C (1996) Biofilms invade microbiology. Science 5283:1795–1797

Pozo MJ, Cordier C, Dumas-Gaudot E, Gianinazzi S, Barea JM, Azcon-Aguilar C (2002) Localized versus systemic effect of arbuscular mycorrhizal fungi on defence responses to *Phytophthora* infection in tomato plants. J Exp Bot 53:525–534

Rawlings GB (1958) Some practical aspects of forest mycotrophy. NZ Soc Soil Sci Proc 3:41–44

Redecker D, Kodner R, Graham LE (2000) Glomalean fungi from the Ordovician. Science 289:1920–1921

Rodriguez H, Fraga R (1999) Phosphate solubilizing bacteria and their role in plant growth promotion. Biotechnol Adv 17:319–339

Roesti D, Ineichen K, Braissant O, Redecker D, Wiemken A, Aragno M (2005) Bacteria associated with spores of the arbuscular mycorrhizal fungi *Glomus geosporum* and *Glomus constrictum*. Appl Environ Microbiol 71:6673–6679

Ruiz-Lozano JM, Bonfante P (2000) A Burkholderia strain living inside the arbuscular mycorrhizal fungus *Gigaspora margarita* possesses the vacb gene, which is involved in host cell colonization by bacteria. Microb Ecol 39:137–144

Salvioli A, Ghignone S, Novero M, Navazio L, Venice F, Bagnaresi P, Bonfante P (2016) Symbiosis with an endobacterium increases the fitness of a mycorrhizal fungus, raising its bioenergetic potential. ISME J 10:130–144

Sarand I, Timonen S, Nurmiaho-Lassila E-L, Koivila T, Haahtela K, Romantschuk M et al (1998) Microbial biofilms and catabolic plasmid harbouring degradative fluorescent pseudomonads in Scots pine ectomycorrhizospheres developed on petroleum contaminated soil. FEMS Microbiol Ecol 27:115–126

Sarand I, Timonen S, Koivula T, Peltola R, Haahtela K, Sen R et al (1999) Tolerance and biodegradation of m-toluate by Scots pine, a mycorrhizal fungus and fluorescent pseudomonads individually and under associative conditions. J Appl Microbiol 86:817–826

Sarand I, Haario H, Jørgensen KS, Romantschuk M (2000) Effect of inoculation of a TOL plasmid containing mycorrhizosphere bacterium on development of Scots pine seedlings, their mycorrhizosphere and the microbial flora in m- toluate-amended soil. FEMS Microbiol Ecol 31:127–141

Scannerini S, Bonfante P (1991) Bacteria and bacteria like objects in endomycorrhizal fungi (Glomaceae). In: Margulis L, Fester R (eds) Symbiosis as source of evolutionary innovation: speciation and morphogenesis. The MIT Press, Cambridge

Scheublin TR, Sanders IR, Keel C, VanDerMeer JR (2010) Characterization of microbial communities colonising the hyphal surfaces of arbuscular mycorrhizal fungi. ISME J 4:752–763

Schreiner R, Mihara K, McDaniel H, Bethlenfalvay G (1997) Mycorrhizal functioning influence plant and soil functions and interactions. Plant Soil 188:199–209
Schüßler A (2000) *Glomus claroideum* forms an arbuscular mycorrhiza-like symbiosis with the hornwort *Anthoceros punctatus*. Mycorrhiza 10:15–21
Secilia J, Bagyaraj DJ (1987) Bacteria and actinomycetes associated with pot cultures of vesicular–arbuscular mycorrhizas. Can J Microbiol 12:1069–1073
Simon L, Bousquet J, Levesque RC, Lalonde M (1993) Origin and diversification of endomycorrhizal fungi and coincidence with vascular land plants. Nature 363:67–69
Smith S, Read D (1997) Mycorrhizal symbiosis. Academic, San Diego
Smith S, Read D (2008) Mycorrhizal symbiosis, 3rd edn. Academic, London
Soderberg KH, Olsson PA, Baath E (2002) Structure and activity of the bacterial community in the rhizosphere of different plant species and the effect of arbuscular mycorrhizal colonisation. FEMS Microbiol Ecol 40:223–231
Sood SG (2003) Chemotactic response of plant-growth-promoting bacteria towards roots of vesicular-arbuscular mycorrhizal tomato plants. FEMS Microbiol Ecol 45:219–227
Staddon PL, Ramsey CB, Ostle N, Ineson P, Fitter AH (2003) Rapid turnover of hyphae of mycorrhizal fungi determined by AMS microanalysis of 14C. Science 300:1138–1140
Taylor TN, Krings M (2005) Fossil microorganisms and land plants: associations and interactions. Symbiosis 40:119–135
Tisserant E, Kohler A, Dozolme-Seddas P, Balestrini R, Benabdellah K, Colard A, Croll Daniel Da Silva C, Gomez SK, Koul R, Ferrol N, Fiorilli V, Formey D, Franken P, Helber N, Hijri M, Lanfranco L, Lindquist E, Liu Y, Malbreil M, Morin E, Poulain J, Shapiro H, van Tuinen D, Waschke A, Azcon-Aguilar C, Bcard G, Bonfante P, Harrison MJ, Kuster H, Lammers P, Paszkowski U, Requena N, Rensing SA, Roux C, Sanders Ian R, Shachar-Hill Y, Tuskan G, JPW Y, Gianinazzi-Pearson V, Martin F (2012) The transcriptome of the arbuscular mycorrhizal fungus *Glomus intraradices* (DAOM 197198) reveals functional tradeoffs in an obligate symbiont. New Phytol 193:755–769
Tisserant E, Malbreil M, Kuo A, Kohler A, Symeonidi A, Balestrini R, Charron P, Duensing N, Frei DFN, Gianinazzi-Pearson V, Gilbert LB, Handa Y, Herr JR, Hijri M, Koul R, Kawaguchi M, Krajinski F, Lammers PJ, Masclaux FG, Murat C, Morin E, Ndikumana S, Pagni M, Petitpierre D, Requena N, Rosikiewicz P, Riley R, Saito K, San CH, Shapiro H, van Tuinen D, Becard G, Bonfante P, Paszkowski U, Shachar-Hill YY, Tuskan GA, Young JP, Sanders IR, Henrissat B, Rensing SA, Grigoriev IV, Corradi N, Roux C, Martin F (2013) Genome of an arbuscular mycorrhizal fungus provides insight into the oldest plant symbiosis. Proc Natl Acad Sci USA 110:20117–20122
Toljander JF, Artursson V, Paul LR, Jansson JK, Finlay RD (2006) Attachment of different soil bacteria to arbuscular mycorrhizal fungal extraradical hyphae is determined by hyphal vitality and fungal species. FEMS Microbiol Lett 254:34–40
Toljander J, Lindahl B, Paul L, Elfstrand M, Finlay R (2007) Influence of arbuscular mycorrhizal mycelial exudates on soil bacterial growth and community structure. FEMS Microbiol Ecol 2:295–304
Torres-Cortes G, Ghignone S, Bonfante P, Schussler A (2015) Mosaic genome of endobacteria in arbuscular mycorrhizal fungi: transkingdom gene transfer in an ancient mycoplasma-fungus association. Proc Natl Acad Sci U S A 112:7785–7790
Torsvik V, Øvreas L (2002) Microbial diversity and function in soil: from genes to ecosystems. Curr Opin Microbiol 5:240–245
Toussaint JP, Kraml M, Nell M, Smith SE, Smith FA, Steinkellner S, Schmiderer C, Vierheilig H, Novak J (2008) Effect of *Glomus mosseae* on concentrations of rosmarinic and caffeic acids and essential oil compounds in basil inoculated with *Fusarium oxysporum* f.sp. basilica. Plant Pathol 57:1109–1116
Tylka G, Hussey R, Roncadori R (1991) Axenic germination of vesicular-arbuscular mycorrhizal fungi: Effects of selected *Streptomyces* species. Phytopathology 81:754–759

Uroz S, Calvaruso C, Turpault MP, Pierrat JC, Mustin C, Frey-Klett P (2007) Effect of the mycorrhizosphere on the genotypic and metabolic diversity of the bacterial communities involved in mineral weathering in a forest soil. Appl Environ Microbiol 73:3019–3027

van der Heijden MG, Martin FM, Selosse MA, Sanders IR (2015) Mycorrhizal ecology and evolution: the past, the present, and the future. New Phytol 205:1406–1423

Vierheilig H (2004) Regulatory mechanisms during the plant arbuscular mycorrhizal fungus interaction. Can J Bot 82:1166–1176

Vivas A, Marulanda A, Ruiz-Lozano JM, Barea JM, Azcon R (2003) Influence of a *Bacillus* sp. on physiological activities of two arbuscular mycorrhizal fungi and on plant responses to PEG-induced drought stress. Mycorrhiza 13:249–256

Vósatka M, Gryndler M (1999) Treatment with culture fractions from *Pseudomonas putida* modifies the development of *Glomus fistulosum* mycorrhiza and the response of potato and maize plants to inoculation. Appl Soil Ecol 11:245–251

Walley FL, Germida JJ (1996) Failure to decontaminate *Glomus clarum* NT4 spores is due to spore wall-associated bacteria. Mycorrhiza 6:43–49

Wang F, Shi N, Jiang R, Zhang F, Feng G (2016) In situ stable isotope probing of phosphate-solubilizing bacteria in the hyphosphere. J Exp Bot 67:1689–1701

Whipps JM (2004) Prospects and limitations for mycorrhizas in biocontrol of root pathogens. Can J Bot 82:1198–1227

Wiemken V (2007) Trehalose synthesis in ectomycorrhizas – a driving force of carbon gain for fungi? New Phytol 174:228–230

Wright SF, Upadhyaya A (1996) Extraction of an abundant and unusual protein from soil and comparison with hyphal protein of arbuscular mycorrhizal fungi. Soil Sci 161:575–586

Wright SF, Franke-Snyder M, Morton JB, Upadhyaya A (1996) Time course study and partial characterization of a protein on hyphae of arbuscular mycorrhizal fungi during active colonization of roots. Plant Soil 181:193–203

Xavier L, Germida J (2003) Bacteria associated with *Glomus clarum* spores influence mycorrhizal activity. Soil Biol Biochem 35:471–478

Zhang L, Fan J, Ding X, He X, Zhang F, Feng G (2014) Hyphosphere interactions between an arbuscular mycorrhizal fungus and a phosphate solubilizing bacterium promote phytate mineralization in soil. Soil Biol Biochem 74:177–183

Zhang L, Xu M, Liu Y, Zhang F, Hodge A, Feng G (2016) Carbon and phosphorus exchange may enable cooperation between an arbuscular mycorrhizal fungus and a phosphate-solubilizing bacterium. New Phytol 210:1022–1032

Chapter 5
Mycorrhizal Helper Bacteria: Sustainable Approach

Devendra K. Choudhary, Ajit Varma, and Narendra Tuteja

Abstract Microorganisms in rhizosphere play an important role in soil processes that determine plant and soil productivity. Tremendous efforts have been made to explore mycorrhizal diversity along with benign role of bacterial population in soil habitats to understand the successful functioning of extraneous microbial bio-inoculants (AMF/PGPR) and their influence on soil health. Improvement in agricultural sustainability requires optimal use and management of soil fertility and soil physical property and relies on soil biological processes and soil biodiversity. Plants play an important role in selecting and enriching the type of microbes by the constituents of their root exudates. The mycorrhizal and bacterial community develops in the rhizosphere which is a result of diverse nature and concentration of organic constituents of exudates and the corresponding ability of them to utilize these as sources of energy. Therefore, rhizosphere microbial community has an efficient system for uptake and catabolism of organic compounds present in root exudates and further transportation in plants mediated through mycorrhizal helper bacteria.

5.1 Introduction

The phrase "mycorrhiza (MRs)" originated from the Greek words wherein myco- is "fungus" and -rhiza means "root" that makes an association with plant root. The occurrence of MRs found in numerous environments showed ecological aspects in form of ecto- and endo-MRs. There are about 6000 species of MRs belonging to the Glomeromycotina, Ascomycotina, and Basidiomycotina and have been characterized based on phenetic and genetic approaches. The types of MRs are decided by the taxonomic position of the plant and fungal partners and it leads to endomycorrhizas (ENMRs) and ectomycorrhizas (ECMRs). It has been reported that ECMRs locate extracellular in plants especially in epidermal or cortical cells (Bonfante 2001). The study conducted suggests that releases of active diffusible

D.K. Choudhary (✉) • A. Varma • N. Tuteja
Amity Institute of Microbial Technology (AIMT), Amity University Uttar Pradesh, Block 'E-3', 4th Floor, Sector-125, Noida, Gautam Buddha Nagar, Uttar Pradesh 201313, India
e-mail: dkchoudhary1@amity.edu

A. Varma et al. (eds.), *Mycorrhiza - Function, Diversity, State of the Art*,
DOI 10.1007/978-3-319-53064-2_5

molecules are important for the establishment of plant-ECMRs interactions, and it leads major signaling mechanisms identified in ENMRs as arbuscular (A)-MRs (Oldroyd and Downie 2008). It is required to make a physical contact between the fungus and the plant which elicit enormous plant responses that lead to fungal colonization (Genre et al. 2005). Upon interaction, the partners release active diffusible molecules that are reciprocally perceived and reflect Ca-mediated responses (Bouwmeester et al. 2007; Navazio et al. 2007). It has been reported that the colonization process are common to all mycorrhizal fungi, whether they form ECMRs and ENMRs including AMRs. As the natural habitat of MRs is soil and wherein fungal symbionts develop in the rhizosphere (RS) those bridges between soil and plant roots (Selosse and Roy 2009). This interaction retains the concept of mutualism, i.e., an interaction of net benefit to both partners (Thompson and Cunningham 2002) that allows nutritional exchange, e.g., AMRs possess active phosphate transporters that take up inorganic phosphate (Pi) from the soil and delivers it to the plant (Harrison and van Buuren 1995).

Researches on MRs also suggest that production of volatile organic compounds (VOCs) from all the members of the underground flora and fauna that may be important for inter- and intra-organismic communication (Tarkka and Piechulla 2007). It has been demonstrated that truffle VOCs have biological activity, causing leaf bleaching and root inhibition of *Arabidopsis thaliana* (Splivallo et al. 2007). Besides, bacterial VOCs may also affect soil fungi including the MRs ones (Tarkka and Piechulla 2007). Such VOCs can be regarded as important elicitors for symbiosis establishment. To see the synergistic effect of VOCs, different approaches like genomics, proteomics, and metabolomics have been deployed to understand the basis of interaction between MHBs and their fungal and plant hosts (Tarkka and Piechulla 2007). In ENMRs, ericoid, and orchid mycorrhizas, the hyphae penetrate the root cells to establish an intracellular symbiosis, irrespective of the plant host wherein AMRs are widespread among various plant taxa, ericoid, and orchid mycorrhizas are restricted to the order Ericales and the family Orchidaceae, respectively (Bonfante and Genre 2008; Smith and Read 2008).

Furthermore, plants also possess phosphate transporters that are mycorrhiza specific. Their role is to receive Pi from the fungus and deliver it to plant cells. A *Medicago truncatula* Pi transporter exclusively expressed during AMRs symbiosis and located in the periarbuscular membrane not only is essential for the acquisition of Pi delivered by the AMRs but is also required to maintain arbuscule vitality and sustain development of the fungus (Harrison 2005; Javot et al. 2007). Pi transport therefore seems to be a signal to sustain fungal growth inside the root and a determinant of arbuscule morphogenesis. Nitrogen is the other important element taken up by most mycorrhizal fungi. Genes involved in organic and inorganic uptake of N have been identified in AMRs and ECMRs fungi (Smith and Read 2008).

Many molecular and physiological data show that plant N transporters are activated during mycorrhization (Guether et al. 2009), suggesting that mycorrhizal fungi release a substantial amount of N to their hosts. While these fungal and plant transporters may be used as clear markers of mycorrhizal function, the reverse nutrient flow is not so clearly characterized. Carbon transfer from plants to mycorrhizal fungi was demonstrated in the 1960s (Smith and Read 2008), but the molecular mechanisms are still unclear. With the exception of the gene described in the glomeromycotan *Geosiphon pyriforme* (Schüßler et al. 2006), which forms symbiosis with a cyanobacterium, and of the *AmMst1* gene from the ECM fungus *Amanita muscaria* (Nehls et al. 1998), no other hexose transporter responsible for the uptake of C released by host cells has so far been characterized in mycorrhizal fungi. In addition, the transfer does not always go in the expected direction; for example, in orchid mycorrhizas or in other heterotrophic plants, C moves from the fungus toward the plant (Selosse and Roy 2009). In this case the nature of the benefit for the fungus is not obvious, although it might gain advantages, for example, by living within a protected niche.

5.2 Associated Bacteria

In natural conditions, bacteria associated with mycorrhizal fungi colonize the surface of extraradical hyphae or, at least in some fungal taxa, live in the cytoplasm as endobacteria. Understanding the interactions between the microorganisms routinely found in the rhizosphere is essential for describing the nature of the soil–plant interface. A brief review of the predominant prokaryotic species thriving in the rhizosphere of mycorrhizal fungi is provided, followed by an analysis of their nutritional interactions. Assessment of bacterial community structure in the soil is based mostly on the use of cultivation-dependent methods as well as cultivation-independent methods including soil metagenomics (van Elsas et al. 1998; Daniel 2005).

Mycorrhiza-associated bacterial communities have been investigated according to established protocols, using microbiological (strain isolation and identification) and molecular screening of 16S rDNA libraries. These investigations, begun in the 1990s, have revealed a wide repertoire of microbes, including several bacterial taxa with a predominance of species from the genera *Pseudomonas*, *Burkholderia*, and *Bacillus* (De Boer et al. 2005). Streptomycetes have been associated with ectomycorrhizal fungi and have been discussed as modulators of plant symbiosis, while archaebacteria thriving in the rhizosphere of mycorrhizal fungi have been reported only in boreal regions, and their limited distribution may be explained by their low preference for nonextreme environments and for rhizospheric soils in particular (Ochsenreiter et al. 2003; Bomberg and Timonen 2007; Schrey and Tarkka 2008). Notably, the detection of rhizosphere bacteria is often dependent on the method of sampling. A significant difference in bacterial community composition associated with *Tuber* sp. fruiting bodies has been found, depending on the origin and preparation of the samples; a predominance of fluorescent pseudomonads in the culture-dependent samples was not confirmed by analysis of environmental samples that, in contrast, showed the predominance of alphaproteobacteria represented by *Sinorhizobium/Ensifer* and *Rhizobium/Agrobacterium* groups as well as by the nitrogen-fixing *Bradyrhizobium* spp. (Barbieri et al. 2005, 2007).

This suggests that environmental sampling more accurately describes the bacterial community, while culturing may be the best approach when selecting interesting mycorrhiza-associated species. Investigation of the microbial diversity sheds light on whether bacterial communities are influenced by mycorrhizal fungi. Electron microscopy has shown that ectomycorrhizal fungi, such as *Suillus bovinus* and *Paxillus involutus*, host distinct populations of bacteria, suggesting spatially and physiologically different habitats in the mycorrhizospheres (Nurmiaho-Lassila et al. 1997). Many other investigations have confirmed this finding using molecular approaches, stating that bacterial community structure depends more on AMRs than on host plant identity (Roesti et al. 2005). In a study observing the negative influence of the exotic tree *Eucalyptus camaldulensis* (Kisa et al. 2007), AMRs fungal symbiosis was found to play a decisive role in the preservation of the soil microbial structure. Similarly, Singh et al. (2008) showed that AMRs are the major factor in determining the bacterial assemblage on grass roots; they also found that this assemblage is influenced by soil pH and is spatially structured.

There is evidence for necrotrophic and extracellular biotrophic activities of bacteria toward mycorrhizal fungi. Pathogenic and nonpathogenic soil bacteria (*Burkholderia* spp.) penetrated the spores of the AM fungus *Gigaspora decipiens* colonizing senescing spores and attaching to fungal hyphae, as detected in a spore lysis assay using GFP-tagged bacteria (Levy et al. 2003). Different kinetics in the attachment to living or nonliving *Glomus* spp. hyphae in five other GFP-tagged soil bacteria from the genera *Paenibacillus*, *Bacillus*, *Pseudomonas*, and *Arthrobacter* were described. These observations suggest that bacterial attachment to hyphae is regulated by two factors: species specificity and fungal vitality (Toljander et al. 2006).

Electron microscopy has revealed bacteria that probably feed on the outer hyaline spore layer of AMRs. As they represent several taxonomic groups of biodegraders such as *Cellvibrio*, *Chondromyces*, *Flexibacter*, *Lysobacter*, and *Pseudomonas*, it is not clear if their activity is beneficial (stimulation of spores germination) or parasitic (Roesti et al. 2005). Although these reports offer good examples of bacterial mycophagy involving both necrotrophy and extracellular biotrophy, the observations are mostly descriptive, and the controlling mechanisms are still unknown. Only in a few cases have molecular and biochemical mechanisms been clarified: Two mutant strains of *Pseudomonas fluorescens* CHA0 with increased capacity to produce extracellular polysaccharides display higher ability to adhere to the surface of AMRs and plant roots, compared with the nonmucoid wild type, showing the importance of cell wall composition for efficient bacterial adhesion (Bianciotto et al. 2001). Another more recent example of a molecular determinant for fungal–bacterial attachment is given by a truffle-secreted protein, a lectin that binds *Rhizobium* sp. (Cerigini et al. 2008).

Interactions between AMRs and bacteria imply both a beneficial effect of the fungi on bacterial development and vice versa (De Boer et al. 2005). An interesting example is *Paenibacillus validus*, which when alone supports the growth and sporulation of *Glomus intraradices* independently of the presence of the plant (Hildebrandt et al. 2002). Two isolates of *P. validus* are highly efficient in sustaining fungal growth up to production of new germinating spores, probably owing to the release of raffinose and a still unidentified trisaccharide (Hildebrandt et al. 2006). This result shows that at least one AMR can grow independently of its plant host if it is in the presence of a bacterium. It might be interesting to see whether the still unknown bacterial compounds are mimicking plant molecules and whether this bacterium is required for establishment of the mycorrhiza (Bouwmeester et al. 2007).

5.3 Functional Attributes of Bacteria in Rhizosphere

Plant growth-promoting rhizobacteria (PGPR) are defined as free-living soil, rhizosphere, rhizoplane, and phyllosphere bacteria that, under some conditions, are beneficial for plants. Plant-associated bacteria can contribute to the health, growth, and development of plants. Examples of bacteria which have been found to enhance plant growth include species of *Pseudomonas*, *Enterobacter*, *Bacillus*, *Erwinia*, *Azospirillum*, and *Arthrobacter* (Rodriguez et al. 2006). The positive effect of many soil bacteria on plants is mediated by a range of mechanism including improvement of mineral nutrition, enhancement of plant tolerance to biotic and abiotic stress, modification of root development, and as well as suppression of soil-borne diseases (Glick 2015). Plant growth promotion by bacteria may result either from direct effects such as the solubilization of soil phosphorus and iron, the production of indole acetic acid (IAA), or indirect effects such as the biocontrol of soil-borne diseases through competition for nutrients and siderophore-mediated competition for iron (Glick 2015).

In the last decade, bacteria belonging to different genera including *Rhizobium*, *Bacillus*, *Pseudomonas*, *Pantoea*, *Paenibacillus*, *Burkholderia*, *Achromobacter*, *Azospirillum*, *Mycobacterium*, *Methylobacterium*, *Variovorax*, and *Enterobacter* etc., have been reported to provide tolerance to host plants under different abiotic stress environments. Abiotic environmental factors include temperature, humidity, light intensity, the supply of water and minerals, and CO_2; these are the parameters and resources that determine the growth of a plant (Choudhary et al. 2016). Soil microorganisms with beneficial activity on plant growth and health represent an attractive alternative to conventional agricultural method. In recent years, several microbial inoculants have been formulated, produced, marketed, and applied successfully by an increasing number of growers (Glick 2015). All the parts of plants are colonized by microorganisms; the rhizosphere represents the main source of bacteria with plant beneficial activity.

5.3.1 Phosphate Solubilization

Phosphorus (P) is one of the major essential macronutrients for plant growth and development. It is present at levels of 400–1200 mg/kg of soil. Most phosphorus in soil is part of insoluble compounds, which makes P unavailable for nutrition. Phosphorus exists in two forms in soil, as organic and inorganic phosphates. The concentration of soluble P in soil is usually very low, only 5% or less of the total amount of P in soil is available for plant nutrition. The phenomenon of fixation and precipitation of P in soil is highly dependent on soil type and pH. Thus, in acid soils, free oxides and hydroxides of aluminum and iron fix P, while in alkaline soils Ca fixes it. Organic acid metabolite production and decrease of medium pH appear to be the major mechanisms for P solubilization. Phosphate-solubilizing bacteria (PSB) are capable of hydrolyzing organic and inorganic phosphorus from insoluble compounds. They are commonly found in rhizoplane and rhizosphere (Estrada et al. 2013). PSB are known to bring a number of transformations of phosphorus, these include:

- Altering the solubility of inorganic compounds of phosphorus.
- Mineralization of organic phosphate compounds into inorganic phosphates.
- Conversion of inorganic, available anion into cell components, i.e., an immobilization process.
- Oxidation or reduction of inorganic phosphorus compounds of these mineralization and immobilization are the most important reactions/processes in phosphorus cycle. Mineralization of most organic phosphorus compounds in most soils is carried out by means of phosphatase enzymes. These catalyze dephosphorylating reactions involving the hydrolysis of phosphoester or phosphoanhydride bonds.

5.3.2 Iron Chelation and Siderophores

Iron is the fourth most abundant element on earth. In aerobic soils, iron is mostly precipitated as hydroxides, oxyhydroxides, and oxides so that amount of iron available for assimilation by living organisms is very low. Both microbes and plants have a quite high iron requirement. To survive with a limited supply of iron, in bacteria, cellular iron deficiency induces the synthesis of low molecular weight siderophores, molecules with an extraordinarily high affinity for Fe^{+3}, as well as membrane receptor able to bind the Fe-siderophore complex (Dimkpa et al. 2009), thereby allowing iron uptake with the help of microorganism. Some important siderophore-producing bacteria include *Escherichia coli*, *Salmonella*, *Klebsiella pneumoniae*, *Vibrio cholerae*, *Vibrio anguillarum*, *Aeromonas*, *Aerobacter aerogenes*, *Enterobacter*, *Yersinia*, and *Mycobacterium* species (Schalk et al. 2011).

5.3.3 IAA Production

PGPB can promote the plant growth by production or changes in the concentration of plant hormone IAA. IAA produced by bacteria improves plant growth by increasing the number of root hairs and lateral roots (Singh et al. 2013). IAA affects plant cell division, extension, and differentiation; stimulates seed and tuber germination; increases the rate of xylem and root development; controls processes of vegetative growth; initiates lateral and adventitious root formation; mediates responses to light, gravity, and fluorescence; and affects photosynthesis, pigment formation, biosynthesis of various metabolites, and resistance to stressful conditions. Microbial biosynthesis of IAA in soil is enhanced by tryptophan from root exudates or decaying cells. It has been suggested that PGPB-synthesizing IAA may prevent the deleterious effects of environment stresses (Tatsuki et al. 2013).

5.3.4 Nitrogen Fixation

Nitrogen (N) is the most vital nutrient for plant growth and productivity. Although, there is about 78% N_2 in the atmosphere, it is unavailable to the growing plants. The atmospheric N_2 is converted into plant-utilizable forms by biological N_2 fixation (BNF) which changes nitrogen to ammonia by nitrogen-fixing microorganisms using a complex enzyme system known as nitrogenase. Nitrogen-fixing organisms are generally categorized as (a) symbiotic N_2-fixing bacteria including members of the family *Rhizobiaceae* which forms symbiosis with leguminous plants (e.g., rhizobia) and non-leguminous trees (e.g., *Frankia*) and (b) nonsymbiotic (free-living, associative, and endophytes) nitrogen-fixing forms such as cyanobacteria (*Anabaena*, *Nostoc*), *Azospirillum*, *Azotobacter* etc. (Yang et al. 2014). However, nonsymbiotic nitrogen-fixing bacteria provide only a small amount of the fixed nitrogen that the bacterially associated host plant requires. Plant growth-promoting bacteria that fix N_2 in non-leguminous plants are also called as diazotrophs capable of forming a non-obligate interaction with the host plants. The process of N_2 fixation is carried out by a complex enzyme, the nitrogenase complex (Zhao et al. 2006). Structure of nitrogenase was elucidated by Zhao et al. (2006) as a two-component metalloenzyme consisting of (1) dinitrogenase reductase which is the iron protein and (2) dinitrogenase which has a metal cofactor. Dinitrogenase reductase provides electrons with high reducing power while dinitrogenase uses these electrons to reduce N_2 to NH_3. Based on the metal cofactor, three different N-fixing systems have been identified: (a) Mo-nitrogenase, (b) V-nitrogenase and (c) Fe-nitrogenase. Structurally, N_2-fixing system varies among different bacterial genera. Most biological nitrogen fixation is carried out by the activity of the molybdenum nitrogenase, which is found in all diazotrophs (Zhao et al. 2006).

5.4 Helper Bacteria

Duponnois and Garbaye (1991) were the first to observe a significant stimulation of ectomycorrhiza formation by *Pseudomonas fluorescens* BBc6. This was experimental evidence for the so-called helper effect. Bacteria involved in mycorrhiza establishment and/or its functioning were therefore defined as mycorrhiza helper bacteria (MHB) by Garbaye (1994) and are currently the most investigated group among bacteria interacting with mycorrhizas (Frey-Klett et al. 2007). Some species are responsible for multiple helper effects, because they influence both plants and associated mycorrhizal fungi. More recently, the stimulating effect of MHB has been evaluated mostly when symbiotic associations are exposed to stresses ranging from drought (Vivas et al. 2003a) to contamination with heavy metals such as Pb (Vivas et al. 2003b), Zn (Vivas et al. 2006), and Cd(II) (Kozdrój et al. 2007). Frey-Klett et al. (2007) proposed some mechanisms that may explain MHB success. These mechanisms involve the production of growth factors that might stimulate fungal spore germination, mycelial growth, increased root branching and greater root colonization, and reduction of soil-mediated stress through detoxification of antagonistic substances and inhibition of competitors and antagonists. A classic example illustrating the helper effect is given by rhizobia producing 1-aminocyclopropane-1-carboxylate (ACC) deaminase; the molecule modulates plant ethylene levels, increasing plant tolerance to environmental stress and stimulating nodulation (Ma et al. 2002).

In agreement with this rationale, the ACC deaminase-producing *Pseudomonas putida* UW4 promotes mycorrhization with the AMRs *Gigaspora rosea* when inoculated into cucumber plants, whereas a bacterial mutant impaired in ethylene production causes decreased mycorrhization (Gamalero et al. 2008). In other cases, the beneficial effects of some MHB are related to changes in gene expression of the mycorrhizal fungus. In a study involving *L. bicolor* S238 N and the MHB *P. fluoresce ns* BBc6R8, bacteria stimulated fungal growth and development as well as altered fungal gene expression. This led to activation of genes potentially involved in recognition processes, transcription regulation, and synthesis of primary metabolism proteins (Deveau et al. 2007). Similarly, the MHB *Streptomyces* sp. AcH 505 promotes hyphal growth and symbiosis formation with spruce in the ECMRs *Amanita muscaria*, acting on basic cell growth processes. Immunofluorescence microscopy in fact revealed morphological changes in the actin cap of fungal hyphae in the presence of bacteria (Schrey et al. 2007). AMRs and PSB could potentially interact synergistically because PSB solubilize phosphate into a form that AMRs can absorb and transport to the plant. However, very little is known about the interactions between these two groups of microorganisms and how they influence the growth of each other. We tested whether different strains of bacteria, which have the capacity to solubilize phosphate, are able to grow along AMRs hyphae and differentially influence the growth of AMRs both outside the roots of carrot in in vitro conditions and inside the roots of potato in the presence of a microbial community. We found strong effects of AMRs on the growth of the different bacterial strains. Different bacterial strains also had very strong effects on the growth of AMRs extraradical hyphae outside the roots of carrot and on colonization of potato roots by AMRs (Ordoñez et al. 2016).

However, the interactions between PSBs and AMRs are poorly understood, and the approach to using both these microbial groups for applications in agriculture is often naive because of variation in soil abiotic and biotic environments in which these organisms have often not been tested (Rodriguez and Sanders 2015). Moreover, very often, only single strains of these microbial groups have been shown in laboratory or greenhouse conditions to have the capacity to solubilize P or to improve plant P acquisition. Indeed, higher plant P uptake capacity has previously been reported when plants are co-inoculated in greenhouse conditions, with AMRs and PSB (Gamalero et al. 2004). These bacteria probably improved the availability of P, which can subsequently be efficiently absorbed by AMRs hyphae (Nazir et al. 2009). Thus, on the basis of results, mostly from artificial experiments conducted in sterilized soil, AMRs and PSB are thought to act synergistically. Recent evidence also points not only to synergistic effects between AMRs and PSB but also to cooperation between these organisms (Zhang et al. 2016). However, most of the beneficial effects of AMRs are observed in experiments conducted in sterile soil (Rodriguez and Sanders 2015). In reality, plants naturally become colonized by the local AMRs community. A more realistic test of their potential is whether adding AMRs inoculum and PSBs to unsterilized soil will give a growth benefit to the plant. Such tests are rarely performed. Isolated beneficial microbes are then used in field applications, where the bacteria and fungi encounter both diverse soil environments and diverse microbial communities, including existing diverse populations of both PSBs and AMRs. It is perhaps unsurprising; therefore, that application of both AMRs and PSBs in agriculture has had very variable success (Owen et al. 2015). Given that both AMRs and PSB must have coexisted in the rhizosphere for millions of years, many possible interactions could have evolved between them. Yet the interaction between AMRs and PSB is not well understood. Firstly, in the mycorrhizosphere, the soil zone influenced by both the roots and the mycorrhizal fungi (Johansson et al. 2004), AMRs exudates create an environment that can influence bacterial growth (Toljander et al. 2007). Attachment of bacteria, with P-solubilizing capacity, to extraradical AMRs hyphae, could ensure that P-solubilizing activities of the bacteria would be located in the zone where they can be most beneficial in allowing the fungi access to additional soluble P. At the same time, attachment to the AMRs hyphae might provide bacteria with a route to efficiently access the mycorrhizosphere (Bianciotto and Bonfante 2002). Some soil bacteria have been shown to attach both to vital and non-vital AMRs hyphae in in vitro conditions (Toljander et al. 2006). However, none of the bacteria in that study were assessed for their P-solubilizing capacity. It is unknown whether any bacteria with phosphate-solubilizing capacity have the ability to attach to AMRs extraradical hyphae (Scheublin et al. 2010). Of those PSB that might associate with AMRs hyphae, it is unknown whether these bacteria might influence either the growth of AMRs inside the roots or of AMRs hyphae outside the roots. A positive effect of PSB on extraradical AMRs hyphal growth could help PSB to access new areas of the mycorrhizosphere and increase access by AMRs hyphae to new sources of solubilized P. Thirdly, populations of PSB are diverse in the soil, and it is unknown whether there is variation among strains in the effects of PSB on AMRs (Collavino et al. 2008; Jorquera et al. 2008; Naik et al. 2008; Meyer et al. 2011).

The finding that PSBs grow along AMRs hyphae is novel. It was already known that bacteria adhere to the surface of AMRs hyphae (Toljander et al. 2006), but the bacteria that adhered were never shown to be P solubilizers. In vitro experiments also show that several different bacterial groups grow around hyphae of *R. irregularis* and that they likely obtain their nutrition from exudates of AMRs hyphae. However, no test was made in that study regarding the capacity of the bacteria to solubilize P. All the bacterial strains we tested and that had P-solubilizing capacity were able to grow on extraradical AMRs hyphae. However, there was significant variation among strains in how much the strains could grow on AMRs hyphae. We have only used one AMRs species in this study, and the ability of different PSB strains to grow on AMRs hyphae could also potentially be AMRs species specific as variable bacterial communities have been shown to colonize spore surfaces of different AMRs species (Agnolucci et al. 2015). The observed growth of some PSBs on AMRs hyphae could be beneficial for the bacteria in two ways. They could use the hyphae as a route to access further areas of the soil, which could be beneficial for the fungus as P solubilizers could grow away from the route along AMRs hyphae into patches containing insoluble P. The bacteria could also use the AMRs hyphae as a route allowing growth in the other direction toward the plant and colonize the rhizosphere, an area that could be rich in resources from plant exudates.

5.5 Conclusions

Understanding the complex microbial community in the rhizosphere environment has proven to be a challenging task because of the vast diversity and the enormity of the population inhabiting this unique habitat. Extensive studies have investigated perturbation of microbial community equilibrium population by changes in environmental conditions and soil management practices. It has long been recognized that the activity of soil microorganisms play an intrinsic role in residue decomposition, nutrient cycling, and crop production. Any shift in microbial community structure can be reflected in implementation of various land use and management systems that lead to development of best management practices for an agroecosystem. This chapter has highlighted the importance of addressing restoration of soil health in the context of multifunctionality of soil along with soil biodiversity. Soil biodiversity indicators can be employed by farmers and governments to assess and monitor soil health and ecosystem functioning under different land use system and management practices. They can help understand impacts of land use change and land degradation processes brought about by various driving forces. More importantly, the assessment and monitoring of soil life and soil health can be used to encourage the development and adaptation by farmers to develop a more sustainable and productive farming system. It is clear that a range of appropriate indicators of soil health are required to account for the multiple dimensions of soil ecosystem functions. Indicators should be selected that show a close link viz., between the soil characteristics and primary production wherein a positive indicator of soil quality could be organic matter and a negative one, incidence of crop damage by soil pathogens.

Acknowledgment Ajit Varma is thankful to Department of Science and Technology and Department of Biotechnology for partial financial funding and to DST-FIST for providing confocal microscope facility.

References

Agnolucci M, Battini F, Cristani C, Giovannetti M (2015) Diverse bacterial communities are recruited on spores of different arbuscular mycorrhizal fungal isolates. Biol Fertil Soils 51:379–389

Barbieri E, Bertini L, Rossi I, Ceccaroli P, Saltarelli R (2005) New evidence for bacterial diversity in the ascoma of the ectomycorrhizal fungus *Tuber borchii* Vittad. FEMS Microbiol Lett 247:23–35

Barbieri E, Guidi C, Bertaux J, Frey-Klett P, Garbaye J (2007) Occurrence and diversity of bacterial communities in *Tuber magnatum* during truffle maturation. Environ Microbiol 9:2234–2246

Bianciotto V, Bonfante P (2002) Arbuscular mycorrhizal fungi: a specialised niche for rhizospheric and endocellular bacteria. Antonie Van Leeuwenhoek 81:365–371

Bianciotto V, Andreotti S, Balestrini R, Bonfante P, Perotto S (2001) Mucoid mutants of the biocontrol strain *Pseudomonas fluorescens* CHA0 show increased ability in biofilm formation on mycorrhizal and nonmycorrhizal carrot roots. Mol Plant-Microbe Interact 14:255–260

Bomberg M, Timonen S (2007) Distribution of Cren- and Euryarchaeota in scots pine mycorrhizospheres and boreal forest humus. Microb Ecol 54:406–416

Bonfante P (2001) At the interface between mycorrhizal fungi and plants: the structural organization of cell wall, plasma membrane and cytoskeleton. In: Hock B (ed) Mycota. IX Fungal Associations. Springer, Berlin, pp 45–91

Bonfante P, Genre A (2008) Plants and arbuscular mycorrhizal fungi: an evolutionary-developmental perspective. Trends Plant Sci 13:492–498

Bouwmeester HJ, Roux C, Lopez-Raez JA, Becard G (2007) Rhizosphere communication of plants, parasitic plants and AM fungi. Trends Plant Sci 12:224–230

Cerigini E, Palma F, Barbieri E, Buffalini M, Stocchi V (2008) The *Tuber borchii* fruiting body-specific protein TBF-1, a novel lectin which interacts with associated *Rhizobium* species. FEMS Microbiol Lett 284:197–203

Choudhary DK, Kasotia A, Jain S, Vaishnav A, Kumari S, Sharma KP, Varma A (2016) Bacterial-mediated tolerance and resistance to plants under abiotic and biotic stresses. J. Plant Growth Regul 35:276–300

Collavino MM, Sansberro PA, Mroginski LA, Aguilar OM (2008) Comparison of in vitro solubilization activity of diverse phosphate-solubilizing bacteria native to acid soil and their ability to promote *Phaseolus vulgaris* growth. Biol Fertil Soils 46:727–738

Daniel R (2005) The metagenomics of soil. Nat Rev Microbiol 3:470–478

De Boer W, Folman LB, Summerbell RC, Boddy L (2005) Living in a fungal world: impact of fungi on soil bacterial niche development. FEMS Microbiol Rev 29:795–811

Deveau A, Palin B, Delaruelle C, Peter M, Kohler A (2007) The mycorrhiza helper *Pseudomonas fluorescens* BBc6R8 has a specific priming effect on the growth, morphology and gene expression of the ectomycorrhizal fungus *Laccaria bicolor* S238 N. New Phytol 175:743–755

Dimkpa CO, Merten D, Svatos A, Buchel G, Kothe E (2009) Metal induced oxidative stress impacting plant growth in contaminated soil is alleviated by microbial siderophores. Soil Biol Biochem 41:154–162

Duponnois R, Garbaye J (1991) Mycorrhization helper bacteria associated with the Douglas fir–*Laccaria laccata* symbiosis: effects in aseptic and in glasshouse conditions. Ann Sci For 48:239–251

Estrada GA, Baldani VLD, de Oliveira DM, Urquiaga S, Baldani JI (2013) Selection of phosphate-solubilizing diazotrophic *Herbaspirillum* and *Burkholderia* strains and their effect on rice crop yield and nutrient uptake. Plant Soil 369:115–129

Frey-Klett P, Garbaye J, Tarkka M (2007) The mycorrhiza helper bacteria revisited. New Phytol 176:22–36

Gamalero E, Trotta A, Massa N, Copetta A, Martinotti MG, Berta G (2004) Impact of two fluorescent pseudomonads and an arbuscular mycorrhizal fungus on tomato plant growth, root architecture and P acquisition. Mycorrhiza 14:185–192

Gamalero E, Berta G, Massa N, Glick BR, Lingua G (2008) Synergistic interactions between the ACC deaminase-producing bacterium *Pseudomonas putida* UW4 and the AM fungus *Gigaspora rosea* positively affect cucumber plant growth. FEMS Microbiol Ecol 64:459–467

Garbaye J (1994) Helper bacteria: a new dimension to the mycorrhizal symbiosis. New Phytol 128:197–210

Genre A, Chabaud M, Timmers T, Bonfante P, Barker DG (2005) Arbuscular mycorrhizal fungi elicit a novel intracellular apparatus in *Medicago truncatula* root epidermal cells before infection. Plant Cell 17:3489–3499

Glick BR (2015) Stress control and ACC aeaminase. In: Lugtenberg B (ed) Principles of plant-microbe interactions. Springer, Cham, pp 257–264

Guether M, Balestrini R, Hannah M, He J, Udvardi M, Bonfante P (2009) Genome-wide reprogramming of regulatory networks, transport, cell wall and membrane biogenesis during arbuscular mycorrhizal symbiosis in *Lotus japonicus*. New Phytol 182:200–212

Harrison MJ (2005) Signaling in the arbuscular mycorrhizal symbiosis. Annu Rev Microbiol 59:19–42

Harrison MJ, van Buuren ML (1995) A phosphate transporter from the mycorrhizal fungus *Glomus versiforme*. Nature 378:626–629

Hildebrandt U, Janetta K, Bothe H (2002) Towards growth of arbuscular mycorrhizal fungi independent of a plant host. Appl Environ Microbiol 68:1919–1924

Hildebrandt U, Ouziad F, Marner FJ, Bothe H (2006) The bacterium *Paenibacillus validus* stimulates growth of the arbuscular mycorrhizal fungus *Glomus intraradices* up to the formation of fertile spores. FEMS Microbiol Lett 254:258–267

Javot H, Varma Penmetsa R, Terzaghi N, Cook DR, Harrison MJ (2007) A *Medicago truncatula* phosphate transporter indispensable for the arbuscular mycorrhizal symbiosis. Proc Natl Acad Sci USA 104:1720–1725

Johansson JF, Paul LR, Finlay RD (2004) Microbial interactions in the mycorrhizosphere and their significance for sustainable agriculture. FEMS Microbial Ecol 48:1–13

Jorquera MA, Hernandez MT, Rengel Z, Marschner P, Mora ML (2008) Isolation of culturable phosphobacteria with both phytate-mineralization and phosphate-solubilization activity from the rhizosphere of plants grown in a volcanic soil. Biol Fertil Soils 44:1025–1034

Kisa M, Sanon A, Thioulouse J, Assigbetse K, Sylla S (2007) Arbuscular mycorrhizal symbiosis can counterbalance the negative influence of the exotic tree species *Eucalyptus camaldulensis* on the structure and functioning of soil microbial communities in a sahelian soil. FEMS Microbiol Ecol 62:32–44

Kozdrój J, Piotrowska-Seget Z, Krupa P (2007) Mycorrhizal fungi and ectomycorrhiza associated bacteria isolated from an industrial desert soil protect pine seedlings against Cd(II) impact. Ecotoxicology 6:449–456

Levy A, Chang BJ, Abbott LK, Kuo J, Harnett G (2003) Invasion of spores of the arbuscular mycorrhizal fungus *Gigaspora decipiens* by *Burkholderia* spp. Appl Environ Microbiol 69:6250–6256

Ma W, Penrose DM, Glick BR (2002) Strategies used by rhizobia to lower plant ethylene levels and increase nodulation. Can J Microbiol 48:947–954

Meyer JB, Frapolli M, Keel C, Maurhofer M (2011) Pyrroloquinoline quinone biosynthesis gene pqqC, a novel molecular marker for studying the phylogeny and diversity of phosphate-solubilizing pseudomonads. Appl Environ Microbiol 77:7345–7354

Naik PR, Sahoo N, Goswami D, Ayyaduraj N, Sakthivel N (2008) Genetic and functional diversity among fluorescent pseudomonads isolated from the rhizosphere of banana. Microb Ecol 56:492–504

Navazio L, Moscatiello R, Genre A, Novero M, Baldan B (2007) A diffusible signal from arbuscular mycorrhizal fungi elicits a transient cytosolic calcium elevation in host plant cells. Plant Physiol 144:673–681

Nazir R, Warmink JA, Boersma H, Van Elsas JD (2009) Mechanisms that promote bacterial fitness in fungal affected soil microhabitats. FEMS Microbiol Ecol 71:161–185

Nehls U, Wiese J, Guttenberger M, Hampp R (1998) Carbon allocation in ectomycorrhizas: identification and expression analysis of an *Amanita muscaria* monosaccharide transporter. Mol Plant-Microbe Interact 11:167–176

Nurmiaho-Lassila EL, Timonen S, Haahtela K, Sen R (1997) Bacterial colonization patterns of intact *Pinus sylvestris* mycorrhizospheres in dry pine forest soil: an electron microscopy study. Can J Microbiol 43:1017–1035

Ochsenreiter T, Selezi D, Quaiser A, Bonch-Osmolovskaya L, Schleper C (2003) Diversity and abundance of Crenarchaeota in terrestrial habitats studied by 16S RNA surveys and real time PCR. Environ Microbiol 5:787–797

Oldroyd GE, Downie JA (2008) Coordinating nodule morphogenesis with rhizobial infection in legumes. Annu Rev Plant Biol 59:519–546

Ordoñez YM, Fernandez BR, Lara LS, Rodriguez A, Uribe-Vélez D, Sanders IR (2016) Bacteria with phosphate solubilizing capacity alter mycorrhizal fungal growth both inside and outside the root and in the presence of native microbial communities. PLoS One 11:e0154438. doi:10.1371/journal.pone.0154438

Owen D, Williams AP, Griffith GW, Withers PJA (2015) Use of commercial bio-inoculants to increase agricultural production through improved phosphorus acquisition. Appl Soil Ecol 86:41–54

Rodriguez A, Sanders IR (2015) The role of community and population ecology in applying mycorrhizal fungi for improved food security. ISME J 9:1053–1061

Rodriguez H, Fraga R, Gonzalez T, Bashan Y (2006) Genetics of phosphate solubilization and its potential applications for improving plant growth-promoting bacteria. Plant Soil 287:15–21

Roesti D, Ineichen K, Braissant O, Redecker D, Wiemken A (2005) Bacteria associated with spores of the arbuscular mycorrhizal fungi *Glomus geosporum* and *Glomus constrictum*. Appl Environ Microbiol 71:6673–6679

Schalk IJ, Hannauer M, Braud A (2011) New roles for bacterial siderophores in metal transport and tolerance. Environ Microbiol 13:2844–2854

Scheublin TR, Sanders IR, Keel C, Van der Meer JR (2010) Characterisation of microbial communities colonising the hyphal surfaces of arbuscular mycorrhizal fungi. ISME J 4:752–763

Schrey SD, Tarkka MT (2008) Friends and foes: streptomycetes as modulators of plant disease and symbiosis. Antonie Van Leeuwenhoek 94:11–19

Schrey SD, Salo V, Raudaskoski M, Hampp R, Nehls U (2007) Interaction with mycorrhiza helper bacterium *Streptomyces* sp. AcH 505 modifies organisation of actin cytoskeleton in the ectomycorrhizal fungus *Amanita muscaria* (fly agaric). Curr Genet 52:77–85

Schüßler A, Martin H, Cohen D, Fitz M, Wipf D (2006) Characterization of a carbohydrate transporter from symbiotic glomeromycotan fungi. Nature 444:933–936

Selosse MA, Roy M (2009) Green plants that feed on fungi: facts and questions about mixotrophy. Trends Plant Sci 14:64–70

Singh BK, Nunan N, Ridgway KP, McNicol J, Young JP (2008) Relationship between assemblages of mycorrhizal fungi and bacteria on grass roots. Environ Microbiol 10:534–541

Singh RK, Malik N, Singh S (2013) Improved nutrient use efficiency increases plant growth of rice with the use of IAA-overproducing strains of endophytic *Burkholderia cepacia* strain RRE25. Microb Ecol 66:375–384

Smith SE, Read DJ (2008) Mycorrhizal symbiosis, 3rd edn. Academic, New York

Splivallo R, Novero M, Bertea CM, Bossi S, Bonfante P (2007) Truffle volatiles inhibit growth and induce an oxidative burst in *Arabidopsis thaliana*. New Phytol 175:417–424

Tarkka MT, Piechulla B (2007) Aromatic weapons: truffles attack plants by the production of volatiles. New Phytol 175:381–383

Tatsuki M, Nakajima N, Fujii H, Shimada T, Nakano M, Hayashi KI, Hayama H, Yoshioka H, Nakamura Y (2013) Increased levels of IAA are required for system 2 ethylene synthesis causing fruit softening in peach (*Prunus persica* L. Batsch). J Exp Bot 64:1049–1059

Thompson JN, Cunningham BM (2002) Geographic structure and dynamics of coevolutionary selection. Nature 417:735–738

Toljander JF, Artursson V, Paul LR, Jansson JK, Finlay RD (2006) Attachment of different soil bacteria to arbuscular mycorrhizal fungal extraradical hyphae is determined by hyphal vitality and fungal species. FEMS Microbiol Lett 254:34–40

Toljander JF, Lindahl BD, Paul LR, Elfstrand M, Finlay RD (2007) Influence of arbuscular mycorrhizal mycelia exudates on soil bacterial growth and community structure. FEMS Microbial Ecol 61:295–304

van Elsas JD, Duarte GF, Rosado AS, Smalla K (1998) Microbiological and molecular biological methods for monitoring microbial inoculants and their effects in the soil environment. J Microbiol Methods 32:133–154

Vivas A, Marulanda A, Ruiz-Lozano JM, Barea JM, Azcón R (2003a) Influence of a *Bacillus* sp. on physiological activities of two arbuscular mycorrhizal fungi and on plant responses to PEG-induced drought stress. Mycorrhiza 13:249–256

Vivas A, Azcón R, Biró B, Barea JM, Ruiz-Lozano JM (2003b) Influence of bacterial strains isolated from lead-polluted soil and their interactions with arbuscular mycorrhizae on the growth of *Trifolium pretense* L. under lead toxicity. Can J Microbiol 49:577–588

Vivas A, Barea JM, Biró B, Azcón R (2006) Effectiveness of autochthonous bacterium and mycorrhizal fungus on *Trifolium* growth, symbiotic development and soil enzymatic activities in Zn contaminated soil. J Appl Microbiol 100:587–598

Yang J, Xie X, Wang X, Dixon R, Wang YP (2014) Reconstruction and minimal gene requirements for the alternative iron-only nitrogenase in *Escherichia coli*. Proc Natl Acad Sci USA 111:3718–3725

Zhang L, Xu M, Liu Y, Zhang F, Hodge A, Feng G (2016) Carbon and phosphorus exchange may enable cooperation between an arbuscular mycorrhizal fungus and a phosphate-solubilizing bacterium. New Phytol. doi:10.1111/nph.13838

Zhao Y, Bian SM, Zhou HN, Huang JF (2006) Diversity of nitrogenase systems in diazotrophs. J Integr Plant Biol 48:745–755

Chapter 6
Mycorrhization of Fagaceae Forests Within Mediterranean Ecosystems

Francisca Reis, Rui M. Tavares, Paula Baptista, and Teresa Lino-Neto

Abstract Mediterranean Fagaceae forests are valuable due to their ecological and socioeconomic aspects. Some profitable plant species, such as *Castanea* (timber and chestnut), *Quercus* (timber and cork), and *Fagus* (timber), encounter in this habitat the excellent edaphoclimatic conditions to develop. All Fagaceae plants are commonly associated to ECM fungal species, which are found in these forests in quite stable communities, mainly enriched in Russulaceae and Telephoraceae species. Currently, the Mediterranean Basin is considered as one of the global biodiversity hotspots, since many of their endemic plant species are not found elsewhere and are now under threat. Due to climate changing and introduction of disease agents, Fagaceae forests are facing an adaptation challenge to both biotic and abiotic threats. Although ECM communities are highly disturbed by climate factors and tree disease incidence, they could play an important role in increasing water availability to the plant and also improving plant tree defense against pathogens. Recent advances, namely, on genomics and transcriptomics, are providing tools for increasing the understanding of Fagaceae mycorrhization process and stress responses to biotic and abiotic stresses. Such studies can provide new information for the implementation of the most adequate management policies for protecting threaten Mediterranean forests.

6.1 Introduction

Plant nutrient acquisition is mainly performed by root symbionts in about 86% of land plant species (Brundrett 2009). From the two most common mycorrhizal associations, arbuscular mycorrhizal (AM) fungi colonize a diverse spectrum of plant species, whereas ectomycorrhizal (ECM) fungi become specialized in trees

F. Reis • R.M. Tavares • T. Lino-Neto (✉)
BioSystems and Integrative Sciences Institute (BioISI), Plant Functional Biology Centre, University of Minho, Campus de Gualtar, 4710-057 Braga, Portugal
e-mail: tlneto@bio.uminho.pt

P. Baptista
REQUIMTE - School of Agriculture, Polytechnic Institute of Bragança, Campus de Santa Apolónia, 5300-253 Bragança, Portugal

A. Varma et al. (eds.), *Mycorrhiza - Function, Diversity, State of the Art*,
DOI 10.1007/978-3-319-53064-2_6

and shrubs colonization playing an essential role in forest sustainability. The physiology of colonization is also different. AM hyphae are capable of enter inside the root cells forming arbuscules, whereas ECM hyphal growth takes place in intercellular spaces of root cells forming an Hartig net, and the root tip is covered by a mantle (Bücking et al. 2012). Boreal, temperate forests (Mediterranean, Northern Hemisphere, South America), rain forests (Africa, India, and Indo-Malay), and seasonal woodlands of Australia are the most important habitats for ECM communities (Tedersoo et al. 2010). Both responsible for seedling establishment and tree growth, ECM are crucial for Pinaceae, Fagaceae, Betulaceae, Nothofagaceae, Leptospermoideae, Dipterocarpaceae, and Amhersteae families in woodland and forest communities (Tedersoo et al. 2010).

The Fagaceae family has a worldwide distribution and is well recognized for comprising the largely widespread beeches (*Fagus*), chestnuts (*Castanea*), and oaks (*Quercus*) species. However, this family comprises a total of about 900 plant species, which are included in nine genera of both deciduous and evergreen trees and shrubs (Kremer et al. 2012). Fagaceae family is currently divided into two subfamilies depending on their floral attributes, fruit morphology and germination: Castaneoideae (comprising *Chrysolepis*, *Castanea*, *Castanopsis*, and *Lithocarpus* genera) and the less consensual subfamily Fagoideae (Manos et al. 2001). The placement of *Fagus* together with *Quercus* and *Trigonobalanoid* genera (*Trigonobalanus*, *Formanodendron*, and *Colombobalanus*, which sometimes are collectively included under *Trigonobalanus*) in Fagoideae is still under debate (Nixon and Crepet 1989; Manos et al. 2001; Oh and Manos 2008; Kremer et al. 2012). Recently, a new genus, *Notholithocarpus*, has been isolated from *Lithocarpus*, since it is more closely related to *Quercus*, *Castanea*, and *Castanopsis* (Manos and Oh 2008). Presenting a high economic value (mostly *Castanea*, *Quercus*, and *Fagus* genera), due to their timber, fruits (chestnuts), and cork, the plantation areas of these plant species have been increasing in the past years (FAO 2013).

6.2 Fagaceae Forest Distribution

Fagaceae forests are mainly distributed in the northern temperate hemisphere, presenting also a biodiversity hotspot in Southeast Asia (reviewed by Kremer et al. 2012). While the temperate, subtropical, and semiarid floras are particularly rich in *Quercus*, *Castanea*, and *Fagus*, the warmer forests of Southeast Asia are comparably diverse in the castaneoid *Lithocarpus* and *Castanopsis* genera (Fig. 6.1). Northern Hemisphere temperate forests are all very similar, presenting high abundance of *Castanea*, *Fagus*, and *Quercus* genera. These temperate forests are characterized by well-defined seasons and moderate climate, comprising at least 4–6 frost-free months with regular rates of precipitation (Manos and Oh 2008). For this reason, European and North America ecosystems are the most closely related (Manos and Oh 2008), being both currently affected by a decrease of native beech

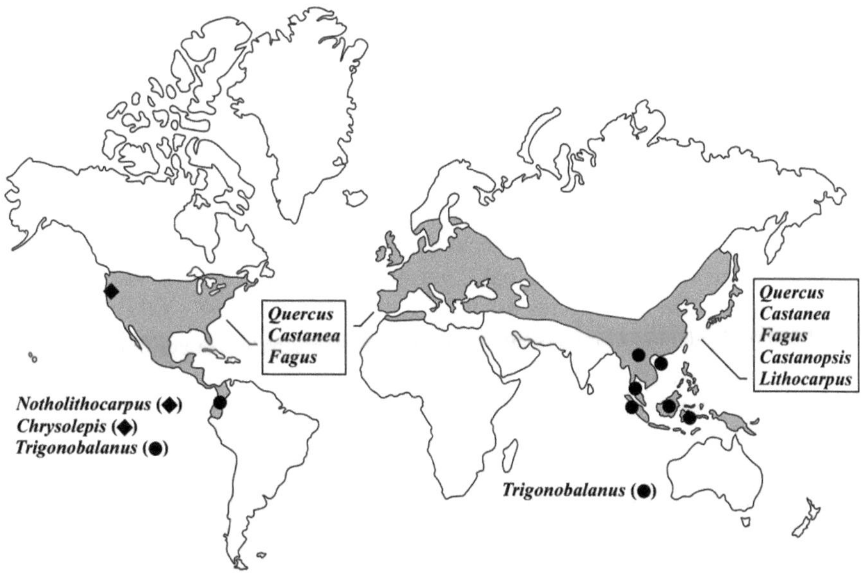

Fig. 6.1 World distribution of Fagaceae genera (adapted from Kremer et al. 2012). *Quercus*, *Castanea*, and *Fagus* genera are the most widespread genera and dominate broadleaf deciduous Mediterranean forests

and oak forests and natural reforestation (Brunet et al. 2010; Dulmer et al. 2014). Anthropogenic influence and disease incidence are two major threats. The Fagaceae forest cut down and forest clearing for activities like agriculture or natural products extraction (e.g., coal mining) has been a major source of income but is degrading forest ecosystems (Bauman et al. 2013). The population awareness for the need of appropriated reforestation programs is thus important to decrease forest erosion and desertification. The knowledge of ECM community of a particular geographic place could contribute for increasing tree adaptation and reforestation survival rate (Ding et al. 2011; Bauman et al. 2013; Dulmer et al. 2014).

Mediterranean climate features have provided unique conditions for the remarkable evolutionary adaptation and divergence of life. Mediterranean Basin only represents 1.5% of earth dry land but comprises about 10% of the total plant species identified (Blondel et al. 2010). From 22,500 plant species found in this region, 11,700 (52%) are endemic to Mediterranean Basin and cannot be found anywhere else in the world (Valavanidis and Vlachogianni 2011). However, the Mediterranean biodiversity has been currently threatened by the habitat loss and degradation, provided by the pollution levels, drought, alien invasive species spread, and overexploitation, among others. For example, from the original Mediterranean forests and shrubs lands, 70% have been destroyed by 1990 (Acácio et al. 2009). This resulted in the recognition of Mediterranean Basin as one of the first 25 Global Biodiversity Hotspots and a hyper-hot candidate for conservation due to the presence of exceptional totals of endemic plants (Myers et al. 2000). For these reasons, the European Union (EU) has classified the Mediterranean Basin as an area of European Community importance and established the "EU Habitats Directive" for the conservation of wild animal and plant species and natural habitats. From the 37 world habitat types identified as priority, 26 occur only in the Mediterranean region (Condé et al. 2005).

Mediterranean natural forests contain about 100 different tree species, whereas only 30 are present in forests of central Europe (four times larger; Valavanidis and Vlachogianni 2011). The Mediterranean forest is mainly composed by broadleaved evergreen tree species, such as oaks and mixed sclerophyllous trees, that alone present more than 20 species in the Mediterranean region (Valavanidis and Vlachogianni 2011). Conifers are also frequently found (Aleppo pine, *Pinus halepensis*; stone pine, *P. pinea*), being the rare conifer species of *Abies*, *Juniperus*, and *Taxus* commonly found in mountains. The most frequent oak species are the cork oak—*Quercus suber*—Fig. 6.2a; the holm oak, *Q. ilex* [considered as two subspecies, *Q. ilex* subsp. *ilex* and *Q. ilex* subsp. *rotundifolia* (Amaral Franco 1990), or as two different species: *Q. ilex* and *Q. rotundifolia* (Lumaret et al. 2002)]; or the Turkey oak, *Q. cerris*. While some oak species, like holm oak and kermes oak (*Q. coccifera*), encircle whole the Mediterranean Sea, others like cork oak and Mediterranean oak (*Q. canariensis*) exhibit a denser distribution in the western region (Condé et al. 2005). Although *Q. robur* is also found in Mediterranean countries, this tree species distribution is more evident in central and northern Europe (Fig. 6.2b), as also reported for *Fagus sylvatica* that has the preferable

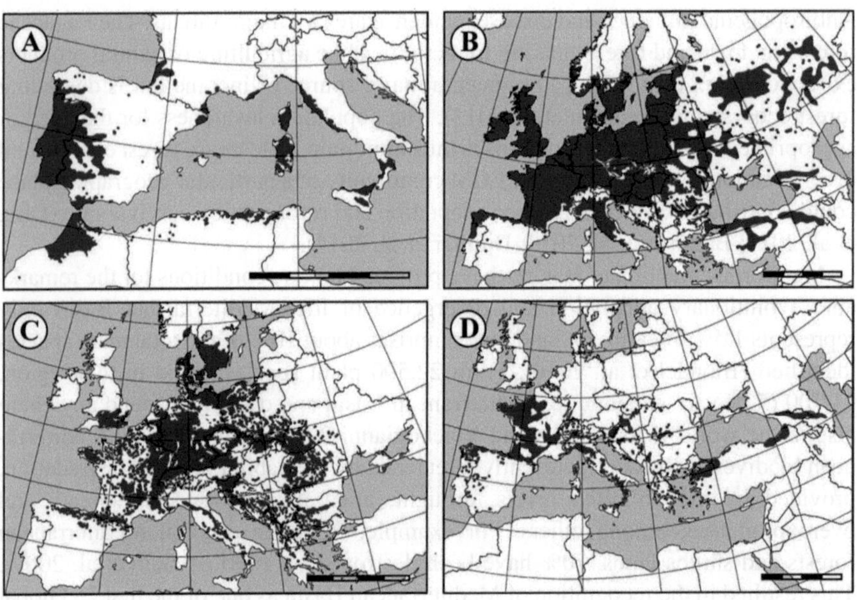

Fig. 6.2 European distribution of the most important Fagaceae species for the economy of Mediterranean Basin countries. *Quercus suber* (**a**), *Fagus sylvatica* (**b**), *Quercus robur* (**c**), and *Castanea sativa* (**d**) (www.euforgen.org)

climate and soil properties in the central Europe (Fig. 6.2c; EUFORGEN 2016). *Castanea sativa* presents the smallest forest area in Europe (predominantly in north of Iberian Peninsula, France, and west of Italy), mainly due to widespread diseases and cultural practices (Fig. 6.2d; Condé et al. 2005). On the other hand, regions with increased water availability are more favorable for downy oak (*Q. pubescens*), Valonia oak (*Q. ithaburensis*), or golden oak (*Q. alnifolia*—Cyprus native) growth (Condé et al. 2005). In all Mediterranean region, the dominant shrubs present in Fagaceae forests are highly aromatic, namely, *Cistus*, *Genista*, *Calluna*, *Arbutus*, thyme, and sage (Condé et al. 2005).

6.3 Mycodiversity in Fagaceae Forest Ecosystems

The interaction between trees and ECM fungi is dependent on many factors, namely, tree species, environmental conditions, and belowground interactions, among others (Öpik et al. 2006). Even season variations have an important role in ECM fungal dynamics in the soil. According to Voříšková et al. (2014), seasonal changes have a significant impact on fungal activity, biomass content and composition, as well as in the relative abundance of different fungal groups in temperate oak forests. A recent work performed in *Castanopsis fargesii*, *Lithocarpus harlandii*, *Pinus armandii*, and *Pinus massoniana* forests revealed that ECM community is much more dependent on the host plant species (33.3%) than soil origin (4.6%) (Ding et al. 2011). This is an important result to take into consideration in reforestation programs, dictating that adequate tree species selection is essential due to ECM host preference.

Fagaceae forests present a quite stable ECM community, mainly consisting of Basidiomycota species, like Russulaceae (Russulales), Thelephoraceae (Thelephorales), *Boletus* (Boletales), Cortinariaceae, Inocybaceae, and Amanitaceae (all Agaricales) species. When analyzing the ECM community diversity of Japanese and Chinese Fagaceae forests, the fungal families Russulaceae and Thelephoraceae were indeed the most abundant, being *Russula*, *Tomentella*, and *Clavulina* the most common ECM fungi (Wang et al. 2011; Toju et al. 2014). Results also showed that these Fagaceae forests, comprising *Castanopsis sieboldii*, *Lithocarpus edulis*, and *Quercus salicina*, present threefold more abundant ECM fungi than non-Fagaceae forests (Lauraceae, e.g., *Machilus japonica* and *Neolitsea sericea*; Toju et al. 2014). A North-American *C. dentata* forest also revealed the same trend as Asian Fagaceae forests, with Russulaceae as the major fungal family identified either by fruit bodies collection (aboveground analysis) or by morphotyping ECM root tips followed by direct sequencing of corresponding rDNA-ITS region (belowground analysis) (Palmer et al. 2008). Although highly abundant, the relative abundance of Boletales, Cortinariaceae, and Thelephoraceae was different in both fungal community views.

The temperate forests from the Mediterranean Basin uncover highly diverse ECM fungal communities, in which several hundreds of fungal species coexist (e.g., Richard et al. 2005; Buée et al. 2009). In a meta-analysis study where fruit bodies surveys were compared in holm oak, cork oak, and mixed forests from Andaluzia (Spain) region, a common dominance of Agaricomycetes species (e.g., Boletales and Russulales) was found (Ortega and Lorite 2007). In this study, a higher diversity and number of exclusive species were reported for cork oak forests. The diversity and structure of other Mediterranean *Quercus*, *Fagus*, and *Castanea* ECM communities have also revealed a high dominance of Russulaceae, Cortinariaceae, Thelephoraceae, and Inocybaceae fruit bodies (Table 6.1).

DNA technologies have improved fungal ecology studies during the recent past years. Fruit bodies as well as root tip descriptions have been greatly enriched by soil-based metabarcoding DNA sequencing (Shokralla et al. 2012). Even though this recent approach revealed a high potential for microbial diversity identification in every ecological guilds, there are still some issues remaining when applying next-generation sequencing (NGS) methods for assessing fungal diversity (Orgiazzi et al. 2015). When studying Fagaceae ECM communities recurring to molecular methods, such as ITS barcoding of ECM tips (e.g., Richard et al. 2005) or ITS metabarcoding of soil sample approaches (Buée et al. 2009), which are methods not dependent on the ability of fungi to produce conspicuous fruit bodies, a different picture of ECM community is obtained. While ECM surveys based exclusively on fruit bodies identification (aboveground approaches) have been hyper-dominant in Basidiomycetes species (mainly Agaricomycetes), a high diversity of Ascomycetes has been detected using belowground approaches based on molecular methods (Peintner et al. 2007; Orgiazzi et al. 2012; Baptista et al. 2015). In spite of that, a higher abundance of Basidiomycota operational taxonomic units (OTUs) has been consistently found. However, from 140 identified taxa among 558 ECM *Q. ilex* root tips, the Ascomycota *Cenococcum geophilum* dominated (35% of ECMs), together with Russulaceae (21.4%), Cortinariaceae (7.1%), and Thelephoraceae (25%) (Richard et al. 2005). The same trend was detected by Azul et al. (2010) when studying the influence of managed oak woodlands dominated by *Q. suber*, under different land use practices, by using the same ECM root tip surveys complemented with ITS rDNA analysis. In this study, the Ascomycota *C. geophilum*, together with Russulaceae and Thelephoraceae, represented 56% of whole ECM fungal community. A positive correlation between ECM fungal richness and silvo-pastoral exploitation regime and low mortality of cork was detected in this study (Azul et al. 2010). In addition, the use of NGS DNA sequencing methods on *Fagus sylvatica* forest soils revealed that the most abundant fungal genera were *Russula*, *Boletus*, but also *C. geophilum* (Coince et al. 2013). Moreover, *C. geophilum* was the main ECM fungus reported in root tip assessment in *Q. rubra* forests, although its abundance has oscillated significantly with tree age (Gebhardt et al. 2007).

Although the ECM association is the dominant symbiotic relationship, Mediterranean Fagaceae species can also be simultaneously colonized by different mycorrhizal fungal types, such as AM and ericoid fungi, among others (Bergero et al. 2000). Accordingly, in oak forests, a higher number of AM fungal spores (mainly *Ambispora gerdemannii*) have been found when compared to other landscapes, such as pine forests, combined forests of pines and oaks, or in several

Table 6.1 ECM communities present in Fagaceae forests in Mediterranean Basin ecosystems. Revision of published studies since 2000

Fagaceae species	Ecosystem	ECM taxa	Approach	Reference
Q. ilex	Corsica Island, France	*Russula*, *Amanita*, *Tricholoma*, *Cortinarius*	Root tips	Richard et al. (2004)
Q. ilex	Mediterranean forests	*Cenococcum geophilum*	Root tips	De Román and De Miguel (2005)
Q. ilex	Mediterranean forests	*Cenococcum geophilum*, Russulaceae, Cortinariaceae, Thelephoraceae	Root tips	Richard et al. (2005)
Q. ilex	Mediterranean forests	Thelephoraceae, Russulaceae, Cortinariaceae	Root tips	Richard et al. (2011)
Q. ilex	Southern France	Thelephoraceae, Pyrenomataceae	Root tips	Taschen et al. (2015)
Q. suber	Moroccan woodlands	*Pisolithus*, *Boletus aureus*	Fruit bodies survey	Yakhlef et al. (2009)
Q. suber	Portuguese montados (savanna-type forests)	*Cenococcum geophilum*, Russulaceae, Thelephoraceae	Root tips	Azul et al. (2010)
Q. suber	Declining forest in northwestern Sardinia, France	Pyronemataceae, Thelephoraceae, Russulaceae, Inocybaceae, Cortinariaceae	Root tips	Lancellotti and Franceschini (2013)
Q. suber	Portuguese forests and landscapes	*Russula*, *Tomentella*, *Cenoccoccum*	Root tips	Reis et al., unpublished results
Q. suber and *Q. canariensis*	South of Spain	*Lactarius chrysorrheus*, *Cenococcum geophilum*	NGS	Aponte et al. (2010)
Q. petraea	Czech Republic	*Russula*, *Lactarius*	NGS	Voříšková et al. (2014)
Q. petraea and *Q. robur*	100-year-old forest in northeastern France	*Tomentella*, *Lactarius*, *Cenococcum*	Root tips	Courty et al. (2008)
C. sativa	Greece	*Amanita caesarea*, *A. rubescens*, *Boletus edulis*, *B. aereus*, *Cantharellus cibarius*, *Craterellus cornucopioides*, *Hydnum repandum*, *H. rufescens*	Fruit bodies surveys	Diamandis and Perlerou (2001)
C. sativa	Italy	*Russula*, *Inocybe*, *Lactarius*, *Tricholoma*, *Cortinarius* and *Amanita*	Fruit bodies surveys	Laganà et al. (2002)
C. sativa	Italy	*Cenococcum geophilum*, *Boletus aestivalis*, *Lactarius chrysorrheus*	Root tips and fruit bodies survey	Peintner et al. (2007)

(continued)

Table 6.1 (continued)

Fagaceae species	Ecosystem	ECM taxa	Approach	Reference
C. sativa	Healthy and *Phytophthora*-infected forests in central Italy	*Cenococcum geophilum, Oidiodendron maius*	Root tips	Blom et al. (2009)
C. sativa	Portuguese orchards	*Russula, Inocybe, Lactarius, Tricholoma, Boletus, Cortinarius, Amanita*	Fruit bodies survey	Baptista et al. (2010)
C. sativa	Portuguese orchards	*Inocybe, Amanita* (above) and *Inocybe, Amanita, Sistotrema* (below)	Fruit bodies survey and NGS	Baptista et al. (2015)

agroecosystems (Chaturvedi et al. 2012). In addition, the symbiotic relationship between plant and ECM fungi can be mediated by other microorganisms or plants (Herrmann 2007; Toju et al. 2014). For example, recent studies on red oak (*Q. rubra*) have showed that soil bacteria can help plants to establish ECM symbiosis by maintaining adequate plant signaling gene levels that will promote mycorrhization (Kurth et al. 2015). Accordingly, as obligatory ECM hosts, *Quercus* are usually sensitive to shifts on microbial communities (Smith et al. 2007).

To conclude, the enriched decaying litter soil from Fagaceae forests is an excellent habitat for fungal development and has been a natural source of many economically important mushrooms (Boa 2004). Those edible ECM fungi naturally associated with Fagaceae trees, mainly in *Castanea* or *Quercus* forests, comprise a main forest subproduct for population food supply, as well as for the production of natural medicines (Boa 2004; Savoie and Largeteau 2011). However, ECM mushroom harvesting has been dramatically decreasing in the past century (Yun and Hall 2004), mainly due to air pollution and litter accumulation in soil surface (Smit et al. 2003). For all these reasons, the preservation of forests including Fagaceae forests has become not only ecologically important but also necessary for maintaining an ECM edible mushroom repository.

6.4 Disturbance and Protection of Fagaceae Forests from Biotic Threats

Beyond ecological and physiological importance to the forests, ECM community is essential for plant tree disease prevention and incidence (Smith and Read 2008). The most devastating diseases of Fagaceae family are caused by *Phytophthora* spp. (ink disease and oaks decline) and *Cryphonectria parasitica* (blight disease). The sudden oak disease caused by *Phytophthora ramorum* has been responsible for the rapid mortality of native oak trees (*Quercus* spp. and *Lithocarpus densiflorus*) in central and northern California (USA) since its first observation in 1995 (DiLeo et al. 2009). More recently, surveys revealed that *P. ramorum* was introduced into Pacific northwest nurseries and into at least eight European countries by movement of stock plants (Brasier et al. 2004). Also, the introduction of the causal agent of chestnut blight disease (*C. parasitica*) by the importation of infected Asian chestnut trees to the USA east coast in the early twentieth century almost led to the extinction of American chestnuts (*C. dentata*; Milgroom et al. 1996). Indeed, this later epidemic has been considered as one of the greatest ecological disasters in US history (Wheeler and Sederoff 2009) and one of the most devastating plant disease epidemics caused by fungi or fungal-like oomycetes (Fisher et al. 2012). Although pedunculate oaks (*Quercus petraea* and *Q. robur*), holm oak (*Q. ilex*), and *Castanopsis* have been also classified as *C. parasitica* host species by the European Plant Protection Organization (EPPO), corresponding plant damages are relatively less when compared with chestnut species. Although susceptible to this fungus, the relatively higher tolerance of European chestnut (*C. sativa*) in comparison to the American chestnut prevented the heavy mortality levels previously observed in the USA (Heiniger and Rigling 1994). However, when *C. parasitica* was first observed in Europe (Genova, Italy, in 1938; reviewed by Anagnostakis 1987), the blight disease rapidly spread all over France, Spain, and Portugal chestnut orchards (Robin and Heiniger 2001).

Within the Mediterranean region, oomycetes from *Phytophtora* spp. are serious threats to Fagaceae forests. Between 1900 and 1950, the main *C. sativa* growing areas of southern Europe, especially Italy, France, and Iberia, suffered heavy mortality due to the chestnut ink disease caused by *Phytophthora cambivora* and *P. cinnamomi* (reviewed by Brasier 2000). After introduction in the late eighteenth century from a center of origin in the Papua New Guinea-Celebes, this disease rapidly spread in France and in all chestnut-growing areas (Vettraino et al. 2002), being the main reason for abandonment of several chestnut orchards. In addition, *P. cinnamomi* has been reported as the agent responsible of ink disease of red oak (*Quercus rubra*; Robin et al. 2012) and as the primary factor of root infection resulting in oak decline and mortality in Mediterranean countries (Brasier et al. 1993). Although cork and holm oak decline have occurred in the Mediterranean Basin since the beginning of the twentieth century, only in the early 1980s, a severe oak decline was reported across the Mediterranean region (Brasier 1996). Oak decline has been described as a complex disease triggered by several interacting environmental constraints, including pathogens (*P. cinnamomi*), as well as drought and other site factors (soil texture and fertility, slope) (Camilo-Alves et al. 2013). The affected oak trees face a progressive defoliation that can go over 75% (Franceschini et al. 2002). Typical symptoms of *Phytophthora* diseases have also been observed in *Fagus* stands of several European countries in the last two decades, which are caused by *P. citricola*, *P. cambivora*, and *P. cactorum* (Schmitz et al. 2006), and in Swedish *Q. robur* stands caused by *P. quercina* (Jönsson-Belyazio and Rosengren 2006).

All *Phytophthora* diseases result in severe leaf loss, which would lead to the reduction of root sugar content and would alter the ECM community of diseased plants. Accordingly, tree crown defoliation has been shown to modify ECM community structure in Scots pine (Kuikka et al. 2003) and increase the frequency of thin mantled ECM morphotypes (Saravesi et al. 2008). Even artificial defoliation has been reported to negatively affect ECM symbionts by reducing the production of fungal biomass in interacting roots (Markkola et al. 2004; Stark and Kytöviita 2005). Comparing healthy and ink-diseased chestnut stands, Blom et al. (2009) found differences in the richness of ECM communities and relative abundance of most important ECM fungi. *C. geophilum* was dominant on both stands, but its relative abundance was 1.5-fold higher in the infected orchard. Also, other Basidiomycota, such as Boletaceae, Paxillaceae, Sistotremataceae, Hydnaceae, and Atheliaceae, showed significantly higher values in infected soils, whereas Thelephoraceae, Cortinariaceae, and Sebacinaceae showed an opposite trend (Blom et al. 2009). As a result of oak decline disease, a reduction of ECM diversity and ECM root colonization has been detected in *Q. ilex* trees (Causin 1996; Montecchio et al. 2004). In contrast, *Q. suber* declined trees do not present differences in ECM community when compared to healthy trees (Lancellotti and Franceschini 2013). But, although no differences in ECM community have been detected in Spanish *Q. ilex* forest trees infected or not with *P. cinnamomi*, non-mycorrhizal root tips seem to be more susceptible to infection than mycorrhizal ones (Corcobado et al. 2014). Although these results indicate that ECM communities are strongly affected in diseased Fagaceae plants, ECM fungal species could also contribute for disease protection. This feature could be provided by the formation of a mantle that serves as a physical barrier to the pathogen, by the production of antibiotics that inhibit pathogen growth and reproduction, by diverging plant exudates that could act as biochemical signals to the disease agent, by providing habitat for antagonistic rhizosphere microorganisms, or by improving plant vigor and protection potential (reviewed by Keen and Vancov 2010). Accordingly, a number of ECM fungi have been already related to *P. cinnamomi* suppression in conifers and eucalyptus forests (Marx 1972; Malajczuk 1979; Malajczuk and McComb 1979), and several ECM fungal isolates (mainly *Suillus brevipes*) have revealed high antagonistic potential against *Phytophtora* sp. (Mohan et al. 2015). The direct protection of ECM fungi against both *P. cambivora* and *P. cinnamomi* infection was achieved after inoculation of *C. sativa* seedlings with *Laccaria laccata*, *Hebeloma crustuliniforme*, *H. sinapizans*, and *Paxillus involutus* (Branzanti et al. 1999). Biocontrol and bioprotection strategies by using ECM could then be the future key for Fagaceae disease prevention and treatment. This kind of information would be important for advising tree nurseries involved in reforestation programs, even though artificial inoculation of *Q. garryana* and *F. sylvatica* seedlings has not been considered necessary in nursery practices (Southworth et al. 2009; Pietras et al. 2013). In any case, the inoculation of *Q. ilex* seedlings with *Hebeloma mesophaeum* revealed to increase the mycorrhizal colonization and plant growth while reducing the need for fertilizers (Oliveira et al. 2010). Also, *Q. ilex* and *Q. faginea* artificial mycorrhization with *Tuber melanosporum* improved seedling growth, water, and phosphorous acquisition (Núñez et al. 2006). Although the growth of cork oak nursery seedlings has not increased by artificial inoculation with

Pisolithus tinctorius, several physiological parameters, such as higher photosynthetic capacity, water use efficiency, and N uptake capacity, benefit from mycorrhization (Sebastiana et al. 2013).

In the recent past years, asymptomatic endophytic fungi have been also regarded as potential biocontrol agents for tree diseases (e.g., Arnold et al. 2003; Blumenstein et al. 2015). The oak decline has been correlated with the diversity and amount of fungal endophytes present on different tissues of *Quercus* spp., and many oak-specific endophytes are specifically described to accelerate the decline of oaks stand (Ragazzi et al. 2001, 2003, 2004). *Q. cerris* exhibited a more diverse endophytic assemblage, but greater infection levels, than *Q. pubescens* suggesting a role of some pathogenic fungal endophytes in Mediterranean oak forests (Moricca et al. 2012).

Other biocontrol agents against Fagaceae diseases are now arising. Strains of the chestnut blight fungus, *C. parasitica*, harboring asymptomatic mycoviruses (CHV1-4; reviewed by Xie and Jiang 2014) are described to induce hypovirulence (virulence attenuation) (Dawe and Nuss 2001). The use of the complex triple interaction (hypovirus, fungal pathogen, and chestnut tree) for controlling chestnut blight in orchards remains a possibility (Xie and Jiang 2014). Antagonistic microbes or metabolites produced by them have been also studied as potential biocontrol agents against *Phytophthora* spp. causing chestnut ink disease (reviewed by Choupina et al. 2014). Most promising results were obtained with *Trichoderma* sp., *Gliocladium* sp., and *Pseudomonas* sp. (Aryantha et al. 2000).

6.5 Fagaceae Mycorrhization in a Mediterranean Changing Climate

The sustainability of forests is extremely dependent on both biotic and abiotic factors, and worldwide climate changes are affecting forests all over the world (Keenan 2015). The effects of drought can be minimized by increasing water uptake through fine root growth, by deep taproot formation, and by osmotic adjustment in water-stressed roots through the accumulation of osmolytes (reviewed in Brunner et al. 2015). Due to their long-term evolutionary adaptation to long periods without rain and high temperatures, typical Mediterranean tree species, particularly evergreen oaks, are particularly adapted to cope with moderate drought without significant losses of production and survival (Ramirez-Valiente et al. 2009, 2011). For example, although not so drought tolerant as *Q. ilex* (described as one of the most drought-resistant oaks), cork oak presents rather drought-tolerant traits such as deep roots (Kurz-Besson et al. 2006). However, Mediterranean forests are now facing problems due to the rapid environmental changes (Lindner et al. 2014). Forests become more likely to be exposed to extreme events, such as the increased risk of fire, extreme drought events, or severe heat waves, which could even lead to the spread of pests and diseases (reviewed by

Bussotti et al. 2013; Moricca and Ragazzi 2008; Moricca et al. 2014). Recurrent episodes of extreme water stress can greatly increase the number of declined trees (also with the contribution of pathogens) and represent a major threat to the survival of Mediterranean plant species (Nardini et al. 2014). Tree plasticity and adaptation to drought is now slower than the increase of stress severity. In *Q. faginea*, a typical Mediterranean tree, the rate of plant adaptive response in xeric environment is significantly lower than drought increase occurring in Spain (Nuche et al. 2014).

As individual plant responses to environmental changes are largely dependent on fungal symbionts (reviewed by Kivlin et al. 2013), the microbial community present in the forest soil is suggested to play an essential role in plant drought stress resistance. The changing environmental conditions are likely to induce changes in plant physiology and root exudation, altering the composition of root exudates in chemoattractants or signal compounds (Kandeler et al. 2006) and thus changing the structure of ECM communities associated with stressed plants (reviewed by Compant et al. 2010). Accordingly, the increased drought imposed by reduction of rainfall induced significant shifts in *Q. ilex* ECM community composition (Richard et al. 2011). The most common taxa identified in these forests are Thelephoraceae, Russulaceae, and Cortinariaceae, but five consecutive years of increased drought have induced a positive response of Cortinariaceae species. In addition, when *F. sylvatica* plants were subjected to drought, no effect was detected in *Lactarius subdulcis* and *Byssocorticium atrovirens* mycorrhizae abundance, but *Xerocomus chrysenteron* mycorrhizae occurrence increased almost twofold (Shi et al. 2002). Furthermore, beech plants mycorrhized with *X. chrysenteron* and *L. subdulcis* were able to better cope with drought stress than others. These observations suggested that distinct ECM taxa differently respond to drought by specifically changing their occurrence/abundance in mycorrhized plants and each plant could be differently affected by drought according to the associated mycorrhizal community. Furthermore, the structure of *F. sylvatica* ECM communities and metabolic activity of each morphotype was reported to be dependent on the season, temperature, and soil moisture, being certain morphotypes more abundant and active in winter than in summer (Buée et al. 2005). The same authors described *C. geophilum* morphotype as being more active during summer, when the increase in temperature and drought could influence its abundance and enzyme activity as reported in oak ecosystems (*Q. robur*, *Q. petraea*, and *Q. pubescens*) (Herzog et al. 2012). Therefore, the overall function of ECM community would result from the occurrence and functional feature of each morphotype. In a complex ecosystem as Fagaceae forests, more than one variable could be influencing ECM communities. European *Q. robur* and *Q. petrea* forests ECM community are influenced by precipitation, pH, and N deposition (Suz et al. 2014).

Diverse drought tolerance levels exhibited by mycorrhized plants are most probably due to the well-recognized differences in drought resistance of specific ECM fungi. *Rhizopogon vinicolor* and *C. geophilum* have been reported as drought-tolerant species, being *C. geophilum* also particularly efficient in protecting forest trees against drought damage, while *L. laccata* is described as a drought-sensitive fungus unable to grow at very low water potentials (Coleman et al. 1989; di Pietro et al. 2007). Since the respiration activity of *C. geophilum* ectomycorrhizae has been reported to be significantly less altered than that of *Lactarius* sp., *C. geophilum* was suggested to better maintain the physiological integrity of beech roots facing drought stress (Jany et al. 2002). In contrast, under high temperatures, a decreased colonization with *C. geophilum* has been detected in *Quercus myrsinaefolia* (Kasai et al. 2000), agreeing with the observation of its reduced respiration under increasing temperature (Malcolm et al. 2008). In any case, *C. geophilum* being a hydrophilic and short-distance exploration fungus has been suggested as a potential indicator of environmental changes (reviewed by Lehto and Zwiazek 2011). However, several problems have been discussed about its use in environmental assessments, including its resistance to other stress factors besides drought and its inability of forming fruit bodies.

The ability for water uptaking in a typical Mediterranean climate is essential for tree resistance to drought scenarios, and ECMs have been recognized as crucial for drought resistance improvement (Kivlin et al. 2013; Brunner et al. 2015). The water status of drought-stressed trees is highly improved by the increased absorbing surface provided by the ECM fungi, through a higher efficient water conduction by mycelial strands, enhanced soil-root hydraulic conductivity, and other hormonal and nutritional effects that modify plant physiology (reviewed by Breda et al. 2006). Moreover, ECM networks can redistribute water from deep soils to roots or move water among roots of drought-stressed plants (Egerton-Warburton et al. 2007; Querejeta et al. 2007). Accordingly, studies performed in *Q. alba* inoculated with *P. tinctorius* revealed higher water potentials and larger root systems than non-inoculated plants (Dixon et al. 1980). Also, *Q. ilex* seedlings inoculated with *T. melanosporum* exhibited half of root hydraulic conductance than non-mycorrhized roots but presented 2.5-fold more fine root surface area (Nardini et al. 2000). The best ECM inoculum for improving drought tolerance is difficult to establish, but their choice should be based on fungal water uptake ability and exploration type. Hydrophilic fungi, such as *Russula*, *Hebeloma*, *Lactarius*, and *Laccaria*, are able to transport water in the apoplast, whereas hydrophobic fungi, like *P. involutus* and *Suillus* spp., need to form mycelia cords to transport water in the symplast (reviewed in Lehto and Zwiazek 2011). On the other hand, contact mycelia or short-distance exploration mycorrhizae are mainly hydrophilic, whereas long-distance exploration are hydrophobic fungal ECMs (Agerer 2001). This particular information would be essential in further research on ECM behavior in drought scenarios or on ECM fungal selection for in vitro and field assays.

Forest fires are common in Mediterranean region during summer period, but fire risk is clearly increasing due to extreme environmental conditions. Indeed, during the last decade, Mediterranean forest fires (especially in Portugal and Greece) have been associated with extreme weather, in particular to extremely long dry periods with hot temperatures and high wind speeds (reviewed by Lindner et al. 2014). Fire events could have significant effects on fungal communities of Mediterranean forests. After a fire event, the complexity of ECM communities tends to be reduced and replaced by a less diverse community, usually composed by resilient fungal species and previously rare species (Pezizales and *Rhizopogon* spp.; reviewed by

Buscardo et al. 2010). Colonization by new fungal species can benefit from a competition decrease, being spores the main structures for postfire natural recolonization. While *Telephora* spp. distribution was strongly affected by fire events in an oak forest, *Tomentella* spp. rapidly raised (Buscardo et al. 2010). When studying the ECM root tips of a *Q. ilex* forest over a 3-year postfire period, the richness of ECM community and the percentage of root tips were also significantly decreased (De Román and De Miguel 2005). *C. geophilum* was the most resilient ECM fungi and maintained its abundance all over the period.

6.6 Advances for Mediterranean Fagaceae ECM Studies

To better understand the symbiotic relationship that occurs between Fagaceae roots and ECM fungi, new molecular tools have been created. Several efforts have been made in order to know the genetic patrimony of several Fagaceae species. To the best of our knowledge, 18 Fagaceae genomes have already been sequenced, eight *Castanea* species and ten *Fagus* species, six of which considered as subspecies (http://www.fagaceae.org/). Other species, such as *Q. alba*, *Q. rubra*, and *Q. suber*, have their genome sequencing ongoing (The Fagaceae Genome Web, http://www.fagaceae.org/home; Genosuber Project—http://www.genosuber.com/). Furthermore, several transcriptomic studies are now allowing the generation of a comprehensive catalog of transcripts from Fagaceae. Recently, a number of transcriptomic studies have been successful at generating expressed sequence tags (ESTs) libraries, mainly from oaks and chestnuts, recurring to NGS approaches (e.g., *Q. robur* and *Q. petraea*, Lesur et al. 2015; *C. sativa* and *C. crenata*, Serrazina et al. 2015). The use of a *Q. robur* gene catalog allowed the discovery of specific molecular mechanisms involved in the regulation of oak ECM symbiosis and the identification of key molecular players involved in ECM formation (Tarkka et al. 2013). Their main findings concern the plant defense gene attenuation and ethylene signaling enhancement during mycorrhization, cell wall remodeling mechanisms, and alteration in several metabolic pathways (e.g., nitrogen, phosphorus, and sugar transporters). Within a national initiative, a Portuguese consortium was created to study cork oak ESTs and thus develop a new genomic resource for studying *Q. suber* (Pereira-Leal et al. 2014). This achievement has been used to better understand processes related with plant development (Rocheta et al. 2014; Teixeira et al. 2014) and adaptation responses to both biotic (Sebastiana et al. 2014) and abiotic factors (Magalhães et al. 2016). The global overview of up- and downregulated genes in cork oak roots following inoculation with the *P. tinctorius* resulted in a better insight of those molecular events that control ECM symbiosis (Sebastiana et al. 2014). ECM colonization resulted in extensive cell wall remodeling, activation of the secretory pathway, alterations in flavonoid biosynthesis, and expression of genes involved in the recognition of fungal effectors. Other identified genes could have putative roles in symbiotic processes such as nutrient exchange with the fungal partner, lateral root formation, or root hair decay (Sebastiana et al. 2014). The

Brunet J, Fritz Ö, Richnau G (2010) Biodiversity in European beech forests—a review with recommendations for sustainable forest management. Ecol Bull 53:77–94

Brunner I, Herzog C, Dawes MA, Arend M, Sperisen C (2015) How tree roots respond to drought. Front Plant Sci 6:547. doi:10.3389/fpls.2015.00547

Bücking H, Liepold E, Ambilwade P (2012) The role of the mycorrhizal symbiosis in nutrient uptake of plants and the regulatory mechanisms underlying these transport processes. Plant Sci 4:107–138. doi:10.5772/52570

Buée M, Vairelles D, Garbaye J (2005) Year-round monitoring of diversity and potential metabolic activity of the ectomycorrhizal community in a beech (*Fagus silvatica*) forest subjected to two thinning regimes. Mycorrhiza 15:235–245. doi:10.1007/s00572-004-0313-6

Buée M, Reich M, Murat C, Morin E, Nilsson RH, Uroz S, Martin F (2009) 454 pyrosequencing analyses of forest soils reveal an unexpectedly high fungal diversity. New Phytol 184:449–445

Buscardo E, Rodriguez-Echeverria S, Martin MP, De Angelis P, Pereira JS, Freitas H (2010) Impact of wildfire return interval on the ectomycorrhizal resistant propagules communities of a Mediterranean open forest. Fungal Biol 114:628–636. doi:10.1016/j.funbio.2010.05.004

Bussotti F, Ferrini F, Pollastrini M, Fini A (2013) The challenge of Mediterranean sclerophyllous vegetation under climate change: from acclimation to adaptation. Environ Exp Bot 103:80–98. doi:10.1016/j.envexpbot.2013.09.013

Camilo-Alves C, da Clara MIE, de Almeida Ribeiro NMC (2013) Decline of Mediterranean oak trees and its association with *Phytophthora cinnamomi*: a review. Eur J For Res 132:411–432. doi:10.1007/s10342-013-0688-z

Causin R, Montecchio L, Accordi SM (1996) Probability of ectomycorrhizal infection in a declining stand of common oak. Ann For Sci 53:743–752. doi:10.1051/forest:19960250

Chaturvedi S, Tewari V, Sharma S, Oehl F, Wiemken A, Prakash A, Sharma AK (2012) Diversity of arbuscular mycorrhizal fungi in oak-pine forests and agricultural land prevalent in the Kumaon Himalayan Hills. Br Microbiol Res J 2:82–96. doi:10.9734/BMRJ/2012/1136

Choupina AB, Estevinho L, Martins I (2014) Scientifically advanced solutions for chestnut ink disease. Appl Microbiol Biotechnol 98:3905–3909. doi:10.1007/s00253-014-5654-2

Coince A, Cael O, Bach C, Lengelle J, Cruaud C, Gavory F, Morin E, Murat C, Marcais B, Buee M (2013) Below-ground fine-scale distribution and soil versus fine root detection of fungal and soil oomycete communities in a French beech forest. Fungal Ecol 6:223–235

Coleman MD, Bledsoe CS, Lopushinsky W (1989) Pure culture response of ectomycorrhizal fungi to imposed water stress. Can J Bot 67:29–39

Compant S, van der Heijden MG, Sessitsch A (2010) Climate change effects on beneficial plant-microorganism interactions. FEMS Microbiol Ecol 73:197–214

Condé S, Richard D, Liamine N (2005) European Environment Agency Europe' s biodiversity The Mediterranean biogeographical region. In: EEA Europe's Biodiversity, Alpine, pp 1–54

Corcobado T, Vivas M, Moreno G, Solla A (2014) Ectomycorrhizal symbiosis in declining and non-declining *Quercus ilex* trees infected with or free of *Phytophthora cinnamomi*. For Ecol Manage 324:72–80. doi:10.1016/j.foreco.2014.03.040

Courty P, Franc A, Pierrat J, Garbaye J, Fore R (2008) Temporal changes in the ectomycorrhizal community in two soil horizons of a temperate oak forest. Appl Environ Microbiol 74:5792–5801. doi:10.1128/AEM.01592-08

Dawe AL, Nuss DL (2001) Hypoviruses and chestnut blight: exploiting viruses to understand and modulate fungal pathogenesis. Annu Rev Genet 35:1–29. doi:10.1146/annurev.genet.35.102401.085929

De Román M, de Miguel AM (2005) Post-fire, seasonal and annual dynamics of the ectomycorrhizal community in a *Quercus ilex* L. forest over a 3-year period. Mycorrhiza 15:471–482. doi:10.1007/s00572-005-0353-6

Di Pietro M, Churin JL, Garbaye J (2007) Differential ability of ectomycorrhizas to survive drying. Mycorrhiza 17:547–555. doi:10.1007/s00572-007-0113-x

Diamandis S, Perlerou C (2001) The mycoflora of the chestnut ecosystems in Greece. For Snow Landsc Res 76:499–504

DiLeo MV, Bostock RM, Rizzo DM (2009) *Phytophthora ramorum* does not cause physiologically significant systemic injury to California bay laurel, its primary reservoir host. Phytopathology 99:1307–1311. doi:10.1094/PHYTO-99-11-1307

Ding Q, Liang Y, Legendre P, He X, Pei K, Du X, Ma K (2011) Diversity and composition of ectomycorrhizal community on seedling roots: the role of host preference and soil origin. Mycorrhiza 21:669–680. doi:10.1007/s00572-011-0374-2

Dixon RK, Wright GM, Behrns GT, Tesky RO, Hinckley TM (1980) Water deficits and root growth of ectomycorrhizal white oak seedlings. Can J For Res 10:545–548

Dulmer KM, Leduc SD, Horton TR (2014) Ectomycorrhizal inoculum potential of northeastern US forest soils for American chestnut restoration: results from field and laboratory bioassays. Mycorrhiza 24:65–74. doi:10.1007/s00572-013-0514-y

Egerton-Warburton LM, Querejeta JI, Allen MF (2007) Common mycorrhizal networks provide a potential pathway for the transfer of hydraulically lifted water between plants. J Exp Bot 58:1473–1483. doi:10.1093/jxb/erm009

EUFORGEN (2016) http://www.euforgen.org/fileadmin/templates/euforgen.org/upload/Documents/Maps/JPG/Quercus_suber.jpg

FAO (2013) http://faostat3.fao.org/download/Q/QV/E

Fisher MC, Henk DA, Briggs CJ, Brownstein JS, Madoff LC, McCraw SL, Gurr SJ (2012) Emerging fungal threats to animal, plant and ecosystem health. Nature 484:186–194. doi:10.1038/nature10947

Franceschini A, Maddau L, Marras F (2002) Osservazioni sull'incidenza di funghi endofiti associati al deperimento di *Quercus suber* e *Q. pubescens*. In: Franceschi-ni A, Marras F (eds) L'Endofitismo di Funghi e Batteri Patogeni in Piante Arboree e Arbustive. Sassari-Tempio Pausania, Italy, pp 313–325

Gebhardt S, Neubert K, Wöllecke J, Münzenberger B, Hüttl RF (2007) Ectomycorrhiza communities of red oak (*Quercus rubra* L.) of different age in the Lusatian lignite mining district, East Germany. Mycorrhiza 17:279–290. doi:10.1007/s00572-006-0103-4

Heiniger U, Rigling D (1994) Biological control of chestnut blight in Europe. Annu Rev Phytopathol 32:581–599. doi:10.1146/annurev.py.32.090194.003053

Herrmann S (2007) Cross talks at the morphogenetic, physiological and gene regulation levels between the mycobiont Piloderma croceum and oak microcuttings (*Quercus robur*) during formation of ectomycorrhizas. Phytochemestry 68:52–67. doi:10.1016/j.phytochem.2006.09.028

Herzog C, Peter M, Pritsch K, Gu MS (2012) Drought and air warming affects abundance and exoenzyme profiles of *Cenococcum geophilum* associated with *Quercus*. Plant Biol 15:230–237. doi:10.1111/j.1438-8677.2012.00614.x

Jany JL, Garbaye J, Martin F (2002) *Cenococcum geophilum* populations show a high degree of genetic diversity in beech forests. New Phytol 154:651–659. doi:10.1046/j.1469-8137.2002.00408.x

Jönsson-Belyazio U, Rosengren U (2006) Can *Phytophthora quercina* have a negative impact on mature pedunculate oaks under field conditions? Ann For Sci 63:661–672. doi:10.1051/forest:2006047

Kandeler E, Mosier AR, Morgan JA, Milchunas DG, King JY, Rudolph S, Tscherko D (2006) Response of soil microbial biomass and enzyme activities to the transient elevation of carbon dioxide in a semi-arid grassland. Soil Biol Biochem 38:2448–2246. doi:10.1016/j.soilbio.2006.02.021

Kasai K, Usami T, Lee J, Ishikawa SI, Oikawa T (2000) Responses of ectomycorrhizal colonization and morphotype assemblage of *Quercus myrsinaefolia* seedlings to elevated air temperature and elevated atmospheric CO2. Microbes Environ 15:197–207

Keen B, Vancov T (2010) *Phytophthora cinnamomi* suppressive soils. In: Curent research, technology and education topics in applied microbiology and microbial biotechnology. FORMATEX, pp 239–250

Keenan RJ (2015) Climate change impacts and adaptation in forest management: a review. Ann For Sci 72:145–167. doi:10.1007/s13595-014-0446-5

Kivlin SN, Emery SM, Rudgers JA (2013) Fungal symbionts alter plant responses to global change. Am J Bot 100:1445–1457. doi:10.3732/ajb.1200558

Kremer A, Abbott AG, Carlson JE, Manos PS, Plomion C, Sisco P, Staton ME, Ueno S, Vendramin GG (2012) Genomics of Fagaceae. Tree Genet Genomes 8:583–610. doi:10.1007/s11295-012-0498-3

Kuikka K, Härmä E, Markkola AM, Rautio P, Roitto M, Saikkonen K, Ahonen-Jonnarth U, Finlay R, Tuomi J (2003) Severe defoliation of Scots pine reduces reproductive investment by ectomycorrhizal symbionts. Ecology 84:2051–2061. doi:10.1890/02-0359

Kurth F, Feldhahn L, Bönn M, Herrmann S, Buscot F, Tarkka MT (2015) Large scale transcriptome analysis reveals interplay between development of forest trees and a beneficial mycorrhiza helper bacterium. BMC Genomics 16:1–13. doi:10.1186/s12864-015-1856-y

Kurz-Besson C, Otieno D, Lobo Do Vale R, Siegwolf R, Schmidt M, Herd A, Nogueira C, David TS, David JS, Tenhunen J, Pereira JS, Chaves M (2006) Hydraulic lift in cork oak trees in a savannah-type Mediterranean ecosystem and its contribution to the local water balance. Plant Soil 282:361–378. doi:10.1007/s11104-006-0005-4

Laganà A, Salerni E, Barluzzi C, Perini C, De Dominicis V (2002) Macrofungi as long-term indicators of forest health and management in central Italy. Cryptogam Mycol 23:39–50

Lancellotti E, Franceschini A (2013) Studies on the ectomycorrhizal community in a declining *Quercus suber* L. stand. Mycorrhiza 23:533–542. doi:10.1007/s00572-013-0493-z

Lehto T, Zwiazek JJ (2011) Ectomycorrhizas and water relations of trees: a review. Mycorrhiza 21:71–90. doi:10.1007/s00572-010-0348-9

Lesur I, Le Provost G, Bento P, Silva C, Leplé JC, Murat F, Ueno F, Bartholomé J, Lalanne C, Ehrenmann C, Plomion C (2015) The oak gene expression atlas: insights into Fagaceae genome evolution and the discovery of genes regulated during bud dormancy release. BMC Genomics 16:112. doi:10.1186/s12864-015-1331-9

Lindner M, Fitzgerald JB, Zimmermann NE, Reyer C, Delzon S, van der Maaten E, Hanewinkel M (2014) Climate change and European forests: what do we know, what are the uncertainties, and what are the implications for forest management? J Environ Manage 146:69–83. doi:10.1016/j.jenvman.2014.07.030

Lumaret R, Mir C, Michaud H, Raynal V (2002) Phylogeographical variation of chloroplast DNA in holm oak (*Quercus ilex* L.) Mol Ecol 11:2327–2336. doi:10.1046/j.1365-294X.2002.01611.x

Magalhães AP, Verde N, Reis F, Martins I, Costa D, Lino-Neto T, Castro PH, Tavares PH, Azevedo H (2016) RNA-Seq and gene network analysis uncover activation of an ABA-dependent signalosome during the cork oak root response to drought. Front Plant Sci 6:1195. doi:10.3389/fpls.2015.01195

Malajczuk N (1979) Biological suppression of *Phytophthora cinnamomi* in eucalyptus and avocados in Australia. In: Schippers B, Gams W (eds) Soil-borne plant pathogens. Academic, London

Malajczuk N, McComb A (1979) The microflora of unsubersied roots of *Eucalyptus calophylla* R. Br. and *Eucalytpus marginata* Donn ex Sm. seedlings grown in soils suppressive and conducive to *Phytophthora cinnamomi* Rands. I. Rhizosphere bacteria, actinomycetes and fungi. Aust J Bot 27:235–254

Malcolm GM, López-Gutiérrez JC, Koide RT, Eissenstat DM (2008) Acclimation to temperature and temperature sensitivity of metabolism by ectomycorrhizal fungi. Glob Chang Biol 14:1169–1180. doi:10.1111/j.1365-2486.2008.01555.x

Manos CH, Oh SH (2008) Phylogenetic relationships and taxonomic status of the paleoendemic Fagaceae of Western North America: recognition of a new genus. Madroño 55:181–190. doi:10.3120/0024-9637-55.3.181

Manos PS, Zhou ZK, Cannon CH (2001) Systematics of Fagaceae: phylogenetic tests of reproductive trait evolution. Int J Plant Sci 162:1361–1379

Markkola AM, Kuikka K, Rautio P, Härmä E, Roitto M, Tuomi J (2004) Defoliation increases carbon limitation in ectomycorrhizal symbiosis of *Betula pubescens*. Oecologia 140:234–240. doi:10.1007/s00442-004-1587-2

Martin F, Aerts A, Ahren D, Brun A, Danchin EG, Duchaussoy F, Gibon J, Kohler A, Lindquist E, Pereda V et al (2008) The genome of *Laccaria bicolor* provides insights into mycorrhizal symbiosis. Nature 452:88–92. doi:10.1038/nature06556

Martin F, Kohler A, Murat C, Balestrini R, Coutinho PM, Jaillon O, Montanini B, Morin E, Noel B, Percudani R et al (2010) Périgord black truffle genome uncovers evolutionary origins and mechanisms of symbiosis. Nature 464:1033–1038. doi:10.1038/nature08867

Marx D (1972) Ectomycorrhizal and nonmycorrhizal shortleaf pine seedlings in soil with *Phytophthora cinnamomi*. Annu Rev Phytopathol 10:1472–1473

Milgroom MG, Wang KR, Zhou Y, Lipari SE, Kaneko S (1996) Intercontinental population structure of the chestnut blight fungus, *Cryphonectria parasitica*. Mycologia 88:179–190. doi:10.2307/3760921

Mohan V, Nivea R, Menon S (2015) Evaluation of ectomycorrhizal fungi as potential bio-control agents against selected plant pathogenic fungi. J Acad Ind Res 3:408–412

Montecchio L, Causin R, Rossi S, Mutto Acordi S (2004) Changes in ectomycorrhizal diversity in a declining *Quercus ilex* coastal forest. Phytopathol Mediterr 43:26–34. doi:10.14601/Phytopathol_Mediterr-1721

Moricca S, Ragazzi A (2008) Fungal endophytes in Mediterranean oak forests: a lesson from *Discula quercina*. Phytopathology 98:380–386. doi:10.1094/PHYTO-98-4-0380

Moricca S, Ginetti B, Ragazzi A (2012) Species- and organ-specificity in endophytes colonizing healthy and declining Mediterranean oaks. Phytopathol Mediterr 51:587–598. doi:10.14601/Phytopathol_Mediterr-11705

Moricca S, Franceschini A, Ragazzi A, Linaldeddu BT, Lancellotti E (2014) Studies on communities of endophytic (end) and ectomycorrhizal (ecm) fungi associated with oaks in pure and mixed stands. In: Pirttilä AM, Sorvari S (eds) Prospects and applications for plant-associated microbes, a laboratory manual: Part B: Fungi

Myers N, Mittermeier RA, Mittermeier CG, Fonseca G, Kent J (2000) Biodiversity hotspots for conservation priorities. Nature 403:853–858. doi:10.1038/35002501

Nardini A, Salleo S, Tyree MT, Vertovec M (2000) Influence of the ectomycorrhizas formed by *Tuber melanosporum* Vitt. on hydraulic conductance and water relations of *Quercus ilex* L. seedlings. Ann For Sci 57:305–312. doi:10.1051/forest:2000121

Nardini A, Lo Gullo MA, Trifilò P, Salleo S (2014) The challenge of the Mediterranean climate to plant hydraulics: responses and adaptations. Environ Exp Bot 103:68–79. doi:10.1016/j.envexpbot.2013.09.018

Nixon KC, Crepet WL (1989) Trigonobalanus (Fagaceae): taxonomy status and phylogenetic relashionship. Am J Bot 76:826–841

Nuche P, Komac B, Camarero JJ, Alados CL (2014) Developmental instability as an index of adaptation to drought stress in a Mediterranean oak. Ecol Indic 40:68–75. doi:10.1016/j.ecolind.2013.12.023

Núñez JAD, Serrano JS, Barreal JAR, de Omeñaca González JAS (2006) The influence of mycorrhization with *Tuber melanosporum* in the afforestation of a Mediterranean site with *Quercus ilex* and *Quercus faginea*. For Ecol Manage 231:226–233. doi:10.1016/j.foreco.2006.05.052

Oh S, Manos PS (2008) Molecular phylogenetics and cupule evolution in Fagaceae as inferred from nuclear CRABS CLAW sequences. Taxon 57:434–451

Oliveira RS, Franco AR, Vosátka M, Castro PML (2010) Management of nursery practices for efficient ectomycorrhizal fungi application in the production of *Quercus ilex*. Symbiosis 52:125–131. doi:10.1007/s13199-010-0092-0

Öpik M, Moora M, Liira J, Zobel M (2006) Composition of root-colonising arbuscular mycorrhizal fungal communities in different ecosystems around the globe. J Ecol 94:778–790. doi:10.1111/j.1365-2745.2006.01136.x

Orgiazzi A, Lumini E, Nilsson RH, Girlanda M, Vizzini A, Bonfante P, Bianciotto V (2012) Unravelling soil fungal communities from different mediterranean land-use backgrounds. PLoS One 7:1–9. doi:10.1371/journal.pone.0034847

Orgiazzi A, Dunbar MB, Panagos P, de Groot GA, Lemanceau P (2015) Soil biodiversity and DNA barcodes: opportunities and challenges. Soil Biol Biochem 80:244–250. doi:10.1016/j.soilbio.2014.10.014

Ortega A, Lorite J (2007) Macrofungi diversity in cork-oak and holm-oak forests in Andalusia (southern Spain); an efficient parameter for establishing priorities for its evaluation and conservation. Cent Eur J Biol 2:276–296. doi:10.2478/s11535-007-0015-0

Palmer J, Lindner D, Volk T (2008) Ectomycorrhizal characterization of an American chestnut (*Castanea dentata*)-dominated community in western Wisconsin. Mycorrhiza 19:27–36. doi:10.1007/s00572-008-0200-7

Peintner U, Iotti M, Klotz P, Bonuso E, Zambonelli A (2007) Soil fungal communities in a *Castanea sativa* (chestnut) forest producing large quantities of *Boletus edulis* sensu lato (porcini): where is the mycelium of porcini? Environ Microbiol 9:880–889. doi:10.1111/j.1462-2920.2006.01208.x

Pereira-Leal JB, Abreu IA, Alabaça CS, Almeida MH, Almeida P, Almeida T et al (2014) A comprehensive assessment of the transcriptome of cork oak (*Quercus suber*) through EST sequencing. BMC Genomics 15:371. doi:10.1186/1471-2164-15-371

Pietras M, Rudawska M, Leski T, Karliński L (2013) Diversity of ectomycorrhizal fungus assemblages on nursery grown European beech seedlings. Ann For Sci 70:115–121. doi:10.1007/s13595-012-0243-y

Plomion C, Bastien C, Bogeat-Triboulot MB, Bouffier L, Déjardin A, Duplessis S, Vacher C (2015) Forest tree genomics: 10 achievements from the past 10 years and future prospects. Ann For Sci 73:77–103. doi:10.1007/s13595-015-0488-3

Querejeta JI, Egerton-Warburton LM, Allen MF (2007) Hydraulic lift may buffer rhizosphere hyphae against the negative effects of severe soil drying in a California Oak savanna. Soil Biol Biochem 39:409–417. doi:10.1016/j.soilbio.2006.08.008

Ragazzi A, Moricca S, Capretti P, Dellavalle I, Mancini F, Turco E (2001) Endophytic fungi in *Quercus cerris*: isolation frequency in relation to phenological phase, tree health and the organ affected. Phytopathol Mediterr 40:165–171. doi:10.14601/Phytopathol_Mediterr-1598

Ragazzi A, Moricca S, Capretti P, Dellavalle I, Turco E (2003) Differences in composition of endophytic mycobiota in twigs and leaves of healthy and declining *Quercus* species in Italy. For Pathol 33:31–38. doi:10.1046/j.1439-0329.2003.3062003.x

Ragazzi A, Moricca S, Dellavalle I (2004) Endophytism in forest trees. Accademia Italiana di Scienze Forestali, Firenze

Ramirez-Valiente JA, Lorenzo Z, Soto A, Valladares F, Gil L, Aranda I (2009) Elucidating the role of genetic drift and natural selection in cork oak differentiation regarding drought tolerance. Mol Ecol 18:3803–3815. doi:10.1111/j.1365-294X.2009.04317.x

Ramirez-Valiente JA, Valladares F, Huertas AD, Granados S, Aranda I (2011) Factors affecting cork oak growth under dry conditions: local adaptation and contrasting additive genetic variance within populations. Tree Genet Genomes 7:285–295. doi:10.1007/s11295-010-0331-9

Richard F, Moreau PA, Selosse MA, Gardes M (2004) Diversity and fruiting patterns of ectomycorrhizal and saprobic fungi in an old-growth Mediterranean forest dominated by *Quercus ilex* L. Can J Bot 82:1711–1729. doi:10.1139/B04-128

Richard F, Millot S, Gardes M, Selosse M-A (2005) Diversity and specificity of ectomycorrhizal fungi retrieved from an old-growth Mediterranean forest dominated by *Quercus ilex*. New Phytol 166:1011–1023. doi:10.1111/j.1469-8137.2005.01382.x

Richard F, Roy M, Shahin O, Sthultz C, Duchemin M, Joffre R, Selosse MA (2011) Ectomycorrhizal communities in a Mediterranean forest ecosystem dominated by *Quercus ilex*: seasonal dynamics and response to drought in the surface organic horizon. Ann For Sci 68:57–68. doi:10.1007/s13595-010-0007-5

Robin C, Heiniger U (2001) Chestnut blight in Europe: diversity of *Cryphonectria parasitica*, hypovirulence and biocontrol. For Snow Landsc Res 76:361–367

Robin C, Smith I, Hansen EM (2012) *Phythophthora cinnamomi*. For Phytophthoras 2(1). doi:10.5399/osu/fp.2.1.3041

Rocheta M, Sobral R, Magalhães J, Amorim MI, Ribeiro T, Pinheiro M, Egas C, Morais-Cecílio L, Costa MM (2014) Comparative transcriptomic analysis of male and female flowers of monoecious *Quercus suber*. Front Plant Sci 6:599. doi:10.3389/fpls.2014.00599

Saravesi K, Markkola AM, Rautio P, Roitto M, Tuomi J (2008) Defoliation causes parallel temporal responses in a host tree and its fungal symbionts. Oecologia 156:117–123. doi:10.1007/s00442-008-0967-4

Savoie J, Largeteau ML (2011) Production of edible mushrooms in forests: trends in development of a mycosilviculture. Appl Microbiol Biotechnol 89:971–979. doi:10.1007/s00253-010-3022-4

Schmitz S, Zini J, Chandelier A (2006) Involvement of Phytophthora species in the decline of beech *Fagus sylvatica* in Wallonia (Belgium). Commun Agric Appl Biol Sci 72(4):879–885

Sebastiana M, Figueiredo A, Acioli B, Sousa L, Pessoa F, Baldi A, Pais MS (2009) Identification of plant genes involved on the initial contact between ectomycorrhizal symbionts (*Castanea sativa*—European chestnut and *Pisolithus tinctorius*). Eur J Soil Biol 45:275–282. doi:10.1016/j.ejsobi.2009.02.001

Sebastiana M, Pereira VT, Alcântara A, Pais MS, Silva AB (2013) Ectomycorrhizal inoculation with *Pisolithus tinctorius* increases the performance of *Quercus suber* L. (cork oak) nursery and field seedlings. New For 44:937–949. doi:10.1007/s11056-013-9386-4

Sebastiana M, Vieira B, Lino-Neto T, Monteiro F, Figueiredo A, Sousa L, Pais MS, Tavares R, Paulo O (2014) Oak root response to ectomycorrhizal symbiosis establishment: RNA-Seq derived transcript identification and expression profiling. PLoS One 9:98376. doi:10.1371/journal.pone.0098376

Serrazina S, Santos C, Machado H, Pesquita C, Vicentini R, Pais MS, Costa R (2015) *Castanea* root transcriptome in response to *Phytophthora cinnamomi* challenge. Tree Genet Genomes 11:1–19. doi:10.1007/s11295-014-0829-7

Shi L, Guttenberger M, Kottke I, Hampp R (2002) The effect of drought on mycorrhizas of beech (*Fagus sylvatica* L.) changes in community structure, and the content of carbohydrates and nitrogen storage bodies of the fungi. Mycorrhiza 12:303–311

Shokralla S, Spall JL, Gibson JF, Hajibabaei M (2012) Next-generation sequencing technologies for environmental DNA research. Mol Ecol 21:1794–1805. doi:10.1111/j.1365-294X.2012.05538.x

Smit E, Veenman C, Baar J (2003) Molecular analysis of ectomycorrhizal basidiomycete communities in a *Pinus sylvestris* L. stand reveals long-term increased diversity after removal of litter and humus layers. FEMS Microbiol Ecol 45:49–57. doi:10.1016/S0168-6496(03)00109-0

Smith SE, Read DJ (2008) Mycorrhizal symbiosis, 3rd edn. Academic, London

Smith ME, Douhan GW, Rizzo DM (2007) Ectomycorrhizal community structure in a xeric *Quercus* woodland based on rDNA sequence analysis of sporocarps and pooled roots. New Phytol 174:847–863. doi:10.1111/j.1469-8137.2007.02040.x

Southworth D, Carrington EM, Frank JL, Gould P, Harrington CA, Devine WD (2009) Mycorrhizas on nursery and field seedlings of *Quercus garryana*. Mycorrhiza 19:149–158. doi:10.1007/s00572-008-0222-1

Stark S, Kytöviita M-M (2005) Evidence of antagonistic interactions between rhizosphere microorganisms and mycorrhizal fungi associated with birch (*Betula pubescens*). Acta Oecol 28:149–155. doi:10.1016/j.actao.2005.03.007

Suz LM, Barsoum N, Benham S, Dietrich H-P, Fetzer KD, Fischer R, Ia PG, Gehrman J, Gofel FK, Mannunger M, Neagu S, Nicolas M, Oldenburger J, Raspe S, Anchez GS, Schrock HW, Schubert A, Verheyen K, Verstraeten A, Bidartondo MI (2014) Environmental drivers of ectomycorrhizal communities in Europe's temperate oak forests. Mol Ecol 23:5628–5644. doi:10.1111/mec.12947

Tarkka MT, Herrmann S, Wubet T, Feldhahn L, Recht S, Kurth F, Mailänder S, Bönn M, Neef M, Angay O, Buscot F et al (2013) OakContigDF159.1, a reference library for studying differential gene expression in *Quercus robur* during controlled biotic interactions: use for quantitative transcriptomic profiling of oak roots in ectomycorrhizal symbiosis. New Phytol 199:529–540. doi:10.1111/nph.12317

Taschen E, Sauve M, Taudiere A, Parlade J, Selosse MA, Richard F (2015) Whose truffle is this? Distribution patterns of ectomycorrhizal fungal diversity in *Tuber melanosporum* brûlés developed in multi-host Mediterranean plant communities. Environ Microbiol 17:2747–2761. doi:10.1111/1462-2920.12741

Tedersoo L, May TW, Smith ME (2010) Ectomycorrhizal lifestyle in fungi: global diversity, distribution, and evolution of phylogenetic lineages. Mycorrhiza 20:217–263. doi:10.1007/s00572-009-0274-x

Teixeira RT, Fortes AM, Pinheiro C, Pereira H (2014) Comparison of good- and bad-quality cork: application of high-throughput sequencing of phellogenic tissue. J Exp Bot 65:4887–4905. doi:10.1093/jxb/eru252

Toju H, Sato H, Tanabe AS (2014) Diversity and spatial structure of belowground plant—fungal symbiosis in a mixed subtropical forest of ectomycorrhizal and arbuscular mycorrhizal plants. PLoS One 9:24–26. doi:10.1371/journal.pone.0086566

Valavanidis A, Vlachogianni T (2011) Ecosystems and biodiversity hotspots in the Mediterranean basin threats and conservation efforts. Sci Adv Environ Toxicol Ecotoxicol Issues 10:1–24

Van der Heijden MGA, Martin FM, Sanders IR (2015) Tansley review Mycorrhizal ecology and evolution: the past, the present, and the future. New Phytol 205:1406–1423. doi:10.1111/nph.13288

Vettraino AM, Barzanti GP, Bianco MC, Ragazzi A, Capretti P, Paoletti E, Luisi N, Anselmi N, Vannini A (2002) Occurrence of *Phytophthora* species in oak stands in Italy and their association with declining oak trees. For Pathol 32:19–28. doi:10.1046/j.1439-0329.2002.00264.x

Voříšková J, Brabcová V, Cajthaml T, Baldrian P (2014) Seasonal dynamics of fungal communities in a temperate oak forest soil. New Phytol 201:269–278. doi:10.1111/nph.12481

Wang Q, Gao C, Guo L (2011) Ectomycorrhizae associated with *Castanopsis fargesii* (Fagaceae) in a subtropical forest. Mycol Prog 10:323–332. doi:10.1007/s11557-010-0705-2

Wheeler N, Sederoff R (2009) Role of genomics in the potential restoration of the *American chestnut*. Tree Genet Genomes 5:181–187. doi:10.1007/s11295-008-0180-y

Xie J, Jiang D (2014) New insights into mycoviruses and exploration for the biological control of crop fungal diseases. Annu Rev Phytopathol 52:45–68. doi:10.1146/annurev-phyto-102313-050222

Yakhlef SB, Kerdouh B, Mousain D, Ducousso M, Duponnois R, Abourouh M (2009) Phylogenetic diversity of Moroccan cork oak woodlands fungi. Biotechnol Agron Soc 13:521–528

Yun W, Hall IA (2004) Edible ectomycorrhizal mushrooms: challenges and achievements. Can J Bot 82(8):1063–1073. doi:10.1139/b04-051

Chapter 7
Ectomycorrhizal Mushrooms: Their Diversity, Ecology and Practical Applications

Rohit Sharma

Abstract Ectomycorrhizal symbiosis is formed by a large number of plants and fungi. It is an association of fungal mycelia and roots of plants, mostly woody trees. Ectomycorrhiza are formed by fungi like *Russula*, *Lactarius*, *Boletus*, *Cantharellus*, etc. which are mostly common edible mushrooms. The trees which form ectomycorrhiza are *Shorea, Pinus,* etc. Ectomycorrhiza has been proved in at least 162 genera and more than 5400 species. Previous studies were based on morphology of ectomycorrhiza, but molecular data were lacking. However, relatively recently molecular studies and identification have confirmed ectomycorrhiza association of various fungi. These are formed by mostly members of *Basidiomycota* and *Ascomycota.* The orders like *Agaricales*, *Boletales*, *Pezizales*, *Helotiales*, and *Cantharellales* include the largest number of ectomycorrhizal lineages. In tropical regions, trees belonging to *Dipterocarpaceae* and *Caesalpiniaceae* form most ectomycorrhiza. There are attempts to study ectomycorrhiza in India but are way behind the studies that are been conducted around the world. Some of the studies conducted in India are related to ectomycorrhizal mushroom diversity and synthesis but none on genomics, ecological, and physiological studies. This chapter discusses from the basics what are ectomycorrhiza and their ecology and also applied aspects of ectomycorrhiza.

7.1 Introduction

Fungi along with other microbes are essential component of forest and grassland ecosystems because of their role as parasites, causal agent of various infections, decomposers of organic matter, and mutualistic symbionts (lichens and mycorrhizae). Fungi are vital for biodiversity and various ecosystem processes thus balancing the ecological system of earth. Fungi forming conspicuous sporocarps or fruiting bodies (popularly called as mushrooms or toadstools) are mostly either plant parasitic, saprobic, or mycorrhizal. Mycorrhizae are highly evolved,

R. Sharma (✉)
Microbial Culture Collection (MCC), National Centre for Cell Science, NCCS Complex, S.P. Pune University, Ganeshkhind, Pune 411 007, Maharashtra, India
e-mail: rsmushroom@gmail.com; rohit@nccs.res.in

A. Varma et al. (eds.), *Mycorrhiza - Function, Diversity, State of the Art*,
DOI 10.1007/978-3-319-53064-2_7

Table 7.1 Types of mycorrhiza and their characteristics

Type of mycorrhiza	Characteristics
Arbuscular mycorrhizas (AM)	Formerly referred to as vesicular-arbuscular mycorrhizas (VAM), they are most prevalent with more than 80% of plants species showing an association involving few fungal genera in the glomeromycota. Most common diagnostic feature is the development of intercellular hyphae, intracellular hyphae and arbuscules in root cortical cells, and production of spores on intra- and extraradical hyphae
Ericoid mycorrhizas	These are found in ericaceae and epacridaceae families of angiosperm. They are important in land ecosystems where soil nitrogen is bound in organic compounds. The nitrogen in these is accessed by plant primarily via fungi associated with fine roots. Epidermal cells of these "hair roots" are colonized by fungal hyphae forming intracellular hyphal bundles or complexes. They belong to ascomycotina and few deuteromycotina
Arbutoid and monotropoid mycorrhizas	These specialized mycorrhizas found in ericales differ structurally from ericoid mycorrhizas in having Hartig net as well as intracellular hyphae and develop a hyphal complex in epidermal cells. The epidermal cells are invaded by a single hypha forming a "peg" around which the host cell elaborates a wall and plasma membrane. Arbutoid mycorrhizas, on the other hand, develop a hyphal complex in epidermal cells
Ectendomycorrhizas	These resemble ectomycorrhiza by having a mantle and Hartig net. However, they are confined to the conifer genera *Pinus* and *Larix* and are formed by a small group of ascomycete fungi
Orchid mycorrhizas	These are restricted to orchidaceae family and are unique in that fungal associations occur with embryo cells of germinating seeds (Peterson et al. 1998) as well as with roots of seedlings and mature plants. In both situations, various fungal species of basidiomycotina form intracellular, short-lived coils called "pelotons," this degenerates and is digested by the host cell

mutualistic associations between soil fungi and plant roots. The symbiotic association is between members of kingdom *Eumycota* (phyla *Basidiomycota*, *Ascomycota*, and *Glomeromycota*) and most vascular plants, especially trees. Fungi form mycorrhizal associations with about 85% of the world's vascular plants (including herbs, shrubs, and trees). Mycorrhizas are generally characterized into seven categories (Table 7.1). Two of the seven types of mycorrhizae are well studied. Endomycorrhizae or Arbuscular Mycorrhizae (AM) is formed by several plants by a limited number of fungal species (ca. 150) in the monophyletic, phylum *Glomeromycota* (Schübler et al. 2001). Ectomycorrhizal (ECM) fungi are much more diverse (>5400 spp.) members of *Basidiomycota* and some *Ascomycota* forming association primarily with woody plants (Read 1991a, b).

Although the term "mycorrhiza" means the association of fungi with roots, relationships called "mycorrhizal associations" are found between hyphal fungi and the organs of higher plants (of different morphological origin) involved in absorption of water and nutrients. In particular, mycorrhizal infection usually

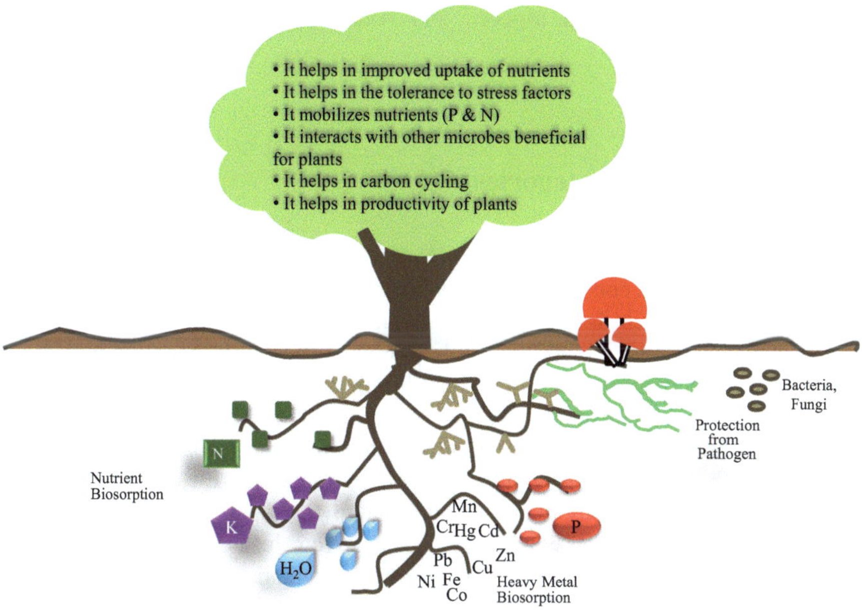

Fig. 7.1 Beneficial effects of ECM fungi

increases the efficiency of nutrient absorption of plants from which the fungus obtains carbon compounds (Fig. 7.1). In mycorrhizas, there is always some penetration of the tissues or a structural modification of roots. Mycorrhizal infected and noninfected roots are clearly distinguishable. The mycorrhizal condition differs from disease as both partners are in normal condition and mutually benefit each other. They are dependent upon one another and interchange of material takes place between their living cells. It was Frank (1885) who recognized and named "mycorrhizen" or "mycorrhiza" for infected roots of temperate forest trees (beech and pine) which are morphologically different from uninfected roots. He later named them as "ectotrophisch" or "ectotrophic" as they possessed conspicuous fungal tissue (sheath or mantle) surrounding plant roots.

Over the past years, research on ECM has shifted from morphological study of ECM to studies on community structure of ectomycorrhizal mushrooms using uncultured approach. Most of the studies focus on the diversity of fungi making ectomycorrhizal association with trees especially those which are edible, understanding the causes and results of an ectomycorrhizal association between plant and fungus, community structure of a forest, biology of ectomycorrhizal mushrooms, synthesis of ECM in laboratory conditions, etc. Recently, Högberg et al. (2001) have studied relative contributions of roots, ECM fungi and free-living microbial heterotrophs to soil respiration using girdling of forest trees.

The present book chapter discusses various aspects of ectomycorrhiza, viz., mycorrhizal categories, morphology of ECM, diversity (plant and fungal), ecology, artificial synthesis, applied aspects, and recent advances in ectomycorrhizal research.

7.2 Mycorrhizal Categories

Mycorrhizal organs may take different forms depending on the nature of higher plant and fungus, and common kinds have been classified and named. But it is now clear that there is much similarity in general physiology with some detailed specialization between different kinds so that their previous bases of separation have been questioned. The pioneer work of Frank (Smith and Read 1997) resulted in the recognition of two broad subdivisions of mycorrhizas, ecto- and endomycorrhizas. ECM forms Mantle and Hartig net of intercellular hyphae on roots of tree species. AM forms arbuscule, vesicles which are more variable than ECM in that it forms symbiosis with herbaceous plants in addition to tree species (Fig. 7.2). Endomycorrhizas are further classified as arbuscular mycorrhizas, ericoid mycorrhizas, arbutoid mycorrhizas, monotropoid mycorrhizas, ectendomycorrhizas, or orchid mycorrhizas (Table 7.1). Each of these categories is characterized by the invasion of plant root cells by fungal hyphae but differ in the nature of intracellular hyphal development (Peterson et al. 2004). The present chapter discusses various aspects of ECM.

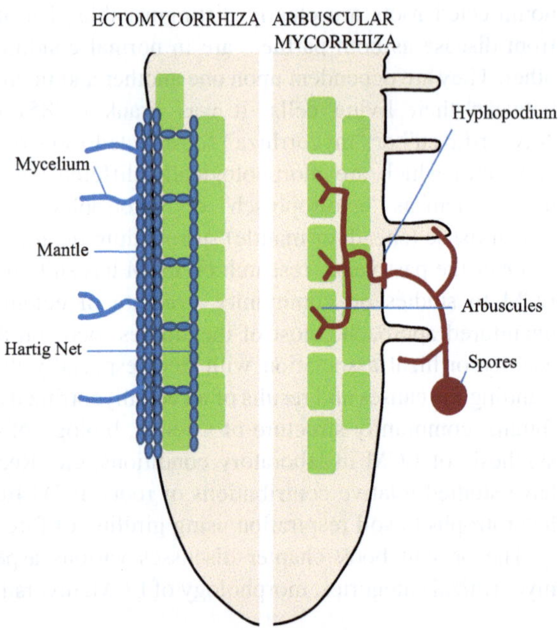

Fig. 7.2 Difference between ectomycorrhiza (ECM) and arbuscular mycorrhiza (AM)

7.3 Ectomycorrhiza

An ECM is a mutualistic symbiotic relationship characterized as a root-fungus association in which the fungus grows on root surface and penetrates the cortex intercellularly to produce a network. ECM mostly occurs in temperate, boreal, and tropical forests (Danell 2002; Dahlberg 2001; Smith and Read 1997; Cairney and Chambers 1999; Verbeken and Buyck 2001; Comandini et al. 2006; Wang and Qiu 2006; Rinaldi et al. 2008). There is a considerable variation in morphological and structural characteristics of ECM. Three features are generally recognized to typify this association:

- Formation of a mantle or sheath of fungal hyphae that covers considerable portions of lateral roots
- Development of hyphae between root cells to form complex branched structure called Hartig net
- Hyphae that come out from the mantle and grow into surrounding soil (extraradical mycelium)

In addition, some ECM develops linear aggregations of hyphae (rhizomorphs and strands) in the extraradical mycelium specialized for rapid transport of nutrients and water. A few ECM fungi develop sclerotia consisting of compact storage hypha surrounded by a ring. Hypogeous or epigeous reproductive bodies are formed periodically from extraradical mycelium. The hyphae do not normally penetrate the cells. Generally, the fungal sheath is usually 20–100 μm thick (often 30–40 μm) and comprises 25–40% of the dry weight of the whole organ. The presence of large fungal component of the absorbing organs suggests that the sheath have selective advantage perhaps in nutrient absorption or storage. This consideration led Lewis (1973) to suggest that the name of ECM be altered to "sheathing mycorrhiza." Hyphal connections run from the sheath between the cells of epidermis and cortex of the plant root, forming the "Hartig Net." Usually there is little hyphal penetration into the cells of plant root of young mycorrhizas, but in senescent parts of a mycorrhizal axis, the cortex becomes colonized by hyphae within the cells.

The presence of Hartig net led some authors to use the term "ectendotrophic" for organs called "ectotrophic." The term "ectendotrophic" (now ectendomycorrhiza) was coined to designate mycorrhizal organs with sheaths/variable development, Hartig net, and extensive intracellular penetration. In addition, considerable research on mycorrhiza of juvenile pines and other conifers especially in nursery conditions has led to the recognition of ectendomycorrhizas, which may merge into the so-called pseudomycorrhizas (believed to be pathological structures) recognized by Melin (1923). There is also considerable variation in the development of various morphological structures. In the extreme, there may be no Hartig net, as in mycorrhiza of *Pisonia grandis* (Ashford and Allaway 1982) or in superficial ECM roots of *Fagus* (Brundrett 2002), or no sheath formation but only a Hartig net as in *Pinus*. ECM associations predominantly forms on fine root tips of host, which are unevenly distributed throughout the soil profile being more abundant in top soil layers containing humus than in underlying layers of mineral soil. The hyphae of ECM fungi are widely distributed through the soil and make large contribution to nutrient uptake and cycling in ecosystems.

Most of us, as biologists, are familiar with the concept of symbiotic association between ECM mushrooms and vascular plants. ECM biomass can account for up to 25% of the total forest root biomass (Pande et al. 2004). The biological role of ECM mushrooms in ecological niche includes uptake of dissolved mineral nutrition, protection from disease-causing pathogens, balancing the ecosystems, etc. These are achieved by various structures of ECM which also make morphological changes in root.

7.4 Morphology of ECM

ECM has received rather separate consideration from the whole root systems which bear them but more emphasis should be given to the root systems of ECM plants and relatively less on mushrooms forming ECM. Besides the ECM of forest trees, there are variants which depart conspicuously in structure from them like the ectendomycorrhizas. They have reduced sheath of surface hyphae, a well-developed Hartig net and intracellular penetration of living hyphae into living cells. Although we have gained knowledge of root systems of temperate ECM trees, very little is known of those in the tropics, subtropics, or semidesert regions. In temperate forests the active roots of ECM trees are intensely developed in surface and subsurface layers of the soil. Humic layers, especially ECM which form the main component of the feeding system, often lie below the second layer in great quantities. The roots tend to grow in a lateral or upward direction colonizing the newly accumulated humus and litter. In the extreme, smallest short roots of *Pinus* have very small meristematic region, grow little, and differentiate mature stellar and cortical tissues close to the apex. In most of the ECM observed in the forest of *Shorea robusta* (Sal) in Madhya Pradesh, India, the ECM roots were below the thick layer of litter. Sometimes, the fruiting body could be excavated along with attached root of *Shorea* confirming the ECM association between the mushroom species and host plant, *S. robusta* (Figs. 7.3, 7.4). In spite of variation, the common ECM types of most trees are similar in general structure. Besides the presence of fungus on surface of mycorrhiza and between the cortical cells, the lesser development of mature tissues behind the meristem is a conspicuous difference. The features of mycorrhiza are, therefore, those of a slow-growing organ, in contrast to the uninfected root. When growth rates of mycorrhizal roots were compared, they grow slower than uninfected roots. Uninfected roots grow approximately five times as fast as mycorrhizas. Moreover, the roots infected with ECM show more branching than noninfected roots. The profuse root branching also help in the forest to identify the ECM infected roots (Figs. 7.3, 7.4).

In ECM, the modifications of branches of root are of two kinds, roots of potentially unlimited extension in length and roots of restricted growth and viable

Fig. 7.3 (**a–b**) Gross morphology of ectomycorrhizae of *Shorea robusta,* dichotomies of short toots. (**b**) Mycorrhizal roots of *Shorea robusta* (sal). (**c**) Ochre brown rough monopodial-pinnate ECM of *Lactarius*. (**d**) Cross section of *Lactarius* ECM. (**e**) Monopodial mycorrhizae showing bulbous and rounded apices covered with hyphal sheath. (**f**) Cross section of ECM of *Russulaceae* member (Sharma 2008)

period. In "heterorhizic" roots, majority of laterals are of limited growth. Apices which are fully infested are usually short and grow slowly, forming racemose branches (Fig. 7.5). Those apices not permanently infested by sheath or uninfected

Fig. 7.4 (**a**) Pale yellow monopodial mycorrhizal roots of *Shorea robusta*. (**b**) Cross section of *Russula* ECM. (**c**) *White Lactarius* arising from *Shorea robusta* ECM roots. (**d**) Cross section of *Lactarius* ECM. (**e**) Monopodial mycorrhizae with bulbous and rounded apex. (**f**) ECM of *Pisolithus* with yellow mycelia (Sharma 2008)

either maintain very active growth and are the leading apices of the system or may abort or become dormant (Lakhanpal 2000). *Pinus* departs from this general form because its short roots are sharply differentiated from the axes which bear them.

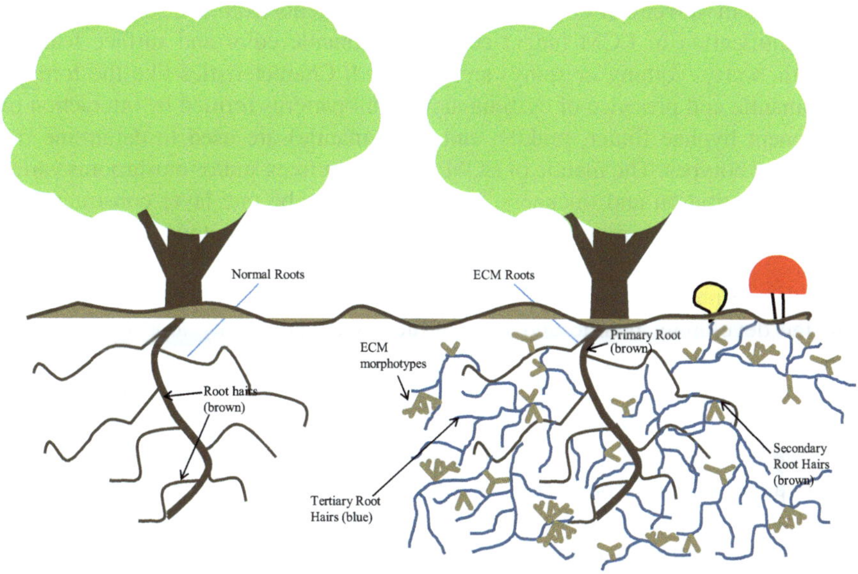

Fig. 7.5 Different types of root morphology in ECM and non-ECM roots

They are simpler in stellar construction, have a restricted apical meristem and root cap, and soon abort if not infected. If infected, they continue to grow, branch dichotomously, and form mycorrhizal systems. "Dichotomy," a characteristic of *Pinus*, is rare or absent in other genera. It is evident that there is an interaction between host and fungi at various levels of ECM development that leads to characteristic features for host-fungus combinations. Agerer (1987, 2002) have described various terms for branching patterns and made into keys. These are also used to identify fungal species associated with host roots. Some common patterns are simple (unramified), monopodial pinnate, monopodial pyramidal, irregularly pinnate, dichotomous, coralloid, and tuberculate. Branching patterns, features of mantle, and extraradical hyphae are used in "morphotyping" field-collected ECM. Tracing hyphal links between sporocarps and roots is a usual way to identify the fungal symbionts associated with roots of particular trees and confirm the symbiosis between both partners. Since the ECM roots and non-ECM roots of a particular tree will differ, it is still followed even in the artificial synthesis experiments. A typical ECM root structure has the following regions: mantle, Hartig net, and extraradical mycelium.

7.4.1 Mantle

The mantle interfaces with root (inner mantle) and soil (outer mantle). It is important in identification of fungal symbionts, interaction with soil microbes, and its role in movement of water and mineral nutrients from soil solution to root. For identification of ECM fungal component, mantle color and surface features (smooth, warty, cottony or spiny) are also used. Characteristics like the form of outer mantle and presence of cystidia along with patterns formed by interaction of component hyphae (inner, middle, and outer mantle) are used to determine the ECM morphotypes. The mantle of ECM of different trees and/or mushrooms varies in both constitution and thickness. It may appear to be one layer constructed of coherent hyphae or it may be two layered pseudoparenchymatous (Peterson et al. 2008).

With continued hyphal growth, root hairs and old root cap cells are incorporated into the developing mantle and root surface is enveloped by loosely organized hyphae. Some ECM may develop cystidia in the outer mantle. The main tissues of sheath may be reminiscent of those of the sporocarps of ECM fungi. Indeed sheath formed by species of *Tuber* have both active and storage hyphae in them, and those of *Lactarius* contained lactifers. The sheath varies in thickness but is typically 20–40 μm thick and comprises 20–30% of the volume of rootlet. The sheath contribution is estimated to be approximately 39% of dry weight of mycorrhiza.

Although Hartig net is the main crossing point of nutrient exchange in most ECM, the repeated branching of inner mantle hyphae suggests their involvement in the bidirectional movement of nutrients. The ECM fungi are capable of absorbing glucose and/or fructose from root cells and converting them into soluble carbohydrates trehalose and mannitol or into the insoluble carbohydrate glycogen which are stored in mantle hyphae. In addition, mantle hyphae accumulate other compounds including lipids, protein, and/or phenolics and polyphosphates. Deleterious metals may be bound to polyphosphates and other vacuolar deposits in the mantle, thereby preventing their uptake into roots. This feature is significant for tree seedlings inoculated with ECM fungi and planted on metal polluted sites. The compact nature of mantles of ECM protects roots from water loss in dry-soil conditions and from ingress of pathogenic organism through roots. Since the mantle interfaces with the soil, it potentially affects the transport of water and nutrient ions into the root. Bacteria may also be located within hyphae and root cells of some ECM, but the function of these has not been studied much. However, some bacterial species located on the surface of mantle hyphae are known to enhance mycorrhiza formation and consequently plant performance. Perhaps, these make available nutrients to the fungal hyphae.

7.4.2 Hartig Net

The Hartig net comprises of hyphae originating from the inner mantle developed between root cells and forms a complex nutrient crossing point in root. In most angiosperms, the Hartig net develops only around epidermal cells and further development is stopped by thick walls of underlying exodermal cells. One notable exception is the genus *Dryas*, where Hartig net hyphae develop up to the cortex.

Here "phi" thickenings (lignified wall thickenings shaped like the Greek letter Φ) appear to block the ingress of the fungus. In *P. grandis* and perhaps a few other species, a Hartig net is not observed.

In conifer trees, Hartig net develops around both epidermal and cortical cells occupying most of the cortex. During the early stage of Hartig net formation, mechanical intrusion of hyphae into middle lamella between epidermal and cortical cells is evident by the tapered hyphal tips. The process is enhanced by hydrolytic enzymes that soften the middle lamella and adjacent cell walls. It is however widely believed that some degree of pectic hydrolysis of the middle lamella also occurs.

An important feature of Hartig net is its labyrinthine hyphal branching where tubular fungal hyphae are replaced by a multi-digitate mode of growth. It increases the surface area for exchange of nutrients. In some types of ECM like *Fagus* tree, the interface is between unmodified hyphal wall and host cell wall. For *Pinus radiata*, a second type of interface is described as consisting of a modified external layer of host and a modified hyphal wall. Scientists have observed this type of interface in synthesized mycorrhizas of *Betula* with *Amanita muscaria* and in *Pseudotsuga menziesii* and *Betula* in association with ascomycetous fungi (Peterson et al. 2004).

The Hartig net is variously described as uniseriate or multiseriate. The development often depends upon the conditions of culture in artificially produced mycorrhizas. The Hartig net is involved in nutrient exchange, and most of the absorbed sugars, mineral nutrients, and water are passed to root cells through these hyphae. Micro-autoradiographic analysis has shown the movement of sugars from root cells to Hartig net and then to mantle and phosphate movement in the vice versa direction (Bücking and Heyser 2001). Hartig net hyphae also act as a reservoir for carbohydrate, lipids, phenolic compounds, and polyphosphates.

7.4.3 Extraradical Mycelium

Extraradical (extramatrical) mycelia are hyphae that develop from the outer mantle into the surrounding soil. These are extensive network penetrating the soil and interconnecting roots of the same plant and/or adjacent plants. It is known with *Cenococcum* ECM that a certain number of hyphal layers are necessary in the mantle for extraradical mycelium initiation. Soil particles adhere to individual hyphae of ECM fungi or groups of hyphae, colonies of bacteria, excrement of earthworms, pollen, litter, and upper layer of organic debris in soil. *Hysterangium* and *Gautieria*, form "mats" of mycelium that bind soil and fine roots encrusted with calcium oxalate crystals. Rhizomorphs can vary considerably in their morphology, color, and internal structure wherein the large number of individual hyphae interconnects each other (Agerer 2001). In complex rhizomorphs, one or more central hyphae (vessel hyphae) are enlarged and modified septa that allow for rapid movement of water and nutrient minerals.

Relatively few ECM fungal species form sclerotia in the extraradical mycelium. Sclerotia (development and structure) have been studied in details for few species: *Pisolithus tinctorius*, *Hebeloma sacchariolens*, *Cenococcum geophilum*, and *Paxillus involutus*. Other mycorrhizal fungi, including species of *Gyrodon*, *Boletus*, *Austropaxillus*, *Cortinarius*, and *Morchella*, are also known to form these structures. At maturity, each sclerotium usually develops a melanized outer covering (rind) that surrounds a cortex (central area) of compact hyphae and a medulla of loosely organized hyphae. Proteins, lipids, polysaccharides, and polyphosphates are stored in cortical region making them ideal propagules. The formation of reproductive bodies (basidiocarps and ascocarps) involves the localized branching of extraradical hyphae, organization of these hyphae into discrete structures, and differentiation of various regions of the sporocarps.

Mobilization, absorption, and translocation of mineral nutrients and water from the soil substrate to plant roots are the main function of extraradical mycelia. In species with rhizomorphs, connecting fine hyphae may pass dissolved nutrients and water for more rapid translocation through the wide diameter central hyphae (vessel hyphae) to the root. Experiments with radioactive isotopes of phosphorus (P) (^{13}P-labeled orthophosphate) have shown that P can be translocated over distances of more than 40 cm through rhizomorphs to roots of colonized plants and subsequently to the shoot system (Finlay and Read 1986a, b). ECM fungi can obtain P from the mycelium network of a saprotrophic fungus and pass P to the host plant. Carbon (C) compounds are translocated in the reverse direction from the host root to the extraradical mycelium for metabolic and growth processes, to developing sclerotia and their storage reserves, and to sporocarps. Production and final biomass of *Laccaria bicolor* basidiocarps are correlated with the rate of photosynthesis of their host *Pinus strobus*. Experiments with radioisotopes of C in the laboratory have confirmed the movement of C from host to fungus as well as from one plant to another through the extraradical mycelium network (Peterson et al. 2004).

7.5 Ecology of ECM Mushrooms

In the past decade, there has been an increasing awareness between ecologist and mycorrhizalogists that mycorrhizal fungi are an integral part of ecosystems and that their ecological function needs to be understood. It has been shown that mycorrhizal fungi contribute to plant diversity, nutrient cycling, acquisition to nutrient sources, and finally to ecosystem functioning. The survey conducted in the forests of Madhya Pradesh and Chhattisgarh states of India yielded several ECM mushrooms. These formed ECM with *S. robusta, Dendrocalamus strictus* trees. These mushrooms are helping in maintaining pure sal forests and the *Shorea* trees in turn are maintaining the fungal diversity. Although the forest slowly turns into mono-tree forest, the ecosystem is healthy due to the tree-fungal symbiosis. This may be one of the reasons why sal forests are mostly mono-tree forest. Moreover, ECM fungi could well contribute to community and ecosystem responses to global changes. As the climate is changing, the ECM fungi will help its symbionts to adapt it. Also, based on the effects of symbiosis on plant fitness, population ecology, dynamics, and evolution of many plants are unlikely to be fully understood without considering their fungal symbionts. ECM are active living components of soil population having some properties of roots and some of microorganisms. The ECM mushrooms make the forest a different ecosystem altogether from the dry forest like that of *Tectona grandis*. Understanding the distribution and ecology of ECM fungi is important for the ecosystem studies as well as selection of plants for plantation (Giachini et al. 2000).

7.5.1 Ecophysiology, Ecosystem Effects, and Global Change

It has been estimated that approximately 80% of all land plant species form associations with mycorrhizal fungi (van der Heijden and Sanders 2002). The abundance of mycorrhizal fungi is enormous as plants that form a symbiosis with these mutualistic fungi dominate most ecosystems. Moreover, most tree plants form mycorrhizal symbiosis with multiple fungal species. Majority of plants in the European calcareous grassland, American tall grass prairie, temperate deciduous forests, tropical rain forest, and shrub land of the threatened South African cape region are associated with mycorrhizal fungi. In contrast, plant communities of arctic tundra and alpine regions often contain a lower percentage of mycorrhizal plant species (Harnett and Wilson 1999; Onguene and Kuyper 2001). In tropical forests like that of Central India, this is mostly occupied by pure sal or mixed evergreen or dry deciduous forest. As per our survey and observation, the dry deciduous forests are mostly dominated by *T. grandis* (teak) and evergreen forests are dominated by *S. robusta* (sal). As per our observation, there were no ECM fungi observed with teak, whereas sal forest contains most of the ECM fungi. Survey conducted in the forests of Madhya Pradesh and Chhattisgarh yielded many ECM-forming fungi. That means single tree plant species can harbor multiple fungal species. The large quantity of litter also helped in the survival of the surface mycelia of the ECM fungi. When the litter is removed, the surface is completely covered by mycelia mat.

It is well known that ECM fungi have beneficial effects on plant growth especially when nutrients availability is low, which is mainly attributed to improved plant nutrition (Jakobsen et al. 2002; Simard et al. 2002). The diameter of ECM hyphae is up to 60 times thinner than plant roots which help to form extensive hyphal networks in soil penetrating the soil particles pores. That is why ECM fungi explore effectively for nutrients in comparison to plant roots. The roots of many ECM trees are completely encapsulated by fungal mantle. A number of traits from both plant and mycorrhizal fungi determine fungal impact on plant growth making it a complex interaction (Smith 2000; Smith et al. 2000).

ECM fungi exude several extracellular enzymes that break complex organic substances and have access to organic N and inorganic P that can be transmitted to their hosts. ECM hyphae are the primary structures that acquire nutrients and are considered as sink of carbon (Jakobsen et al. 2002; Simard et al. 2002). It is estimated that 10–50% of assimilated carbon is translocated to ECM roots (Voke 2012). The influence of global changes such as elevated carbon dioxide, N deposition, ozone, UV radiation, and climate may change mycorrhizal associations (Rillig et al. 2002). Recently, several groups are working on the sequestration of carbon in ECM fungi.

7.5.2 Biodiversity, Plant, and Fungal Communities

One of the major goals in ecology is to search for mechanisms that determine biological diversity (Grime 2001). Apart from the presence of ECM fungi, species composition and diversity of ECM fungal communities also affect plant diversity and productivity (Hart and Klironomos 2002; van der Heijden 2002). Moreover, the influence of ECM on plant diversity depends on plant species composition. ECM diversity might also play a role in seedling establishment of forest trees (Jonsson et al. 2001). Evidence shows that the reverse is also true, that plants affect populations and community composition of fungal symbionts. Bever et al. (2002) have taken a population and community approach to look at how fungi affect plant fitness but at the same time how plant species affect ECM fungal fitness.

Because of differential effects of ECM fungal species on plant growth, it is essential to know which factors determine the diversity and composition of ECM communities (Erland and Taylor 2002). Both abiotic soil factors and biotic factors (such as plant species composition) affect the composition of ECM fungal communities. Enhanced levels of available soil N as caused by atmospheric N deposition change the composition of ECM fungal communities. It also alters levels of root colonization of plants associated with ECM fungi (Rillig et al. 2002). It appears that N deposition and soil acidification reduce ECM diversity. This is often accompanied by a shift in community structure so that dominance by one fungus/few species increases. The negative impact of N deposition on ECM fungal diversity leads to reduced viability of many temperate and boreal forests.

The diversity study of ECM fungi in India is less. They are mostly focused in the Himalayan region. In the Himalayan region, an ECM fungus like *Morchella* is found to form ECM with *Pinus*. *Suillus sibiricus*, an edible fungus, has been found to be associated with plantations of *Pinus wallichiana* in northwestern Himalayan region (Sagar and Lakhanpal 2005). However, in the study conducted in the forests of Central India primarily from the extensive surveyed Madhya Pradesh, we collected 61 species of ECM mushrooms belonging to nine genera (Sharma 2008; Sharma et al. 2008a, b, 2009a, c, 2010b). Most of the mushrooms which belong to *Russula*, *Lactarius*, *Boletus*, *Leccinum*, and *Amanita* were found to form ECM symbiosis with the *Shorea* tree (Fig. 7.6). Some species of *Scleroderma* and *Geastrum* were also found to be ECM with the *Shorea* tree. Apart from *Shorea*, we also observed one species of *Pisolithus* on *Eucalyptus*. Recording of important

Fig. 7.6 Diversity of ECM mushrooms found in forests of *Shorea robusta* forming ECM. (**a–e**) Species of *Russula*. (**f**) Species of *Lactarius*. (**g**, **h**) Species of *Boletus*. (**i**) Species of *Strobilomyces*. (**j–l**) Species of *Amanita* (Sharma et al. 2008a, 2009a, 2010d, e)

characters of ECM mushrooms is important for their identification (Fig. 7.7). A table mentioning all isolates is given in Table 7.2. It is observed that the forests of *Shorea* harbor a lot of ECM fungal diversities. ECM fungal diversity has also been reported on plants of *Dipterocarpaceae* (*Dipterocarpus indicus*, *Kingiodendron pinnatum*, *Humboldtia brunonis*) from the southern part of India (Natarajan et al. 2005b). He reported about 30 ECM fungi from the forests of Uppangala, the dominant one being members of *Russula*. Reddy et al. (2005) also reported a new species of *Pisolithus indicus* associated with *Dipterocarpaceae*. A comparison of three prominent ECM mushroom diversity studies of different regions of India is compiled in Table 7.3.

The members of *Dipterocarpaceae* are not only dominant in Indian evergreen forests but also in other Southeast Asian countries. Watling and Lee (1995) reported about 24 ECM fungi which included agarics, boletes, and earth balls. The dominant forms were members of *Amanita*, *Boletus*, and *Russula*. Similarly, many ectomycorrhizal fungal species associated with dipterocarps have been reported from the Philippines, Thailand, Sri Lanka, and Indonesia (Natarajan et al. 2005b). Recently Thomas et al. (2002) have reported a new genus *Anamika* (*A. indica*) under *Hopea* sp. in Wayanad District, India. Relatively recently, the use of molecular methods to identify the ECM fungi has helped in the authentic characterization of ECM fungi and identification of cryptic species. The species richness values for ECM in oak and conifer forest were 43 and 55, respectively, which were close to a midpoint range for similar other forests studied globally. Taxonomic studies of ECM mushrooms in the forest of Western Himalayas regions have been documented by Lakhanpal (1993, 1996). Similar work has also been undertaken in

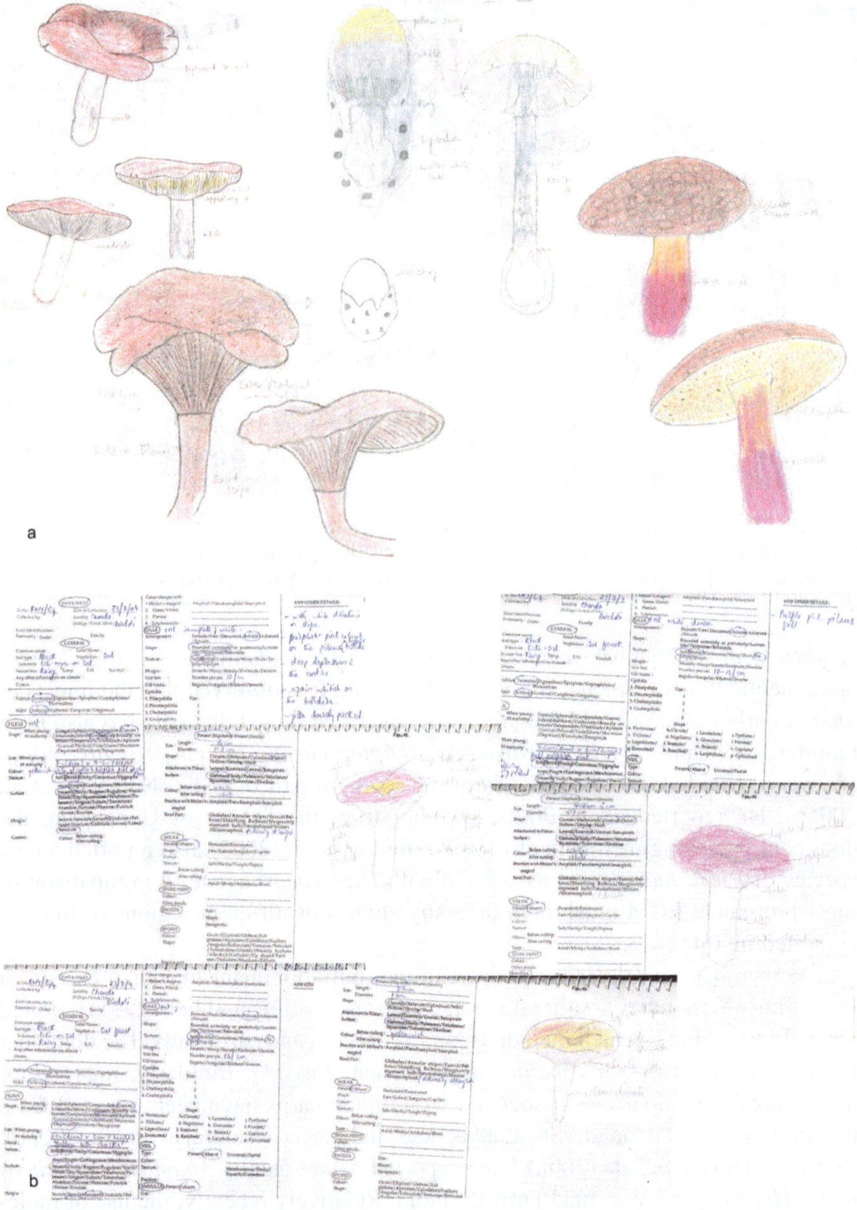

Fig. 7.7 Field notebook helpful in the identification of ECM mushrooms. (**a**) Hand drawings of ECM mushrooms on field. (**b**) Field data sheet filled on field (Sharma 2008)

Western Ghats by Natarajan et al. (2005a, b) and Natarajan and Ravindran (2003a, b). Limitations of estimates include the fragmentary nature of data and consideration of only species associated with dominant trees. Several workers have contributed to the taxonomic and biodiversity status of mushroom (not specifically ECM

Table 7.2 Important genera (in terms of species) of ECM fungi in three India studies (Sharma et al. 2009c)

		Percent association with			
Genus	Total mycorrhizal species	Sal	Teak	Bamboo	Eucalyptus
Amanita	5	+	–	–	–
Russula	24	+	–	–	–
Lactarius	8	+	–	–	–
Boletus	9	+	–	–	–
Leccinum	2	+	–	–	–
Geaster	5	+	–	–	–
Pisolithus	2	–	–	–	+
Scleroderma	5	+	–	–	–
Cantharellus	1	–	–	+	–

mushrooms)—Sharma and Sidhu (1991), Kaul (2002), Sharda (1991), Rattan and Khurana (1978), Khoshoo (1996), Verma et al. (1995), Lakhanpal (1993), and Bhagwat et al. (2000) in the Eastern Himalayan region; Natarajan and Ravindran (2003a, b) and Natarajan et al. (2005a, b) in the Western Ghats of Kerela and Maharashtra; Khoshoo (1991) and Kaushal (1991) in the Northwest Himalaya; Natarajan et al. (2005a, b) in South India; and Saini and Atri (1993) and Purkayastha and Chandra (1976, 1985) in Indo-Gangetic plains (Punjab, Uttar Pradesh and Bengal).

Besides the recognized hotspots in Western Ghats and the North Eastern Himalayan region, Central India is also home to the world's important tropical deciduous forest. These are rich in large unexplored microbial diversity yet to be exploited and conserved. The climatic conditions also make natural habitat for a large number of mushrooms. Central India is home to one of the world's important tropical deciduous rain forests. During the past one decade, regional mycologists launched fungal species richness monitoring studies in central forests dominated by sal, teak, and bamboo, tree species that form the base of region economy. Many studies have concentrated on mushroom diversity but none on ECM mushroom diversity. However, the fact that ECM fungi produce easily surveyed macroscopic sporocarps makes them particularly appropriate for large-scale monitoring studies. In Central - India and nearby regions, the Department of Forestry, Government of India, manages or holds an unusually large number of tracts of young, natural, and ancient teak, sal, and bamboo stands. The presence of variously aged and differently managed stands provided a unique opportunity to compare and contrast fungal communities through time as well as to investigate the impact of timber removal on fungal diversity. In this region, there are few reports on the diversity of ECM mushrooms (Sharma et al. 2008a, 2009a, 2010b, d, e).

Table 7.3 Important genera (in terms of species) of ECM fungi in three India studies (Sharma et al. 2009c)

S. No.	Mushroom species	Western Ghats	Western Ghats	Central India
		Natarajan et al. (2005a, b)	Pande et al. (2004)	Sharma et al. (2009a, b, c)
1	*Russula*	12	13	24
2	*Lactarius*	–	9	6
3	*Amanita*	5	15	5
4	*Anamika*	1	–	–
5	*Boletus*	–	12	10
6	*Leccinum*	–	6	2
7	*Strobilomyces*	1	1	1
8	*Suillus*	2	7	–
9	*Cantharellus*	–	–	1
10	*Astrosporina*	3	–	–
11	*Laccaria*	2	1	–
12	*Cortinarius*	1	4	–
13	*Hygrophorus*	–	4	–
14	*Clitocybe*	–	3	–
15	*Hebeloma*	–	2	–
16	*Volvariella*	–	1	–
17	*Inocybe*	–	1	–
18	*Galaria*	–	1	–
19	*Tubaria*	–	1	–
20	*Lacrymaria*	–	1	–
21	*Astraeus*	–	–	–
22	*Scleroderma*	–	–	5
23	*Geaster*	–	–	5
24	*Pisolithus*	1	–	2
25	*Lepista*	–	1	–
26	*Leucopaxillus*	–	1	–
27	*Oudemansiella*	–	1	–
28	*Tricholoma*	–	2	–
29	*Agaricus*	–	1	–

7.5.3 Geographical Distribution and Host Specificity

Despite the long evolutionary history, many mycorrhizal fungi appear to be non-host specific, and available evidence suggests that there is an ongoing parallel evolution of the partners in response to environmental change (Sanders 2002). An interesting phylogenetic analysis suggests that ECM fungi have evolved repeatedly from saprotrophic precursors (Cairney 2000). Unlike AM fungi which are completely dependent on their host for C, many ECM fungi can be cultured separately on agar plates without host roots. Despite their obligate nature, AM fungi are not thought to be host specific.

Plant individuals can be colonized by different species of AM fungi or ECM fungi. Dual colonization by both ECM and AM fungi also occurs in some plant species pointing to the absence of any host specificity. Plant trees show a certain degree of host specificity like only *S. robusta* (sal) forms ECM with mushrooms even in mixed forest where other trees are also present. Unlike AM fungi, not all tree plants form ECM symbiosis. The stability of host specificity is unclear considering that reversals from mycorrhizal mutualists to saprotrophic fungi occur. There is one major exception with regard to specificity in mycorrhizal symbioses, the myco-heterotrophic plants. These plants depend on fungi for the supply of C and photosynthesis is completely or partly lost. Spatial distribution of plants and fungi is often tightly linked due to host specificity. In India, most of the ECM fungi are reported from the Himalayas, Western Ghats, Eastern Himalayas, Central India, and forests of Southern India (Natarajan et al. 2005b; Pande et al. 2004; Sharma et al. 2009c). The same species of mushrooms are found to be forming ECM with different plant tree species in different portions of Earth.

7.5.4 Population and Dynamics

The quantitative measurement of the relations between populations, infection, and plant growth is an important factor to understand responses to efficiency of inoculum and inoculation for adequate ECM formation. Quantitative evaluations in relation to the soil environment are necessary to manage ECM population for maximum effect. The capacity of hyphae to which it can grow from spores, sclerotia, and mycelial strands through soil will affect infection dynamics. Larger propagules like sclerotia and mycelial strands are capable of growth through soil then small propagules (such as basidiospores) for considerable distances obtaining high infection levels.

The longevity of different types of propagules is an important factor for reforestation, but has received little experimental study till now. Basidiospores and sclerotia have considerable longevity, whereas mycelial strands appear fragile and hence have poor inocula in soil. Inoculum potential/infection relations for any one-fungus/plant combination will vary with environmental factors, viz., chemical status, pH, heavy metals, moisture, temperature, organic matter, and micro flora of soil. It is expected that pollutants and acidification of soils can strongly affect ECM formation and fungal spore on roots (Kjøller and Clemmensen 2008). The number of ECM formed under various conditions results from effects of environment on fungal factors in soil and rhizosphere and on root susceptibility to infection and spread.

It is commonly observed that as the forest ages, fungal species which dominated in early stages may be replaced by other fungi and is referred to as “late stage.” The separation of early and late species by no means is clear cut, as some fungal species persist throughout the life of tree plantations, and it is prominent in a situation where an exotic fungal species has been introduced. Moreover, where a natural situation pertains, as the environment changes with stand age, various fungi will be advantaged and will replace some of the prominent early-stage fungi. Almost all the basidiospores responding to seedling roots are of those species which are regarded as "early-stage" fungi, e.g., species of *Hebeloma, Paxillus*, *Suillus*, *Pisolithus, Scleroderma*, and *Inocybe*. ECM fungi may be associated with several tree species (Sterkenburg 2016). Growth of ECM fungi in rhizospheric soil may be a key factor to understand ECM composition on a root as it ages and to selection of fungi for tolerance of deleterious factors. It has been recognized that studies in laboratory media have little relevance to growth in the rhizosphere, where the physicochemical environment is different. Slow growth of fungi which appear at late stage on young roots as well as poor spore germination may explain their failure to dominate the rhizosphere of young roots. Moreover, some signals (metabolite, chemicals) may be secreted by older tree roots. Fungal communities in general are affected by soil fertility. The development of extraradical hyphae varies considerably between fungus (and hosts) and environment. An understanding of factors affecting growth of extraradical hyphae is important, not only to uptake of nutrient and water but also to other ECM functions. Some ECM fungi, viz., *Russula luteolus*, *Suillus bovinus*, and *Hebeloma crustuliniforme*, form highly branched mycelial strands up to 40 cm (Skinner and Bowen 1974a, b; Finlay and Read 1986a, b).

In recent years, molecular methods have been used to identify the genotypes. Various molecular methods like random amplified polymorphic DNA (RAPD), simple sequence repeat (SSR), amplified fragment length polymorphism (AFLP), single-strand conformational polymorphism (SSCP) and inter-retrotransposon amplified polymorphism (IRAP) markers, and repeat SSR markers have helped in knowing the distribution of various genotypes of ECM fungi (Murata et al. 2005; Zhou et al. 2001; Dunham et al. 2003; Kretzer et al. 2004; Wu et al. 2005; Bergemann et al. 2006). They may not completely resolve the genotypes but help in knowing the distribution (Redecker et al. 2001; Dunham et al. 2003; Bagley and Orlovich 2004). The fruiting body collection also helps to know that multiple genotypes of a genus are present in a particular location, like in genera *Russula, Lactarius, Amanita,* etc. Sometimes the spatial distribution is restricted because the distribution of host tree species in a mixed forest influences the distribution of ECM genotypes (Zhou et al. 2000).

7.5.5 Nutrient Circulation in Ecosystem

As a rule, mycorrhizal infection enhances plant growth by increasing nutrient uptake via increasing in the absorbing surface area, by mobilizing sparingly available nutrients sources, or by excretion of chelating compounds or ectoenzymes. Bidirectional transfer of nutrients between plant and fungus is typical of ECM (including other types) and is the basis for the prolonged compatible interactions typical of these symbioses. Depending on the tree species, different amounts of mineral nutrients from the soil to the root occur via fungal hyphae (Marschner

1995). Except where mantle is diffuse, most nutrients delivered to Hartig net are likely to be transported through the living external hyphae. Some ECM have mycelial cords which transport water and nutrients over long distances and in some cases via specialized nonliving hyphae.

At cellular level, interfaces in all types of mycorrhizas are composed of membranes of both partners which are separated by apoplastic region (Smith et al. 1994). The interface is simple intercellular wall to wall contact in ECM, but the fungal partner remains in apoplast space outside the plant protoplast and cell membrane. Fungal colonization of host root tissues is entirely extracellular in ECM fungi. In Hartig net region, the hyphal branch and septa formation profusely becomes uneven or incomplete giving characteristic labyrinthine system (Martin and Nehls 2009). Surface fibrils and acid phosphatase activity present in mantle vanish as hyphae become tightly pressed against host cell walls. Adjacent walls of fungal and host become indistinguishable from each other forming a homogenous interfacial matrix.

They are important for the delivery of carbon to soil and are responsible for a substantial component of forest-soil carbon fluxes (Nehls 2008; Anderson and Cairney 2007; Högberg et al. 2001; Högberg and Högberg 2002; Godbold et al. 2006; Hobbie 2006). However, extracellular material is deposited around hyphae of ECM and also accumulates in intercellular spaces of the fungal sheath in ECM of *Eucalyptus* and *Pisonia* (Pritsch and Garbaye 2011). Evidence (or lack of it) for the nature of the compounds transferred between the symbionts has been reviewed previously by many workers. Sugars are important in carbohydrate transfer with hydrolysis of sucrose (or trehalose in orchid mycorrhizas) along with synthesis of characteristic "nonrecyclable" carbohydrates (e.g., mannitol in ECM), as important steps in polarizing transport in favor of one symbiont (Dearnaley et al. 2012). ECM has a major influence on N and lesser effect on P nutrition wherein inorganic orthophosphate is the major form in which phosphorus is transferred (Wiemken 2007; Baxter and Dighton 2005). Moreover in ECM, the coexistence of ATPase activity on plant and fungal plasma membrane at Hartig net interface suggests that both the systems work cooperatively in bidirectional nutrient exchange. P_i (inorganic phosphorus) transfer to host tissue in excised beech ECM was around 10–20% of P_i absorbed by sheath hyphae, and in mycelia of *P. tinctorius* (Pers.), Coker and Couch net efflux was also 10% of P_i absorbed (Smith et al. 1994).

Hydrolysis of sucrose to hexoses by invertase in the apoplast would allow a net sugar transport to the fungus. The fact that individual ECM roots act as greater sinks for photosynthetically fixed C than non-mycorrhizal roots strongly suggests greater concentration at the interface and the probability of increased loss from the root cells. Release of phytohormones at the interface might also influence nutrient transport. Auxin ,e.g., can increase ATPase activity in plant tissue and may have a role in ECM formation, and the cytokinin N^{6-2} isopentenyl–adenosine (2iPA) has been shown to influence both absorption and loss of ions from mycelium of the ECM fungus *Suillus variegatus*.

Many studies have shown that ECM fungi can utilize organic N sources through the production of extracellular acid proteinases. Mycorrhizal infection can provide host plants with access to N sources, which are normally unavailable to non-mycorrhizal roots (Gobert and Plassard 2008). Further, proteolytic capacity may vary greatly between fungal isolates. Possibly all fungi can assimilate ammonia by a combination of the glutamate dehydrogenase and glutamine synthetase pathways, whereas a smaller number of species can efficiently reduce nitrate. Ammonia is rapidly assimilated in the extraradical hyphae and N is transferred to the host primarily as glutamine. ECM symbiosis alters metabolic pathways of N assimilation in the fungal symbiont. Not much is known on the role of mycorrhiza in uptake of K, Ca, Mg, and S. Majority studies on ECM and micronutrient uptake are focused on the protection from excessive uptake of Cu and Zn on soils high in heavy metals. Production of siderophores is widespread between ECM. However, boron is essential for the growth of fungi and ECM may increase concentrations in the host plants. Plant growth or reproduction is not always increased by mycorrhizal infection due to high efficiency of P acquisition or low P requirement of plants. The acquisition of P is reduced when external hyphae are destroyed by grazing soil animals, soil disturbance, or fungicides.

Identification of efficient genotypes of both fungus and host and understanding the way their function is integrated depend on the identification and quantification of the key processes involved in nutrient uptake and use. Despite worldwide research interest, progress is slow in understanding the mechanisms involved, the differences between ECM fungi in their capacity to deliver P to the host plant, and in quantification of the benefit ECM plants have under field and in natural ecosystems. However, better matching of ECM fungi with host and site conditions is required for the full potential of large-scale inoculation programs to be realized. Moreover, recent studies of genome sequencing are giving new insight to the mycorrhizal symbiotic association (Martin et al. 2008).

7.5.6 Interactions with Other Microorganisms

In addition to increasing absorptive surface area of their host plant root systems, the hyphae of ECM fungi provide an increased surface area for interactions with other microorganisms and provide an important pathway for translocation of energy-rich plant assimilates (products of photosynthesis) to the soil (Finlay 2004). These ECM interactions are synergetic, competitive, or antagonistic and have applied significance in areas such as biological control, bioremediation, and sustainable forestry. Bacteria with a potential to fix nitrogen have been discovered which grow symbiotically with tuberculate roots of ECM plants forming rood nodules and also applied as plant growth-promoting rhizobacteria (PGPR) (Cumming et al. 2015). Exudation and reabsorption of fluid droplets at ECM hyphal tips have been earlier demonstrated. The extent to which interactions between ECM mycelia and other microbes affect organic and mineral substrates is unclear, and further experiments are needed to distinguish between the activity of ECM hyphae and activity facilitated by ECM for the uptake of compounds (Finlay 2008; Churchland and Grayston 2014).

7.5.7 Multifunctional Role of ECM

The effects of mycorrhizal fungi have traditionally been considered within narrow perspective of their effects on the mineral nutrition of individual plants. Research during the past 20 years has increasingly viewed symbiotic ECM associations between plants and fungi within a wider, multifunctional perspective. New molecular methods have been applied to investigate ECM fungal communities, and greater attention has been paid to their possible effects at the level of plant community. And we have acquired greater knowledge about fungal species diversity and become more aware of potential functional diversity of ECM fungi. The new multifunctional perspective includes:

> mobilization of N and P from organic polymers; release of nutrients from mineral particles or rock surfaces via weathering; interactions with myco-heterotrophic plants; effects on carbon cycling; mediation of plant responses to stress factors such as soil acidification, toxic metals, drought and plant pathogens; as well as a range of possible interactions with other soil microorganisms.

The role of ECM fungi in shaping terrestrial ecosystems is fundamental. Many of the characteristic plant communities that dominate major terrestrial biomes are due to selection which has favored different types of symbiotic associations that are adapted to soil, vegetation, and climatic conditions characterizing these different environments. Comparative analysis of different systems will improve our understanding of responses to environmental and climatic changes. It is an important prerequisite for future sustainable management of terrestrial ecosystems (Anderson and Cairney 2007; Finlay 2004).

7.6 Practical Applications

7.6.1 Artificial Synthesis of ECM

Interest in cultivating edible ECM mushrooms has grown in the past few decades with the realization that there had been dramatic falls in the availability of mushroom species and increased market demand exceeded their supply. Melin (1922, 1923, 1936) developed the pure culture synthesis technique to experimentally demonstrate the ability of known fungus isolates to form ECM with specific hosts under pure culture conditions. Use of these techniques has also led to the discovery of important physiological aspects of symbiosis. These include uptake of nutrients and water by fungus and translocation to the host, movement of photosynthate from host to fungus, interactions of growth-regulating substances, host-to-host transfer of carbohydrate via a shared fungal symbiont, protection against root pathogens, effects of temperature on mycorrhiza development, specificity and compatibility between fungus and host, and several other processes.

One must recognize the artificiality of pure culture synthesis and limit extrapolation of results to natural situations. Positive synthesis results confirm the ability of that particular host-fungus combination to form ECM. Negative results suggest that union of organisms in question seems unlikely. Melin (1923) primarily used flasks containing sterile sand moistened with a nutrient solution. Several investigators have modified Melin's technique often trying more complex arrangements to reduce the artificial nature of the enclosed system. Hacskaylo (1953) greatly improved the system by using vermiculite instead of sand as the substrate. Vermiculite provides better aeration and moisture holding capacity than sand. Marx and Zak (1965) further improved the substrate by stabilizing the acidity with an addition of finely ground sphagnum peat moss. Molina and Palmer (1982) reports excellent seedling growth and ECM development in the glass test tube system filled with vermiculite and peat moss moistened with nutrient solution which is helpful in running numerous syntheses in a relatively small area. Pachlewski and Pachlewski (1974) also reported good mycorrhiza synthesis in a large test tube but used a solid agar substrate rather than peat moss and vermiculite. Danell (1994) reported artificial ECM formation between *Cantharellus cibarius* with a host tree plant. In another study from India, Sharma et al. (2008b, 2009b, 2010a, c) formed in vitro synthesis of ECM between *D. strictus* and *Cantharellus* sp. (Fig. 7.8). The roots showed root modification and clear interaction between host plant and ECM fungus. The ECM fungal inoculum was produced on used tea leaves and sand (Sharma and Rajak 2011).

7.6.2 Applied Aspects of ECM

Much of our understanding on the functions of ECM has come from research directed toward practical application in forestry. Repeated failures in the establishment of exotic pine plantations in the tropics and other areas where ECM hosts do not naturally occur clearly demonstrated the dependence of these trees on their fungal symbionts. Only after inoculation with forest soil containing ECM fungus propagules could these trees survive and function properly. This information provides many necessary tools and concepts for strengthening forestry programs around the world. Today, wide-scale inoculation of forest nurseries with selected ECM fungi appears imminent. Commercial interest in producing pure culture of ECM fungi inoculum expands the possibilities of worldwide application. The success of these inoculation programs hinges on selection of effective and beneficial fungal symbionts. New inoculation programs must be strongly research oriented from the outset.

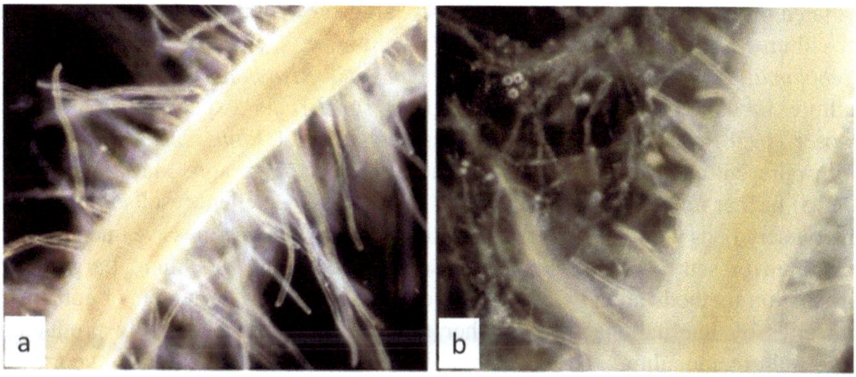

Fig. 7.8 Aseptic ECM synthesis. (**a**) Uninoculated *D. strictus* roots with long root hairs. (**b**) *I*noculated roots of *Dendrocalamus strictus* with *C. tropicalis* with small, less root hairs, and mycelial coverage (Sharma et al. 2009b)

It has been experienced during the introduction of exotic pines into the Southern Hemisphere and Tropical Island that mycorrhizal fungi should accompany host trees. Afforestation attempts in treeless grasslands of the USA and steppes of Russia have also required inoculation for success. Although successful inoculation of tree seedlings (already planted) in field have been known, nursery inoculation is more common. Seedlings inoculated in nursery can establish a healthy ECM system before out planting. Tree seedlings lacking ECM suffer severe nutrient deficiencies early in their first growing season; the deficiencies persist until mycorrhizae are formed. The most commonly used and probably most reliable ECM inoculum is soil taken from beneath ECM hosts. Soil inoculum may also be added to the planting hole when seedlings are out planted. Soil inoculation has been instrumental in the establishment of exotic pine plantations in the Southern Hemisphere and continues as a regular practice there today.

Planting mycorrhizal "nurse" seedlings or incorporating chopped roots of ECM hosts into nursery beds as a source of fungi for neighboring young seedlings has been successful (Sim and Eom 2006). Chevaliar and Grente (1973) were able to inoculate seedlings with the prized truffle fungus *Tuber melanosporum* by use of nurse seedlings already mycorrhizal with this fungus. Basidiospores and ascospores or crushed sporocarps have been used occasionally as inoculum, usually in small experiments. Some investigators have reported good success with this technique. Asexual spores and sclerotia are further sources of inoculum. The gasteromycetes (puffballs and related fungi) with abundant spore masses offer better source of spores than gills. Most of the recent research has been with *P. tinctorius*. Inoculation with spores of *Rhizopogon* species also appears promising. Abundant *Rhizopogon* mycorrhizae formed on seedlings produced from the coated seed of *P. radiata* D. Don with basidiospores of *Rhizopogon luteolus*.

Although many difficulties remain in using pure culture as source, techniques for wide-scale application are now being developed. Unfortunately, many ECM fungi isolated from either fruiting body or ECM itself grow extremely slowly or not at all in culture. Still, many do grow well in culture, e.g., most species of *Suillus, Hebeloma, Laccaria, Amanita, Rhizopogon*, and *Pisolithus*. Most of the pure culture inoculation has been restricted to small-scale experiments, although Moser (1958) successfully inoculated nursery beds of *Pinus cembra* in Austria with pure cultures of *Suillus plorans* more than 20 years ago. Marx and Bryan (1975) further refined Moser's technique and reported excellent results in inoculating nursery beds with *P. tinctorius*. The logistics of producing massive quantities of inoculum presently limits wide-scale use of pure culture inoculum. Large-scale production methods are now being developed in industrial fermenters or container nurseries by international firms. Other firms are also experimentally producing pure culture inoculum of ECM fungi. Industry representatives and mycorrhiza researchers are optimistic that effective commercial inoculum will soon be available in the market, and it can be used for reforestation of waste and degraded sites (Fig. 7.9).

The promising outlook for pure culture inoculation raises still another important question; which fungus is best for a particular host or habitat? The effectiveness of the various ECM fungi on different host species has to be repeatedly emphasized. With thousands of ECM fungi and numerous hosts, careful selection of the best fungi for particular host is critical. Many important criteria must be considered when selecting fungus candidates for nursery inoculation. Careful experimentation and good record keeping are essential throughout evaluations of each isolate. One must first be able to isolate the particular fungus and grow it reasonably well in culture.

Relatively fast-growing fungi are generally preferred for inoculation because of their short incubation period. Unfortunately, many otherwise desirable ECM fungi grow slowly. According to Marx (1980) fresh cultures are preferred to cultures repeatedly transferred and stored for several years. He further suggests passing important fungus cultures through a host inoculation and mycorrhiza formation followed by re-isolation, every few years to maintain mycorrhiza-forming capacity. Moreover, fungi, which produce large hyphal stands of rhizomorphs in culture of soil, may be superior in soil exploration and mineral uptake to those which lack rhizomorphic growth.

There have been a lot of studies conducted on ECM. In recent times most of the studies are focusing on genome sequencing, genomics, and metagenomics and several other aspects (Martin et al. 2016; Redeker et al. 2004; Tedersoo et al. 2016). There are some interesting articles which have discussed the importance of ECM studies highlighting their role in distribution, evolution, and phylogenetic studies (Tedersoo et al. 2010). Studies are focusing on the one-to-one fungus-plant symbiosis, i.e., between individual partners and benefits shared by them (Kennedy et al. 2015). Some researchers are also focusing on links between community structure and function for ectomycorrhizal fungus (Walker et al. 2014). There is increased to study various aspects of ECM synthesis in detail.

Fig. 7.9 Forest regeneration and reclamation of mine sites. Exploitation of potential ECM mushroom by inoculation of nursery seedlings and tissue culture plantlets of bamboo, sal, and other tree species can help in forest reclamation (Sharma 2008)

References

Agerer R (1987, 2002) Colour atlas of ectomycorrhizae. Einhorn-Verlag, Schwabisch Gmiid, Germany

Agerer R (2001) Exploration types of ectomycorrhizas: a proposal to classify ectomycorrhizal mycelial systems according to their patterns of differentiation and putative ecological importance. Mycorrhiza 11:107–114

Anderson IC, Cairney JWG (2007) Ectomycorrhizal fungi: exploring the mycelial frontier. FEMS Microbiol Rev 31:388–406

Ashford AE, Allaway WG (1982) A sheathing mycorrhiza on *Pisonia grandis* R. Br. (*Nyctaginaceae*) with development of transfer cells rather than a Hartig net. New Phytol 90:511–517

Bagley SJ, Orlovich DA (2004) Genet size and distribution of *Amanita muscaria* in a suburban park, Dunedin, New Zealand. N Z J Bot 42:939–947

Baxter JW, Dighton J (2005) Phosphorus source alters host plant response to ectomycorrhizal diversity. Mycorrhiza 15:513–523

Bergemann SE, Douhan GW, Garbaletto M, Miller SL (2006) No evidence of population structure across three isolated subpopulations of *Russula brevipes* in an oak/pine woodland. New Phytol 170:177–184

Bever JD, Pringle A, Bchults PA (2002) Dynamics with in the plant-arbuscular mycorrhizal fungal mutualism: testing the Nahire of community feedback. In: van der Heijden MGA, Sanders IR (eds) Mycorrhizal ecology. Springer, Berlin, pp 267–294

Bhagwat SA, Brown ND, Watkinson SC, Savill PS, Jennings SB (2000) Macrofungal diversity in three forested land use types, a case study from the Western Ghats of India. In: Tropical mycology. Liverpool John Moores University, Liverpool, pp 25–29

Brundrett MC (2002) Coevolution of roots and mycorrhizas of land plants. New Phytol 154: 275–304

Bücking H, Heyser W (2001) Microautoradiographic localization of phosphate and carbohydrates in mycorrhizal roots of *Populus tremula* × *Populus alba* and the implications for transfer processes in ectomycorrhizal associations. Tree Physiol 21:101–107

Cairney JWG (2000) Evolution of mycorrhiza systems. Naturwissenschaften 87:467–475

Cairney JWG, Chambers SM (1999) Ectomycorrhizal fungi—key genera in profile. Springer, Berlin, Heidelberg, p 369

Chevaliar G, Grente J (1973) Propagation de la mycorrhization par la truffle a partir de racines excisees et de plantules inseminatrices. Ann Phytopathol 4:317–318

Churchland C, Grayston SJ (2014) Specificity of plant-microbe interactions in the tree mycorrhizosphere biome and consequences for soil C cycling. Front Microbiol 5:1–20

Comandini O, Contu M, Rinaldi AC (2006) An overview of Citrus ectomycorrhizal fungi. Mycorrhiza 16:381–395

Cumming JR, Zawaski C, Desai S, Collart FR (2015) Phosphorus disequilibrium in the tripartite plant ectomycorrhiza-plant growth promoting rhizobacterial association. J Soil Sci Plant Nutr 15:464–485

Dahlberg A (2001) Effects of fire on ectomycorrhizal fungi in Fennos Canadian boreal forests. Silva Fennica 36:69–80

Danell E (1994) Formation and growth of the ectomycorrhiza of *Cantharellus cibarius*. Mycorrhiza 5:88–97

Danell E (2002) Current research on chanterelle cultivation in Sweden. In: Hall I, Wang Y, Danell E, Zambonelii A (eds) Edible mycorrhizal mushrooms and their cultivation. Crop and Food Research, Christ Church, pp 1–4

Dearnaley JDW, Martos F, Selosse M-A (2012) Orchid mycorrhizas: molecular ecology, physiology, evolution and conservation aspects. In: Hock B (ed) Fungal associations. Springer, Berlin, pp 207–230

Dunham SM, Kretzer A, Pfrender ME (2003) Characterization of Pacific golden chanterelle (*Cantharellus formosus*) genet size using co-dominant microsatellite markers. Mol Ecol 12: 1607–1618

Erland S, Taylor AFS (2002) Diversity of ectomycorrhizal fungal communities in relation to the abiotic-environment. In: van der Heijden MGA, Sanders JR (eds) Mycorrhizal ecology. Springer, Berlin, Heidelberg, p 465

Finlay RD (2004) Mycorrhizal fungi and their multifunctional role. Mycologist 18:91–96

Finlay RD (2008) Ecological aspects of mycorrhizal symbiosis: with special emphasis on the functional diversity of interactions involving the extraradical mycelium. J Exp Bot 59: 1115–1126

Finlay RD, Read DJ (1986a) The structure and function of the vegetative mycelium of ecto-mycorrhizal plants-I, translocation of 14C-labeled carbon between plants interconnected by a common mycelium. New Phytol 120:105–115

Finlay RD, Read DJ (1986b) The structure and function of the vegetative mycelium of ecto-mycorrhizal plants-II, the uptake and distribution of phosphorus by mycelial strands interconnecting host plants. New Phytol 103:157–165

Frank AB (1885) Uber die auf wurzelsbiose beruhends eranhrung gewisser baume durch unter-irdische pilze. Ber Deut Bot Ges 3:128–145

Giachini AJ, Oliviera VI, Castellano MA, Trappe JM (2000) Ectomycorrhizal fungi in *Eucalyptus* and *Pinus* plantations in southern Brazil. Mycologia 92:1166–1177

Gobert A, Plassard C (2008) The beneficial effect of mycorrhizae on N utilization by the host plant: myth or reality? In: Varma A (ed) Mycorrhiza: state of the art, genetics and molecular biology, eco-function, biotechnology, eco-physiology, structure and systematics. Springer, Berlin, Heidelberg, p 797

Godbold DL, Hoosbeek MR, Lukac M et al (2006) Mycorrhizal hyphal turnover as a dominant process for carbon input into soil organic matter. Plant Soil 281:15–24

Grime JP (2001) Plant strategies, vegetation processes and ecosystem properties, 2nd edn. Wiley, Chichester

Hacskaylo E (1953) Pure culture synthesis of prime mycorrhizae in terralite. Mycologia 45: 971–975

Harnett DC, Wilson WT (1999) Mycorrhizae influence plant community structure and diversity in tall grass prairie. Ecology 80:1187–1195

Hart MM, Klironomos JN (2002) Diversity of arbuscular mycorrhizal fungi and ecosystem functioning. In: van der Heijden MGA, Sanders IR (eds) Mycorrhizal ecology. Springer, Heilderberg, p 465

Hobbie EA (2006) Carbon allocation to ectomycorrhizal fungi correlates with below ground allocation in culture studies. Ecology 87:563–569

Högberg MN, Högberg P (2002) Extramatrical ectomycorrhizal mycelium contributes one-third of microbial biomass and produces, together with associated roots, half the dissolved organic carbon in a forest soil. New Phytol 154:791–795

Högberg P, Nordgren A, Buchmann N, Taylor AFS, Ekblad A, Högberg MN, Nyberg G, Ottosson-Löfvenius M, Read DJ (2001) Large-scale forest girdling shows that current photosynthesis drives soil respiration. Nature 411:789–792

Jakobsen I, Smith SE, Smith FA (2002) Function and diversity of arbuscular mycorrhizae in carbon and mineral nutrition. In: van der Heijden MGA, Sanders IR (eds) Mycorrhizal ecology. Springer, Berlin, Heidelberg, p 465

Jonsson LM, Nilsson M-C, Wardle DA, Zachrisson O (2001) Context dependent effects of ectomycorrhizal species richness on tree seedling productivity. Oikos 93:353–364

Kaul TN (2002) In: Watling R, Frankland JC, Ainsworth AM, Isaac S, Robinson CH (eds) Conservation of mycodiversity in India: an appraisal. CABI Publishing, New York, p 191

Kaushal SC (1991) Systematics of NW Himalayan species of Helvella (operculate discomycete). In: Khoshoo TN, Sharma M (eds) Himalayan botanical researches. Ashish Publishing House, New Delhi, pp 61–75

Kennedy P, Walker JKM, Bogar L (2015) Interspecific mycorrhizal networks and non-networking hosts: exploring the ecology of the host genus Alnus. In: Horton TR (ed) Mycorrhizal networks, Ecological studies 224, Chapter 8. Springer, p 227

Khoshoo TN (1991) Conservation of biodiversity in biosphere. In: Khoshoo TN, Sharma M (eds) Indian geosphere biosphere programme: some aspects. Har-Anand Publications, Vikas Publishing House Private, New Delhi, pp 183–233

Khoshoo TN (1996) Biodiversity in the Indian Himalayas: conservation and utilization. In: Banking on biodiversity—report on the regional consultation on biodiversity assessment in the Hindukush Himalayas

Kjøller R, Clemmensen KE (2008) The impact of liming on ectomycorrhizal fungal communities in coniferous forests in Southern Sweden. Skogsstyrelsen februari Publications, Jönköping

Kretzer AM, Dunham S, Molina R, Spatafora JW (2004) Microsatellite markers reveal the below ground distribution of genets of two species of *Rhizopogon* forming tuberculate ectomycorrhizas on Douglas fir. New Phytol 161:313–320

Lakhanpal TN (1993) The Himalayan agaricales status of systematics. Mush Res 2:1–10

Lakhanpal TN (1996) Mushrooms of India: Boletaceae. In: Mukerji KG (ed) Studies in cryptogamic botany, vol I. APH Publishing Corporation, Delhi

Lakhanpal TN (2000) Ectomycorrhiza—an overview. In: Mukerji (ed) Mycorrhizal biology. Kluwer Academic Plenum Publishers, New York, pp 101–118

Lewis DH (1973) Concepts in fungal nutrition and the origin of biotrophy. Biol Rev 48:261–273

Marschner H (1995) Mineral nutrition of higher plants. Academic Press, London

Martin F, Nehls U (2009) Harnessing ectomycorrhizal genomics for ecological insights. Curr Opin Plant Biol 12:508–515

Martin F, Aerts A, Ahrén D (2008) The genome of *Laccaria bicolor* provides insights into mycorrhizal symbiosis. Nature 452:88–92

Martin F, Kohler A, Murat C, Veneault-Fourrey C, Hibbett DS (2016) Unearthing the roots of ectomycorrhizal symbioses. Nat Rev Microbiol 14:760–773

Marx DH (1980) Ectomycorrhiza fungus inoculations, a tool for improving forestation practices. In: Mikola P (ed) Tropical mycorrhiza research. Oxford University Press, Oxford, pp 13–71

Marx DH, Bryan WC (1975) Growth and ectomycorrhizal development of loblolly pine seedlings in fumigated soil infested with the fungal symbiont *Pisolithus tinctorius*. For Sci 21:245–254

Marx DH, Zak B (1965) Effect of pH on mycorrhizal formation of slash pine in aseptic culture. For Sci 11:66–75

Melin E (1922) Untersuchungen iiber die Larix Mycorrhiza I. Synthese der Mykorrhiza in Rein culture. Sven Bot Tidskr 16:161–196

Melin E (1923) Experimentelle Unters uchungen uber die Okologie der Mykorrhizen von Pinus sylvestris and Pinus abies. Mycol Unters 2:72–331

Melin E (1936) Metroden der experimentelle untersuchung mylcotropher pflanzen. Handb Biol Arbeitsmety 11:1015–1108

Molina R, Palmer JG (1982) Isolation, maintenance and pure culture manipulation of ectomycorrhizal fungi. In: Schenck NC (ed) Methods and principles of mycorrhizal research. APS, Saint Paul, pp 115–119

Moser M (1958) Die Kiinstliche Mycorrhizaimp fung an Forstpflanzen. I. Erfahrungen bei der Reinkulture von Mycorrhizapilzen. For Wiss Centralbl 77:32–40

Murata H, Ohta A, Yamada A, Narimatsu M, Futamura N (2005) Genetic mosaics in the massive persisting rhizosphere colony "shiro" of the ectomycorrhizal basidiomycete *Tricholoma matsutake*. Mycorrhiza 15:505–512

Natarajan K, Ravindran C (2003a) Two new species of the genus *Entoloma* from south India. Mycotaxon 85:143–146

Natarajan K, Ravindran C (2003b) Two new species of the genus *Pholiota* from south India. Mycotaxon 85:271–275

Natarajan K, Narayanan K, Ravindran C, Kumaresan V (2005a) Biodiversity of agarics from Nilgiri Biosphere Reserve, Western Ghats, India. Curr Sci 88:1890–1893

Natarajan K, Senthilarasu G, Kumaresan V, Riviere T (2005b) Diversity in ectomycorrhizal fungi of a dipterocarp forest in Western Ghats. Curr Sci 88:1893–1895

Nehls U (2008) Mastering ectomycorrhizal symbiosis: the impact of carbohydrates. J Exp Bot 59: 1097–1108
Onguene NA, Kuyper TW (2001) Mycorrhizal associations in the rain forest of South Cameroon. For Ecol Manage 140:277–287
Pachlewski R, Pachlewski J (1974) Studies on symbiotic properties of mycorrhizal fungi of Pine (*Pinus silvertris* L.) with the aid of the method of mycorrhizal synthesis in pure cultures on agar. For Res Inst, Warsaw, Poland, p 228
Pande V, Palni UT, Singh SP (2004) Species diversity of ectomycorrhizal fungi associated with temperate forest of Western Himalaya: a preliminary assessment. Curr Sci 86:1619–1623
Peterson RL, Uetake Y, Zelmer C (1998) Fungal symbioses with orchid protocorms. Symbiosis 25:29–55
Peterson RL, Massicotte HB, Melville LH (2004) Mycorrhizas: anatomy and cell biology. CABI Publishing, CAB International, Wallingford, Oxon
Peterson RL, Wagg C, Pautier M (2008) Associations between microfungal endophytes and roots: do structural features indicate function? Botany 86:445–456
Pritsch K, Garbaye J (2011) Enzyme secretion by ECM fungi and exploitation of mineral nutrients from soil organic matter. Ann For Sci 68:25–32
Purkayastha RP, Chandra A (1976) Indian edible mushrooms. Firma KLM Pvt. Ltd., Calcutta
Purkayastha RP, Chandra A (1985) Manual of Indian edible mushrooms. Today and Tomorrow's Printers and Publishers, New Delhi
Rattan SS, Khurana IPS (1978) The clavariaceae of the Sikkim Himalayas. Bibliotheca Mycologia, vol 66. Cramer in der A.R. Gantner Verlag Kommanditgesellschaft FL-9490 Vadauz. Liechtenstein 66:1–68
Read DJ (1991a) Mycorrhizas in ecosystems. Experimentia 47:376–391
Read DJ (1991b) Mycorrhizal in ecosystems natures' response to the "Law of the minimum". In: Hawksworth DL (ed) Frontiers in mycology. CAB International, Wallingford, pp 101–130
Reddy MS, Singla S, Natarajan K, Senthilrasu G (2005) *Pisolithus indicus*, a new species of ectomycorrhizal fungus associated with Dipetrocarps in India. Mycologia 97:838–843
Redecker D, Szaro TM, Bowman RJ, Bruns TD (2001) Small genets of *Lactarius xanthogalactus*, *Russula cremoricolor* and *Amanita francheti* in late-stage ectomycorrhizal successions. Mol Ecol 10:1025–1034
Redeker KR, Treseder KK, Allen MF (2004) Ectomycorrhizal fungi: a new source of atmospheric methyl halides. Glob Chang Biol 10:1009–1016
Rillig MC, Wright SF, Eviner V (2002) The role of arbuscular mycorrhizal fungi and glomalin in soil aggregation: comparing effects of five plant species. Plant Soil 238:325–333
Rinaldi AC, Comandini O, Kuyper TW (2008) Ectomycorrhizal fungal diversity: separating the wheat from the chaff. Fungal Divers 33:1–45
Sagar A, Lakhanpal TN (2005) Pure culture synthesis of *Pinus wallichiana* ectomycorrhizal with *Suillus sibiricus*. Indian Phytopathol 58:323–325
Saini SS, Atri NS (1993) Studies on genus *Lactarius* from India. Indian Phytopathol 46:360–364
Sanders IR (2002) Specificity in the arbuscular mycorrhizal symbiosis. In: van der Heijden MGA, Sanders IR (eds) Mycorrhizal ecology. Springer, Berlin, Heidelberg
Schübler A, Schwarzott D, Walker C (2001) A new fungal phylum, the *Glomeromycota*: phylogeny and evolution. Mycol Res 105:1413–1421
Sharda RM (1991) Clavaroid homobasidiomycetes in the Himalaya, a check list. In: Khullar SP, Sharma MP (eds) Himalayan botanical researches. Ashish publishing House, New Delhi, pp 31–60
Sharma R (2008) Studies on ectomycorrhizal mushrooms of M.P. and Chhattisgarh. PhD thesis, R.D. University, Jabalpur, India
Sharma R, Rajak RC (2011). Ectomycorrhizal Interaction between *Cantharellus* and *Dendrocalamus*. In: Rai M, Varma V (eds) Diversity and biotechnology of ectomycorrhizae. Soil Biol 25. Springer, Berlin, Heidelberg, pp 405–428

Sharma PM, Sidhu D (1991) Notes on Himalayan *Geoglossaceae*. In: Khullar SP, Sharma MP (eds) Himalayan botanical researches. Ashish Publishing House, New Delhi, pp 13–29
Sharma R, Rajak RC, Pandey AK (2008a) Some ectomycorrhizal mushrooms of Central India—I. *Russula*. J Mycopathol Res 46:201–212
Sharma R, Rajak RC, Pandey AK (2008b) Growth response of *Dendrocalamus* seedlings by inoculation with ectomycorrhizal fungi. Middle East J Sci Res 3:200–206
Sharma R, Rajak RC, Pandey AK (2009a) Some ectomycorrhizal mushrooms of Central India—II. *Lactarius*. J Mycopathol Res 47:43–47
Sharma R, Rajak RC, Pandey AK (2009b) Simple technique for ectomycorrhizal formation between *Cantharellus* and *Dendrocalamus strictus*. Taiwan J For Sci 24:141–148
Sharma R, Rajak RC, Pandey AK (2009c) Ectomycorrhizal mushrooms in Indian tropical forests. Biodiversity 10:25–30
Sharma R, Rajak RC, Pandey AK (2010a) Mass multiplication of ectomycorrhizal *Cantharellus* inoculum for large scale tailoring nursery inoculations of bamboo seedlings. Asian J Sci Res 4:84–89
Sharma R, Rajak RC, Pandey AK (2010b) Some ectomycorrhizal mushrooms of Central India-V. *Pisolithus, Scleroderma, Geastrum, Cantharellus*. J Mycopathol Res 48:337–342
Sharma R, Rajak RC, Pandey AK (2010c) Evidence of antagonistic interactions between rhizosphere and mycorrhizal fungi associated with *Dendrocalamus strictus* (Bamboo). J Yeast Fungal Res 1:112–117
Sharma R, Rajak RC, Pandey AK (2010d) Some ectomycorrhizal mushrooms of Central India–III. *Amanita*. J Mycopathol Res 48:81–84
Sharma R, Rajak RC, Pandey AK (2010e) Some ectomycorrhizal mushrooms of Central India-IV. *Boletus, Leccinum*. J Mycopathol Res 48:329–335
Sim M-Y, Eom A-H (2006) Effects of ectomycorrhizal fungi on growth of seedlings of *Pinus densiflora*. Mycobiology 34:191–195
Simard SW, Durall D, Jones M (2002) Carbon and nutrient fluxes within and between mycorrhizal plants. In: van der Heijden MGA, Sanders IR (eds) Mycorrhizal ecology. Springer, Berlin, Heidelberg
Skinner MF, Bowen GD (1974a) The uptake and translocation of phosphate by mycelial strands of pine mycorrhizas. Soil Biol Biochem 6:53–56
Skinner MF, Bowen GD (1974b) The penetration of soil by mycelial strands of ectomycorrhizal fungi. Soil Biol Biochem 6:57–61
Smith FA (2000) Measuring the influence of mycorrhizas. New Phytol 148:4–6
Smith SE, Read DJ (1997) Mycorrhizal symbiosis, 2nd edn. Academic Press, London
Smith SE, Gianinazzi-Pearson V, Koide R, Cairney JWG (1994) Nutrient transport in mycorrhizas: structure, physiology and consequences for efficiency of the symbiosis. Plant Soil 159: 103–113
Smith FA, Timonen S, Smith SE (2000) In: Blom WPM, Visser EJW (eds) Mycorrhizas. Springer, Berlin, Heidelberg, New York
Sterkenburg E (2016) Drivers of soil fungal communities in boreal forests–feedbacks on soil fertility and decomposition. Doctoral Thesis, Swedish University of Agricultural Sciences, Uppsala
Tedersoo L, May TW, Smith ME (2010) Ectomycorrhizal lifestyle in fungi: global diversity, distribution, and evolution of phylogenetic lineages. Mycorrhiza 20:217–263
Tedersoo L, Liiv I, Kivistik PA, Anslan S, Kõljalg U, Bahram M (2016) Genomics and metagenomics technologies to recover ribosomal DNA and single-copy genes from old fruit-body and ectomycorrhiza specimens. MycoKeys 13:1–20
Thomas KA, Peintner U, Moser MM, Manimohan P (2002) *Anamika*, a new mycorrhizal genus of *Cortinariaceae* from India and its phylogenetic position based on ITS and LSU sequences. Mycol Res 106:245–251
van der Heijden MGA (2002) Arbuscular mycorrhizal fungi as a determinant of plant diversity: in search of underlying mechanisms and general principles. In: van der Heifden MGA, Sanders IR (eds) Mycorrhizal ecology. Springer, Berlin, Heilderberg
van der Heijden MGA, Sanders IR (2002) Mycorrhizal ecology. Springer, Berlin

Verbeken A, Buyck B (2001) Diversity and ecology of tropical ectomycorrhizal fungi of Africa. In: Watling R, Frankland JC, Ainsworth AM, Isaac S, Robinson CH (eds) Tropical mycology, vol I. CABI Publishing, UK, pp 11–24

Verma RN, Singh GB, Mukta S (1995) Mushroom flora of north-eastern hills. In: Advances in horticulture-13, mushroom. Malhotra Publishing House, New Delhi, pp 329–349

Voke NR (2012) The effect of roots and ectomycorrhizal fungi on carbon cycling in forest soils. The University of York, York

Walker JKM, Cohen H, Higgins LM, Kennedy PG (2014) Testing the link between community structure and function for ectomycorrhizal fungi involved in a global tripartite symbiosis. New Phytol 202:287–296

Wang B, Qiu YL (2006) Phylogenetic distribution and evolution of mycorrhizas in land plants. Mycorrhiza 16:299–363

Watling R, Lee SS (1995) Ectomycorrhizal fungi associated with members of the *Dipterocarpaceae* in Peninsular Malaysia. J Trop For Sci 7:657–669

Wiemken V (2007) Trehalose synthesis in ectomycorrhizas—a driving force of carbon gain for fungi. New Phytol 174:228–230

Wu B, Nara K, Hogetsu T (2005) Genetic structure of *Cenococcum geophilum* populations in primary successional volcanic deserts on Mount Fuji as revealed by microsatellite markers. New Phytol 165:285–293

Zhou Z, Miwa M, Hogetsu T (2000) Genet distribution of ectomycorrhizal fungus *Suillus grevillei* populations in two *Larix kaempferi* stands over two years. J Plant Res 113:365–374

Zhou Z, Miwa M, Hogetsu T (2001) Polymorphism of simple sequence repeats reveals gene flow within and between ectomycorrhizal *Suillus grevillei* populations. New Phytol 149:339–348

Chapter 8
Plant Flavonoids: Key Players in Signaling, Establishment, and Regulation of Rhizobial and Mycorrhizal Endosymbioses

Priyanka Singla and Neera Garg

Abstract Plants belonging to family fabaceae play an imperative role in restoring soil fertility, with the remarkable ability to engage endosymbiotically with both rhizobia and arbuscular mycorrhiza (AM). Establishment of both symbioses is based on a finely regulated molecular dialogue between two partners. Plant roots secrete an assortment of flavonoids, competent to shape rhizosphere microflora by amplifying chemotactic surface motility of beneficial microorganisms while combating pathogenic ones. Flavonoids potentially regulate transcriptional activity of many microbial genes, e.g. *nod* genes, and fungal hyphal branching and initiate the production of microsymbiont signal molecules (Nod/Myc factor). The perception of these lipo-chito-oligosaccharides at epidermis stimulates partly analogous downstream signal transduction cascade to activate symbiosis-related genes and consequently enable successful penetration of both microsymbionts in the host. In response to host-specific microbe, selective accumulation of flavonoids drives suppression of plant innate immunity as well as cortical cell dedifferentiation into symbiosome. High degree of coordination between root cortical cell machinery and rhizobia/AM results in the formation of symbiotic interfaces—nodules/arbuscules respectively, where harboring bacteroids and arbuscules deliver macronutrients (nitrogen and phosphorus) to host in exchange for photosynthates. Flavonoids cross-link with plant proteins to form an O_2- diffusion barrier in the symbiosome membrane and serve as a checkpoint for nitrogenase efficiency. Under nutrient-rich conditions, plants regulate flavonoid fluxes to prevent an excessive establishment of metabolically expensive symbioses. Therefore, understanding these selective forces that govern host selection of beneficial rhizomicrobiome, followed by underlying establishment and regulation of symbioses in legumes, is crucial for agrobiologists to achieve sustainable agriculture.

P. Singla • N. Garg (✉)
Department of Botany, Panjab University, Chandigarh 160014, India
e-mail: gargneera@gmail.com

A. Varma et al. (eds.), *Mycorrhiza - Function, Diversity, State of the Art*,
DOI 10.1007/978-3-319-53064-2_8

8.1 Introduction

Among plant nutrients, nitrogen (N) and phosphorus (P) are the most limiting nutrients worldwide whose continued supply as fertilizers is necessary if world food needs are to be fulfilled. Modern agriculture has been highly reliant on industrial fertilizers, and the vast amount of resources spent on their production is leading to a substantial carbon footprint (exploiting ~50% of fossil fuel) of this industrial sector (Jensen et al. 2012). However, the rising fossil fuel cost is making chemical fertilizers dramatically expensive; CO_2 emission during fossil fuel combustion is contributing to the greenhouse effect, and leaching of applied fertilizers (~30–50%) is leading to major environmental problems, e.g. eutrophication (Crutzen et al. 2007). Thus, efforts must be directed to enhance the exploitation of rhizospheric soil microorganisms which can effectively ensure substantial uptake of essential nutrients by plants and thus recuperate agricultural fields with nutrient-poor soils (Bonfante and Genre 2008; Parniske 2008; Vieira et al. 2010). Two of the most widespread rhizospheric interactions, the *Rhizobium*–legume (RL) and the arbuscular mycorrhiza (AM)–plant symbioses, have particular significance as natural mini-fertilizer factories in land ecosystems (Venturi and Keel 2016).

Rhizobium–legume (RL) symbiosis is almost entirely limited to economically important family Fabaceae which evolved only about 60 Mya (Manchanda and Garg 2007; Delaux et al. 2015). On the other hand, since 500 million years ago (Mya), ubiquitous soil AM fungi establish a symbiotic relationship with the roots of more than 90% of all higher plants (Smith and Read 2008; Wilde et al. 2009). RL interaction leads to the formation of symbiotic root nodules, where *Rhizobium* bacteria acquires vast fitness output from host carbon and energy (Peix et al. 2015) and legumes gain access to otherwise unavailable soil nitrogen (replenishing approximately 200 million tons of N_2 annually) (Ferguson et al. 2010; Kondorosi et al. 2013). In mycorrhizal association, vegetative growth and reproductive spore production of biotropic fungal symbionts rely on reduced carbon of living plant tissue (4–20% of their photosynthate) and in return provide various benefits, including—but not limited to—nutrient and water uptake (Parniske 2008).

Legumes have the ability to host N_2-fixing bacteria and AM at the same time (Antunes et al. 2006), and mycorrhizal symbiosis associated with legumes is an essential link for effective phosphorus (P) nutrition leading to enhanced N_2 fixation that advocates a synergistic tripartite association (Geneva et al. 2006). In most cases investigated, especially when both nitrogen and phosphate are limiting factors, rhizobia and AM fungi appear to act synergistically since combined inoculation with rhizobia and mycorrhiza enhances plant growth and reproduction more than inoculation with either microsymbiont alone and also mutually increases each other's establishment (Gould and Lister 2005). Striking similarities between two symbionts have been reported with respect to mutual recognition, infection process, and genetic and hormonal regulation (Mukherjee and Ané 2011). Exchange of molecular signals between the host plant and microsymbiont in the form of cell-to-cell inter-organismal communication is an important stage for the initiation of effective symbiosis (Mapope and Dakora 2013), and during N_2 fixation, a wide variety of host FLAVONOIDS have been shown to attract compatible rhizobia for symbiosis, by their *nod* gene inducing activity (Shaw et al. 2006; Mandal et al. 2010). Apart from the function of flavonoids in RL interaction, they also act as signaling compounds in the host communication with AM (Catford et al. 2006; Steinkellner et al. 2007; Shaw and Hooker 2008). Moreover, plant flavonoids vary qualitatively and quantitatively with root endosymbiont colonization, indicating dependability of nodulation and mycorrhization on flavonoids beyond signaling (Carlsen et al. 2008; Zhuang et al. 2013). Thus, this prompts in-depth review on diverse role of flavonoids in the establishment of legume symbioses with rhizobia and AM.

8.2 Flavonoid: A Versatile Compound

8.2.1 Structural Diversity and Subcellular Distribution

Flavonoids (derived from the Latin word for yellow, favus), the low-molecular-weight secondary metabolites, are biologically active polyphenolic compounds and are widespread throughout the plant kingdom, ranging from mosses to angiosperms (Williams and Grayer 2004). Diverse flavonoid molecules share the same core carbon framework of phenyl-benzopyran functionality, the flavan nucleus (Fig. 8.1), consisting of two aromatic rings with six carbon atoms (rings A and B) interconnected by a heterocyclic benzopyrano (chromano) C ring with three carbon atoms (Saito et al. 2013; Cheng et al. 2014). The position of B aromatic ring linkage to the benzopyrano moiety allows a broad separation of these compounds into flavonoids (2-phenyl-benzopyrans, e.g., kaempferol, apigenin), isoflavonoids (3-phenyl-benzopyrans, e.g., genistein), and neoflavonoids (4-phenyl-benzopyrans) (Winkel-Shirley 2001; Dixon and Pasinetti 2010). These groups usually share a common chalcone precursor and therefore are biogenetically and structurally related (Marais et al. 2006). A number of divergent chalcones and flavonoid structures are formed from the extensive modification (rearrangement, alkylation, oxidation, and glycosylation) of the basic molecules (Halbwirth 2010). To date, more than 10,000 different flavonoids have been identified in plants (broadly

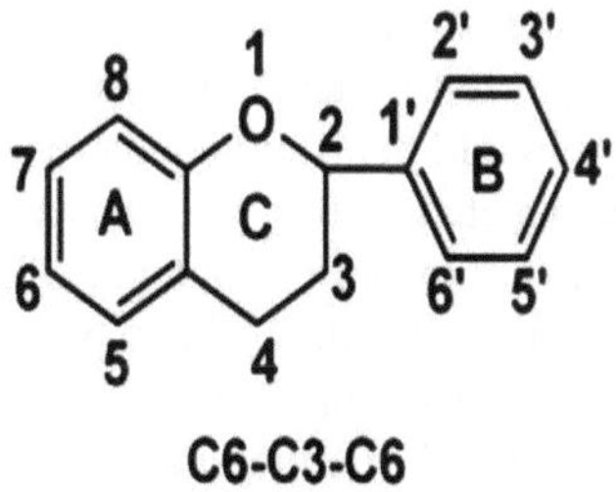

Fig. 8.1 Flavan nucleus of flavonoids

classified into flavonols, flavones, flavan-3-ols, flavanones, anthocyanins, isoflavones) and the number is still increasing (D'haeseleer et al. 2010; Hassan and Mathesius 2012; Cheynier et al. 2013) and even within the same species a number of different flavonoids have been identified (Martens and Mithöfer 2005). Typically, flavonoids are stored in their glycoside forms (more stable than the free form); however, some flavones and flavonols can be found naturally as the aglycone (Birt and Jeffery 2013). The chemical diversity, size, three-dimensional shape, and physical and biochemical properties of flavonoids allow them to interact with targets in different subcellular locations to influence biological activity in plants, animals, and microbes (Taylor and Grotewold 2005; Buer et al. 2010).

Consistent with their diverse physiological functions, flavonoids are found in most plant cell compartments, including cytosol, vacuole (anthocyanins, flavonol and flavone glycosides), ER, chloroplast (quercetin and kaempferol glycosides), nucleus (isoflavonoids coumestrol and 4′,7-dihydroxyflavone), pollen surface and small vesicles, as well as the extracellular space (Saslowsky et al. 2005; Lepiniec et al. 2006; Hsieh and Huang 2007; Naoumkina et al. 2007; Hernández et al. 2009). In the plant, flavonoids function as developmental regulators, antioxidants, pigments, UV sunscreens, nutrient acquisitor, energy escape valve, auxin transport regulators, defense compounds against pathogens (signal for jasmonate-induced mobilization of vacuolar isoflavonoid glucosides for phytoalexin biosynthesis), and signals during symbiosis (Buer and Muday 2004; Naoumkina and Dixon 2008; Cheynier et al. 2013). However, their importance in plant biology goes beyond their specific functions within the plant (Dixon and Pasinetti 2010). Flavone and flavonolaglycones have been detected in root exudates from numerous species, where isoflavones are particularly secreted by legume roots into the rhizosphere (Zhao and Dixon 2010).

8.2.2 Flavonoids in the Rhizosphere

In response to elicitors, both aglycone and glycoside flavonoids are often exuded into the rhizosphere in order to fulfill some of their ecological roles as mediators of belowground interactions, for example, in order to attract compatible rhizosphere-dwelling rhizobia, stimulate or inhibit rhizobial *nod* gene expression, inhibit root pathogens, stimulate mycorrhizal spore germination and hyphal branching, affect quorum sensing, and chelate soil nutrients (Broughton et al. 2003; Cooper 2004; Martens and Mithöfer 2005). Exudation of flavonoids is ATP dependent and catalyzed by an ABC-type transporter as has been suggested by the study of isoflavonoid genistein exudation from soybean root plasma membrane vesicles (Sugiyama et al. 2007). This was also supported by the study of Badri et al. (2009) where ABC transporter mutants of *Arabidopsis* like *abcg30* had altered root exudate profiles. Besides ABC-type transpoters, flavonoids can also be released passively in the rhizosphere from decomposing root cap and border cells (Shaw et al. 2006). Another important mechanism for releasing active flavonoid aglycones during root–microbe interactions is through apoplastic β-glucosidases, which release isoflavones from their conjugates in soybean roots (Suzuki et al. 2006). Recently, Sugiyama et al. (2016) reported that the expression of gene encoding isoflavone conjugates hydrolyzing beta-glucosidase (ICHG) coordinately peaked at vegetative stages with the higher secretion of daidzein. However, under nitrogen-deficient conditions, besides daidzein, genistein was also highly secreted, with no induction of ICHG. Thus, their study suggested that two pathways for isoflavone secretion in soybean roots are expected to have distinct physiological roles, i.e., ICHG-mediated secretion during vegetative growth and ATP-dependent transport during nitrogen deficiency.

Once in the rhizosphere, the fate of flavonoid persistence in the rhizosphere (likely to be from hours to days) varies with environmental conditions, flavonoid structure, and the presence of soil microbes, some of which can metabolize or modify flavonoids. Flavonoids can be absorbed to the cell wall and to soil particles with cationic binding sites, thus becoming unavailable (Shaw and Hooker 2008). Depending on their structural modifications, the solubility and mobility of flavonoids in the soil varies. For example, genistein is described as "practically insoluble in water," whereas genistin, the glucoside, is "sparingly soluble in water" (O'Neil et al. 1996). Thus, conjugated forms are expected to be less adsorbed to the soil matrix, more mobile, and therefore more bioavailable than the free aglycone form. While glycosylation improves their solubility in water, it is likely that flavonoid glycosides are quickly deglycosylated by microorganisms and plant exoenzymes, leaving the more hydrophobic aglycone (Sosa et al. 2010; Weston and Mathesius 2013). Once present in the aglycone form, new flavonoid structures may be produced during biodegradation of a parent flavonoid, which can be more efficient inducers of *nod* genes than the flavonoids themselves (Anders and Huber 2010; Rose et al. 2012). Moderate bioavailablity and mobility of flavonoids (organic carbon normalized partition coefficients, i.e., logKoc of 3.12 for formononetin and 3.19 for naringenin), makes ecological sense, as it would be evolutionarily favorable for a plant root to produce a chemical signal that is sufficiently bioavailable to allow interaction with its intended microbial target, but, at the same time, not so mobile that it will diffuse rapidly to the outer reaches of the rhizosphere. Thus, bioavailability, mobility and persistence of a flavonoid will determine the degree and outcome of flavonoid–microbe interaction with important consequences for plant nutrition (Shaw and Hooker 2008; Weston and Mathesius 2013).

8.3 Flavonoids in *Rhizobium*–Legume (RL) Endosymbiosis

Among the wide range of bacteria that have the ability to reduce N_2 to ammonia (NH_3), the most important are soil-dwelling prokaryotic bacteria collectively called rhizobia, where each legume species has its own cognate *Rhizobium* partner (s) (Gibson et al. 2008; Soyano and Kawaguchi 2014). Legume genomes are at least 50 times larger than those of their microsymbionts; nevertheless, their respective contributions are probably not vastly different (Irving et al. 2000). Thus, RL symbioses are cross-kingdom collaboration between two vastly different genomes (Yang et al. 2010). The endosymbiosis generally commences within a specific "susceptible" root zone close to the root tip, where initial bacteria–host recognition takes place (Laloum et al. 2014). The bacteria invade the roots of compatible legume plants leading to the development of specialized root structures called nodules, providing a unique ecological niche in which bacteria differentiate into bacteroids and fix N_2. Although the interaction is beneficial to both partners, it comes with rigid rules that are strictly enforced by both the partners (Oldroyd et al. 2011). Here, we review the various signaling pathways by which the plant allows bacterial infection and promotes the construction of the nodule as well as how intricate transaction of metabolites (especially flavonoids) directs the bacteria into a nitrogen-fixing organelle-like state.

8.3.1 Signaling in Rhizobium*–Legume Symbiosis*

The succession to the symbiotic affirmation by two originally autonomous, free-living partners is governed by reciprocal generation and perception of signals, which has been described as "molecular dialogue" (De'narie et al. 1993). At least three different sets of symbiotic signals are exchanged between legumes and rhizobia during nodule development: flavonoids, Nod factors, EPS, and extra-cellular proteins (NOPs = *n*odulation *o*uter *p*roteins) (Broughton et al. 2003).

8.3.1.1 Signals from the Host Plants

A diverse array of compounds is exuded into the rhizospere, including sugars, aliphatic as well as aromatic acids, amino acids, amines, and many other low-molecular-weight compounds such as flavonoids, isoflavonoids, steroids, alkaloids, vitamins, and growth regulators (Skorupska et al. 2010; Lareen et al. 2016). The "rhizosphere effect," first described by Hiltner (1904), assumes that many micro-organisms are attracted to sugars, acids, and amino acids exuded by plant roots, which serve as C and energy sources for microorganisms. However, in addition to providing a C-rich environment, plant roots initiate cross talk with soil microbes by producing signals that are recognized by the microbes, which in turn produce signals that initiate colonization (Bais et al. 2004). Among the myriad of rhizodepositions, normally and continuously exuded by plants into the rhizosphere are phenolic compounds especially flavonoids that mediate signal traffic between roots and beneficial rhizosphere-dwelling rhizobia (Gibson et al. 2008; Skorupska et al. 2010; McNear 2013; Mandal et al. 2016). Although small quantities are excreted continuously, flavonoid concentrations in the rhizosphere increase in response to compatible rhizobia. Niches in the legume rhizosphere are tailored to rhizobial inoculation, as exudation of flavonoids is mostly restricted to the elongating root hair zone from which most nodules later develop (Zuanazzi et al. 1998). The role of flavonoids as bacterial chemoattractant and transcriptional activator of bacterial *nod* genes points out their central position in modifying rhizobial phenotypes possibly in relation to plant–root association and then symbiotic interaction (Brencic and Winans 2005; Spini et al. 2016). Once exuded, flavonoids, especially aglycone forms, are recognized to diffuse into the rhizobial membrane (Kobayashi et al. 2004; Wang et al. 2012), possibly through porins (Taylor and Grotewold 2005). Further, it activates rhizobia *nod* (nodulation) D gene expression (Ferguson et al. 2010), and this successful liaison between flavonoids and NodD proteins signals the beginning of association (Broughton et al. 2000). The first flavonoid to be discovered to act as *nod* gene inducers was luteolin, isolated from *Medicago sativa* and 7,4′ dihydroxyflavone (DHF) from *Trifolium repens* (white clover) (Redmond et al. 1986). Since then, about 30 *nod* gene-inducing flavonoids (especially flavanones and isoflavonoids) have been isolated from nine legume genera (Limpens and Bisseling 2008; Gholami et al. 2014). Most of these flavonoids are active as *nod* gene inducers at nanomolar to low micromolar concentrations (Begum et al. 2001a; Gholami et al. 2014) and stimulate bacterial *nod* gene expression within minutes. The specific exudation of flavonoid (mixtures) from legume hosts together with the specific perception of flavonoids by NodD proteins of different rhizobia is partially responsible for the host specificity of the symbiosis (Gibson et al. 2008; Skorupska et al. 2010; Rose et al. 2012). Hence, point mutations in *nodD* affect recognition of inducing flavonoids and cause extension of host range (Broughton et al. 2000). Generally, the NodDs of broad host range rhizobia, e.g., NGR234, respond to a wider range of flavonoid species (including phenolics that are inhibitors in other rhizobia, e.g., vanillin, *iso*-vanillin, as well as several estrogenic compounds) than those present in restricted host range rhizobia (Peck et al. 2006; Wang et al. 2012). *nodD* products of various *Rhizobium* species respond in different ways to flavonoids, and NodD homologues from the same strain may have various flavonoid preferences. In *R. meliloti*, NodD1 was activated when cells were supplied with a complex plant seed extract or the flavonoid luteolin. NodD2 only derepressed transcription when supplied with the complex extract, not with purified luteolin, while NodD3 apparently modulates the expression of *nod* genes even in the absence of any plant factor (Smit et al. 1992). Recently, Peck et al. (2013) presented a structural model of wild-type NodD1 identifying residues important for inducer binding, protein multimerization, and interaction with RNA polymerase at *nod* gene promoters. Species-specific flavonoids interact with a class of transcriptional activators of the LysR family which have an N-terminal ligand-binding domain that regulates the activity of the associated C-terminal DNA-binding domain, and NodD ligand-binding domain is thought to function as a flavonoid receptor. The perception of flavonoids by rhizobia is linked to elevation in concentrations of intracellular calcium in rhizobia that subsequently induces NodD proteins for Nod factor expression (Moscatiello et al. 2010; Hassan and Mathesius 2012).

Of the large number of available flavonones, flavones, isoflavones, and other related compounds, capability of flavonoid to act as *nod* gene inducers varies with the variation in host varieties, bacterial strains, and/or signal compounds (Begum et al. 2001a, b). Within the variety of flavonoids, isoflavonoids, and other compounds secreted by *Lupinus albus* roots into the rhizosphere, major proportion is composed of aldonic acids which act as natural *nod* gene inducers of *Rhizobium lupini*, *Mesorhizobium loti*, and *Sinorhizobium meliloti* (Gagnon and Ibrahim 1998). Luteolin, genistein, naringenin, hesperetin, and apigenin are the principal flavonoids involved in *nod* gene expression in *S. meliloti*, *Bradyrhizobium japonicum*, *Rhizobium*, and *R. tibeticum*, respectively (Kapulnik et al. 1987; Graham 1991; Begum et al. 2001b; Belkheir et al. 2001; Novák et al. 2002; Tsvetkova et al. 2006; Brechenmacher et al. 2010; Abd-Alla et al. 2014). Genistein, coumestrol, and daidzein have been reported as important inducers of rhizobial nodulation genes in the early stages of symbiosis between soybean and *B. japonicum* (Antunes and Goss 2005; Miransari and Smith 2009; Tian et al. 2014; Sugiyama et al. 2016). Begum et al. (2001a) suggested that the attachment of the B-ring to C-2 of flavonoids, as found in flavones and flavanones, is of crucial importance for induction. In this regard, Zhang et al. (2007) provided genetic evidence that RNA interference-mediated suppression of *MtFNSII* genes in *Medicago truncatula* resulted in flavone-depleted roots and led to significantly reduced nodulation when inoculated with *S. meliloti*. In addition, hydroxylation at the C-4 and the C-7 positions of flavones is important for this activity (Brencic and Winans 2005; Subramanian et al. 2007).

Interestingly, some flavonoids also show *nod* gene repressing activity for certain rhizobia. Isoflavonoids, medicarpin and coumestrol, have been reported to negatively control Nod factor production in *S. meliloti* (Zuanazzi et al. 1998). Jain and Nainawatee (1999) studied that except quercetin, alfalfa exudates decreased growth and protein content of *R. meliloti* cells. Naringenin induces the expression of *nod* genes in *R. leguminosarum–Pisum sativum*; however, quercetin is an inhibitor of nodulation (Novák et al. 2002). Inducers in one species or strain of *Rhizobium* are frequently anti-inducers in another species; thus, one type of flavonoid can have opposing effects on different bacteria; for example, the isoflavone diadzein induces *nod* gene expression in *B. japonicum* (nodulating soybean), whereas it inhibits that of those from *R. leguminosarum* (nodulating clover or peas) and thus contributes to host specificity (Andersen and Markham 2006). It has been suggested that a mixture of flavonoids is more effective in inducing *nod* genes as opposed to a single compound (Mandal et al. 2010). Both functions (induction or anti-induction) can co-occur in the exudates of the same plant, where different flavonoids in root exudates can act synergistically as nod gene inducers, but also in an antagonistic manner as anti-inducers (Cooper 2007; Makarova et al. 2015). Luteolin and 7,40-dihydroxyflavone are inducers, whereas genistein inhibits expression of *S. meliloti nod* genes (Kosslak et al. 1987; Hartwig et al. 1990; Peck et al. 2006). Thus, the ratio of inducers to anti-inducers in root exudates may be involved in determination of host recognition, maintaining an optimal level of Nod factor production and preventing elicitation of defense responses by the plant (Zuanazzi et al. 1998; Cesco et al. 2010). Li et al. (2016) provided a previously unidentified mechanism by which flavonoids in exudates of one crop root can promote N_2 fixation in another crop in a two-crop intercropping system. In their study, maize root exudates contained significant flavonoids and promoted flavonoid synthesis in faba bean, thus triggering N_2 fixation.

8.3.1.2 Signals from the Microsymbiont

Second set of signals is synthesized when cytoplasmic membrane-bound NodD–flavonoid complexes activate transcription from conserved 49-bp DNA motifs, i.e., "*nod*-box" promoters found in the promoter regions of many nodulation loci (Gibson et al. 2008). Thus, NodD proteins act as both sensors of the plant signal and transcriptional activators of *nod* loci on symbiotic plasmids (Redmond et al. 1986; Downie 2010). The concerted transcriptional activation of common and host-specific *nod* genes leads to the synthesis of Nod factor (NF) which is a key to legume doors (Mulder et al. 2005; Cooper 2007; Remigi et al. 2016). NFs belong to lipo-chito-oligosaccharides (LCOs) family, having an oligosaccharide backbone of four or five β-1-4-linked *N*-acetyl-D-glucosamine units with a terminal nonreducing sugar *N*-acylated by a 16–18 carbon fatty acid (Fauvart and Michiels 2008; Hamel and Beaudoin 2010). Assembly of the chitin backbone is performed by an *N*-acetylglucosaminyltransferase encoded by *nodC*, and the deacetylase NodB removes the *N*-acetyl moiety from the nonreducing terminus of the *N*-acetyl-glucosamine oligosaccharides. NodG has the enzymatic activity of a 3-oxoacyl-acyl carrier protein reductase and is involved in fatty acid elongation (López-Lara and Geiger 2001). Finally, an acyltransferase coded by *nodA* links the acyl chain to the acetyl-free carbon C-2 of the nonreducing end of the oligosaccharide (Brencic and Winans 2005). In addition to the common *nod-DABC* genes that are essential for symbiosis, the rhizobia harbor different combinations of other *nod*, *nol*, and *noe* genes which may have been recruited from paralogues in the course of the evolution, allowing the diversification of NF structures and host ranges (Taurian et al. 2008; Vieira et al. 2010). For instance, *nod*V and *nod*W of *B. japonicum* are essential for the nodulation of *Macroptilium atropurpureum*, *Vigna radiata*, and *V. unguiculata* but contribute only marginally to the symbiosis with *G. max*. *nodH* encodes a sulfotransferase that transfers a sulfate group to the reducing end of NFs of *R. meliloti* and elicits Ca^{2+} spiking (Wais et al. 2002). A *nodF* mutant produces a NF with a modified *N*-acylation on the terminal nonreducing *N*-acetyl glucosamine residue (Demont et al. 1993). NF produced by *nodL* mutants lacks a C6-*O*-acetylation on the terminal nonreducing glucosamine (Ardourel et al. 1995). *nodFL* double mutants trigger Ca^{2+} spiking and root hair deformation but are unable to infect their hosts (Wais et al. 2002; Haney et al. 2011). Thus, a major determinant of host-symbiont specificity is attributed to the different NF substituents (sulphuryl, methyl, carbamoyl, acetyl, fucosyl, arabinosyl, and other groups) attached to the oligosaccharide backbone as well as differences in the structure of the acyl chain (Downie 2010; Ferguson et al. 2010; Kouchi et al. 2010). Nod factors can trigger plant responses like root hair deformation and calcium oscillations (called calcium spiking) at astonishingly low concentrations, i.e., as little as 10^{-13} M Nod factor

(Rose et al. 2012). As Nod factors also stimulate the synthesis and release of flavonoids from legume roots, the response to inoculation is amplified (Broughton et al. 2003).

8.3.2 Nod Factor (NF) Signaling Pathway in the Root Epidermis

Symbiosis initiates if the above-stated chemical cross talk between the interacting partners successfully culminates in the production of NFs (Bek et al. 2010). The mechanism, by which plants regulate the intracellular uptake of symbiotic bacteria, depends on physiological and molecular reprogramming that is associated with the perception of NF and resultant downstream signaling (Xie et al. 2012; Liang et al. 2013). Rhizobia have two main ways of entering the plant root: *via* the root hair or through cracks in root epidermal tissue (Oldroyd and Downie 2008; Ribeiro et al. 2015); however, rhizobial entry along the infection threads in root hairs is largely common (Mathesius 2009; Downie 2010). Two receptor-like kinases (RLK) of chitin-binding LysM RLK family, located on epidermal cells, are involved in nod factor binding: LjNFR1 and LjNFR5 in *L. japonicus*, PsSYM2A and PsSYM10 in *P. sativum*, MtLYK3/MtLYK4 and MtNFP in *M. truncatula*, and GmNFR1α/β and GmNFR5α/β in soybean (Ferguson et al. 2010; Indrasumunar et al. 2010; Broghammer et al. 2012; Wang et al. 2012). These receptors consist of an intracellular kinase domain, a transmembrane domain, and an extracellular portion having LysM domains (Gough 2003; Mathesius 2009). Interestingly, LjNFR1/PsSYM2A/MtLYK3/MtLYK4/GmNFR1α/β has a typical serine/threonine kinase domain, while LjNFR5/PsSYM10/MtNFP/GmNFR5α/β lacks the activation loop (Indrasumunar et al. 2010). The absence of an activation loop in one of the kinase domains suggests that the two LysM RLKs may assemble into a heterodimeric receptor, with the active kinase (NFR kinase) domain triggering downstream signal transduction through phosphorylation (Markmann et al. 2008; Radutoiu et al. 2008; Hamel and Beaudoin 2010; Indrasumunar et al. 2015).

Rhizobial infection has many similarities with pathogenic infection and induction of defense responses accompanies both interactions, but defense responses are induced to a lesser extent during rhizobial infection. Recently, it was evidenced by Ivanova et al. (2015) that a range of plant defense responses like suberization, callose and unesterified pectin deposition, as well as activation of defense genes can be triggered by different single mutations in symbiotic genes (*sym33, sym40, sym42*) that cause perception of an otherwise beneficial strain of *Rhizobium* as a pathogen. Besides symbiosis, LysM-RLKs have a role in immune signaling, indicating that NF signaling and pathogen chitin-based immune signaling are intertwined. However, Nod factor signal, unlike microbe-associated molecular pattern (MAMP) derived from microbes, suppresses an innate immune response in the host (Tóth and Stacey 2015). Two putative models have been put forward to explain this evolutionarily conserved dual function: (a) perception of LCOs factors modulates the balance between different LysM-RK receptor complexes, favoring a symbiotic complex at the expense of complexes required for immune responses, or (b) tight regulation of the receptor complexes at the posttranslational level, involving rapid endocytotic turnover, subsequently prevents activation of defense responses (Limpens et al. 2015). NF perception leads to root hair deformation and to changes in the root hair cytoskeleton (within 3–5 min) that are required for root hair curling (the so-called Shepherd's crooks within 1–3 h) and invasion (Weerasinghe et al. 2005; Yokota et al. 2009). Mutations in genes coding for the NF LRR RLK (Leucine-rich repeat receptor-like kinases), the putative ion channels, or the nucleoporins abolish Ca^{2+} spiking and continued nodule development events; however, they maintain the Ca^{2+} fluxes and root hair deformation events (Kanamori et al. 2006; Miwa et al. 2006; Saito et al. 2007; Capoen et al. 2011; Morieri et al. 2013). In contrast, mutations in genes encoding for a calcium and calmodulin-dependent kinase called CCaMK or DMI3 do not affect Ca^{2+} fluxes and Ca^{2+} spiking events but block continued nodule development (Lévy et al. 2004; Miwa et al. 2006). This suggests that the NF LRR RLK, the ion channels, and the nucleoporins act downstream of NF perception, but upstream of Ca^{2+} spiking, whereas the CCaMK acts downstream of Ca^{2+} spiking (Limpens and Bisseling 2008). Within the nucleus, this sustained calcium spiking is decoded by CCaMK, which phosphorylates a transcriptional regulator CYCLOPS/IPD3 (Yano et al. 2008; Kouchi et al. 2010; Singh et al. 2014). CYCLOPS, together with other TFs belonging to the GRAS (NSP1 and NSP2) (Smit et al. 2005; Oldroyd and Downie 2008), ERF (ERN1) (Cerri et al. 2012; Rose et al. 2012), and the nodule inception (NIN) activator (Marsh et al. 2007) families, modulates early symbiotic gene expression (like *ENOD2*, *ENOD40*, *ENOD11*) for infection thread (IT) formation and polar tip growth (Yano et al. 2008; Madsen et al. 2010). Downstream of DMI3 and NIN, members of Nuclear Factor Y family, i.e., NF-YA1 and NF-YA2 (a CCAAT-box-binding heterotrimeric TF complex), act as early symbiotic regulators of ENOD11 (Cerri et al. 2012; Laloum et al. 2014). The secondary induction of Nod signals by flavonoids inside the roots is thought to be responsible for an additional level of host specificity. Thus, flavonoids play a multitude of roles during the process of nodulation (Subramanian et al. 2007).

8.3.3 Formation of Nodules: The Conjugal Lodging

NF perception in the epidermis activates a series of events, including polarized root hair tip growth, invagination associated with bacterial infection, and the promotion of cell division in the cortex leading to the nodule meristem (Oldroyd et al. 2011; Gourion et al. 2015; Laplaze et al. 2015). Cytoskeletal rearrangements have been reported in pericycle cells of *M. truncatula* within just 16 h of rhizobia inoculation (Timmers et al. 1999), and *ENOD40* expression is reported in cortical cells within just 24 h of rhizobia inoculation (Mulder et al. 2005; Murray 2011). Thus, coordination between epidermal and nodule organogenesis seems to be crucial for successful nitrogen-fixing nodule formation (Oldroyd and Downie 2008). Although initial bacterial infection events do not require nodule primordia formation, it is subsequently required to direct infection thread growth (Oldroyd et al. 2011; Rose et al. 2012). To achieve such rapid mitotic activity in the underlying cortical cells after exposing the outer root to rhizobia/NF, role of auxin and cytokinin is imperative (Ryu et al. 2012).

8.3.3.1 Rhizobial Invasion

The host plant permits rhizobium to enter root tissues through plasma membrane-derived conduits called infection threads (ITs) (Jones et al. 2007; Fournier et al. 2008). The new growth can result in the root hair curling/bending, which results in NF-producing bacteria becoming entrapped between appressed cell walls, forming so-called infection pocket (Murray 2011; Wang et al. 2012). The rhizobia entrapped in these infection pockets continue to divide, forming colonies that are referred to as infection foci from which root hair ITs start to develop (Oldroyd et al. 2011). This results in elevated Nod factor concentrations, which are thought to be required to reach a Nod factor threshold concentration (Oldroyd and Downie 2008). The invaginating plant cell wall, along with the extended plasma membrane, grows as a hollow tube within the root hair cell and the bacteria multiply within the polar centripetally growing infection threads, where new cell wall and membrane material are being synthesized at their tip (Xie et al. 2012; Haag et al. 2013).

A third set of signals are represented by other rhizobial products necessary for continued infection thread development and/or preventing defense mechanisms (López-Baena et al. 2016). Among them are extracellular lipopolysaccharides (LPS), extracellular polysaccharides (EPS) and related compounds, as well as proteins exported by the type III secretion system (T3SS) (Limpens and Bisseling 2008; Haag et al. 2013). EPS facilitate attachment of bacterial cells to both biotic and abiotic surfaces and biofilm formation due to hydrophobic interactions and heterogeneity of the envelope surface (Janczarek et al. 2015), affect different stages of the organogenesis of nodules (Kelly et al. 2012), and along with LPS also facilitate suppression of defense response (Dalla Via et al. 2016). Further, bacterial effector proteins delivered from rhizobia into the plant cytosol through a T3SS or T4SS can act to either negatively or positively modulate nodulation, i.e., alter the symbiotic state toward pathogenesis or vice versa (Nelson and Sadowsky 2015; Tóth and Stacey 2015). The suppression of the MAMP-triggered immunity through a T3SS constituted the first evolutionary step toward symbiosis (Gourion et al. 2015; Yamazaki and Hayashi 2015; Okazaki et al. 2016). Interestingly, rhizobial NF, T3SS, and T4SS depend on a common regulator activated by legume-secreted flavonoids (Janczarek and Skorupska 2011; Gourion et al. 2015; Smith et al. 2015). Role of flavonoids in protein secretion in ITs suggests that the same keys can unlock different doors (Broughton et al. 2003). Growth in the ITs is critical stage for selection of competitive rhizobia, and this stage provides checkpoint for the plant because bacteria mutated in cell wall components such as LPS, EPS, as well as BacA either fail to be released from ITs or fail to form mature symbiosomes (Gibson et al. 2008; Janczarek et al. 2015).

Concomitantly, certain cortical cells divide to form nodule primordia, and it is toward these primordia that the infection thread grows. After initiating the symbiotic dialog, flavonoids function as positional signals for cell division and/or growth in nodulating roots. This function was inferred because both the induction of a chalcone synthase-GusA (CHS–GusA) fusion and the accumulation of flavonoids occurred at the site where either purified Nod factor or nodulating rhizobia strains (but not of non-nodulating strains) were applied (Mathesius et al. 1998). Genes encoding phenyl propanoid biosynthesis enzymes including *Chalcone-O-Methyltransferase* (required for the production of the potent *nod* gene inducer 4,4-dihydroxy-2-methoxychalcone) not only express in rhizobially infected root hairs but also in nodule infection zone (not in the nitrogen fixation zone) (Chen et al. 2015). Some intriguing effects of plant phenolics are the ones associated with long-distance polar auxin transport (PAT) streams (Taylor and Grotewold 2005; Peer et al. 2011). Evidence that flavonoids regulate auxin accumulation in vivo was obtained using the flavonoid-deficient mutant, *tt4*, where accumulation of [^{14}C] indole-3-acetic acid in whole seedling was defective as a considerable amount of auxin escaped from the roots. Treatment of the *tt4* mutant with the missing intermediate naringenin restored normal auxin distribution and accumulation by the root (Murphy et al. 2000). Silencing of lignin biosynthetic gene in *Arabidopsis thaliana* led to redirection of metabolic flux into flavonoid synthesis through chalcone synthase activity. The level of plant growth reduction of HCT-deficient plants was correlated with the inhibition of auxin transport, while suppression of flavonoid accumulation by chalcone synthase repression in HCT-deficient plants restored normal auxin transport and wild-type plant growth. Thus, reduced size phenotype of HCT-silenced plants is not due to the alteration of lignin synthesis but to flavonoid accumulation (Besseau et al. 2007). The molecular interplay of flavonoids and Nod factors is likely to occur at several stages during nodule ontogeny (Taylor and Grotewold 2005). Nod factors bring about an immediate and transient inhibition of PAT in a highly localized fashion, possibly through the action of specific flavonoids, resulting in a change in auxin concentrations and subsequent stimulation of cell divisions at the site of nodule initiation (de Billy et al. 2001; Laplaze et al. 2015). Wasson et al. (2006) identified that accumulation of auxin and nodule formation was restored in the naringenin and liquiritigenin-supplemented chalcone synthase (CHS)-silenced root cultures, thereby indicating that the ability of nodule-forming rhizobia to inhibit auxin transport was flavonoid dependent. Zhang et al. (2009) silenced different flavonoid biosynthesis enzymes to generate transgenic *M. truncatula* roots with different flavonoid profiles. Silencing of chalcone synthase led to flavonoid-deficient roots, while silencing of isoflavone synthase and flavone synthase led to roots deficient for a subset of flavonoids, isoflavonoids (formononetin and biochanin A), and flavones (7,4-dihydroxyflavone), respectively. When tested for nodulation by *S. meliloti*, flavonoid-deficient roots had a near complete loss of nodulation, whereas flavone-deficient roots had reduced nodulation. Isoflavone-deficient roots nodulated normally, suggesting that isoflavones might not play a critical role in *M. truncatula* nodulation, even though they are the most abundant root flavonoids. Supplementation of flavone-deficient roots with 7,4-dihydroxyflavone, a major inducer of *S. meliloti nod* genes, completely restored nodulation. However, the same treatment did not restore nodulation in flavonoid-deficient roots, suggesting that other non-*nod* gene-inducing flavonoid compounds are also critical to nodulation. Supplementation of roots with the flavonol kaempferol (an inhibitor of auxin transport), in combination with the use of flavone-pretreated *S. meliloti* cells, completely restored nodulation in flavonoid-deficient roots. These observations indicated that flavones might act as internal inducers of rhizobial *nod* genes and that flavonols might act as auxin transport regulators during nodulation. However, all flavonoids have not been shown to exhibit auxin transport inhibition. The flavonol subclass in particular, such as kaempferol and quercetin, shows the strongest inhibitory activity (Ng et al. 2015).

Several possible mechanisms by which flavonoids modulate auxin transport have been reported. Acropetal auxin transport in the root, as well as basipetal auxin transport in the inflorescence, hypocotyl, and root, is all elevated in the absence of flavonoids. Flavonoids, such as quercetin, apigenin, and kaempferol, do not directly compete with IAA but are implicated as inhibitors of IAA transport across the plasma membrane by binding to a plasma membrane protein known as NPA receptor (Cooke et al. 2002). Rhizobia-induced flavonoids could bind to the auxin transporters AtMDR (Multidrug resistance), AtAPM (Aminopeptidase M1), and AtPINs (Pin-formed) (Peer et al. 2004) and prevent their intracellular trafficking, i.e., negatively regulate PAT in vivo (Buer and Muday 2004; Taylor and Grotewold 2005). Another possible mechanism is that flavonoids alter either the amount of synthesis or localization of auxin transport proteins, perhaps by phosphorylation of transcription factors that control the synthesis of these auxin carriers (Buer and Muday 2004). In addition to the participation of flavonoids in the polar transport of auxins, flavonoids are involved in the inhibition of auxin breakdown by peroxidases. Monohydroxy B-ring flavonoids are suggested as cofactors of peroxidase functioning as an IAA oxidase (destroys the hormone), whereas dihydroxy B-ring forms act as inhibitors of IAA-degrading activity (Mathesius 2001). Similarly, Agati and Tattini (2010) postulated that the high light-induced biosynthesis of flavonoids may have a role in regulating whole-plant and individual organ architecture. Pollastri and Tattini (2011) confirmed this as high sunlight-induced synthesis of quercetin derivatives fine-tuned auxin gradients as well as local auxin concentrations which represent the actual determinants for different morphological responses.

As IT reaches the base of the root hair cell, the nucleus in the adjacent cortical cells starts to reposition itself and a centrally located cytoplasmic bridge forms to establish pre-IT in this cell (Timmers et al. 1999). This process is repeated at each cell junction, thereby extending transcellularly, through a tip growth-like mechanism, from the epidermis or subepidermal cortex toward a subtending region of dividing cortical cells that have initiated the formation of a nodule primordium

(NP) (Fournier et al. 2008; Haag et al. 2013). NSP1/2 and NIN not only act downstream of CCaMK in the epidermis, but they also act downstream of CCaMK and the cytokinin receptor in the cortex (Marsh et al. 2007; Suzaki and Kawaguchi 2014; van Zeijl et al. 2015). *ENOD11* and *ENOD12* encode hydroxyproline-rich glycoproteins (HyPRPs), with relatively few tyrosine residues, which enhance cell wall plasticity or in components of the infection thread matrix. Hence, very rapidly and even before contact, legume root cells are paving the way for the accommodation of their symbiotic partner by remodeling the cell wall barrier (Rose et al. 2012; comprehensively reviewed by Rich et al. 2014). Cytokinin receptor LjLHK1/MtCRE1 is essential for Nod factor-induced gene expression and subsequent nodule primordium formation. Miri et al. (2016) postulated that cytokinin participates in orchestrating signaling events that promote rhizobial colonization of the root cortex and limit the extent of subsequent infection at the root epidermis by systemic autoregulation of nodulation (AON), thus maintaining homeostasis of the symbiotic interaction. Recent studies have revealed that exogenously applied cytokinin can bypass *Rhizobium* Nod factor signaling and the symbiotic phenotype of the *Mtcre1* mutant can be complemented, at least in part, by the exogenous application of flavonoids, suggesting that flavonoids mimic cytokinin functioning or even act as small secondary molecules downstream of cytokinin (Ng et al. 2015).

8.3.3.2 Nodule Organogenesis into Symbiosome: The Unifying Feature of Endosymbioses

Eventually rhizobia are released from the growing tip of ITs into an infection droplet in the cytoplasm [a subset of nascent nodule primodia (NP) cells], where they undergo differentiation to bacteroids in confined facultative organelle-like compartments called symbiosomes (SMs) (Kereszt et al. 2011). The rapid growth of bacteria at the tip tends to select out a single bacterial strain even if more than one strain becomes entrapped initially. This selection even works with two near-identical rhizobial strains in the infection (Gage 2002) and can select against potential “cheaters,” which will attempt to coinfect the nodule niche without conferring any benefit to the plant (Downie 2010). Through a process resembling endocytosis, the bacteria are surrounded by a plant-derived membrane, called the peribacteroid membrane, which forms SMs (McNear 2013). To accommodate a high number of endosymbionts, extreme plant cell enlargement is the consequence of repeated endoreduplication (ER) of the genome without mitosis (Kondorosi and Kondorosi 2004; Ribeiro et al. 2015). The bacteroids multiply in the growing host nodule cells to a certain cell density, adapt to the endosymbiotic lifestyle and microaerobic conditions, and mature to nitrogen-fixing bacteroids which convert atmospheric nitrogen gas into ammonia, using the nitrogenase enzyme complex (Ferguson et al. 2010). The couple jointly needs to modify their house especially to insulate the microsymbiont from the plant cytoplasm and to lower oxygen tensions for nitrogen fixation. One essential modification to the corridor is seen as the peribacteroid or symbiosome membrane in infected cortical cells (Broughton et al. 2000). Thus, rhizobia remain outside the plant cytoplasm and are engulfed in a symbiosome membrane, which functions to regulate nutrient exchange between the partners (Mathesius 2009). In nodulating plants, flavonoids have been observed to delimit the border between symbiotic sites and non-symbiotic plant tissue. Another notable feature in these barrier regions is the accumulation of histidine-rich glycoproteins, such as ENOD2, which are capable of cross-linking with flavonoids (Stafford 1997). The existence of this barrier in legumes has been postulated to act as an O_2− diffusion barrier that helps to prevent inactivation of nitrogenase in bacteroids (Cohen et al. 2001).

Two major morphological types of nodules are determined by the host legumes: determinate and indeterminate. Differences between the two nodule types are the site of first internal cell divisions, maintenance of a meristematic region, and the form of the mature nodules (Oldroyd et al. 2011; Hichri et al. 2016). Different roles of flavonoids in regulating auxin homeostasis for development of determinate and indeterminate nodules have been reported (Wasson et al. 2006; Subramanian et al. 2007; Zhang et al. 2009). Silencing the flavonoid pathway in *M. Truncatula* (forming indeterminate nodules) prevented localized auxin transport inhibition, and thus, flavonoid-deficient roots were unable to initiate nodules, even though normal root hair curling was observed (Wasson et al. 2006). However, Subramanian et al. (2005) silenced isoflavone synthase, the entry point enzyme for isoflavone biosynthesis in soybean (forming determinate nodules). Isoflavonoid-depleted roots of the determinate nodule forming legume soybean were also deficient in both auxin transport inhibition and nodulation. However, nodulation in isoflavone-deficient soybean roots could be restored by using isoflavone-hypersensitive rhizobia cells, suggesting that induction of Nod signal biosynthesis by isoflavones is crucial, whereas auxin transport inhibition by isoflavones is not. Consistently, during indeterminate nodule formation, flavonoid-regulated auxin transport inhibition occurs at the site of rhizobial infection, whereas during determinate nodule formation there is no significant auxin transport inhibition (Subramanian et al. 2007).

8.3.4 Role of Flavonoids in Autoregulation of Nodulation

At times of sufficient nitrogen supply to the legumes, rhizobial symbiosis can be limited by both local and systemic feedback regulatory mechanisms (Mathesius 2009). Since nodulation and the subsequent nitrogen fixation are energy-intensive processes, host plant maintains a balance between cost and benefit by tightly controlling the number of nodules it forms, *via* a complex root-to-shoot-to-root signaling loop called AON (Herder and Parniske 2009; Magori and Kawaguchi 2009). There are various root [e.g., CLAVATA3/ESR-related (CLE) peptide] and shoot (e.g., SDI) derived inhibitors identified in autoregulation of nodulation (Reid et al. 2011). Rossen et al. (1987) recognized four classes of mutations in the *nodD* gene of *Rhizobium leguminosarum* biovar viciae that can affect its ability to autoregulate and/or activate other *nod* genes in the presence of flavonoid inducers. Class I mutations led to defects in both autoregulation and activation of other *nod* genes in the presence of inducers, class II mutations left activation of the transcription of *nodABC* and *nodEF* unaffected but virtually abolished autoregulation, and class III mutations affected the induction of other *nod* genes but not autoregulation. More adversely, class IV mutations showed the activation of other nod genes even in the absence of inducer molecules. Some of the anti-inducers also acted as inducers in the strains with class IV *nodD* mutations, thereby pointing toward the direct interaction of nodD proteins with the flavonoids. All the mutations led to decrease in nodule number; however, class IV mutations even failed to fix N_2 (even when a wild-type copy of nodD was present in the strain). This suggests that continued transcription of *nodABCIJ* and/or *nodEF* by bacteria in the nodule interferes with their ability to develop in N_2-fixing bacteroids (Rossen et al. 1987). Catford et al. (2006) identified that levels of formononetin and its 7-*O*-glucoside ononin systemically suppressed in response to *S. meliloti* and NFs, suggesting that these isoflavonoids are probably involved in the systemic suppression of nodulation as their application to autoregulated roots again promoted nodulation (Fig. 8.2).

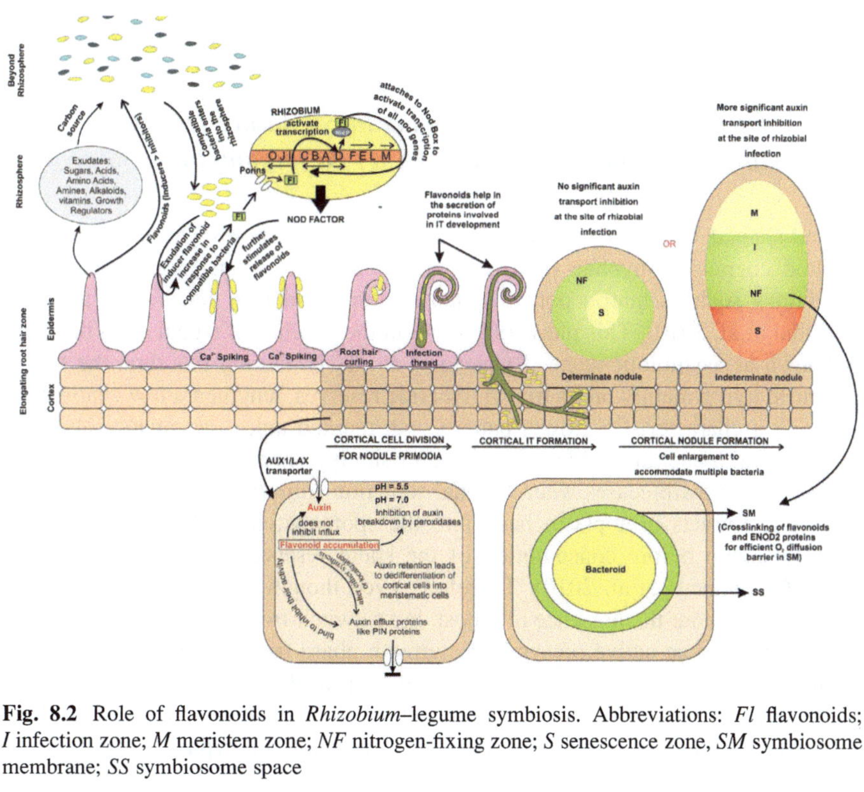

Fig. 8.2 Role of flavonoids in *Rhizobium*–legume symbiosis. Abbreviations: *Fl* flavonoids; *I* infection zone; *M* meristem zone; *NF* nitrogen-fixing zone; *S* senescence zone, *SM* symbiosome membrane; *SS* symbiosome space

8.4 Flavonoids in Arbuscular Mycorrhizal Symbiosis

Phosphorus (P) is the second most important macronutrient, next to nitrogen; however, P is the least accessible and hence most frequently deficient macronutrient limiting growth of crop plants (Balemi and Negisho 2012). Further, excess P supply in the soil, as chemical fertilizers, is another major environmental concern as it increases the risk of P movement to surface and groundwaters (Grant et al. 2005). To this end, arbuscular mycorrhiza (AM) fungi could be of great benefit in enhancing soil P utilization efficiency of the crop and improve nutritional status of the crop (Aggarwal et al. 2011). Unlike rhizobial symbiosis, mycorrhizae are the rule in nature, not the exception. In this mutualistic relationship, mycorrhizal hyphae (up to 100 times longer than root hairs) act as an extension of the root system, extending to soil beyond the P depletion zone (Mohammadi et al. 2011; Balemi and Negisho 2012). Thus, mycorrhization aids roots by providing a more widespread nutrient foraging system to plants, and this foraging provides fungus, an avenue to find new hosts (Denison and Kiers 2011; Bücking et al. 2012). A better understanding of this microbial interaction is therefore a potential key to determine the profitability and sustainability of agricultural systems (Mohammadi et al. 2011). Thus, here we review the mechanisms by which the plant signals (especially flavonoids) allow fungal infection and promote mutualistic symbiosis.

8.4.1 Unfolding Role of Flavonoids in Signaling Imputed to Arbuscular Mycorrhiza (AM) Development

Round-shaped thick cell-walled multinucleate fungal spores germinate to form haploid coenocytic hyphae (Hijri and Sanders 2005), when environmental soil conditions such as matric potential, temperature, and CO_2 levels are favorable (Mandal et al. 2010). This phase of growth in the absence of signal from the plant is known as asymbiotic stage (Requena et al. 2007). During this time the fungal colonies, living mainly on its triacylglyceride reserves, lengthen a few centimeters showing a characteristic growth pattern with marked apical dominance and sporadic hyphal branching. AM spores are surprisingly dynamic, and in the absence of a host root, growth ceases where hyphal septation from the apex arrests development and retracts protoplasm back into the spore within 8–20 days, becoming dormant again before consumption of the spore reserves (Bonfante and Genre 2010; Giovannetti et al. 2010). Therefore, even though spores are competent to germinate without the presence of a host, but a switch from the aymbiotic stage of development to an active presymbiotic growth phase triggers only in response to initial recognition by multifaceted fine-tuned reciprocal signaling events (Gachomo et al. 2009). Prior to colonization, significant developmental step in the life cycle of mycorrhizal fungi is host signal interceded increase in hyphal branching and metabolic activity, which ensures directional growth of hyhal branches toward host root (Pinior et al. 1999).

Germ tubes, with limited growth potential, respond to the presence of "branching factor" in host root exudates in their vicinity, where fungal morphology shifts toward enhanced hyphal growth and extensive hyphal branching (Buee et al. 2000). The branch-inducing factor is present in root exudates of all the mycotrophic plants but absent in those of nonhost plants. Although no directional growth has been observed toward the root, several experiments showed that host-derived signals intensify hyphal ramification, thereby increasing the probability of contact with a host root (Requena et al. 2007). By analogy to the *Rhizobium*–legume symbiosis, some of the potential host exudates that stimulate spore germination, hyphal branching in the soil, and fungal invasion and arbuscule formation inside the root, often in a symbiont-specific manner, have been identified as *flavonoids* (Scervino et al. 2007; Steinkellner et al. 2007). These chemical signals exuded by the plant and the thigmotropic signals from the rhizodermis are possibly recognized by receptor proteins associated with the fungal plasma membrane (Requena et al. 2007). Interestingly, Siqueira et al. (1991b) suggested that flavonoids can stimulate fungal growth directly or remove AM fungal self-inhibition. Pyranoisoflavones produced by white lupin, which is not a host for mycorrhizal fungi, inhibited hyphal branching of mycorrhizal fungi, suggesting that flavonoids could play both stimulating and inhibitory roles on fungal symbionts in the soil (Akiyama et al. 2010). Martens and Mithöfer (2005) considered the absence of flavone biosynthesis pathway in Brassicaceae as one of the reasons of this family lacking mycorrhizae. Thus, it is likely that host and nonhost plants can modulate the establishment of symbiosis by altering the profile of flavonoid exudates (Hassan and Mathesius 2012).

A variety of flavanones (hesperetin and naringenin), flavones (apigenin), and isoflavones (formononetin) have been reported to stimulate spore germination and hyphal growth of AM fungi *in vitro* (Gianinazzi-Pearson et al. 1989; Tsai and Phillips 1991; Nair et al. 1991). Effect of these compounds on AM fungi is flavonoid type and concentration, AM fungal species and genera, and host specific (reviewed by Vierheilig et al. 1998). At low concentration, different flavonoids can increase AM fungal spore germination, hyphal growth, and hyphal branching, while at high concentration, the same flavonoid turns inhibitory (Nair et al. 1991; Baptista and Siqueira 1994). Siqueira et al. (1991b) reported that isoflavonoids (formononetin and biochanin A) at concentrations of 5 mg l^{-1} stimulated AM colonization in *T. repens* L., while flavone chrysin increased root colonization at concentration 40–60 mg l^{-1}. Morandi et al. (1992) tested the effect of 2 isoflavonoids (glyceollin I and coumestrol at concentration 0, 0.05, 0.5, 5, and 50 μM) and 1 flavonoid (quercetin at concentration 0, 0.1, 1, and 10 μM) on *in vitro* spore germination of *Gigaspora margarita*. After 5 and 7 days, number of germ tubes per spore was slightly increased by glyceollin I; mycelium length from germinated spores was increased by low concentrations of glyceollin I but was significantly decreased at the highest concentration. A positive correlation was found between coumestrol concentration and mycelium length, while vesicle number decreased by coumestrol, quercetin, and the highest concentration of glyceollin (but was increased by glyceollin at 0.5 μM). Thus, similar to the case of *nod* gene induction, it is probably not simply the absence or presence of a specific flavonoid in root exudates which determines the signal properties, but rather a tightly controlled concentration-specific release pattern of inducers and anti-inducers.

Bécard et al. (1992) reported that only the flavonols stimulated *in vitro* growth of germinated spores of *Gi. margarita*, while the flavones, flavanones, and isoflavones tested were generally inhibitory. Quercetin (10 μM) prolonged hyphal growth from germinated spores of *Gi. margarita*, while the glycosides of quercetin, rutin and quercitrin, were not stimulatory, thereby indicating the role of structure of flavonoid in AM colonization. In general, at least one hydroxyl group on the B aromatic ring has been found to be necessary for stimulatory effect. Besides this, a hydroxyl on position 3 is also essential to confer stimulatory activity to the flavonol molecule, as the flavones luteolin and apigenin, lacking this hydroxyl, showed no effect (Bécard et al. 1992). Glycosylation at position 3 also promoted the loss of the stimulatory activity, as in quercitrin and rutin. Moreover, it has been suggested that saturation of the 2,3 double bond in flavonols, as seen in flavanones, promoted the loss of activity on hyphal growth of *Gi. margarita* (Chabot et al. 1992). This loss of activity was attributed to the loss of the planar configuration of the flavonol molecule when the double bond disappears; e.g., dihydroquercetin and dihydrokaempferol showed no effect on hyphal growth on *Gi. margarita*. However, these results are in contrast with those of Gianinazzi-Pearson et al. (1989) and Baptista and Siqueira (1994). Naringenin, lacking the 2,3 double bond, was stimulatory with *Gi. margarita* and *Gi. gigantea*, whereas apigenin, differing only by the presence of the 2,3 double bond, showed no effect with *Gi. gigantea*. In general, *Glomus* spp. are stimulated by flavonols as well as by isoflavones (Vierheilig et al. 1998). Whereas some flavonoids such as quercetin exhibit a general stimulatory effect on the hyphal growth of different AMF genera (Chabot et al. 1992, Baptista and Siqueira 1994), data with biochanin A showed a stimulation of *Glomus* (Nair et al. 1991; Vierheilig et al. 1998) but not *Gigaspora* species (Chabot et al. 1992, Baptista and Siqueira 1994).

8.4.2 Myc Factor Signaling Pathway in the Root Epidermis

Branched fungal hyphae secrete fungal chitin elicitor to be called "Myc factor" that drive various morphological and physiological changes in the hosts (Requena and Breuninger 2004), to orchestrate the AM infection process by counteracting the plant immune program and upregulating the expression of symbiosis-related genes (Kloppholz et al. 2011). At an early stage of mycorrhizal formation, plant constitutive chitinases may partially cleave these elicitors and thus inducing transient plant defense response (Antunes and Goss 2005; Maillet et al. 2011). However, unlike fungal pathogens, diffusible Myc factors from AM increase lateral root formation, elicit a transient cytosolic calcium elevation within a few minutes, and induce expression of specific genes in only specific root cells in direct contact with the penetrating fungus (Navazio et al. 2007), while a suppressor activity is induced in non-colonized neighboring cells (Kosuta et al. 2003; Genre et al. 2005).

Several mutants in legumes like *M. truncatula*, *Lotus japonicus*, and pea are deficient for nodulation and AM, indicating the existence of a conserved symbiotic (Sym) pathway required for the establishment of both symbioses (Mitra et al. 2004b; Kistner et al. 2005). One common signaling component is the receptor-like leucine-rich kinase *SymRK (MtDMI2/MtSYM2)* that is involved in the direct or indirect transduction of fungal or rhizobial signals through its intracellular kinase domain to the cytoplasm (Stracke et al. 2002). Evolutionarily more recent Nod factors also overlap with Myc symbiotic signals, both structurally and functionally (Laparre et al. 2014; Limpens et al. 2015). Three nucleopore-associated proteins: NUP85, NUP133, and NENA (Saito et al. 2007) act downstream and could be involved in the transport of CASTOR and/or *DMI1/POLLUX* (ion channels) to the inner nuclear envelope (Riely et al. 2007). In both the symbioses, these channels lead to calcium oscillations in the nucleus and perinuclear cytoplasm for activating calcium-calmodulin-dependent protein kinase *MtSYMI3*/DMI3/CCAMK (Mitra et al. 2004a). *CCAMK* is known to phosphorylate the last identified SYM gene *CYCLOPS,* encoding IPD3/CYCLOPS protein which interacts with DMI3 (Messinese et al. 2007). In order to activate the appropriate symbiotic program, legumes have to discriminate the two types of symbiotic signals. *CYCLOPS* represents a branch point in the common SYM pathway, as infection threat formation and arbuscular development are *CYCLOPS* dependent, but nodule organogenesis is *CYCLOPS* independent (Yano et al. 2008). Microbial activation of the common symbiosis signaling pathway can also modulate innate immune responses through its effect on the hormonal landscape, where Limpens et al. (2015) speculated a central role for DELLA proteins, in part by influencing the salicylic acid–jasmonic acid balance. However, profound mycorrhization in roots of *cre1* mutant of *M. truncatula* suggested that MtCRE1 (Cytokinin Response 1 required for N_2 fixation) does not belong to the ancestral common symbiotic pathway (Laffont et al. 2015). In addition to the Sym pathway, a parallel pathway exists and mediates AM signaling in non-nodulating eudicots and monocots (Mukherjee and Ané 2011).

Further, hyphal contact with host roots has been reported to induce the flavonoid pathway in a number of host species, in particular in infected cells (Steinkellner et al. 2007; Abdel-Lateif et al. 2012). Harrison and Dixon (1993) reported that in *M. truncatula* + *G.versiforme* interactions, the most striking changes in identified root metabolites included the overall accumulation of formononetin malonyl glucoside (FGM), medicarpin malonyl glucoside (MGM), and daidzein and a transient increase in free medicarpin (the major phytoalexin from MGM) in the early stages of the interaction, as a defense response to fungal elicitors. In mycorrhizal state, either the fungal symbiont may be consuming theses phytoalexin precursors as carbon precursors or conjugation of medicarpin may inactivate or remove a potentially toxic metabolite. In contrast to the levels of phenylalanine ammonia lyase (PAL) and CHS transcripts, which remained elevated throughout the interaction, the level of Isoflavone reductase (IFR) transcripts decreased in later stages of the interaction to 2.5-fold below the level in uninoculated control roots, suggesting that the established mycorrhizal association in *M. truncatula* brought about a specific suppression of this transcript to decline medicarpin levels in these roots. However, in the interaction between myc$^-$ line and *G. versiforme* the levels of medicarpin and related transcripts remained elevated throughout the interaction suggesting that a defense response was occurring. Further, the inability of myc$^-$ *M. sativa* line to form a complete interaction may be attributed to the absence of coumestrol and 4′,7-dihydroxyflavone (Harrison and Dixon 1993). Studies of Guenoune et al. (2001) and Akiyama et al. (2002) on flavonoids and AM reopened the discussion about the involvement of flavonoids as regulatory compounds during signaling of mycorrhization. Guenoune et al. (2001) demonstrated that medicarpin-3-*O*-glucoside, an isoflavonoid phytoalexin, accumulated both in roots colonized by the pathogenic fungus and in AM-treated roots receiving high P and prevented these roots from being colonized by AMF. Increases in the steady-state levels of chalcone isomerase and IFR mRNAs, as defense responses of alfalfa roots to the pathogenic fungus *Rhizoctonia solani*, were reduced significantly in roots simultaneously infected with the arbuscular mycorrhizal (AM) fungus *Glomus intraradices*. These data suggested that during early stages of colonization by *G. intraradices*, suppression of defense-related properties is associated with the successful establishment of AM symbiosis. Akiyama et al. (2002) detected the C-glycosylflavone—isovitexin 2″-*O*-β-glucoside in non-mycorrhizal, P-deficient melon roots but not in roots with a high P status or in mycorrhizal roots (inoculated with *G. caledonium*). Application of this flavonoid (at concentrations of 20 and 50 μM) to AM plants not only enhanced root colonization in plants grown under low P conditions, but also in plants grown under high P conditions, thus clearly showing that a high P status or the mycorrhizal status of a plant can reduce the accumulation of a flavonoid in roots which stimulates mycorrhization.

Tsai and Phillips (1991) reported that after 21 days of application, 1.0–2.5 μM quercetin-3-D-galactoside, 4′,7-dihydroxyflavone, and 4′,7-dihydroxyflavanone promoted spore germination, hyphal elongation, and hyphal branching of *Glomus etunicatum* and *Glomus macrocarpum in vitro*. On the other hand, formononetin, an isoflavone that is released from stressed alfalfa roots, inhibited germination of both *Glomus* species. However, flavonoids tested in this study were identified as products from non-AM alfalfa seedlings only, strongly advocating that AM fungi alter flavonoid metabolism in plants and compounds produced as a result of infection have a role in differentiation of vesicles, arbuscules, and/or spores (Tsai and Phillips 1991). Again, application of biochanin A and formononetin in the rhizosphere of AM nonhost plants—*Lupinus polyphyllus* and *Spinacia oleracea* (grown in the presence of *G. mosseae*)—resulted in hyphal attachment and more hyphae around these roots (Vierheilig and Piché 1995). These results could indicate a simple hyphal growth stimulation to be responsible for the enhanced root colonization in AM host plants; however, as soil application also resulted in a colonization of *L. polyphyllus* roots by G. *mosseae* (Vierheilig and Piché 1995; Vierheilig et al. 1996), a role of these compounds in the plant–fungus communication during the colonization stage is suggested (Vierheilig et al. 1998). Vierheilig and Piché (2002) presented a complex model for the role of flavonoids during the AM association. They proposed that a fundamental molecular dialogue occurs that regulates not only the early development of AM symbiosis but also subsequent colonization which has to be balanced for establishment of genuine mycorrhizal symbiosis.

8.4.3 AM Hyphal Invasion and Differentiation of Intraradical Hyphae: The Conjugal Lodging

After hyphal docking to cell surface by means of an appresorium, the fungus enters into root epidermis *via* AM-specific prepenetration apparatus (Bonfante and Genre 2010). AM fungi first grow intercellularly or cross outer cells with linear or simple coiled hyphae (Gianinazzi and Gianinazzi-Pearson 1988). Reaching the inner cortex the fungal symbionts switch to different mode of colonization and penetrate the plant cell, where between cell wall and plant plasma membrane fungal hyphae extensively ramify to form a dichotomously branched haustorium, called "arbuscule" with a high surface to volume area (Gutjahr and Parniske 2013). Although the arbuscule eventually expands to largely fill the cortical cell, the hyphal tree does not enter the plant symplast. This plant–fungal interface consists of the fungal cell wall, the perihyphal or periarbuscular membrane (PAM), and a matrix between the two organisms—periarbuscular space (Pumplin and Harrison 2009; Rich et al. 2014). Hofferek et al. (2014) reported that in the root cortex nodulation and mycorrhization are regulated by *NSP2* whose spatiotemporal expression negatively regulates both types of root endosymbioses through perception of the nutritional status (such as high Pi or N) of the host plant. VAPYRIN also promotes intracellular accommodation of endosymbionts by interacting with membranes/cytoskeleton and is thus required for intracellular accommodation of AM fungi and rhizobia in epidermal as well as cortical cells (Feddermann et al. 2010; Gobbato 2015). Shtark et al. (2016) indicated that as compared to wild type (wt), though non-nodulating pea mutants (*sym7, sym11, sym14*) had considerable increase in root surface colonization by *Rhizophagus irregularis*, a substantial decrease in internal colonization (especially arbuscule formation) was recorded. *sym34* mutants displayed strongly reduced root surface colonization, but 10 days later internal colonization did not differ from wt. While *sym38* mutant did not differ from wt indicating that except SYM38, all analyzed genes were essential for both nodule and AM development (Shtark et al. 2016). Various homologues of *MtCBS1/2* gene, to encode protein containing a cystathionine-b-synthase (CBS) domain in ITs, act downstream NIN in rhizobial infection, in nodule organogenesis, as well as in mycorrhization, again suggesting common infection mechanisms (Sinharoy et al. 2016).

Harrison and Dixon (1993) reported that roots of the mycorrhizal resistant *M. sativa* mutant (Myc^-) do not become colonized, but hyphae grew attached to the root surface forming appresoria. In the presence of *G. versiforme,* levels of several flavonoids (diadzein, formononetin, FGM, medicarpin, MGM) increased in the *M. sativa* (Myc^-) roots in the same way as in colonized root of the wild-type (Myc^+) *M. sativa* (Harrison and Dixon 1993). Volpin et al. (1994) also found an increase of formononetin in inoculated but still uncolonized roots of *M. sativa* (Myc^+). Thus, these two researches suggested that these flavonoids play a role in the plant–fungus signaling during precolonization and cell-to-cell stage. Further, Harrison and Dixon (1993) also demonstrated that some flavonoids are newly induced after AM fungal root colonization, where coumestrol and 4′,7-dihydroxyflavone were only present in colonized roots of (Myc^+) *M. sativa* and *M. truncatula*, but were not detected in uncolonized roots of the two plants and in roots of inoculated (Myc^-) *M. sativa*. This further suggested that these flavonoids play a signaling role during the intraradical phase of the hyphae in the AM symbiosis and might be an indication for its role in the formation of highly ramified arbuscules. Naringenin found in root exudates of bean (Hungria et al. 1991) enhanced in axenic culture of *Gi. margarita*, where the number of vesicle clusters per germinated spore increased. Earlier *in vitro* experiments by Gianinazzi-Pearson et al. (1989) evidenced similar effects with apigenin and hesperetin. Luteolin, found in the seed rinse of alfalfa (Hartwig et al. 1990), although showing no stimulation of hyphal growth, stimulated the production of auxiliary cells in *Gi. margarita* (Bécard et al. 1992). Thus, hyphal differentiation possibly requires a different stimulatory mechanism or induction than do hyphal growth. Some flavonoids may be involved in only one of these mechanisms and not in the other. However, quercetin and myricetin, which exhibit a hyphal growth stimulating effect on different AM fungal genera, also enhanced the formation of auxiliary cells (Bécard et al. 1992), suggesting again a more general role of these two compounds compared to other flavonoids. Sometimes similar changes could be observed within a plant family. The level of coumestrol and daidzein increased in colonized roots of *Glycine max*, *M. sativa*, and *M. truncatula* (Morandi et al. 1984; Harrison and Dixon 1993), all belonging to the leguminosae family, and the level of blumenin was enhanced in colonized roots of several members of the Gramineae (Maier et al. 1995). Change of flavonoids level also depends on the AM fungal species, i.e., different AM fungi have different requirements for their development and thus induce different levels of these compounds or that the plant recognizes the two fungi differently. Ponce et al. (2004) reported that 5,6,7,8,9-hydroxy chalcone (NM7), 3,7-hydroxy-4′-methoxy flavone, 5,6,7,8-hydroxy-4′-methoxy flavones (RR4), and 3,5,6,7,4′-hydroxy flavones (RR4-2) could be detected only in non-mycorrhizal roots of white clover, whereas the flavonoids acacetin, quercetin, and rhamnetin were only present in roots inoculated with *G. intraradices*.

Arbuscules act as the epicenter of mutualism as they represent an extreme form of intimacy and compatibility and are site of nutrient transfer from the fungus to the host plant (Hughes et al. 2008; Denison and Kiers 2011; Mohammadi et al. 2011). In return, host-derived carbon is transferred to the fungi and stored in energy-rich vesicles to support vegetative growth or spore formation (Genre and Bonfante 2010). Whereas a high frequency of arbuscules usually indicates cost-effective nutrient exchange in both directions, high vesicular colonization is a potential

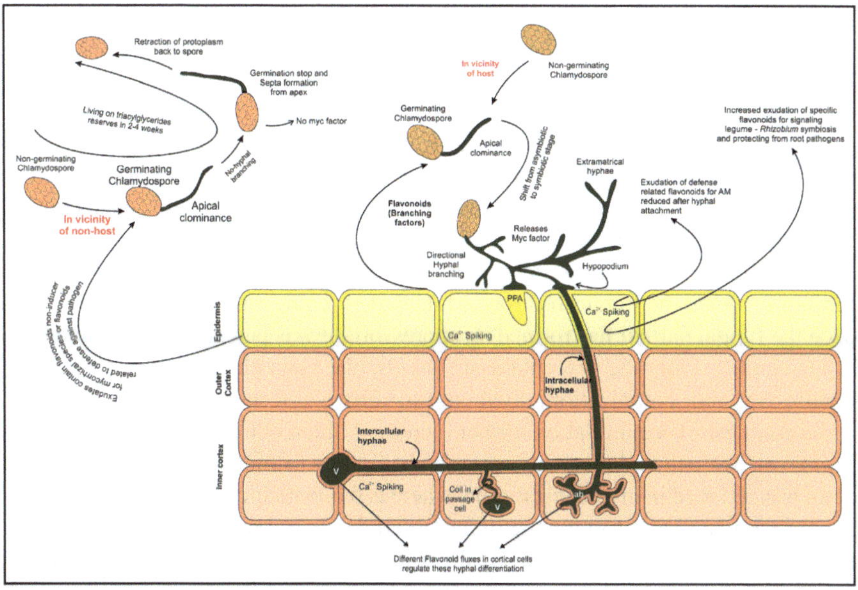

Fig. 8.3 Role of flavonoids in Arbuscular mycorrhizal symbiosis. Abbreviations: *ab* arbuscule; Ca^{2+} Calcium ion; *PPA* prepenetration apparatus; *v* vesicle

indicator of fungal resource hoarding prominently under high external nutrient conditions, when hosts are less dependent on fungal partners (Johnson 2010; Nijjer et al. 2010). Further, host molecules like lysophosphatidylcholine (LPc) potentially allow them to evaluate the amount of P delivered via the mycorrhizal pathway (Bucher et al. 2009). Host will decrease C provision or directly digest arbuscules when there is insufficient P being transferred to the host across the colonized cell (Kobae and Hata 2010). Resource exchange is followed by rapid arbuscule collapse in 4–5 days, with structures degenerating within 2.5–5.5 h (Bonfante and Genre 2010), much more rapidly than the decline in N_2 fixation in nodules (Wong and Evans 1971). The fungal life cycle is completed after formation of asexual chlamydospores on the external mycelium, which allow them to propagate and survive in the absence of a host, for more than 10 years (Giovannetti et al. 2010) (Fig. 8.3).

8.4.4 Role of Flavonoids in Autoregulation of Mycorrhization

An already mycorrhizal plant develops one mechanism to repulse further colonization by fungi, i.e., to not differentiate between AMF and soilborne pathogenic fungi (Vierheilig and Piché 2002; Mechri et al. 2015). Flavonoids again play a role in autoregulating mycorrhization, as depending on the flavonoids and their concentration, they can exhibit an inhibitory and/or a stimulatory effect on fungi (Vierheilig et al. 1998; Subramanian et al. 2007; reviewed by Lira et al. 2015). Larose et al. (2002) found that after an application of a mixture of spores and extraradical hyphae of *G. intraradices* to *M. sativa* roots, coumestrol, ononin, and daidzein levels were increased, whereas formononetin was detected at lower levels, indicating that at this stage of symbiosis, i.e., during appressoria formation, fungal partner, with signals released by its mycelium, is actually asking for a change of the concentration of certain flavonoids, which exhibit stimulatory effects on fungi. The accumulation of medicarpin (defense phytoalexin) only at the stage of appressoria formation, when the fungus has formed its first structures in the root, indicates that once the fungal partner has formed its appressoria it was perceived in a similar way as a pathogen. Later the perception of fungi by the plant changed and at a stage of intense AM root colonization, no altered medicarpin accumulation was observed, while coumestrol accumulation further increased. Interestingly, formononetin/daidzein levels were linked inversely, as at the beginning of root colonization formononetin was reduced in mycorrhizal roots whereas daidzein increased and at the end of the experiment, when formononetin increased daidzein levels seemed to decrease. Besides coumestrol, medicarpin again accumulated at the end of colonization, thereby pointing toward the role of coumestrol at any stage of root colonization; medicarpin probably plays a role at the beginning of root colonization and at a later stage. There was a positive link between the accumulation of coumestrol, medicarpin, and/or formononetin at the later stages and the autoregulation of mycorrhization (Vierheilig and Piché 2002). Some species-specific quantitative changes were observed, as ononin was highly accumulated in roots colonized by *G. mosseae,* but only weakly accumulated in roots colonized by *G. intradices* while coumestrol was high in roots colonized by *G. intradices* but lower in roots colonized by *G. mosseae*. The only compound that accumulated to similar levels in all AMF-colonized roots was genistein, and the similar accumulation pattern of genistein in mycorrhizal roots independent of the root-colonizing AMF indicated a general, non-species- or non-genus-specific regulatory role of this compound during mycorrhization (Larose et al. 2002). Later on, Scervino et al. (2005a) reported that acacetin and rhamnetin inhibited the formation of AM penetration structures as well as colonization and pointed toward a possible involvement of these two flavonoids in the autoregulation process of the AM symbiosis. Even, flavonoid NM7 inhibited the number of entry points and the percentage of root length colonization of the test plant by *Gigaspora* and *Glomus* species. The presence of 5,6,7,8,9-hydroxy chalcone in non-AM colonized clover roots and the disappearance of this flavonoid in mycorrhizal clover roots indicated that the AM fungus *G. intraradices* induced a mechanism, which reduces this inhibitory molecule in the mycorrhizal root. Quercetin, 5,6,7,8-hydroxy-4′-methoxy flavones, and 3,5,6,7,4-′-hydroxy flavone increased the number of entry points and the AM colonization of tomato roots by *Gigaspora* (not *Glomus*). Further, Scervino et al. (2005b) linked flavonoids changes in mycorrhizal roots directly to regulatory processes of the AM symbiosis for the first time. Interestingly, RR4 and RR4-2 stimulated most of the presymbiotic stages of *Gigaspora* (but not of the two *Glomus* species), supporting the implication of Akiyama et al. (2002) that a flavonoid, stimulating root colonization by AM fungi, was only present in non-mycorrhizal but not in mycorrhizal roots. They further suggested that the compound might be involved in one stage of the fungal regulation, but not in another, as RR4 increased the percentage of spore germination of *Gi. margarita*, however, inhibited the hyphal growth, and showed no effect on hyphal branching and the cluster formation of auxiliary cells. NM7 was detectable only in non-mycorrhizal white clover roots and inhibited the presymbiotic development of all AM fungi tested. There was stimulatory effect of newly synthesized quercetin on spore germination and/or hyphal growth of *Gigaspora* species. However, acacetin and rhamnetin exhibited an inhibitory effect on nearly all fungal parameters of both AM genera (*Gigaspora* and *Glomus*) tested, pointing toward an implication in the autoregulation of mycorrhization (Scervino et al. 2005b). Scervino et al. (2009) reported that 2 μM concentration of 3-methoxy-5,6,7,8-hydroxy-4′hydroxy flavone (NMHTV) isolated from shoots of non-arbuscular mycorrhizal (AM) inoculated clover did not affect the percentage of germination of spores but significantly increased the formation of entry points and symbiotic stage of *Gi. margarita* spores on colonization of tomato (*Lycopersicum esculentum*), while at higher concentration of 8 μM concentration inhibited the hyphal length of *Gi. rosea*. The absence of stimulation of the AM presymbiotic and symbiotic stages in tomato by exogenous application of the newly synthesized flavonoids 5,6,7,8-hydroxy-3-methoxy flavone (MH-1), 5,6,7,8-hydroxy-4′-hydroxy flavone (MH-2), and 5,7-hydroxy-3,4′-methoxy flavone (MH-3), isolated from AM clover (*T. repens*) shoots, indicated that the autoregulation of the AM symbiosis can be, at least partially, due to the disappearance of flavonoids in AM-colonized plants that stimulated the AM symbiosis.

Working with split-root systems of barley, alfalfa, and soybean, it was reported that root colonization of one side of a split-root system strongly suppressed mycorrhization of "autoregulated roots" on the other side (Vierheilig 2004b; Meixner et al. 2005). Meixner et al. (2005) reported that the soybean supernodulating mutant *nts1007* [mutated in the receptor kinase gene *GmNARK* (Searle et al. 2003)], which lacks the autoregulatory mechanism to control nodulation, did not autoregulate AMF root colonization. Thus, these reports indicated a similar mechanism of autoregulation in both symbioses. Alike *Rhizobium*–legume symbiosis, flavonoids have also been suggested to be involved in the regulation of mycorrhization as roots treated with certain flavonoids exhibited increased AM fungal root colonization (Vierheilig and Piché 2002; Vierheilig 2004a; Scervino et al. 2005a, b). Catford et al. (2006) studied split-root systems (SRS) of alfalfa plants, where one side of SRS was inoculated with *S. meliloti* or *Glomus mosseae* or was treated with Nod factor. All the three applications resulted in accumulation of daidzein but reduced levels of formononetin, ononin, and medicarpin. These treatments given on one side altered isoflavonoids of the medicarpin synthesis pathway on the other side, thereby suggesting that levels of formononetin and its glycoside ononin were not only locally reduced but also systemically downregulated in parts of the root system that were not in contact with the invading symbionts. Thus,

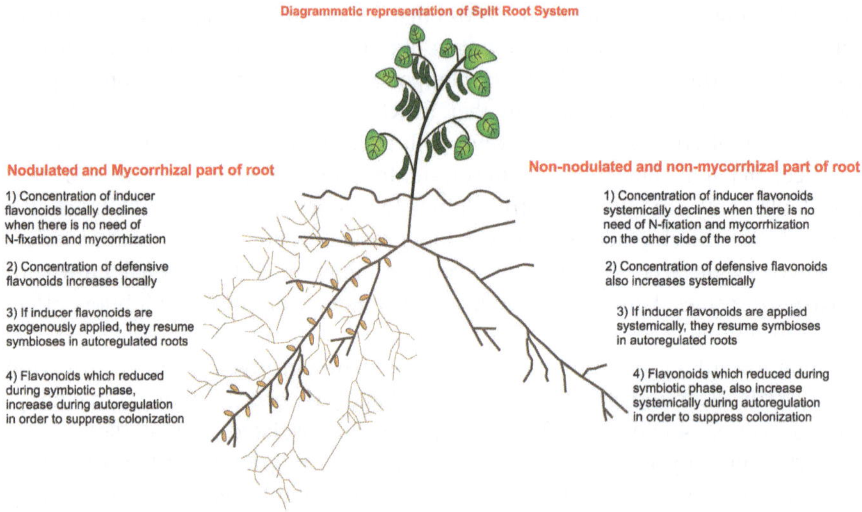

Fig. 8.4 Role of flavonoids in autoregulation of nodulation and mycorrhization

Catford et al. (2006) suggested that systemic downregulation of formononetin and ononin levels in non-treated root parts is a metabolic reponse to the putative autoregulation signals that mediate suppression of symbioses. Exogenous application of isoflavonoids to roots inactivated or compensated the symbiosis-suppressing effects induced by long-distance autoregulation signals and promoted nodule formation and stimulated establishment of the AM fungal symbiosis. Ononin (but not formononetin) promoted AMF root colonization, whereas formononetin stimulated nodulation in autoregulated parts of the root system. Hence, although autoregulation of nodulation and mycorrhization seem to share some common signaling events (Vierheilig 2004a; Meixner et al. 2005; Vierheilig et al. 2008), the effects of systemically regulated isoflavonoids on establishment of symbiosis were different (Fig. 8.4).

Many legume plants benefit from the tripartite symbiosis of arbuscular mycorrhizal fungi and rhizobia. Siqueira et al. (1991b) reported that in isoflavonoid formononetin and biochanin A (considered "anti-nod inducers") treated plants, *Rhizobium* nodulation was enhanced in the presence of AM fungi which might have resulted from improved mycorrhizal colonization and plant growth. Soil applications of these compounds had also shown to stimulate mycorrhizal formation and growth of non-nodulated plants in the presence of AM fungi (Siqueira et al. 1991a), thus indicating that isoflavonoid growth stimulation was not nodulation mediated. However, Xie et al. (1995) found that the inoculation of *G. max* with *B. japonicum* increased the colonization by *G. mosseae*, and an increased flavonoid concentration in root exudates was suggested to be responsible for this effect. Xie et al. (1997) reported the increase in mycorrhizal colonization (from <30 to 65%) and sporocarp formation of *G. mosseae* (Nicol. & Gerd.) Gerdemann & Trappin in roots of *Lablab purpureus* (L.) Sweet treated with Nod factors purified from

Rhizobium sp. NGR234. Antunes et al. (2006) found that the concentration of daidzein was at least four times greater in soybean root than in the seed, whereas coumestrol, which was absent in the seed, was newly synthesized. The significant increase in diadzein and coumestrol appeared to result mainly from the development of AM, which might have helped in bradyrhizobial symbiosis (diadzein and coumestrol are involved in bradyrhizobial symbiosis) and thus played a key role in the early stages of the tripartite symbiosis. Morandi et al. (2009) reported a new *M. truncatula* mutant B9, defective for nodulation but was hypermycorrhizal. It represented a new tool for the study of plant metabolites differentially regulating mycorrhiza and nodulation symbioses, in particular those related to autoregulation mechanisms. In contrast to wild A17, mutant was characterized by considerably higher root concentrations of the phytoestrogen coumestrol (*nod* gene inhibitor for *S. meliloti* and hyphal growth inducer of *Gi. margarita*) and coumestrol conjugate malonyl glycoside, thus leading to low and late nodulation but high mycorrhizal colonization in B9 mutant. Fokom et al. (2010) pointed out significant interaction between arbuscular mycorrhizal fungi and 3-*o*-glucoside kaempferol in enhancing P nutrition, N_2 fixation, as well as phenolic metabolism in *Vigna unguiculata* (L.) Walp growing in low-P soil of southern Cameroon. Many flavonoids like quercetin, luteolin, and other substituted flavones and flavanones are released by germinating seeds and living roots of legume crops over time and their interactions with fungal pathogens suggest that while stimulating mycorrhizal fungi they may be equally important for plant protection against soilborne pathogens (Hassan and Mathesius 2012). Cordeiro et al. (2015) evidenced from field experiment that formononetin can improve mycorrhizal colonization, nodules number, and reduce the negative effects of fungicides Carbendazim + Thiram in soybean production. Thus, utilizing flavonoids can be an excellent opportunity to utilize and manipulate *Rhizobium* and AM fungi in order to enhance crop productivity in a cost-effective manner and with reduced agricultural chemical inputs. Currently, seed exudates, which are mixture of flavonoids, are being used commercially to promote *Rhizobium*–legume symbiosis and N_2 fixation in agricultural practices (Skorupska et al. 2010). A commercial product "SoyaSignal" (mixture of genistein and daidzein) in Northern America is being applied either directly to the seed or in furrows in soils that contain adequate populations of *Bradyrhizobium* (Smith and Zhang 1999). SoyaSignal significantly improved nodulation and nitrogen fixation which resulted in an average 7% increase in grain yield (Leibovitch et al. 2001; Broughton et al. 2003). This study linked the function of different flavonoids to the establishment of the tripartite symbiosis and suggested that these compounds are produced and released into the rhizosphere as a function of the colonization process.

8.5 Conclusion

Thus, this chapter highlighted the significance of flavonoids as common architects in the establishment of two agriculturally important root symbioses—nodulation and mycorrhizal interactions. Information attained from these studies illustrated that besides stimulating microbial swarming motility toward host and activating symbiosis-related microbial genes, multifaceted flavonoids contribute in various stages of symbiosome developmental program. Their differential endogenous accumulation regulates modulation of plant innate defense reactions, infection thread/hyphal growth, intraradical hyphal differentiation, cortical cell morphogenesis for providing optimal niche to the microbes, and prevention of metabolically inefficient symbioses. Deciphering the complexity of flavonoids induced responses and their interplays with other major regulators of symbioses such as hormones make flavonoids a vital player in plant–microbe symbioses. However, gaps still remain in the knowledge base and further researches involving the endogenous regulation of flavonoid biosynthesis during these mutualistic associations as well as use of flavonoids as potential soil amendments are needed to optimize their implementation in improving rhizobial and mycorrhizal symbioses.

References

Abd-Alla MH, El-enany A-WE, Bagy MK, Bashandy SR (2014) Alleviating the inhibitory effect of salinity stress on *nod* gene expression in *Rhizobium tibeticum*—fenugreek (*Trigonella foenum* graecum) symbiosis by isoflavonoids treatment. J Plant Interact 9:275–284

Abdel-Lateif K, Bogusz D, Hocher V (2012) The role of flavonoids in the establishment of plant roots endosymbioses with arbuscular mycorrhiza fungi, rhizobia and *Frankia* bacteria. Plant Signal Behav 7:636–641

Agati G, Tattini M (2010) Multiple functional roles of flavonoids in photoprotection. New Phytol 186:786–793

Aggarwal A, Kadian N, Tanwar A, Yadav A, Gupta KK (2011) Role of arbuscular mycorrhizal fungi (AMF) in global sustainable development. J Appl Nat Sci 3:340–351

Akiyama K, Matsuoka H, Hayashi H (2002) Isolation and identification of a phosphate deficiency-induced C-glycosyl-flavonoid that stimulates arbuscular mycorrhiza formation in melon roots. Mol Plant Microbe Interact 15:334–340

Akiyama K, Tanigawa F, Kashihara T, Hayashi H (2010) Lupin pyranoisoflavones inhibiting hyphal development in arbuscular mycorrhizal fungi. Phytochemistry 71:1865–1871

Anders S, Huber W (2010) Differential expression analysis for sequence count data. Genome Biol 11:R106

Andersen OM, Markham KR (2006) Flavonoids: chemistry, biochemistry and applications. CRC Press, Boca Raton

Antunes PM, Goss MJ (2005) Communication in the tripartite symbiosis formed by arbuscular mycorrhizal fungi, *Rhizobia* and legume plants: a review. Am Soc Agron 48:199–221

Antunes PM, de Variennes A, Rajcan I, Goss MJ (2006) Accumulation of specific flavonoids in soybean as a function of the early tripartite symbiosis with arbuscular mycorrhizal fungi and *Bradyrhizobium japonicum*. Soil Biol Biochem 38:1234–1242

Ardourel M, Lortet G, Maillet F, Roche P, Truchet G, Promé JC, Rosenberg C (1995) In *Rhizobium meliloti*, the operon associated with the nod box n5 comprises *nodL, noeA* and *noeB*, three host-range genes specifically required for the nodulation of particular *Medicago* species. Mol Microbiol 17:687–699

Badri DV, Quintana N, El Kassis EG, Kim HK, Choi YH, Sugiyama A, Verpoorte R, Martinoia E, Manter DK, Vivanco JM (2009) An ABC transporter mutation alters root exudation of phytochemicals that provoke an overhaul of natural soil microbiota. Plant Physiol 151: 2006–2017

Bais HP, Park SW, Weir TL, Callaway RM, Vivanco JM (2004) How plants communicate using the underground information superhighway. Trends Plant Sci 9:26–32

Balemi T, Negisho K (2012) Management of soil phosphorus and plant adaptation mechanisms to phosphorus stress for sustainable crop production: a review. J Soil Sci Plant Nutr 12:547–562

Baptista MJ, Siqueira JO (1994) Efeito de flavonoides na germinação de esporos e no crescimento assimbiótico do fungo micorrízico arbuscular *Gigaspora gigantea*. Rev Bras Fisiol Veg 6: 127–134

Bécard G, Douds DD, Pfeffer PE (1992) Extensive *in vitro* hyphal growth of vesicular arbuscular mycorrhizal fungi in the presence of CO_2 and flavonols. Appl Environ Microbiol 58:821–825

Begum AA, Leibovitch S, Migner P, Zhang F (2001a) Specific flavonoids induced nod gene expression and pre-activated nod genes of *Rhizobium leguminosarum* increased pea (*Pisum sativum* L.) and lentil (*Lens culinaris* L.) nodulation in controlled growth chamber environments. J Exp Bot 52:1537–1543

Begum AA, Leibovitch S, Migner P, Zhang F (2001b) Inoculation of pea (*Pisum sativum* L.) by *Rhizobium leguminosarum* bv. viceae preincubated with naringenin and hesperetin or application of naringenin and hesperetin directly into soil increased pea nodulation under short season conditions. Plant Soil 237:71–80

Bek AS, Sauer J, Thygesen MB, Duus JO, Petersen BO, Thirup S, James E, Jensen KJ, Stougaard J, Radutoiu S (2010) Improved characterization of nod factors and genetically based variations in LysM receptors domain identify amino acids expendable for nod factor recognition in lotus spp. Mol Plant Microbe 23:58–66

Belkheir AM, Zhou X, Smith DL (2001) Variability in yield and yield component responses to genistein pre-incubated *Bradyrhizobium japonicum* by soybean [*Glycine max* (L.) Merr] cultivars. Plant Soil 229:41–46

Besseau S, Hoffmann L, Geoffroy P, Lapierre C, Pollet B, Legranda M (2007) Flavonoid accumulation in *Arabidopsis* repressed in lignin synthesis affects auxin transport and plant growth. Plant Cell 19:148–162

Birt DF, Jeffery E (2013) Flavonoids. American Society for Nutrition. Adv Nutr 4:576–577

Bonfante P, Genre A (2008) Plants and arbuscular mycorrhizal fungi: an evolutionary-developmental perspective. Trends Plant Sci 13:492–498

Bonfante P, Genre A (2010) Mechanisms underlying beneficial plant-fungus interactions in mycorrhizal symbiosis. Nat Commun 1:48. doi:10.1038/ncomms1046

Brechenmacher L, Lei Z, Libault M, Findley S, Sugawara M, Sadowsky MJ, Sumner LW, Stacey G (2010) Soybean metabolites regulated in root hairs in response to the symbiotic bacterium *Bradyrhizobium japonicum*. Plant Physiol 153:1808–1822

Brencic A, Winans SC (2005) Detection of and response to signals involved in host-microbe interactions by plant-associated bacteria. Microbiol Mol Biol Rev 69:155–194

Broghammer A, Krusell L, Blaise M, Sauer J, Sullivan JT, Maolanon N, Vinther M, Lorentzen A, Madsen EB, Jensen KJ, Roepstorff P, Thirup S, Ronson CW, Thygesen MB, Stougaard J (2012) Legume receptors perceive the rhizobial lipochitin oligosaccharide signal molecules by direct binding. Proc Natl Acad Sci USA 109:13859–13864

Broughton WJ, Jabbouri S, Perret X (2000) Keys to symbiotic harmony. J Bacteriol 182: 5641–5652

Broughton WJ, Zhang F, Perret X, Staehelin C (2003) Signals exchanged between legumes and *Rhizobium*: agricultural uses and perspectives. Plant Soil 252:129–137

Bucher M, Wegmuller S, Drissner D (2009) Chasing the structures of small molecules in arbuscular mycorrhizal signaling. Curr Opin Plant Biol 12:500–507

Bücking H, Liepold E, Ambilwade P (2012) The role of the mycorrhizal symbiosis in nutrient uptake of plants and the regulatory mechanisms underlying these transport processes. In: Dhal NK, Sahu SC (eds) Plant science. Intech, Janeza Trdine, p 107

Buee M, Rossignol M, Jauneau A, Ranjeva R, Becard G (2000) The pre-symbiotic growth of arbuscular mycorrhizal fungi is induced by a branching factor partially purified from plant root exudates. Mol Plant Microbe Interact 13:693–698

Buer CS, Muday GK (2004) The *transparent testa4* mutation prevents flavonoid synthesis and alters auxin transport and the response of *Arabidopsis* roots to gravity and light. Plant Cell 16:1191–1205

Buer CS, Imin N, Djordjevic MA (2010) Flavonoids: new roles for old molecules. J Integr Plant Biol 52:98–111

Capoen W, Sun J, Wysham D, Otegui MS, Venkateshwaran M, Hirsch S, Miwa H, Downie JA, Morris RJ, Ane JM, Oldroyd GE (2011) Nuclear membranes control symbiotic calcium signaling of legumes. Proc Natl Acad Sci USA 108:14348–14353

Carlsen SCK, Understrup A, Fomsgaard IS, Mortensen AG, Ravnskov S (2008) Flavonoids in roots of white clover: interaction of arbuscular mycorrhizal fungi and a pathogenic fungus. Plant Soil 302:33–43

Catford JG, Stehelin C, Larose G, Piché Y, Vierheilig H (2006) Systemically suppressed isoflavonoids and their stimulating effects on nodulation and mycorrhization in alfalfa split-root systems. Plant Soil 285:257–266

Cerri MR, Frances L, Laloum T, Auriac MC, Niebel A, Oldroyd GE, Barker DG, Fournier J, de Carvalho-Niebel F (2012) *Medicago truncatula* ERN transcription factors: regulatory interplay with NSP1/NSP2 GRAS factors and expression dynamics throughout rhizobial infection. Plant Physiol 160:2155–2172

Cesco S, Neumann G, Tomasi N, Pinton R, Weisskopf L (2010) Release of plant-borne flavonoids into the rhizosphere and their role in plant nutrition. Plant Soil 329:1–25

Chabot S, Bel-Rhlid R, Chênevert R, Piché Y (1992) Hyphal growth promotion *in vitro* of the VA mycorrhizal fungus, *Gigaspora margarita* Becker et Hall, by the activity of structurally specific flavonoid compounds under CO_2-enriched conditions. New Phytol 122:461–467

Chen D-S, Liu C-W, Roy S, Cousins D, Stacey N, Murray JD (2015) Identification of a core set of rhizobial infection genes using data from single cell-types. Front Plant Sci 6:575. doi:10.3389/fpls.2015.00575

Cheng A-X, Han X-J, Wu Y-F, Lou H-X (2014) The function and catalysis of 2-oxoglutarate-dependent oxygenases involved in plant flavonoid biosynthesis. Int J Mol Sci 15:1080–1095

Cheynier V, Comte G, Davies KM, Lattanzio V, Martens S (2013) Plant phenolics: recent advances on their biosynthesis, genetics, and ecophysiology. Plant Physiol Biochem 72:1–20

Cohen MF, Sakihama Y, Yamasaki H (2001) Roles of plant flavonoids in interactions with microbes: from protection against pathogens to the mediation of mutualism. In: Pandalai SG (ed) Recent research developments in plant physiology 2. Research Signpost, Trivandrum, pp 157–173

Cooke TJ, Poli D, Sztein AE, Cohen JD (2002) Evolutionary patterns in auxin action. Plant Mol Biol 49:319–338

Cooper JE (2004) Multiple responses of rhizobia to flavonoids during legume root infection. Adv Bot Res 41:1–62

Cooper JE (2007) Early interactions between legumes and rhizobia: disclosing complexity in a molecular dialogue. J Appl Microbiol 103:1355–1365

Cordeiro MAS, Ferreira DA, Paulino HB, Souza CRF, Siqueira JO, Carneiro MAC (2015) Mycorrhization stimulant based in formononetin associated to fungicide and doses of phosphorus in soybean in the Cerrado. Biosci J 31:1062–1070

Crutzen P, Mosier AR, Smith KA, Winiwarter W (2007) N_2O release from agro-fuel production negates global warming reduction by replacing fossil fuels. Atmos Chem Phys Discuss 7: 11191–11205

D'haeseleer K, Keyser AD, Goormachtig S, Holsters M (2010) Transcription factor MtATB2: about nodulation, sucrose, and senescence. Plant Cell Physiol 51:1416–1424

Dalla Via V, Zanetti ME, Blanco F (2016) How legumes recognize rhizobia. Plant Signal Behav 11(2):e1120396. doi:10.1080/15592324.2015.1120396

de Billy F, Grosjean C, May S, Bennett M, Cullimore JV (2001) Expression studies on *AUX1*-like genes in *Medicago truncatula* suggest that auxin is required at two steps in early nodule development. Mol Plant Microbe Interact 14:267–277

De'narie J, Debelle F, Truchet G, Prome JC (1993) *Rhizobium* and legume nodulation: a molecular dialogue. In: Palacios R, Moira J, Newton WE (eds) New horizons in nitrogen fixation. Kluwer, Dordrecht, pp 19–30

Delaux P-M, Radhakrishnan G, Oldroyd G (2015) Tracing the evolutionary path to nitrogen-fixing crops. Curr Opin Plant Biol 26:95–99

Demont N, Debellé F, Aurelle H, Dénarié J, Promé JC (1993) Role of the *Rhizobium meliloti nodF* and *nodE* genes in the biosynthesis of lipo-oligosaccharidic nodulation factors. J Biol Chem 268:20134–20142

Denison RF, Kiers ET (2011) Life histories of symbiotic rhizobia and mycorrhizal fungi. Curr Biol 21:775–785

Dixon RA, Pasinetti GM (2010) Flavonoids and Isoflavonoids: from plant biology to agriculture and neuroscience. Plant Physiol 154:453–457

Downie JA (2010) The roles of extracellular proteins, polysaccharides and signals in the interactions of rhizobia with legume roots. FEMS Microbiol Rev 34:150–170

Fauvart M, Michiels J (2008) Rhizobial secreted proteins as determinants of host specificity in the *Rhizobium*-legume symbiosis. FEMS Microbiol Lett 285:1–9

Feddermann N, Finlaya R, Bollerb T, Elfstrand M (2010) Functional diversity in arbuscular mycorrhiza—the role of gene expression, phosphorous nutrition and symbiotic efficiency. Fungal Ecol 3:1–8

Ferguson BJ, Indrasumunar A, Hayashi S, Lin M-L, Lin Y-H, Reid DE, Gresshoff PM (2010) Molecular analysis of legume nodule development and autoregulation. J Integr Plant Biol 52:61–76

Fokom R, Nana WL, Tchameni S, Nwaga D (2010) Arbuscular mycorrhizal fungi (AMF) colonisation and rhizobia nodulation of cowpea as affected by flavonoid application. Res J Agric Biol Sci 6:1015–1021

Fournier J, Timmers AC, Sieberer BJ, Jauneau A, Chabaud M, Barker DG (2008) Mechanism of infection thread elongation in root hairs of *Medicago truncatula* and dynamic interplay with associated rhizobial colonization. Plant Physiol 148:1985–1995

Gachomo E, Allen JW, Pfeffer PE, Govindarajulu M, Douds DD, Jin HR, Nagahashi G, Lammers PJ, Shachar-Hill Y, Bücking H (2009) Germinating spores of *Glomus intraradices* can use internal and exogenous nitrogen sources for de novo biosynthesis of amino acids. New Phytol 184:399–411

Gage DJ (2002) Analysis of infection thread development using Gfp- and DsRed-expressing *Sinorhizobium meliloti*. J Bacteriol 184:7042–7046

Gagnon H, Ibrahim RK (1998) Aldonic acids: a novel family of *nod* gene inducers of *Mesorhizobium loti*, *Rhizobium lupini*, and *Sinorhizobium meliloti*. Mol Plant Microbe Interact 11:988–998

Geneva M, Zehirov G, Djonova E, Kaloyanova N, Georgiev G, Stancheva I (2006) The effect of inoculation of pea plants with mycorrhizal fungi and *Rhizobium* on nitrogen and phosphorus assimilation. Plant Soil Environ 52:435–440

Genre A, Bonfante P (2010) The making of symbiotic cells in arbuscular mycorrhizal roots. In: Koltai H, Kapulnik Y (eds) Arbuscular mycorrhizas: physiology and function. Springer, New York, pp 57–71. doi:10.1007/978-90-481-9489-6_3

Genre A, Chabaud M, Timmers T, Bonfante P, Barker DG (2005) Arbuscular mycorrhizal fungi elicit a novel intracellular apparatus in *Medicago truncatula* root epidermal cells before infection. Plant Cell 17:3489–3499

Gholami A, Geyter ND, Pollier J, Goormachtig S, Goossens A (2014) Natural product biosynthesis in *Medicago* species. Nat Prod Rep 31:356–380

Gianinazzi S, Gianinazzi-Pearson V (1988) Mycorrhizae: a plant's health insurance. Chim Oggi 10:56–68

Gianinazzi-Pearson V, Branzanti B, Gianinazzi S (1989) *In vitro* enhancement of spore germination and early hyphal growth of a vesicular-arbuscular mycorrhizal fungus by host root exudates and plant flavonoids. Symbiosis 7:243–255

Gibson KE, Kobayashi H, Walker GC (2008) Molecular determinants of a symbiotic chronic infection. Annu Rev Genet 42:413–441

Giovannetti M, Avio L, Sbrana C (2010) Fungal spore germination and presymbiotic mycelial growth—physiological and genetic aspects. In: Koltai H, Kapulnik Y (eds) Arbuscular mycorrhizas: physiology and function. Springer, New York, pp 3–32

Gobbato E (2015) Recent developments in arbuscular mycorrhizal signaling. Curr Opin Plant Biol 26:1–7

Gough C (2003) *Rhizobium* symbiosis: insight into Nod factor receptors. Curr Biol 13:973–975

Gould KS, Lister C (2005) Flavonoid functions in plants. In: Anderson OM, Markham KR (eds) Flavonoids: chemistry, biochemistry, and applications. CRC, Boca Raton, pp 397–441

Gourion B, Berrabah F, Ratet P, Stacey G (2015) *Rhizobium*-legume symbioses: the crucial role of plant immunity. Trends Plant Sci 20:186–194

Graham TL (1991) Flavonoid and isoflavonoid distribution in developing soybean seedling tissues and in seed and root exudates. Plant Physiol 95:594–603

Grant C, Bittman S, Montreal M, Plenchette C, Morel C (2005) Soil and fertilizer phosphorus: effects on plant P supply and mycorrhizal development. Can J Plant Sci 85:3–14

Guenoune D, Galili S, Phillips DA, Volpin H, Chet I, Okon Y, Kapulnik Y (2001) The defense response elicited by the pathogen *Rhizoctonia solani* is suppressed by colonization of the AM-fungus *Glomus intraradices*. Plant Sci 160:925–932

Gutjahr C, Parniske M (2013) Cell and developmental biology of arbuscular mycorrhiza symbiosis. Annu Rev Cell Dev Biol 29:593–617

Haag AF, Arnold MFF, Myka KK, Kerscher B, Dall'Angelo S, Zanda M, Mergaert P, Ferguson GP (2013) Molecular insights into bacteroid development during *Rhizobium*-legume symbiosis. FEMS Microbiol Rev 37:364–383

Halbwirth H (2010) The creation and physiological relevance of divergent hydroxylation patterns in the flavonoid pathway. Int J Mol Sci 11:595–621

Hamel LP, Beaudoin N (2010) Chitooligosaccharide sensing and downstream signaling: contrasted outcomes in pathogenic and beneficial plant-microbe interactions. Planta 232: 787–806

Haney CH, Riely BK, Tricoli DM, Cook DR, Ehrhardt DW, Long SR (2011) Symbiotic rhizobia bacteria trigger a change in localization and dynamics of the *Medicago truncatula* receptor kinase LYK3. Plant Cell 23:2774–2787

Harrison M, Dixon RA (1993) Isoflavonoid accumulation and expression of defense gene transcripts during the establishment of vesicular-arbuscular mycorrhizal associations in roots of *Medicago truncatula*. Mol Plant Microbe Interact 6:643–654

Hartwig UA, Maxwell CA, Joseph CM, Phillips DA (1990) Chrysoeriol and luteolin released from alfalfa seeds induce *nod* genes in *Rhizobium meliloti*. Plant Physiol 92:116–122

Hassan S, Mathesius U (2012) The role of flavonoids in root-rhizosphere signalling: opportunities and challenges for improving plant-microbe interactions. J Exp Bot 63:3429–3444

Herder GD, Parniske M (2009) The unbearable naivety of legumes in symbiosis. Curr Opin Plant Biol 12:491–499

Hernández I, Alegre L, Van Breusegem F, Munné-Bosch S (2009) How relevant are flavonoids as antioxidants in plants? Trends Plant Sci 14:125–132

Hichri I, Meilhoc E, Boscari A, Bruand C, Frendo P, Brouquisse R (2016) Nitric oxide: jack-of-all-trades of the nitrogen-fixing symbiosis? Adv Bot Res 77. doi:10.1016/bs.abr.2015.10.014

Hijri M, Sanders IR (2005) Low gene copy number shows that arbuscular mycorrhizal fungi inherit genetically different nuclei. Nature 433:160–163

Hiltner L (1904) Uber neure erfahrungen und probleme auf dem gebeit der bodenbackteriologie und unter besonderer berucksichtigung der grundungung und brache. Arb Deut Landwirsch Ges 98:59–78

Hofferek V, Mendrinna A, Gaude N, Krajinski F, Devers EA (2014) MiR171h restricts root symbioses and shows like its target *NSP2*a complex transcriptional regulation in *Medicago truncatula*. BMC Plant Biol 14:199. doi:10.1186/s12870-014-0199-1

Hsieh K, Huang AH (2007) Tapetosomes in *Brassica* tapetum accumulate endoplasmic reticulum-derived flavonoids and alkanes for delivery to the pollen surface. Plant Cell 19:582–596

Hughes JK, Hodge A, Fitter AH, Atkin OK (2008) Mycorrhizal respiration: implications for global scaling relationships. Trends Plant Sci 13:583–588

Hungria M, Joseph CM, Phillips DA (1991) *Rhizobium nod*-gene inducers exuded naturally from roots of common bean (*Phaseolus vulgaris* L.). Plant Physiol 97:759–764

Indrasumunar A, Kereszt A, Searle I, Miyagi M, Li D, Nguyen CD, Men A, Carroll BJ, Gresshoff PM (2010) Inactivation of duplicated nod factor receptor 5 (*NFR5*) genes in recessive loss-of-function non-nodulation mutants of allotetraploid soybean (*Glycine max* L. Merr.) Plant Cell Physiol 51:201–214

Indrasumunar A, Wilde J, Hayashi S, Li D, Gresshoff PM (2015) Functional analysis of duplicated symbiosis receptor kinase (*SymRK*) genes during nodulation and mycorrhizal infection in soybean (*Glycine max*). J Plant Physiol 176:157–168

Irving HR, Boukli NM, Kelly MN, Broughton WJ (2000) Nod-factors in symbiotic development of root hairs. In: Ridge RW, Emons AMC (eds) Root hairs: cell and molecular biology. Springer, Tokyo, pp 241–265

Ivanova KA, Tsyganova AV, Brewin NJ, Tikhonovich IA, Tsyganov VE (2015) Induction of host defences by *Rhizobium* during ineffective nodulation of pea (*Pisum sativum* L.) carrying symbiotically defective mutations *sym40* (PsEFD), *sym33* (PsIPD3/PsCYCLOPS) and *sym42*. Protoplasma 252:1505–1517

Jain V, Nainawatee HS (1999) Flavonoids influence growth and saccharide metabolism of *Rhizobium meliloti*. Folia Microbiol 44:311–316

Janczarek M, Skorupska A (2011) Modulation of *rosR* expression and exopolysaccharide production in *Rhizobium leguminosarum* bv. trifolii by phosphate and clover root exudates. Int J Mol Sci 12:4132–4155

Janczarek M, Rachwał K, Cieśla J, Ginalska G, Bieganowski A (2015) Production of exopolysaccharide by *Rhizobium leguminosarum* bv. trifolii and its role in bacterial attachment and surface properties. Plant Soil 388:211–227

Jensen ES, Peoples MB, Boddey RM, Gresshoff PM, Hauggaard-Nielsen H, Alves BJR, Morrison MJ (2012) Legumes for mitigation of climate change and feedstock in a bio-based economy—a review. Agron Sustain Dev 32:329–364

Johnson NC (2010) Resource stoichiometry elucidates the structure and function of arbuscular mycorrhizas across scales. New Phytol 185:631–647

Jones KM, Kobayashi H, Davies BW, Taga ME, Walker GC (2007) How rhizobial symbionts invade plants: the *Sinorhizobium-Medicago* model. Nat Rev Microbiol 5:619–633

Kanamori N, Madsen LH, Radutoiu S, Frantescu M, Quistgaard EM, Miwa H, Downie JA, James EK, Felle HH, Haaning LL, Jensen TH, Sato S, Nakamura Y, Tabata S, Sandal N, Stougaard J (2006) A nucleoporin is required for induction of Ca^{2+} spiking in legume nodule development and essential for rhizobial and fungal symbiosis. Proc Natl Acad Sci USA 103:359–364

Kapulnik Y, Joseph CM, Phillips DA (1987) Flavone limitation to root nodulation and symbiotic nitrogen fixation in alfalfa. Plant Physiol 84:1193–1196

Kelly SJ, Muszynski A, Kawaharada Y, Hubber AM, Sullivan JT, Sandal N, Carlson RW, Stougaard J, Ronson CW (2012) Conditional requirement for exopolysaccharide in the *Mesorhizobium–Lotus* symbiosis. Mol Plant Microbe Interact 26:319–329

Kereszt A, Mergaert P, Kondorosi E (2011) Bacteroid development in legume nodules: evolution of mutual benefit or of sacrificial victims? Mol Plant Microbe Interact 24:1300–1309

Kistner C, Winzer T, Pitzschke A, Mulder L, Sato S, Kaneko T, Tabata S, Sandal N, Stougaard J, Webb KJ, Szczyglowski K, Parniske M (2005) Seven *Lotus japonicus* genes required for transcriptional reprogramming of the root during fungal and bacterial symbiosis. Plant Cell 17:2217–2229

Kloppholz S, Kuhn H, Requena N (2011) A secreted fungal effector of *Glomus intraradices* promotes symbiotic biotrophy. Curr Biol 21:1204–1209

Kobae Y, Hata S (2010) Dynamics of periarbuscular membranes visualized with a fluorescent phosphate transporter in arbuscular mycorrhizal roots of rice. Plant Cell Physiol 51:341–353

Kobayashi H, Naciri-Graven Y, Broughton WJ, Perret X (2004) Flavanoids induce temporal shifts in gene-expression of nod-box controlled loci in *Rhizobium* sp. NGR234. Mol Microbiol 51: 335–347

Kondorosi E, Kondorosi E (2004) Endoreduplication and activation of the anaphase-promoting complex during symbiotic cell development. FEBS Lett 567:152–157

Kondorosi E, Mergaert P, Kereszt A (2013) A paradigm for endosymbiotic life: cell differentiation of rhizobium bacteria provoked by host plant factors. Annu Rev Microbiol 67:611–628

Kosslak RM, Bookland R, Barkei J, Paaren EH, Appelbaum ER (1987) Induction of *Bradyrhizobium japonicum* common *nod* genes by isoflavones isolated from *Glycine max*. Proc Natl Acad Sci USA 84:7428–7432

Kosuta S, Chabaud M, Lougnon G, Gough C, Dénarie J, Barker DG, Bécard G (2003) A diffusible factor from arbuscular mycorrhizal fungi induces symbiosis-specific *MtENOD11* expression in roots of *Medicago truncatula*. Plant Physiol 131:1–11

Kouchi H, Imaizumi-Anraku H, Hayashi M, Hakoyama T, Nakagawa T, Umehara Y, Suganuma N, Kawaguchi M (2010) How many peas in a pod? Legume genes responsible for mutualistic symbioses underground. Plant Cell Physiol 51:1381–1397

Laffont C, Rey T, André O, Novero M, Kazmierczak T, Debellé F, Bonfante P, Jacquet C, Frugier F (2015) The CRE1 cytokinin pathway is differentially recruited depending on *Medicago truncatula* root environments and negatively regulates resistance to a pathogen. PLoS One 10(1):e0116819. doi:10.1371/journal.pone.0116819

Laloum T, Baudin M, Frances L, Lepage A, Billault-Penneteau B, Cerri MR, Ariel F, Jardinaud M-F, Gamas P, de Carvalho-Niebel F, Niebel A (2014) Two CCAAT-box-binding transcription factors redundantly regulate early steps of the legume-rhizobia endosymbiosis. Plant J 79:757–768

Laparre J, Malbreila M, Letissec F, Portaisc JC, Rouxa C, Bécarda G, Puech-Pagèsa V (2014) Combining metabolomics and gene expression analysis reveals that propionyl- and butyryl carnitines are involved in late stages of arbuscular mycorrhizal symbiosis. Mol Plant 7: 554–566

Laplaze L, Lucas M, Champion A (2015) Rhizobial root hair infection requires auxin signaling. Trends Plant Sci 20:332–334

Lareen A, Burton F, Schäfer P (2016) Plant root-microbe communication in shaping root microbiomes. Plant Mol Biol 90:575–587

Larose G, Chênevert R, Moutoglis P, Gagné S, Piché Y, Vierheilig H (2002) Flavonoid levels in roots of *Medicago sativa* are modulated by the developmental stage of the symbiosis and the root colonizing arbuscular mycorrhizal fungus. J Plant Physiol 159:1329–1339

Leibovitch S, Migner P, Zhang F, Smith DL (2001) Evaluation of the effect of Soyasignal technology on soybean yield [*Glycine max* (L.)] Merr under field conditions over 6 years in Eastern Canada and the Northern United States. J Agron Crop Sci 187:281–292

Lepiniec L, Debeaujon I, Routaboul JM, Baudry A, Pourcel L, Nesi N, Caboche M (2006) Genetics and biochemistry of seed flavonoids. Annu Rev Plant Biol 57:405–430

Lévy J, Bres C, Geurts R, Chalhoub B, Kulikova O, Duc G, Journet EP, Ané JM, Lauber E, Bisseling T, Dénarié J, Rosenberg C, Debellé F (2004) A putative Ca^{2+} and calmodulin-dependent protein kinase required for bacterial and fungal symbioses. Science 303:1361–1364

Li B, Li Y-Y, Wu H-M, Zhang F-F, Lia C-J, Li X-X, Lambers H, Li L (2016) Root exudates drive interspecific facilitation by enhancing nodulation and N_2 fixation. PNAS. doi:10.1073/pnas.1523580113

Liang Y, Cao Y, Tanaka K, Wan J, Choi J, ho ho Kang C, Qui J, Stacey G (2013) Nonlegumes respond to rhizobial nod factors by suppressing the innate immune response. Science 34: 1384-1387

Limpens E, Bisseling T (2008) Nod factor signal transduction in the Rhizobium-legume symbiosis. In: AMC E, Ketelaar T (eds) Root hairs, Plant cell monographs, vol 12. Springer, Berlin, pp 249–276

Limpens E, van Zeijl A, Geurts R (2015) Lipochitooligosaccharides modulate plant host immunity to enable endosymbioses. Annu Rev Phytopathol 53:311–334

Lira MA Jr, Nascimento LRS, Fracetto GGM (2015) Legume-rhizobia signal exchange: promiscuity and environmental effects. Front Microbiol 6:945. doi:10.3389/fmicb.2015.00945

López-Baena FJ, Ruiz-Sainz JE, Rodríguez-Carvajal MA, Vinardell JM (2016) Bacterial molecular signals in the *Sinorhizobium fredii*-soybean symbiosis. Int J Mol Sci 17:755. doi:10.3390/ijms17050755

López-Lara IM, Geiger O (2001) The nodulation protein NodG shows the enzymatic activity of an 3-oxoacyl-acyl carrier protein reductase. Mol Plant Microbe Interact 14:349–357

Madsen LH, Tirichine L, Jurkiwicz A, Sullivan JT, Heckmann AB, Bek AS, Ronson CW, James EK, Stougaard J (2010) The molecular network governing nodule organogenesis and infection in the model legume *Lotus japonicus*. Nat Commun 1:1–12

Magori S, Kawaguchi M (2009) Long-distance control of nodulation: molecules and models. Mol Cells 27:129–134

Maier W, Peipp H, Schmidt J, Wray V, Strack D (1995) Levels of a terpenoid glycoside (blumenin) and cell wallbound phenolics in some cereal mycorrhizas. Plant Physiol 109: 465–470

Maillet F, Poinsot V, André O, Puech-Pagès V, Haouy A, Gueunier M, Cromer L, Giraudet D, Formey D, Niebel A, Martinez EA, Driguez H, Bécard G, Dénarié J (2011) Fungal lipochitooligosaccharide symbiotic signals in arbuscular mycorrhiza. Nature 469:58–63

Makarova LE, Dudareva LV, Petrova IG (2015) The content of phenolic compounds in the pea seedling root exudates depends on the size of their roots and inoculation of bacteria mutualistic and antagonistic type of interactions. J Stress Physiol Biochem 11:94–103

Manchanda G, Garg N (2007) Endomycorrhizal and rhizobial symbiosis: how much do they share? J Plant Interact 2:79–88

Mandal SM, Chakraborty D, Dey S (2010) Phenolic acids act as signaling molecules in plant-microbe symbioses. Plant Signal Behav 5:359–368

Mandal SM, Chakraborty D, Dutta SR, Ghosh AK, Pati BR, Korpole S, Paul D (2016) Induction of *nodD* gene in a *Betarhizobium* isolate, *Cupriavidus* sp. of *Mimosa pudica*, by root nodule phenolic acids. Curr Microbiol 72:733–737

Mapope N, Dakora FD (2013) Rrole of flavonoid and isoflavonoid molecules in symbiotic functioning and host-plant defence in the leguminosae. In: Gurib-Fakim A, Eloff JN (eds) Chemistry for sustainable development in Africa. Springer, Berlin, Heidelberg, pp 33–48

Marais JPJ, Deavours B, Dixon RA, Ferreira D (2006) The stereochemistry of flavonoids. In: Grotewold E (ed) The science of flavonoids. The Ohio State University, Columbus, OH, pp 1–46

Markmann K, Giczey G, Parniske M (2008) Functional adaptation of a plant receptorkinase paved the way for the evolution of intracellular root symbioses with bacteria. PLoS Biol 6:e68

Marsh JF, Rakocevic A, Mitra RM, Brocard L, Sun J, Eschstruth A, Long SR, Schultze M, Ratet P, Oldroyd GE (2007) *Medicago truncatula* NIN is essential for rhizobial-independent nodule

organogenesis induced by autoactive calcium/calmodulin-dependent protein kinase. Plant Physiol 144:324–335
Martens S, Mithöfer A (2005) Flavones and flavone synthases. Phytochemistry 66:2399–2407
Mathesius U (2001) Flavonoids induced in cells undergoing nodule organogenesis in white clover are regulators of auxin breakdown by peroxidase. J Exp Bot 52:419–426
Mathesius U (2009) Comparative proteomic studies of root-microbe interactions. J Proteomics 72:353–366
Mathesius U, Bayliss C, Weinman JJ, Schlaman HRM, Spaink HP, Rolfe BG, Mccully ME, Djordjevic MA (1998) Flavonoids synthesized in cortical cells during nodule initiation are early developmental markers in white clover. Mol Plant Microbe Interact 11:1223–1232
McNear DH Jr (2013) The rhizosphere—roots, soil and everything in between. Nat Educ Knowl 4:1
Mechri B, Tekaya M, Cheheb H, Attia F, Hammami M (2015) Accumulation of flavonoids and phenolic compounds in olive tree roots in response to mycorrhizal colonization: a possible mechanism for regulation of defense molecules. J Plant Physiol 185:40–43
Meixner C, Ludwig-Müller J, Miersch O, Gresshoff P, Staehelin C, Vierheilig H (2005) Lack of mycorrhizal autoregulation and phytohormonal changes in the supernodulating soybean mutant *nts1007*. Planta 222:709–715
Messinese E, Mun JH, Yeun LH, Jayaraman D, Rougé P, Barre A, Lougnon G, Schornack S, Bono JJ, Cook DR, Ané JM (2007) A novel nuclear protein interacts with the symbiotic DMI3 calcium- and calmodulin-dependent protein kinase of *Medicago truncatula*. Mol Plant Microbe Interact 20:912–921
Miransari M, Smith DL (2009) Alleviating salt stress on soybean (*Glycine max* (L.) Merr.)-*Bradyrhizobium japonicum* symbiosis, using signal molecule genistein. Eur J Soil Biol 45: 146–152
Miri M, Janakirama P, Held M, Ross L, Szczyglowski K (2016) Into the root: how cytokinin controls rhizobial infection. Trends Plant Sci. doi:10.1016/j.tplants.2015.09.003
Mitra RM, Gleason CA, Edwards A, Hadfield J, Downie JA, Oldroyd GED, Long SR (2004a) A Ca^{2+}/calmodulin-dependent protein kinase required for symbiotic nodule development: gene identification by transcript-based cloning. Proc Natl Acad Sci USA 101:4701–4705
Mitra RM, Shaw SL, Long SR (2004b) Six nonnodulating plant mutants defective for Nod factor-induced transcriptional changes associated with the legume-rhizobia symbiosis. Proc Natl Acad Sci USA 101:10217–10222
Miwa H, Sun J, Oldroyd GE, Downie JA (2006) Analysis of Nod-factor-induced calcium signaling in root hairs of symbiotically defective mutants of *Lotus japonicus*. Mol Plant Microbe Interact 19:914–923
Mohammadi K, Khalesro S, Sohrabi Y, Heidari G (2011) A review: beneficial effects of the mycorrhizal fungi for plant growth. J Appl Environ Biol Sci 1:310–319
Morandi D, Bailey JA, Gianinazzi-Pearson V (1984) Isoflavonoid accumulation in soybean roots infected with vesicular-arbuscular mycorrhizal fungi. Physiol Plant Pathol 24:357–364
Morandi D, Branzanti B, Gianinazzi-Pearson V (1992) Effect of some plant flavonoid on *in vitro* behavior of an arbuscular mycorrhizal. Agronomie 12:811–816
Morandi D, Le Signor C, Gianinazzi-Pearson V, Duc G (2009) A *Medicago truncatula* mutant hyper-responsive to mycorrhiza and defective for nodulation. Mycorrhiza 19:435–441
Morieri G, Martinez EA, Jarynowski A, Driguez H, Morris R, Oldroyd GED, Downie JA (2013) Host-specific nod-factors associated with *Medicago truncatula* nodule infection differentially induce calcium influx and calcium spiking in root hairs. New Phytol 200:656–662
Moscatiello R, Squartini A, Mariani P, Navazio L (2010) Flavonoid-induced calcium signalling in *Rhizobium leguminosarum* bv. viciae. New Phytol 188:814–823
Mukherjee A, Ané JM (2011) Germinating spore exudates from arbuscular mycorrhizal fungi: molecular and developmental responses in plants and their regulation by ethylene. Mol Plant Microbe Interact 24:260–270

Mulder L, Hogg B, Bersoult A, Cullimore JV (2005) Integration of signalling pathways in the establishment of the legume-rhizobia symbiosis. Physiol Plant 123:207–218
Murphy A, Peer WA, Taiz L (2000) Regulation of auxin transport by aminopeptidases and endogenous flavonoids. Planta 211:315–324
Murray JD (2011) Invasion by invitation: rhizobial infection in legumes. Mol Plant Microbe Interact 24:631–639
Nair MG, Safir GR, Siqueira JO (1991) Isolation and identification of vesicular-arbuscular mycorrhiza-stimulatory compounds from clover (*Trifolium repens*) roots. Appl Environ Microbiol 57:434–439
Naoumkina M, Dixon RA (2008) Subcellular localization of flavonid natural products: a signalling function? Plant Signal Behav 3:573–575
Naoumkina M, Farag MA, Sumner LW, Tang Y, Liu C-J, Dixon RA (2007) Different mechanisms for phytoalexin induction by pathogen and wound signals in *Medicago truncatula*. Proc Natl Acad Sci USA 104:17909–17915
Navazio L, Moscatiello R, Genre A, Novero M, Baldan B, Bonfante P, Mariani P (2007) A diffusible signal from arbuscular mycorrhizal fungi elicits a transient cytosolic calcium elevation in host plant cells. Plant Physiol 144:673–681
Nelson MS, Sadowsky MJ (2015) Secretion systems and signal exchange between nitrogen-fixing rhizobia and legumes. Front Plant Sci 6:491. doi:10.3389/fpls.2015.00491
Ng JLP, Perrine-Walker F, Wasson AP, Mathesius U (2015) Flavonoids and auxin transport inhibitors rescue symbiotic nodulation in the *Medicago truncatula* cytokinin perception mutant *cre1*. Plant Cell 27:2210–2226
Nijjer S, Rogers WE, Siemann E (2010) The impacts of fertilization on mycorrhizal production and investment in western gulf coast grasslands. Am Midl Nat 163:124–133
Novák K, Chovanec P, Skrdleta V, Kropacova M, Lisa L, Nemcova M (2002) Effect of exogenous flavonoids on nodulation of pea (*Pisum sativum* L.) J Exp Bot 53:1735–1745
O'Neil M, Heckelman P, Koch C, Roman K, Kenny C, D'Arecca M (1996) The merck index: encyclopedia of chemicals, drugs and biological. Merck Publishing, Rahway, NJ
Okazaki S, Tittabutr P, Teulet A, Thouin J, Fardoux J, Chaintreuil C, Gully D, Arrighi J-F, Furuta N, Miwa H, Yasuda M, Nouwen N, Teaumroong N, Giraud E (2016) *Rhizobium*-legume symbiosis in the absence of Nod factors: two possible scenarios with or without the T3SS. ISME J 10:64–74
Oldroyd GE, Downie JA (2008) Coordinating nodule morphogenesis with rhizobial infection in legumes. Annu Rev Plant Biol 59:519–546
Oldroyd GED, Murray JD, Poole PS, Downie JA (2011) The rules of engagement in the legume-rhizobial symbiosis. Annu Rev Genet 45:119–144
Parniske M (2008) Arbuscular mycorrhiza: the mother of plant root endosymbioses. Nat Rev Microbiol 6:763–775
Peck MC, Fisher RF, Long SR (2006) Diverse flavonoids stimulate NodD1 binding to *nod* gene promoters in *Sinorhizobium meliloti*. J Bacteriol 188:5417–5427
Peck MC, Fisher RF, Bliss R, Long SR (2013) Isolation and characterization of mutant *Sinorhizobium meliloti* NodD1 proteins with altered responses to luteolin. J Bacteriol 195: 3714–3723
Peer WA, Bandyopadhyay A, Blakeslee JJ, Makam SN, Chen RJ, Masson PH, Murphy AS (2004) Variation in expression and protein localization of the PIN family of auxin efflux facilitator proteins in flavonoid mutants with altered auxin transport in *Arabidopsis thaliana*. Plant Cell 16:1898–1911
Peer WA, Blakeslee JJ, Yang HB, Murphy AS (2011) Seven things we think we know about auxin transport. Mol Plant 4:487–504
Peix A, Ramírez-Bahena MH, Velázquez E, Bedmar EJ (2015) Bacterial associations with legumes. Crit Rev Plant Sci 34:17–42
Pinior A, Wyss U, Piché Y, Vierheilig H (1999) Plants colonized by AM fungi regulate further root colonization by AM fungi through altered root exudation. Can J Bot 77:891–897

Pollastri S, Tattini M (2011) Flavonols: old compounds for old roles. Ann Bot 108:1225–1233

Ponce M, Cervino M, Erra-Balsells R, Ocampo JA, Godeas A (2004) Flavonoids from shoots and roots of *Trifolium repens* (white clover) grown in presence or absence of the arbuscular mycorrhizal fungus *Glomus intraradices*. Phytochemistry 65:3131–3134

Pumplin N, Harrison MJ (2009) Live-cell imaging reveals periarbuscular membrane domains and organelle location in *Medicago truncatula* roots during arbuscular mycorrhizal symbiosis. Plant Physiol 151:809–819

Radutoiu S, Madsen LH, Madsen EB, Felle HH, Umehara Y, Grønlund M, Sato S, Nakamura Y, Tabata S, Sandal N, Stougaard J (2008) Plant recognition of symbiotic bacteria requires two LysM receptor-like kinases. Nature 425:585–592

Redmond JW, Batley M, Djordjevic MA, Innes RW, Kuempel PL, Rolfe BG (1986) Flavones induce expression of nodulation genes in *Rhizobium*. Nature 323:632–635

Reid DE, Ferguson BJ, Hayashi S, Lin Y-H, Gresshoff PM (2011) Molecular mechanisms controlling legume autoregulation of nodulation. Ann Bot 108:789–795

Remigi P, Zhu J, Young JPW, Masson-Boivin C (2016) Symbiosis within symbiosis: evolving nitrogen-fixing legume symbionts. Trends Microbiol 24:63–75

Requena N, Breuninger M (2004) The old arbuscular mycorrhizal symbiosis in the light of the molecular era. In: Esser K, Lüttge U, Beyschlag W, Murata J (eds) Progress in botany. Springer, Berlin, Heidelberg, pp 323–356

Requena N, Serrano E, Ocón A, Breuninger M (2007) Plant signals and fungal perception during arbuscular mycorrhizal establishment. Phytochemistry 68:33–40

Ribeiro CW, Alloing G, Mandon K, Frendo P (2015) Redox regulation of differentiation in symbiotic nitrogen fixation. Biochim Biophys Acta 1850:1469–1478

Rich MK, Schorderet M, Reinhardt D (2014) The role of the cell wall compartment in mutualistic symbioses of plants. Front Plant Sci 5:238. doi:10.3389/fpls.2014.00238

Riely BK, Lougnon G, Ané JM, Cook DR (2007) The symbiotic ion channel homolog DMI1 is localized in the nuclear membrane of *Medicago truncatula* roots. Plant J 49:208–216

Rose CM, Venkateshwaran M, Volkening JD, Grimsrud PA, Maeda J, Bailey DJ, Park K, Howes-Podoll M, den Os D, Yeun LH, Westphall MS, Sussman MR, Ane J-H, Coon JJ (2012) Rapid phosphoproteomic and transcriptomic changes in the rhizobia-legume symbiosis. Mol Cell Proteomics 11:724–744

Rossen L, Davis EO, Johnston AWB (1987) Plant-induced expression of Rhizobium genes involved in host specificity and early stages of nodulation. Trends Biochem Sci 12:430–433

Ryu H, Cho H, Choi D, Hwang I (2012) Plant hormonal regulation of nitrogen-fixing nodule organogenesis. Mol Cells 34:117–126

Saito A, Keda S, Ezura H, Minamisawa K (2007) Microbial community analysis of the phytosphere using culture-independent methodologies. Microbes Environ 22:93–105

Saito K, Yonekura-Sakakibara K, Nakabayashi R, Higashi Y, Yamazaki M, Tohge T, Fernie AR (2013) The flavonoid biosynthetic pathway in *Arabidopsis*: structural and genetic diversity. Plant Physiol Biochem 72:21–34

Saslowsky DE, Warek U, Winkel BSJ (2005) Nuclear localization of flavonoid enzymes in Arabidopsis. J Biol Chem 280:23735–23740

Scervino JM, Ponce MA, Erra-Bassells R, Vierheilig H, Ocampo JA, Godeas A (2005a) Arbuscular mycorrhizal colonization of tomato by *Gigaspora* and *Glomus* species in the presence of root flavonoids. J Plant Physiol 162:625–633

Scervino JM, Ponce MA, Erra-Bassells R, Vierheilig H, Ocampo JA, Godeas A (2005b) Flavonoids exhibit fungal species and genus specific effects on the presymbiotic growth of *Gigaspora* and *Glomus*. Mycol Res 109:789–794

Scervino JM, Ponce MA, Erra-Bassells R, Bompadre J, Vierheilig H, Ocampo JA, Godeas A (2007) The effect of flavones and flavonols on colonization of tomato plants by arbuscular mycorrhizal fungi of the genera *Gigaspora and Glomus*. Can J Microbiol 53:702–709

Scervino JM, Ponce MA, Monica ID, Vierheilig H, Ocampo JA, Godeas A (2009) Development of arbuscular mycorrhizal fungi in the presence of different patterns of *Trifolium repens* shoot flavonoids. J Soil Sci Plant Nutr 9:102–115

Searle IR, Men AE, Laniya TS, Buzas DM, Iturbe-Ormaetxe I, Carroll BJ, Gresshoff PM (2003) Long-distance signaling in nodulation directed by a CLAVATA1-like receptor kinase. Science 299:109–112

Shaw LJ, Hooker JE (2008) The fate and toxicity of the flavonoids naringenin and formononetin in soil. Soil Biol Biochem 40:528–536

Shaw LJ, Morris P, Hooker JE (2006) Perception and modification of plant flavonoid signals by rhizosphere microorganisms. Environ Microbiol 8:1867–1880

Shtark OY, Sulima AS, Zhernakov AI, Kliukova MS, Fedorina JV, Pinaev AG, Kryukov AA, Akhtemova GA, Tikhonovich IA, Zhukov VA (2016) Arbuscular mycorrhiza development in pea (*Pisum sativum* L.) mutants impaired in five early nodulation genes including putative orthologs of *NSP1* and *NSP2*. Symbiosis 68:129–144

Singh DP, Prabha R, Meena KK, Sharma L, Sharma AK (2014) Induced accumulation of polyphenolics and flavonoids in cyanobacteria under salt stress protects organisms through enhanced antioxidant activity. Am J Plant Sci 5(5):43916. doi:10.4236/ajps.2014.55087

Sinharoy S, Liu C, Breakspear A, Guan D, Shailes S, Nakashima J, Zhang S, Wen J, Torres-Jerez I, Oldroyd G, Murray JD, Udvardi MK (2016) A *Medicago truncatula* cystathionine-b-synthase-like domain-containing protein is required for rhizobial infection and symbiotic nitrogen fixation. Plant Physiol 170:2204–2217

Siqueira JO, Safir GR, Nair MG (1991a) Significance of phenolic compounds in plant-soil-microbial systems. Crit Rev Plant Sci 10:63–121

Siqueira JO, Safir GR, Nair MG (1991b) Stimulation of vesicular arbuscular mycorrhiza formation and growth of white clover by flavonoid compounds. New Phytol 118:87–93

Skorupska A, Wielbo J, Kidaj D, Marek-Kozaczuk M (2010) Enhancing *Rhizobium*-legume symbiosis using signaling factors. In: Khan MS (ed) Microbes for legume improvement. Springer, Wien, pp 27–54

Smit G, Puvanesarajah V, Carlson RW, Barbour WM, Stacey G (1992) *Bradyrhizobium japonicum nodD*1 can be specifically induced by soybean flavonoids that do not induce the *nodYABCSUIJ* operon. J Biol Chem 267:310–318

Smit P, Raedts J, Portyanko V, Debellé F, Gough C, Bisseling T, Geurts R (2005) NSP1 of the GRAS protein family is essential for rhizobial nod factor-induced transcription. Science 308:1789–1791

Smith SE, Read DJ (2008) Mycorrhizal symbiosis. Academic Press, Inc., San Diego, CA

Smith DL, Zhang F (1999) Composition for enhancing grain yield and protein yield of legumes grown under environmental conditions that inhibit or delay nodulation thereof. US Patent-5922316

Smith DL, Praslickova D, Ilangumaran G (2015) Inter-organismal signaling and management of the phytomicrobiome. Front Plant Sci 6:722. doi:10.3389/fpls.2015.00722

Sosa T, Valares C, Alias JC, Lobon NC (2010) Persistence of flavonoids in *Cictus landanifer* soils. Plant Soil 337:51–63

Soyano T, Kawaguchi M (2014) Systemic regulation of root nodule formation. Division of Symbiotic Systems, National Institute for Basic Biology, Okazaki, Aichi. Available via doi:10.5772/56991

Spini G, Decorosi F, Cerboneschi M, Tegli S, Mengoni A, Viti C, Giovannetti L (2016) Effect of the plant flavonoid luteolin on *Ensifer meliloti* 3001 phenotypic responses. Plant Soil 399: 159–178

Stafford HA (1997) Roles of flavonoids in symbiotic and defense functions in legume roots. Bot Rev 63:27–39

Steinkellner S, Lendzemo V, Langer I, Schweiger I, Khaosaad T, Toussaint J-P, Vierheilig H (2007) Flavonoids and strigolactones in root exudates as signals in symbiotic and pathogenic plant-fungus interactions. Molecules 12:1290–1306

Stracke S, Kistner C, Yoshida S, Mulder L, Sato S, Kaneko T, Tabata S, Sandal N, Stougaard J, Szczyglowski K, Parniske M (2002) A plant receptor-like kinase required for both bacterial and fungal symbiosis. Nature 417:959–962

Subramanian S, Graham MY, Yu O, Graham TL (2005) RNA interference of soybean isoflavone synthase genes leads to silencing in tissues distal to the transformation site and to enhanced susceptibility to phytophthora sojae. Plant Physiol 137:1345–1353

Subramanian S, Stacey G, Yu O (2007) Distinct, crucial roles of flavonoids during legume nodulation. Trends Plant Sci 12:282–285

Sugiyama A, Shitan N, Yazaki K (2007) Involvement of a soybean ATP-binding cassette-type transporter in the secretion of genistein, a signal flavonoid in legume-*Rhizobium* symbiosis. Plant Physiol 144:2000–2008

Sugiyama A, Yamazaki Y, Yamashita K, Takahashi S, Nakayama T, Yazaki K (2016) Developmental and nutritional regulation of isoflavone secretion from soybean roots. Biosci Biotechnol Biochem 80:89–94

Suzaki T, Kawaguchi M (2014) Root nodulation: a developmental program involving cell fate conversion triggered by symbiotic bacterial infection. Curr Opin Plant Biol 21:16–22

Suzuki H, Takahashi S, Watanabe R, Fukushima Y, Fujita N, Noguchi A, Yokoyama R, Nishitani K, Nishino T, Nakayama T (2006) An isoflavone conjugate-hydrolyzing beta-glucosidase from the roots of soybean (*Glycine max*) seedlings. Purification, gene cloning, phylogenetics, and cellular localization. J Biol Chem 281:30251–30259

Taurian T, Morón B, Soria-Díaz ME, Angelini JG, Tejero-Mateo P, Gil-Serrano A, Megías M, Fabra A (2008) Signal molecules in the groundnut-*Bradyrhizobia* interaction. Arch Microbiol 189:345–356

Taylor LP, Grotewold E (2005) Flavonoids as developmental regulators. Curr Opin Plant Biol 8: 317–323

Tian F, Jia T, Yu B (2014) Physiological regulation of seed soaking with soybean isoflavones on drought tolerance of *Glycine max* and *Glycine soja*. Plant Growth Regul 74:229–237

Timmers ACJ, Auriac MC, Truchet G (1999) Refined analysis of early symbiotic steps of the *Rhizobium-Medicago* interaction in relationship with microtubular cytoskeleton rearrangements. Development 126:3617–3628

Tóth K, Stacey G (2015) Does plant immunity play a critical role during initiation of the legume-*Rhizobium* symbiosis? Front Plant Sci 6:401. doi:10.3389/fpls.2015.00401

Tsai SM, Phillips DA (1991) Flavonoids released naturally from alfalfa promote development of symbiotic *Glomus* spores *in vitro*. Appl Environ Microbiol 57:1485–1488

Tsvetkova G, Teofilova T, Georgiev GI (2006) Effect of naringein and quercetin on activity of nodabc genes of strain d293 and following nodulation and nitrogen fixation response of inoculated pea plants (*Pisum sativum* l.). Gen Appl Plant Physiol Special Issue:67–71

van Zeijl A, Op den Camp RHM, Deinum EE, Charnikhova T, Franssen H, Op den Camp HJM, Bouwmeester H, Kohlen W, Bisseling T, Geurts R (2015) Rhizobium lipo-chitooligosaccharide signaling triggers accumulation of cytokinins in *Medicago truncatula* roots. Mol Plant 8:1213–1226

Venturi V, Keel C (2016) Signaling in the rhizosphere. Trends Plant Sci 21:187–198

Vieira RF, Mendes IC, Reis-Junior FB, Hungria M (2010) Symbiotic nitrogen fixation in tropical food grain legumes: current status. In: Khan MS, Musarrat J, Zaidi A (eds) Microbes for legume improvement. Springer, Dordercht, Heidelberg, London, New York, pp 427–472. doi:10.1007/978-3-211-99753-6

Vierheilig H (2004a) Regulatory mechanisms during the plant-arbuscular mycorrhizal fungus interaction. Can J Bot 82:1166–1176

Vierheilig H (2004b) Further root colonization by arbuscular mycorrhizal fungi in already mycorrhizal plants is suppressed after a critical level of root colonization. J Plant Physiol 161: 339–341

Vierheilig H, Piché Y (1995) Facteurs biochimiques potentiellement impliqués dans les interactions entre les champignons endomycorhiziens et leurs plantes non-hôtes. In: Fortin JA,

Charest C, Picheé Y (eds) La symbiose mycorhizienne. Orbis Frelighsburg, Québec, pp 109–124

Vierheilig H, Piché Y (2002) Signalling in arbuscular mycorrhiza: facts and hypotheses. In: Buslig B, Manthey J (eds) Flavonoids in cell functions. Kluwer/Plenum, New York, pp 23–39

Vierheilig H, Albrecht C, Bago B, Piché Y (1996) Do flavonoids play a role in root colonization by AM fungi? First International Conference on Mycorrhizae, Berkeley, CA

Vierheilig H, Bago B, Albrecht C, Poulin MJ, Piché Y (1998) Flavonoids and arbuscular mycorrhizal fungi. In: Manthey J, Buslig B (eds) Flavonoids in the living system. Plenum Press, New York, pp 9–33

Vierheilig H, Steinkellner S, Khaosaad T, Garcia-Garrido JM (2008) The biocontrol effect of mycorrhization on soilborne fungal pathogens and the autoregulation of the AM symbiosis: one mechanism, two effects? In: Varma A (ed) Mycorrhiza: state of the art, genetics and molecular biology, eco-function, biotechnology, eco-physiology, structure and systematics. Springer, Berlin, Heidelberg, pp 307–320

Volpin H, Elkind Y, Okon Y, Kapulnik Y (1994) A vesicular-arbuscular mycorrhizal fungus (*Glomus intraradix*) induces a defense response in alfalfa roots. Plant Physiol 104:683–689

Wais RJ, Keating DH, Long SR (2002) Structure-function analysis of nod factor-induced root hair calcium spiking in *Rhizobium*-legume symbiosis. Plant Physiol 129:211–224

Wang D, Yang S, Tang F, Zhu H (2012) Symbiosis specificity in the legume-rhizobial mutualism. Cell Miocrobiol 14:334–342

Wasson AP, Pellerone FI, Mathesius U (2006) Silencing the flavonoid pathway in *Medicago truncatula* inhibits root nodule formation and prevents auxin transport regulation by rhizobia. Plant Cell 18:1617–1629

Weerasinghe RR, Bird DMK, Allen NS (2005) Root-knot nematodes and bacterial Nod factors elicit common signal transduction events in *Lotus japonicus*. Proc Natl Acad Sci USA 102:3147–3152

Weston LA, Mathesius U (2013) Flavonoids: their structure, biosynthesis and role in the rhizosphere, including allelopathy. J Chem Ecol 39:283–297

Wilde P, Manal A, Stodden M, Sieverding E, Hilderbrandt U, Bothe H (2009) Biodiversity of arbuscular mycorrhizal fungi in roots and soils of two salt marshes. Environ Microbiol 11:1548–1561

Williams CA, Grayer RJ (2004) Anthocyanins and other flavonoids. Nat Prod Rep 21:539–573

Winkel-Shirley B (2001) It takes a garden. How work on diverse plant species has contributed to an understanding of flavonoid metabolism. Plant Physiol 127:1399–1404

Wong PP, Evans HJ (1971) Poly-ß-hydroxybutyrate utilization by soybean (*Glycine max* Merr.) nodules and assessment of its role in maintenance of nitrogenase activity. Plant Physiol 47:750–755

Xie Z-P, Staehelin C, Vierheilig H, Wiemken A, Jabbouri S, Broughton WJ, Vögeli-Lange R, Boller T (1995) Rhizobial nodulation factors stimulate mycorrhizal colonization of nodulating and nonnodulating soybeans. Plant Physiol 108:1519–1525

Xie ZP, Muller J, Wiemken A, Broughton WJ, Boller T (1997) Nod factors and tri-iodobenzoic acid stimulate mycorrhizal colonization and affect carbohydrate partitioning in mycorrhizal roots of *Lablab purpureus*. New Phytol 139:361–366

Xie F, Murray JD, Kim J, Heckmann AB, Edwards A, Oldroyd GE, Downie JA (2012) Legume pectate lyase required for root infection by rhizobia. Proc Natl Acad Sci USA 109:633–638

Yamazaki A, Hayashi M (2015) Building the interaction interfaces: host responses upon infection with microorganisms. Curr Opin Plant Biol 23:132–139

Yang S, Tang F, Gao M, Krishnan HB, Zhu H (2010) R gene-controlled host specificity in the legume-rhizobia symbiosis. Proc Natl Acad Sci USA 107:18735–18740

Yano K, Yoshida S, Muller J, Singh S, Banba M, Vickers K, Markmann K, White C, Schuller B, Sato S, Asamizu E, Tabata S, Murooka Y, Perry J, Wang TL, Kawaguchi M, Imaizumi-Anraku H, Hayashi M, Parniske M (2008) CYCLOPS, a mediator of symbiotic intracellular accommodation. Proc Natl Acad Sci USA 105:20540–20545

Yokota S, Mizuo K, Moriya N, Oshio S, Sugawara I, Takeda K (2009) Effect of prenatal exposure to diesel exhaust on dopaminergic system in mice. Neurosci Lett 449:38–41

Zhang JA, Subramanian S, Zhang YS, Yu O (2007) Flavone synthases from *Medicago truncatula* are flavanone-2-hydroxylases and are important for nodulation. Plant Physiol 144:741–751

Zhang L, Tian LH, Zhao JF, Song Y, Zhang CJ, Guo Y (2009) Identification of an apoplastic protein involved in the initial phase of salt stress response in rice root by two-dimensional electrophoresis. Plant Physiol 149:916–928

Zhao J, Dixon RA (2010) The 'ins' and 'outs' of flavonoid transport. Trends Plant Sci 15:72–80

Zhuang X, Gao J, Ma A, Fu S, Zhuang G (2013) Bioactive molecules in soil ecosystems: masters of the underground. Int J Mol Sci 14:8841–8868

Zuanazzi JAS, Clergeot PH, Quirion JC, Husson HP, Kondorosi A, Ratet P (1998) Production of *Sinorhizobium meliloti nod* gene activator and repressor flavonoids from *Medicago sativa* roots. Mol Plant Microbe Interact 11:784–794

Chapter 9
Mycorrhizas in Forest Tree Health

Vivek Kumar, Manoj Kumar, Ram Prasad, Narendra Tuteja, and Ajit Varma

Abstract Mycorrhizas impart fungus-root associations and a true symbiotic relation between fungi and plant roots, which is very close to that of nodular microbes (actinomycetes and bacteria) in legume crops. This type of association boosts both the allies (i.e. the fungus and its specific host plant) to be reciprocally benefited by each other, as the fungus absorbs hydrogenous nutrients and edaphic factors from soil and channelizes the same to the plant system, and in turn synthesis its nutrition (carbohydrates and photosynthates) from the host plant for its growth and multiplication.

There is now plenty of proof to support the mutual declaration that most tree species in normal ecosystems have mycorrhizal associations. Evidence about the global distribution of tree plants with diverse populations of mycorrhizal relations is used to establish associations with the major abiotic factors (water, temperature) which control the distribution pattern of forest trees, and also to extend more restricted edaphic factors. Environmental inferences of mycorrhizal associations in forest ecosystems and the role of soil or ecological influences, mycorrhizal fungus physiognomies or host plant belongings in an individual manner or in groupings are considered accordingly. Factors which can affect the existence mycorrhizal associations are (a) root characteristics, (b) edaphic or ecological factors, (c) soil biodynamics, (d) soil commotion and (e) tree host-fungus compatibility, thus, we address the overall ecological dynamics in this review. Environmental themes on ecological aspects have been discussed in this chapter which include (a) mycorrhizal phenology, (b) influences responsible for variable grades of mycorrhizal reliance in host trees, (c) the implication of mycorrhizal hyphae in soil, (d) nutritional rivalry connecting mycorrhizal and non-mycorrhizal tree species and (e) mycorrhizal connections associating effluence and additional stresses, the rhizosphere, soil possessions and so on. The population dynamics of mycorrhizal fungi and the impact of their relations on forest tree ecology also form a comprehensive discussion.

V. Kumar • M. Kumar (✉) • R. Prasad • N. Tuteja • A. Varma
Amity Institute of Microbial Technology (AIMT), Amity University Uttar Pradesh, Block 'E-3', 4th Floor, Sector-125, Noida, Gautam Buddha Nagar, Uttar Pradesh 201313, India
e-mail: manojjnu@gmail.com

A. Varma et al. (eds.), *Mycorrhiza - Function, Diversity, State of the Art*,
DOI 10.1007/978-3-319-53064-2_9

9.1 Introduction

Fungi play essential roles in many microbiological and environmental processes, swaying soil richness and putrefaction, the driving of minerals and organic matter, as well as tree health and nourishment. Mycota are heterotrophs, necessitating exterior sources of carbon for vigor and cellular synthesis, and they have espoused three different trophic approaches to obtaining this carbon, occurring as saprotrophs, necrotrophs and biotrophs. Mycorrhizal symbiosis is the most primitive, extensive form of fungal cooperation with plants and trees. The term 'mykorrhiza' was first used in 1885 by Frank (2005) to define the adapted root structures of forest trees, and has since been protracted to cover a range of mutualistic, symbiotic relations between fungi and tree plant roots (Smith and Read 2008). A diverse category of mycorrhizal symbiosis has been classified on the basis of morphological features and the fungal and tree species involved.

Arbuscular mycorrhiza is the recorded primitive and widespread form. Paleobotanical and molecular sequence data proposed that the first terrestrial plants formed links with Glomalean fungi from the Glomeromycota about 460 million years ago (Redecker et al. 2000). This is projected to be some 300–400 million years before the advent of root nodule symbioses with nitrogen-fixing bacteria. Arbuscular mycorrhizal (AM) associations can be molded with a very wide range of tree/plant species—as many as 250,000. 150–200 individual species of AM fungi have so far been noted on the basis of morphology, but DNA-based revisions recommend that the true variety of these symbionts may be very much advanced (Fitter 2005). The cooperation is characterized by highly pronged fungal assemblies—*arbuscules*—which cultivate intracellularly without penetrating the host plasmalemma.

9.1.1 Mycorrhizal Contribution and Ecological Aspects

Mycorrhizas form extremely developed, mutualistic relations between earth (soil), fungi and plant roots. The associates in this connotation are adherents of the fungus kingdom (Zygomycetes, Ascomycetes and Basidiomycetes, but not protoctistan fungi such as Oomycetes) and maximum vascular plants (Harley and Smith 1983; Kendrick 1985). In the mycorrhizal literature, the term 'symbiosis' is often used to describe these highly interdependent mutualistic relationships whereby the host plant receives mineral nutrients while the fungus obtains photosynthetically derived carbon compounds (Harley 1989; Harley and Smith 1983). A minimum of seven diverse forms of mycorrhizal association have been documented, relating different groups of fungi and host plants and distinct morphology shapes (Hadley 1982; Harley 1989; Harley and Smith 1983; Read 1983). The frequent associations are (a) arbuscular mycorrhizas (AM), in which zygomycetous fungi yield arbuscules, hyphae, as well as vesicles that are embedded in root cortex cells;

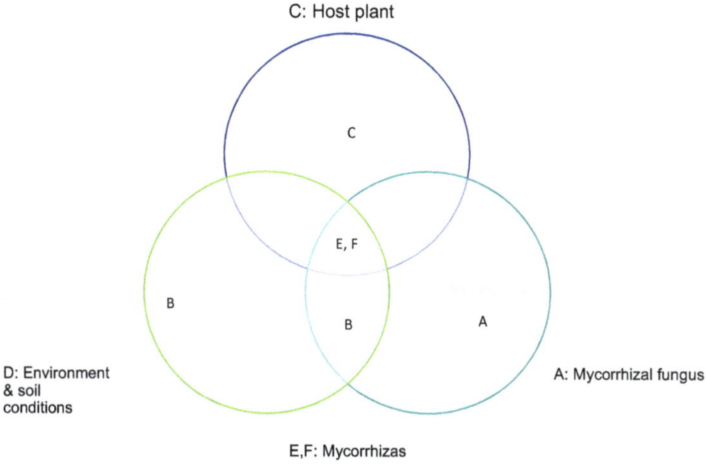

Fig. 9.1 The categorized regions refer to mycorrhizal ecological areas

(b) ectomycorrhizas (ECM), where Basidiomycetes and other fungi constitute a blanket around roots and a savory net within root cells; (c) orchid mycorrhizas, where fungi release loops of hyphae within roots (or stems) of orchidaceous plants; and (d) ericoid mycorrhizas, involving hyphal loops/coils in exterior cells of the slimy 'hair roots' of plants in the Ericales.

Associations of mycorrhizas are always dependent on the selective features of the host plant trees and conversing mycorrhizal fungi, where edaphic and ecological factors play important roles (Harley and Smith 1983; Mosse and Hayman 1980). Existing reported factors have been viewed for categorized regions worldwide which highlight variable climatic conditions for mycorrhizal ecology. Since the investigator (Högberg et al. 1999) reported that ectomycorrhizal fungi have an established relation with numerous tree species with common approach to infect (Fig. 9.1).

9.1.2 Mycorrhizas for Tree Root Architecture

Mycorrhizal fungi are understood as natural enhancers for plant root architecture and capacity builders of a plant by up to a thousand times in just a few months. It is an evidence from natural yard that the mycorrhizal roots of a mature tree which is laid out end to end and stretched for miles due to the mycorrhizal hyphae and its response at the feeding habit level (Fig. 9.2).

Use of high throughput microscopic facility has revolved the anatomy of AM, further it reveals the structural orientation at histological level (Abbott 1982; Abdul Karim 2004).

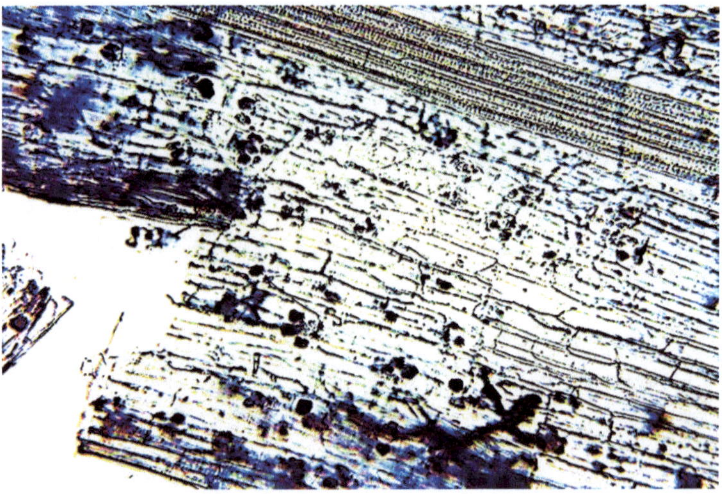

Fig. 9.2 Hyphae of *Rhizopogon mycorrhizas* attached to conifer root system (courtesy of Dr. Mike Amaranthus, Oregon State University)

Arbuscular mycorrhizal fungi (AMF) constitute mutualistic connotations with most native plants, including trees (Abbott et al. 1992). They are known to be beneficial partners and they get to induce prevalent changes in host tree physiology. It has long been considered that colonization by AMF cannot change root architecture, but significant alteration has now been demonstrated unequivocally for several plants, including trees with AMF-colonized plants usually possessing a more highly branched root system. Additionally, alteration of root system endurance has also been measured with AMF-colonized root arrangements (Abu-Zeyad et al. 1999). The mechanisms by which these alterations arise are not fully understood, but positive impacts are not entirely due to altered phosphorus nutrition or are likely to involve AMF interface with plant cell cycles. Upcoming research should focus on signifying the extent of AMF variation in root structure and development, and further the importance of AMF interactions with plant cell cycles in determining these changes (Barnola and Montilla 1997; Barroetaveña et al. 2007).

9.1.3 Mycorrhizas in Woodland Environments

To understand the scaling system on which mycorrhizal associations occur, an exemplary piece of research in Swedish woodlands found that between 0.6 and 1.2 million ECM were present in one square meter of forest land! In ancient forests, it has been reported that individual fungi can be predominantly huge—the mycelia of some extend over 100 square meters (Harley and Smith 1983).

The effects of mycorrhizas are not limited to the fungus and its host (Batty et al. 2002). One of the numerous ways in which they preserve soil health is by providing a 'safety net', averting nutrients from being leached away (Chilvers et al. 1987). The complete bionetwork can profit from tree plant diversity and enhanced soil edifices. Research in forest land in Australia and South Africa has uncovered that the networking systems of mycorrhizas in the forest permit the transfer of important amounts of carbon *between* trees, even those of diverse species (Chilvers et al. 1987; Batty et al. 2002).

Carbon is a proven 'energy currency' within ecological facets, and it has been proven that when Douglas fir (*Pseudotsuga menziesii*) is shaded, there is an increase in the amount of carbon it receives from birch (*Betula* spp.). This insight indicates for a shift change from the importance on rivalry to one on reservoir distribution in tree groups, and exemplifies how mycorrhizas can indorse coexistence and biodiversity (Clarkson and Robards 1975; Christensen 1989; Claridge and May 1994; Claridge 2002).

Furthermore, though the trees in forests are recognized as being an important carbon bowl, the fact that mycorrhizas also hoard large amounts of carbon resources shows that they may have a vital role to play in dealing with global warming.

9.1.4 Carbon Flow

Energy-rich carbon complexes from tree roots to soil microbiota establishes an essential supply procedure to the soil ecosystem (Finlay and Söderström 1992). Substantial intake of carbon through mycorrhizal mycelia is to promote the diverse machineries of soil ecosystem, at the same time a value system of upholding ectomycorrhizal relations with gratified sequestration of atmospheric carbon is recognized. A few recent trials by Högberg et al. (2001) employed the girdling of forest trees to differentiate the comparative characteristics of roots and ectomycorrhizal fungi and free-living microheterotrophs to soil breath. A decline of approximately 50% in soil respiration was proven following girdling, signifying that the flow of current assimilates is the main chauffeur of soil respiration. Separately, straight respiratory damage, energy-rich carbon complexes are mandatory for the majority of the biological activities, and additional information is required regarding the quantities and forms of diverse mixtures and the apparatuses regulating their translocation and eventual segregation. Latent effects comprise the production of enzymes, organic acids and additional compounds, swaying the dilapidation of organic substrates of mineral substrates (Rosling et al. 2004a, b) and the biosynthesis of antibiotic substances involved in chemical defence or antagonism. Synthesis of glycol-proteins (i.e. glomalin) that are indicative in the development and constancy of soil masses may also have a significant influence on other microbiota connected with the arbuscular mycorrhizal mycelium (Johansson et al. 2004). Though distribution of carbon may be augmented in ectomycorrhizal mycelia inhabiting reinforcements of disintegrating plant litter, Leake et al. (2001) revealed that the provision of carbon to the mycorrhizal mycelia inter-relating with the mycelium of the wood-cracking agent *Phanerochaete velutina* was reduced. It has been intensively discussed and controversies have arisen on the eventual fate of carbon attainment to the mycorrhizal mycelium, and it has also been found that there are significant evidence about transmission between trees connected by shared hyphal systems. Finlay and Söderström (1992) revealed that association of plants to a mutual mycelial network could be of significance without any arguments concerning net interplant transmission of carbon, subsequently plantlets could gain access to a large absorptive network of hyphae with minimal investment of carbon. The research outcomes of Högberg et al. (1999) are reliable with the above disagreement, since the investigators found that promiscuous ectomycorrhizal fungi creating relations with numerous tree species had natural signatures of 13C nearer to those of overstorey trees, signifying that the overstorey trees moderately or exclusively support the carbon strains of the nutrient-absorbing mycelia of their suspected entrants, the understorey trees. Transmission of carbon may also happen between green trees/conifers and non-photosynthetic, mycoheterotrophic plants sharing the mycelium. Distribution of topical photoassimilates through mycorrhizal mycelia affects nutrient subtleties and microbiota inhabitants in the mycorrhizosphere (Jones et al. 2004; Finlay and Rosling 2006), and Högberg and Read (2006) have demonstrated that such physiological endurance and dynamic interdependence of the plant–microbe–soil system encounters the extensive view that soil movement is subjugated by decomposer organisms using adult detrital substances, and that rhizosphere litter contributes the equivalent of above-ground litter.

9.1.5 Bioremediation of Forest Soil Pollutants

Evidence of mycorrhizal mycelia might show in the bioremediation of edaphic pollutants, which is yet to go for confirmatory testing. Meharg and Cairney (2000) revised potential ways in which ectomycorrhizal fungi might support rhizosphere remediation of persistent organic pollutants (POPs). Numerous fungi which are able to convert these compounds have been studied for degradation of POPs—such as polyhalogenated biphenyls, polyaromatic hydrocarbons, chlorinated phenols and pesticides—but moderately few mycorrhizal taxa have been verified. Meharg et al. (1997a) showed that degradation of 2,4-dichlorophenol by the two ectomycorrhizal fungi *Paxillus involutus* and *Suillus variegatus* was higher when the fungi were growing in symbiosis with *Pinus sylvestris* than when they were grown in pure culture. In other experiments (Meharg et al. 1997b), *S. variegatus* has been publicized to be effective in degrading 2,4,6-trinitrotoluene. A potential benefit of using mycorrhizal fungi in bioremediation is that they accept a direct supply of carbon from their plant hosts to sustain growth in polluted substrates. Some of this ingested carbon may consequently be accessible to bacteria related to the mycorrhizal mycelium (Sun et al. 1999) and this may have consequences for bioremediation in the mycorrhizosphere. On the contrary, efforts to familiarize micro-organisms with bioremediation properties often fail because the inoculants fail to establish themselves. Mycorrhizal hyphae may enable the establishment of some bacteria, and Sarand et al. (1998) recommended that mycorrhizal hyphae were able to sustain the microbial biofilms of catabolic plasmid Tol^{+}-harbouring bacteria which could be vigorous in the bioremediation of petroleum-contaminated soil (under-earth pipeline) in forest. In additional experiments, these authors (Sarand et al. 2000) confirmed that the number of Tol^{+} bacteria was advanced in mycorrhizospheric soil as compared with loose soil, and inoculation with bacteria had a positive effect on tree and fungal growth. The occurrence of easily available plant-derived carbon sources did not impede the degradation of the *m-toluate* by the bacteria (Sarand et al. 1999). However, in other trials Genney et al. (2004) indicated that degradation of the polycyclic aromatic hydrocarbon fluorene was blocked in a Scots pine ecto-mycorrhizosphere. Joner et al. (2006) also confirmed obstructed phytoremediation of polycyclic aromatic hydrocarbons (PAHs) by the ectomycorrhizal mycelium of *S. bovinus* that was accredited to nutrient exhaustion by the foraging fungus. Overall, the implication of AMF is least investigated with respect to bioremediation, but experiments by Joner et al. (2001) recommend that the dissipation of PAHs may be improved in the presence of arbuscular mycorrhizas, and that vicissitudes in the composition of the mycorrhiza-associated microflora may be accountable for the observed decreases in PAH concentrations.

Acknowledgment Ajit Varma is thankful to the Department of Science and Technology and the Department of Biotechnology for partial financial funding, and to DST-FIST for providing a confocal microscope facility.

References

Abbott LK (1982) Comparative anatomy of vesicular-arbuscular mycorrhizas formed on subterranean clover. Aust J Bot 30:485–499

Abbott LK, Robson AD, Gazey C (1992) Selection of inoculant vesicular-arbuscular mycorrhizal fungi. In: Norris JR, Read DJ, Varma AK (eds) Techniques for the study of mycorrhiza, Methods in microbiology, vol 24. Academic Press, London, pp 1–21

Abdul Karim NB. (2004) Molecular and enzymatic groupings of fungi from tropical orchids of Western Australia and their patterns of tissue colonisation. PhD Thesis, The University of Western Australia

Abu-Zeyad R, Khan AG, Khoo C (1999) Occurrence of arbuscular mycorrhiza in *Castanospermum australe* A. Cunn. & Fraser and effects on growth and production of castanospermine. Mycorrhiza 9:111–117

Barnola LG, Montilla MG (1997) Vertical distribution of mycorrhizal colonization, root hairs, and belowground biomass in three contrasting sites from the tropical high mountains, Mérida, Venezuela. Arct Alp Rese 29:206–212

Barroetaveña C, Cázares E, Rajchenberg M (2007) Ectomycorrhizal fungi associated with ponderosa pine and Douglas-fir: a comparison of species richness in native western North American forests and Patagonian plantations from Argentina. Mycorrhiza 17:355–373

Batty AJ, Dixon KW, Brundrett MC, Sivasithamparam K (2002) Orchid conservation and mycorrhizal associations. In: Sivasithamparam K, Dixon KW, Barrett RL (eds) Microorganisms in plant conservation and biodiversity. Kluwer Academic Publishers, The Netherlands, pp 195–226

Chilvers GA, Lapeyrie FF, Horan DP (1987) Ectomycorrhizal vs endomycorrhizal fungi within the same root system. New Phytol 107:441–448

Christensen M (1989) A view of fungal ecology. Mycologia 81:1–19

Claridge AW (2002) Ecological role of hypogeous ectomycorrhizal fungi in Australian forests and woodlands. Plant Soil 244:291–305

Claridge AW, May TW (1994) Mycophagy among Australian mammals. Aust J Ecol 19:251–275

Clarkson DT, Robards AW (1975) The endodermis, its structural development and physiological role. In: Macfadyen A, Ford ED (eds) The development and function of roots. Academic Press, New York, pp 415–436

Finlay RD, Rosling A (2006) Integrated nutrient cycles in forest ecosystems, the role of ectomycorrhizal fungi. In: Gadd GM (ed) Fungi in biogeochemical cycles. Cambridge University Press, Cambridge, pp 28–50

Finlay RD, Söderström B (1992) Mycorrhiza and carbon flow to soil. In: Allen MF (ed) Mycorrhizal functioning. Chapman and Hall, London, pp 134–160

Fitter AH (2005) Darkness visible, reflections on underground ecology. J Ecol 93:231–243

Frank B. (2005) On the nutritional dependence of certain trees on root symbiosis with belowground fungi. Mycorrhiza 15:267–275 (Translation of Frank's original 1885 paper.).

Genney DR, Alexander IJ, Killham K, Meharg AA (2004) Degradation of the polycyclic aromatic hydrocarbon (PAH) fluorene is retarded in a Scots pine ectomycorrhizosphere. New Phytol 163:641–649

Hadley G. (1982) Orchid mycorrhiza. In: Orchid biology reviews and perspectives II Arditti J. pp. 85-118. Cornell University Press, Ithaca

Harley JL (1989) The significance of mycorrhiza. Mycol Res 92:129–139

Harley JL, Smith SE (1983) Mycorrhizal symbiosis. Academic Press, Toronto

Högberg P, Read DJ (2006) Towards a more plant physiological perspective on soil ecology. Trends Ecol Evol 21:548–554

Högberg P, Plamboeck AH, Taylor AFS, Fransson PMA (1999) Natural C-13 abundance reveals trophic status of fungi and host-origin of carbon in mycorrhizal fungi in mixed forests. Proc Natl Acad Sci USA 96:8534–8539

Högberg P, Nordgren A, Buchmann N, Taylor AFS, Ekblad A, Högberg MN, Nyberg G, Ottosson-Löfvenius M, Read DJ (2001) Large-scale forest girdling shows that current photosynthesis drives soil respiration. Nature 2001(411):789–792

Johansson J, Paul L, Finlay RD (2004) Microbial interactions in the mycorrhizosphere and their significance for sustainable agriculture. FEMS Microbiol Ecol 48:1–12

Joner EJ, Johansen A, Loibner AP, Cruz MAD, Szolar OHJ, Portal J-M, Leyval C (2001) Rhizosphere effects on microbial community structure and dissipation and toxicity of polycyclic aromatic hydrocarbons (PAHs) in spiked soil. Environ Sci Tech 35:2773–2777

Joner EJ, Leyval C, Colpaert JV (2006) Ectomycorrhizas impede phytoremediation of polycyclic aromatic hydrocarbons (PAHs) both within and beyond the rhizosphere. Environ Pollut 142: 34–38

Jones DL, Hodge A, Kuzyakov Y (2004) Plant and mycorrhizal regulation of rhizodeposition. New Phytol 163:459–480

Kendrick B (1985) The Fifth Kingdom. Mycologue Publications, Waterloo, ON

Leake JR, Donnelly DP, Saunders EM, Boddy L, Read DJ (2001) Rates and quantities of carbon flux to ectomycorrhizal mycelium following 14C pulse labeling of *Pinus sylvestris* seedlings: effects of litter patches and interaction with a wood-decomposer fungus. Tree Physiol 21: 71–82

Meharg AA, Cairney JWG (2000) Ectomycorrhizas: extending the capacities of rhizosphere remediation? Soil Biol Biochem 32:1475–1484

Meharg AA, Cairney JWG, Maguire N (1997a) Mineralization of 2,4-dichlorophenol by ectomycorrhizal fungi in axenic culture and in symbiosis with pine. Chemosphere 34:2495–2504

Meharg AA, Dennis GR, Cairney JWG (1997b) Biotransformation of 2,4,6-trinitrotoluene (TNT) by ectomycorrhizal basidiomycetes. Chemosphere 35:513–521

Mosse B, Hayman DS (1980) Mycorrhiza in agricultural plants. In: Mikola P (ed) Tropical mycorrhizal research. Claredon Press, Oxford, pp 213–230

Read DJ (1983) The biology of mycorrhiza in the Ericales. Can J Bot 61:985–1004

Redecker D, Kodner R, Graham LE (2000) Glomalean fungi from the Ordovician. Science 289: 1920–1921

Rosling A, Lindahl BD, Finlay RD (2004a) Carbon allocation in intact mycorrhizal systems of *Pinus sylvestris* L. seedlings colonizing different mineral substrates. New Phytol 162:795–802

Rosling A, Lindahl BD, Taylor AFS, Finlay RD (2004b) Mycelial growth and substrate acidification of ectomycorrhizal fungi in response to different minerals. FEMS Microbiol Ecol 47: 31–37

Sarand I, Timonen S, Nurmiaho Lassil EL, Koivula T, Haahtela K, Romantschuk M, Sen R (1998) Microbial biofilms and catabolic plasmid harbouring degradative fluorescent pseudomonads in Scots pine mycorrhizospheres developed on petroleum contaminated soil. FEMS Microbiol Ecol 27:115–126

Sarand I, Timonen S, Koivula T, Peltola R, Haahtela K, Sen R, Romantschuk M (1999) Tolerance and biodegradation of m-toluate by Scots pine, a mycorrhizal fungus and fluorescent pseudomonads individually and under associative conditions. J Appl Microbiol 86:817–826

Sarand I, Haario H, Jørgensen KS, Romantschuk M (2000) Effect of inoculation of a TOL plasmid containing mycorrhizosphere bacterium on development of Scots pine seedlings, their mycorrhizosphere and the microbial flora in m-toluate-amended soil. FEMS Microbiol Ecol 31: 127–141

Smith SE, Read DJ (2008) Mycorrhizal symbiosis, 3rd edn. Academic Press, Amsterdam

Sun Y-P, Unestam T, Lucas SD, Johanson KJ, Kenne L, Finlay RD (1999) Exudation–reabsorption in mycorrhizal fungi, the dynamic interface for interaction with soil and other microorganisms. Mycorrhiza 9:137–144

Chapter 10
Ectomycorrhizal Fungi: A Major Player in Early Succession

Izabela L. Kałucka and Andrzej M. Jagodziński

Abstract Ectomycorrhizal (ECM) fungi are a key organism group enabling and enhancing the process of open land colonization by ECM-dependent trees and shrubs. Through their functional traits, interactions with both abiotic and biotic environment, and their own successional dynamics, they significantly affect woody vegetation succession coupled with soil and ecosystem development. In this chapter, we review the role of ECM fungi in the processes of early primary and secondary succession, including non-anthropogenic natural systems, like glacier forefronts, volcanic deserts, and sand dunes, as well as major sites disturbed by intensive human activity, such as mine spoils, fire-affected sites, clear-cuts and timber harvesting areas, and post-agricultural lands. Successional traits of ECM fungal community reflecting their life histories and species composition, dispersal, spatial and temporal structure, host preferences, and sensitivity to environmental filters underpin key ecosystem services provided by ECM fungi in the processes of forest development, management, and restoration. While the rapidly increasing influence of climate change, environmental damage, species invasions, and biodiversity reduction become obvious, ECM fungi and their successional traits must be considered in afforestation and carbon sequestration polices, in sustainable forest management, as well as in biodiversity conservation and rehabilitation practices.

I.L. Kałucka (✉)
Faculty of Biology and Environmental Protection, Department of Algology and Mycology, University of Łódź, Banacha 12/16, 90-237 Łódź, Poland
e-mail: izabela.kalucka@biol.uni.lodz.pl

A.M. Jagodziński
Institute of Dendrology, Polish Academy of Sciences, Parkowa 5, 62-035 Kórnik, Poland

Faculty of Forestry, Department of Game Management and Forest Protection, Poznań University of Life Sciences, Wojska Polskiego 71c, 60-625 Poznań, Poland
e-mail: amj@man.poznan.pl

A. Varma et al. (eds.), *Mycorrhiza - Function, Diversity, State of the Art*,
DOI 10.1007/978-3-319-53064-2_10

10.1 Introduction

Although typical of a relatively small number of plant species—about 8000 taxa (<3% of phanerogams), ectomycorrhiza nevertheless plays an important role on a global scale. The association develops mainly with trees of the families *Betulaceae*, *Pinaceae*, *Fagaceae*, and *Salicaceae*, which form extensive boreal and temperate forests covering much of Eurasia and North America, the *Dipterocarpaceae* and some tribes of the *Fabaceae* in the tropics, and the *Myrtaceae* and *Fagaceae* in the temperate and subtropical regions of the southern hemisphere (Taylor and Alexander 2005; Smith and Read 2008; Brundrett 2009). Although not that numerous, some shrubs and even herbaceous plant species important in early-successional sites can also form ectomycorrhizas (e.g., *Arctostaphylos*, *Bistorta*, *Cistus*, *Helianthemum*, *Kobresia*, *Salix herbacea*). Ectomycorrhizal (ECM) associations have evolved in different lineages of plants and fungi several times (Hibbett and Matheny 2009; Tedersoo et al. 2010; Tedersoo and Smith 2013). They involve a great variety of fungi, estimated for 8000–10,000 species (Taylor and Alexander 2005; Rinaldi et al. 2008). Most of these belong to Basidiomycota; there are also a significant number of Ascomycota and a few zygomycetous fungi in this group. Among them, about 4500 species produce epigeous sporocarps while others form hypogeous fruit bodies or do not form them at all.

ECM fungi are vital for biogeochemical cycles and for nutrient acquisition by their plant partners. Radiating from the rhizosphere, vegetative mycelia also provide a source of mycorrhizal inoculum for neighboring host plants and form common mycorrhizal networks that connect a number of different plant individuals (van der Heijden and Horton 2009; Simard et al. 2012). As they interact with soil environment and influence inter-plant relations, they have been proved to be important drivers of the structure and dynamics of ECM plant populations and forest communities (van der Heijden et al. 1998; Klironomos et al. 2011). The majority of ECM fungi have a broad host range; however, they display different levels of preference for host plant species (Ishida et al. 2007; Tedersoo et al. 2008). These assemblages of fungi can change with time both on individual roots and at the scale of the plant communities and ecosystems, providing different ecological functions and services depending on the successional stage and the legacies of the ecosystem history (Dickie et al. 2013). In early-successional systems, ECM fungi are crucial for the ability of their plant partners to colonize novel sites, help open successional processes, and develop plant communities, driving primary and secondary succession initiated by natural events (e.g., glacier retreat, volcanic eruption, windthrow, or natural fire) or anthropogenic disturbances (e.g., mining, prescribed burning, timber harvesting, or agricultural activities). They are important for natural establishment and regeneration of woody vegetation, in forest reclamation of areas destroyed by human activities, and in managing timber forests.

10.2 Major Conceptual Framework

In natural conditions, mycorrhizal fungi rely almost exclusively on their plant partners for carbon needs (Treseder et al. 2006) and may be treated as extensions of plant roots for their physiological capabilities (Berendsen et al. 2012). However, it should also be recognized that they are free-living microorganisms which respond to and affect their environment components and variables (Dickie et al. 2013). Mycorrhizal fungi have their own species sequence trajectories in successional systems, which in long time spans seem unpredictable but indicate some clear patterns that mirror spatial and temporal changes of the host plants and ecosystems. Although the element of chance renders each succession unique, making it the outcome of a number of different circumstances and complicated interactions, recognizable fungal communities do develop as a result of underlying ecological mechanisms and adaptations (Allen and Allen 1992; Twieg et al. 2007; Dickie et al. 2009, 2014).

Succession of both plants and fungi can be defined as "a directional change in the composition, relative abundance and spatial pattern of species comprising communities" (Frankland 1992). Rayner and Todd (1979) described it more precisely in the mycological sense as "the sequential occupation of the same site by thalli (normally mycelia), either of different fungi, or of different associations of fungi." A substantial difference exists between succession of saprotrophic and mycorrhizal fungi. The former are dependent on the substrate they occupy and the resource which they eventually deplete so that the community can be reduced to zero and another suitable resource has to be found for colonization (Dix and Webster 1995). A succession of mycorrhizal fungi associated with host roots is more akin to that of vegetation since the resource, in the short term, is renewable (Frankland 1992).

As summarized by Frankland (1992, 1998), Park's (1968) division of fungal successions into the substratum and seral types has been widely adapted. Substratum succession refers to time-related changes in the community structure of fungi which are bound to the substrate usually limited in space and which offers an unrenewable resource. Examples concern mainly saprotrophic fungi colonizing various types of organic materials such as plant litter, wood, dung, or burns. Seral succession is a series of fungal assemblages following successional changes in vegetation and usually concerns mycorrhizal fungal communities. However, the present knowledge on the successional ecosystems, including not only vegetation and seral plant communities but also incorporating other important constituents, like soil formation and nutrient cycling, microbial communities, and invertebrates (Bever et al. 2010; Wardle et al. 2012; Mueller et al. 2015), suggests that mycorrhizal fungi successions must be thought about in a more complex way. As pointed out by Dickie et al. (2013), mycorrhizal fungal communities are not simply passengers of ecological processes. They do not just follow successional changes in vegetation—they can also act as important drivers, both of pedogenesis and plant community development, as well as foster the association networks in ecosystems (Montesinos-Navarro et al. 2012; Afkhami et al. 2014).

The terms "primary" or "secondary succession" are frequently used to describe mycorrhizal successions. They come directly from the initial theories of vegetation succession developed by pioneers in plant ecology, e.g., Clements (1916) and Gleason (1939).

Primary succession is essentially the process of colonization and establishment of vegetation on newly exposed substrates, usually scarce in nutrients, barren and devoid of autochthonous organic matter, although a considerable amount of allochthonous, windborne biological materials, in the form of living organisms or organic detritus coming from abrasion of surrounding established ecosystems, may be deposited (Hodkinson et al. 2002). Secondary succession occurs after a disruption of primary succession (disturbance) and involves the replacement of preexisting vegetation on developed soil; the soil has an organic component and is usually enriched with nutrients produced by mineralization of residues left by the previous community (Glenn-Lewin et al. 1992; Begon et al. 1996).

The terms characterize the whole ecosystem process and not just the vegetation or fungal communities alone. Therefore, the expressions "primary" or "secondary succession of ectomycorrhizal fungi" which are sometimes used, while convenient, may be inaccurate. Rather, "succession of ectomycorrhizal fungi in primary or secondary successional site" appears more appropriate. Some authors go even further. According to Jones et al. (2003), ECM fungi undergo primary succession when they colonize the land that previously supported only non-ECM plants, and spores are the most important sources of initial inoculum. They undergo secondary succession when trees had been harvested prior to the development of a new stand, and the full range of inoculum types is present. This leads the authors to the curious conclusion that ECM fungi associated with trees developing in post-agricultural fields (typical secondary successional site) undergo primary succession: "... fungi undergoing primary succession as they colonized birch planted in an agricultural field." To avoid such confusions, succession of ECM fungi should not be addressed separately from the rest of the ecosystem.

Disturbance is a fundamental phenomenon in the concept of secondary succession. White and Pickett (1985) defined it as "any relatively discrete event in time that disrupts ecosystem, community, or population structure and changes resources, substrate availability, or the physical environment." As a result of alterations in the structure and composition of the component communities, the efficiency of nutrient cycling and resource utilization declines at the ecosystem level, with a resultant change in the succession trajectory (Zak 1992). In the most drastic cases, after a catastrophic disturbance like a volcanic eruption, vast rock- or mudslide, large-scale open-cast mining, or mine spoil formation, primary successional sites are formed. When the disturbance alters the preexisting community or partially destroys it so that it is not completely devoid of any vegetation and organic matter, secondary succession can start after the disturbance has ceased, for example, in old fields, burnt areas, large-scale clear-cuts, or forest openings, with diaspores coming from the outside of the disturbed system and dormant seed/spore bank. Depending on the duration, spatial extent, and severity scale of disturbance, one can expect different responses of species richness and composition, overall diversity, and spatial heterogeneity of the system (Faliński 1986; Pickett et al. 1989). Moreover, possible wide variation in size, intensity, frequency, and spatial heterogeneity of disturbances leads to uneven patterns of surviving organisms and their propagules; this may result in different initial successional patterns and fading distinction between primary and secondary succession (van der Maarel 1993; Turner et al. 1998).

Subsequent changes of the fungal community linked closely with the undisturbed development of the vegetation during primary or secondary succession essentially define seral succession. Disturbances less severe with respect to extent can leave a part of ECM fungi intact (Jonsson et al. 1999; Hewitt et al. 2013). The inoculum coming from the previous assemblage of fungi associated with the preexisting plant community enables the process of ECM community regeneration. As reviewed by Jones et al. (2003), extramatrical hyphae and dying mycorrhizae, as well as viable sclerotia and spores still present in the soil for some time after disturbance, can act as effective inoculum for seedlings in regenerating stands. This is particularly important in fire-affected areas as well as in clear-cuts and forest openings of different origins.

Another ecological concept that is referred to especially in earlier studies on fungal populations and successional communities is the concept of life history (Jumpponen and Egerton-Warburton 2005; Douhan et al. 2011). Life history of an organism is its pattern of obtaining food (foraging), growth, differentiation, and reproduction from birth to death (Andrews 1992). This pattern, also called a strategy, is a key factor in the environmental selection of species which are best fitted for certain habitat conditions. In the evolutionary sense, it allows for adaptation to the changing environment, thus underlying the process of succession.

The concept of r- and K-selection was introduced to ecology by MacArthur and Wilson (1967); its applications to fungi in the context of life history have been discussed by Andrews and Harris (1986) and Andrews (1992). The term r-selection refers to "uncrowded" environments and species with a high population growth rate, high productivity, and large reproductive effort. Individuals have a short life expectancy, are smaller in size, mature earlier, and have more and smaller progeny; r-selection is advantageous during colonization events. The term K-selection refers to habitats "saturated" with species which have to compete for limited resources and use them more efficiently; they channel energy into maintenance and their reproductive effort is small. Individuals have longer life expectancy, are larger, show delayed reproduction, allocate more resources to growth (size), and have fewer but larger offspring. K-selection is advantageous in established situations. As most species fall between the extremes of r- and K-selection, the concept was further developed to the R-C-S concept (Grime 1977, 1979), which was also adapted for fungi (Cooke and Rayner 1984; Pugh and Boddy 1988; Andrews 1992). It assumes that there are three major forms of selection: R-selection for a "ruderal strategy" (selection for short life span and high-speed production in disturbed but productive environments or pioneer situations), C-selection for a "competitive strategy" (selection for highly competitive ability and long life span in productive, relatively undisturbed conditions), and S-selection for "stress-tolerant strategy" (selection for adaptation to sustained environmental stress or resource depletion). As fungi may switch their behavior during different phases of growth or show different strategies (or their combinations) under various circumstances, the three primary ecological strategies can be used to define or compare behaviors shown by species rather at a particular time and are not always useful for classification of individual fungi. As it was reviewed by Douhan et al. (2011), these concepts were successfully applied in numerous ECM population studies showing that species with numerous small genets reproduce mainly from meiotic spores and establish new individuals each year, whereas species forming rather few, large genets develop mainly perennial mycelium.

The assumption that the composition of ECM fungal community is determined by r-selection at the beginning of the successional process while K-selection prevails when trees age and organic matter accumulates underpins the concept of "early-" and "late-stage fungi" that emerged from a series of investigations carried out in first-generation forests planted in secondary successional sites (Mason et al. 1983; Dighton and Mason 1985; Dighton et al. 1986). Species of fungi occurring at the early stages of woodland development ("early-stage fungi") are, in due course, superseded or joined by others, which are typical of older trees and mature forests ("late-stage fungi"). Also, early-stage species tend to persist on young, peripheral tree roots, whereas they are replaced by late-stage fungi on roots nearer the tree trunk (Ford et al. 1980). This ordered succession was first ascribed to the age of host trees and the temporal stage of woodland development including organic matter accumulation and to changes in host physiology, photosynthate allocation in particular, as a result of aging. Apart from the "early-stage" and "late-stage" strategies, which were also ascribed as generally corresponding to R and C types (Deacon and Fleming 1992), a "multistage" strategy of intermediate character was also proposed (Danielson 1991; Visser 1995). The concept assumed not only changes in the dominant ECM species composition but also in the species richness and diversity, which, according to the concept's authors, increase at least until the stage of tree canopy closure (Dighton and Mason 1985; Dighton et al. 1986; Last et al. 1987).

The "early-" and "late-stage" model has played a very inspiring role in mycorrhizal ecology and stimulated a lot of interesting research on spatiotemporal changes in ECM communities. It has earned support (Jansen 1991; Visser 1995; Helm et al. 1996; Jumpponen et al. 2002; Nara et al. 2003a, b) as well as criticism (Keizer and Arnolds 1994; Termorshuizen 1991). Growing evidence suggested that successional changes in an ECM community should be referred more to resource changes and abilities of early- and late-stage fungi to colonize roots in soils with different accumulations of litter, development of humus layer, and content of organic matter, as well as to the stand structure (Last et al. 1987; Blasius and Oberwinkler 1989; Baar 1996; Jumpponen et al. 1999a). At present, the shifts in the vegetation composition and function, including photosynthesis efficiency and nutrient requirements, are still considered strong determinants of ECM fungi successional traits (Buscot 2015). However, the role of biogeographic factors, dispersal, host identity, niche partitioning, interspecific competition, environmental and biotic filters are also cited (Bruns 1995; Jumpponen and Egerton-Warburton 2005; Bahram et al. 2015; Tedersoo et al. 2014; Mujic et al. 2016), and instead of the terms "early-" and "late-stage" fungi, the more appropriate terms "early-successional" and "late-successional" fungi are used (Dickie et al. 2013). However, there is still no clear delimitation of using these terms in the timescale context. Early- and late-successional ECM fungi can be described both in the time span of woodland development from open community colonized by first ECM host plants to first-generation mature forest (Ashkannejhad and Horton 2006), through a few types of the subsequent woodland communities (Helm et al. 1996; Jumpponen et al. 1999a), and in the time span of the whole successional trajectory including progressive, mature, and retrogressive stages of woodland ecosystem (Dickie et al. 2013).

Early-successional ECM species usually colonize saplings. Most of these fungi have a low carbohydrate demand (supply by host tree) and exploit nutrients within the inorganic pool. They are also capable of rapid mycelial growth, spread mainly by sexual spores, and exhibit features of "ruderal strategy" typical of pioneering species. These fungi colonize tree roots in early-successional stages, in primary or disturbed habitats. Features like these were confirmed, for instance, in *Laccaria bicolor* (de La Bastide et al. 1994) and *Suillus luteus* (Muller et al. 2004). Also, it has been shown that the spores of pioneer ECM fungi germinate easily, especially in the presence of host roots, and colonize them effectively, such as with species of *Inocybe*, *Laccaria*, *Hebeloma*, *Suillus*, and *Rhizopogon* (Theodorou and Bowen 1987; Ashkannejhad and Horton 2006; Ishida et al. 2008; Nara 2009). Late-successional fungi occupy undisturbed habitats, associate with older/larger trees or older forests, have a high carbohydrate demand, derive nutrients mainly from the organic pool, exhibit high competitive abilities due to the production of mycelial strands, and spread mainly via vegetative mycelium. Thus, they clearly exhibit features of "competitive strategy." Examples include *Cortinarius rotundisporus* (Sawyer et al. 1999) and *Russula brevipes* (Bergemann and Miller 2002). Spores of the fungi that usually do not appear in pioneer habitats but are often dominant in mature forests, such as species of *Russula*, *Lactarius*, *Amanita*, and *Cortinarius*, do not germinate well in pioneer conditions, even in the presence of host roots (Nara 2009). In late-successional (mature) stands, seedlings are usually colonized with late-successional fungi as they get an access to the already established common mycorrhizal network and potentially can even gain nutritional support from mature trees (Teste and Simard 2008; Richard et al. 2009; van der Heijden and Horton 2009; Simard et al. 2012).

10.3 Chronosequence Studies

Succession of woody vegetation and associated fungi is a long-term ecological process. Development of mature forest community lasts from a few decades in coniferous forests subject to natural cyclic fires (Larsen 1997) to hundreds of years in old-growth temperate and boreal forests (Faliński 1986). If the whole life span of some subtropical forest ecosystems is considered, it can be hundreds of thousand or even a few million years (Wardle et al. 2004). Monitoring temporal changes in such systems would be well beyond the capacity of a single researcher. The space-for-time replacement and chronosequence studies are the most widely used and appreciated methods that overcome this difficulty in field research (Pickett 1989; Walker et al. 2010). As summarized by Jumpponen et al. (2012), chronosequence studies suffer from potential correlations between position in the chronosequence and substrate chemistry, fluctuations in weather or climatic conditions, distance to propagule sources, or other environmental parameters. But they benefit from providing a single location where substrates of different ages can be observed in a relatively homogeneous environment—similar soil parent material, climate, and regional species pools (Jenny 1980). A chronosequence set of sites selected across a range of various-aged vegetation communities provides opportunities to evaluate communities of mycorrhizal fungi on host plant species representative for subsequent successional stages and to follow the fungal seral succession. Mycorrhizal chronosequence studies usually assess changes resulting from three causes: changes in plant species, age of plants, and the successional stage itself (Helm et al. 1996). Apart from the observations of mycorrhizal fungi associated with plants comprising the host communities in such real, complicated systems, field experiments and glasshouse bioassays are conducted to test hypotheses referring to selected variables or simplified conditions (Helm et al. 1999; Nara 2006a; Ashkannejhad and Horton 2006). As pointed out by Hobbie et al. (2005), one valuable technique has been to take the symbiosis out of the field and study it in the laboratory or greenhouse, but inferences derived from the laboratory should always be tested against field data (Read and Perez-Moreno 2003). Chronosequence studies, especially these using molecular techniques of mycorrhizal species identification, have yielded a large body of data, considerably increasing our knowledge of patterns and mechanisms involved in ECM fungal successions.

10.4 Primary Successional Sites

In primary successional sites, vegetation develops on newly exposed substrates void of any autochthonic organic matter where the supply of ECM fungi inoculum is limited and depends mostly on air- and animal-mediated arrival of propagules from "outside" the community (Trappe and Luoma 1992). Compatible fungal associates are vital for successful establishment and persistence of mycorrhiza-dependent plants. Their establishment is driven by the general assembly rules and ecological filters accounting for both stochastic (e.g., distribution of suitable establishment sites and propagules) and deterministic factors (facilitative and competitive processes after community establishment) (Jumpponen and Egerton-Warburton 2005; Jumpponen et al. 2012).

10.4.1 Glacier Forefronts

Glacier forefronts are unique areas of primary succession whereby non-vegetation terrain exposed from beneath glacial ice develops toward a complex plant community and can be regarded as model study systems (Jumpponen et al. 2012). Glaciers leave a harsh environment of extreme conditions, including strong nutrient limitations. Glacier forelands are relatively young, settling of plants and fungi occurs within comparatively short, usually known time span, diversity of both plant and fungal communities is low, and supply of ECM inoculum is relatively limited.

Glacier chronosequence studies on ECM communities (or including them among other groups of organisms) focused on successional changes in occurrence of sporocarps (Helm et al. 1996; Jumpponen et al. 1999a, 2002; Alfredsen and Høiland 2001), ECM roots (Helm et al. 1996, 1999; Trowbridge and Jumpponen 2004; Cázares et al. 2005; Blaalid et al. 2012), and soil microbial DNA (Jumpponen 2003; Brown and Jumpponen 2014).

The presence of potential ECM fungi in the form of aerially deposited, dormant spore bank was detected in recently exposed, early-successional areas prior to the presence of susceptible hosts (Jumpponen 2003; Trowbridge and Jumpponen 2004). The propagules were sparse in young soil substrates, but their numbers and diversity increased with time since deglaciation; the ECM propagule bank seemed to establish in areas deglaciated for at least 40 years (Jumpponen et al. 2002). Spore dispersal of ECM fungi is facilitated not only by wind: many species, particularly of hypogeous fungi, are introduced in feces of visiting mycophagous animals (Cázares and Trappe 1994). In the accumulation of propagules, stochastic patterns are important (Jumpponen et al. 2002; Jumpponen 2003). Similarly to colonizing plants, spores may be more frequently deposited in so-called safe sites which are more likely to accumulate large numbers of viable propagules and thus favor higher rates of fungal colonization of host roots (Jumpponen et al. 1999b; Trowbridge and Jumpponen 2004). Despite potential inoculum limitations, all ECM plants occurring naturally at glacier forelands, even those in early-successional sites, had their roots colonized with ECM fungi (Helm et al. 1996; Cázares et al. 2005). These areas, especially at alpine but not high latitude locations, are usually not considerably far from established woodlands, which are a rich and continuous source of fungal propagules. Thus, at longer timescale, it is unlikely that ECM colonization could be really a limiting factor for successional processes here. However, aerial random spore dispersal may be much slower and limiting in arctic situations (Hodkinson et al. 2003; Blaalid et al. 2012).

Ectomycorrhizal hosts are among first plant colonizers of glacier barrens. They associate relatively rapidly with fungal partners and become symbiotic (Hobbie et al. 2005). Mycorrhizal colonization of seedlings occurs within 2 years of germination, sometimes within a few weeks (Helm et al. 1996), whereas total ECM formation percentages increase from about 25% in the barren stage to over 50% in the next and later stages. All the *Pinaceae* and *Salicaceae* sampled at the glacier foreland nearly 100 years old had ectomycorrhizal roots (Cázares et al. 2005; Jumpponen et al. 2012), even the first *Abies lasiocarpa* seedlings appearing on the substrate deglaciated for 20 years. Degree of ECM root tip infection in early-successional plants—*Kobresia myosuroides* (Mühlmann and Peintner 2008a), *Polygonum viviparum* = *Bistorta vivipara* (Mühlmann et al. 2008), and *S. herbacea* (Mühlmann and Peintner 2008b) reached 95, 100, and 93%, respectively, in the area deglaciated for 150 years. Plant species with ectomycorrhizal associations are among earliest vascular colonizers also in arctic-alpine glacier forelands (Alfredsen and Høiland 2001; Hodkinson et al. 2003). However, in High Arctic, their ECM colonization may proceed much slower (Fujiyoshi et al. 2011).

Successional changes were also observed in fruiting patterns. Sporocarps of fungal genera such as *Inocybe*, *Thelephora*, and *Laccaria* and some species of *Cortinarius* and *Hebeloma* were found predominantly in the early stages (usually <50 years old), with a few common, non-host-specific ruderal species being the most frequent. Later, they were supplemented with a number of species of *Lactarius*, *Suillus*, *Leccinum*, *Russula*, and *Amanita* and other species of *Cortinarius*, with several infrequently fruiting or host-specific taxa, especially in areas of litter accumulation (Helm et al. 1996; Alfredsen and Høiland 2001; Jumpponen et al. 2002). In some cases, Ascomycetes were a significant component of the community, especially in the inter-canopy areas (Trowbridge and Jumpponen 2004). The prolonged presence, and not replacement, of the initial fungal colonizers could result from the heterogeneity in the host age distribution (coexistence of hosts of various age) and differing soil conditions in patchy plant communities, thus facilitating the presence of early, ruderal ECM fungi in the oldest parts of the chronosequence (Jumpponen et al. 2002). Increasing number of ECM species showed positive correlation with the frequency of the hosts (Alfredsen and Høiland 2001). Some of the earliest species soon disappeared, some species were frequent through the whole chronosequence and showed no preference toward the successional stage, and some preferred more established vegetation. The highest number of ECM species was usually found at the end of the chronosequences examined, although relatively high fungal richness may occur already in the recently exposed areas (Blaalid et al. 2012). The increase in the diversity of ECM fungi is attributed to the increase in both the number of host species and the range of their ages, which in turn increased the diversity of available habitats (Jumpponen et al. 2002).

On the Lyman Glacier forefront (the North Cascades mountain range, Washington, USA), Trowbridge and Jumpponen (2004) found that individual ECM taxa showed preferences for canopy or inter-canopy microenvironments. They argued that such patterns in ECM species were related to organic matter accumulation and N availability associated with soil development. Under-canopy soils promoted the occurrence of fungi relying on host photosynthates and facilitated selection of symbionts based on their compatibility with a plant host. Thus, plant establishment in this early-successional system, decoupled from the substrate age, may lead to convergence of ECM fungal communities (Jumpponen et al. 2012) and is a key factor affecting ECM fungal succession (Jumpponen and Egerton-Warburton 2005). Despite usual low species number and high heterogeneity, functional diversity of ECM community allows colonization of early-successional plants (Jumpponen 2003), facilitates subsequent establishment of late-successional plants through sharing non-host-specific fungal symbionts (Helm et al. 1999), and supports high nutritional demands of developing plant communities, especially through the presence of efficient host-specific ECM taxa (Hobbie et al. 2005). However, discrepancies between the ECM fungal communities in the glacial forefronts and in the late-successional stages outside the terminal moraine or in adjacent mature secondary stands may last for centuries, as the harsh environment makes the succession process slow, resulting in patchy vegetation structure and lack of well-developed organic layers (Jumpponen et al. 1999a; Blaalid et al. 2012). Moreover, although the ECM fungal community may follow specific trajectory different from other groups of fungi, niche preferences, pedogenesis, and successional age were also proved to drive community shifts along primary successional chronosequence (Brown and Jumpponen 2014).

10.4.2 Volcanic Deserts

Volcanic eruptions are natural catastrophic disturbances creating primary successional sites both in the past and at present. Flows of lava and thick deposits of pumice and ash constitute a new, barren substrate ready for recolonization. The process of vegetation succession and soil development is usually very slow—300 years after eruption plant cover makes up 5% of the land area (Nara et al. 2003a), while formation of 6 cm deep soil profile may take almost 2000 years (Peña-Ramírez et al. 2009). Since the former vegetation may be completely destroyed and covered with scoria, any fungal inoculum must be brought in from far outside the area by wind dispersal or animals. As majority of spores in this environment decrease their activity with time (Ishida et al. 2008), after hundreds of years a new dormant spore bank may be hardly present or active (Nara and Hogetsu 2004).

Germination patterns and infectivity of ECM fungal species correspond well with their ecological traits and can critically affect initial colonization of non-mycorrhizal habitats and development of ECM communities (Ishida et al. 2008). All ECM fungi known to occur as first-stage colonizers in a Mt Fuji volcanic desert (Japan), such as *Laccaria* spp. and *Inocybe lacera*, exhibited high germination rates in the presence of roots of susceptible plant hosts. They readily formed mycorrhizal associations, whereas majority of late-stage fungi, e.g., of the *Cortinarius* and *Russula*, exhibited weak responses to host roots.

Because of limited propagule availability, non-mycorrhizal seedlings that were transplanted into most parts of the volcanic desert on Mt Fuji remained uncolonized by ECM fungi, although all established *Salix* shrubs were associated with ECM symbionts (Nara et al. 2003b; Nara and Hogetsu 2004). However, seedlings could easily develop ECM associations when growing beside *Salix* shrubs, and their roots could join the mycelial network of hyphae emanating from these plants (Nara 2006a). This resulted in better nutrient acquisition and improved growth, indicating that early-established *Salix* shrubs facilitate the subsequent establishment of conspecific seedlings by providing ECM fungal symbionts. Moreover, the analysis of spatial coincidence between secondary colonizing timber species (*Betula ermanii* and *Larix kaempferi*), which are ECM host plants that follow *Salix* in the initial stage of forest formation, showed that these individuals were all accompanied by *Salix* with no exception (Nara 2006b). The ECM fungal communities of these timber species showed high similarity to that of *Salix* and were dominated by species common to this host even a decade after their establishment. The seedlings of *Betula* and *Larix* may have been connected to *Salix* shrubs by common mycelia of the same ECM fungi (Nara and Hogetsu 2004). These were generalist species that were compatible with two or more plant families (*Betulaceae*, *Pinaceae*, *Salicaceae*). Thus, pioneer *Salix* may contribute to tree succession by providing adjacent late colonizers with compatible ECM fungal symbionts.

The succession process can run much faster when the extent of disturbance or its severity are not equally devastating and when the adjacent forest edges, islands of preexisting vegetation, or its remnants underneath the present deposits serve as ready propagule sources (Allen et al. 1992; Obase et al. 2005). It can also be facilitated by the activity of small mammals (Allen et al. 1992).

In such circumstances, the invasion of ECM host plants on volcanic deserts has been observed after a few to several years of substrate exposure (Allen et al. 1992; Obase et al. 2007). The *Salicaceae* are dominant woody pioneers in volcanic deserts. Formation of ectomycorrhizas on the roots of *Salix* was found as early as 2 years after the eruption (Obase et al. 2007); 5 years after this disturbance all seedlings of other ECM woody plants had their roots colonized with the ratio reaching 75% (Obase et al. 2008).

No ECM sporocarps were observed until 10 years after the eruption of Mount St Helens (USA), although some ECM trees had already appeared (Allen et al. 1992). In the volcanic desert on Mount Fuji (Japan), the sporocarp abundance and the species composition of sporocarps changed with host size enlargement (Nara et al. 2003a); the size of the ECM host, *Salix reinii*, was taken as an indicator of the age after the first colonization. *Laccaria laccata*, *L. amethystina*, and *I. lacera* were the first colonizers (first-stage species) and were subsequently joined by additional taxa as the host grew: *L. murina* and *Scleroderma bovista* (second-stage species) and later by other species of *Hebeloma*, *Cortinarius*, *Russula*, and *Inocybe* (late-stage species). As the ECM species richness increased, none of the species disappeared or were replaced in this early-successional sere. As sporocarps of the second-stage species were situated on the bare ground on the outsides of the *Salix* patches (where no litter had accumulated), their occurrence was attributed to the increased host age.

The occurrence of many late-stage species was clearly related to changes in soil conditions, mainly to the accumulation of organic material. The sporocarps of these species were always situated inside the vegetation patches, where litter was continuously deposited. Species of *Laccaria*, *Inocybe*, *Hebeloma*, *Suillus*, and *Scleroderma* were also found near mature trees that survived the eruption of Mt Usu (Japan) (Obase et al. 2005).

A well-corresponding pattern of species composition was found below ground, particularly in the earlier successional stages (Nara et al. 2003b). Species recorded as sporocarps dominated the underground ECM community. The sere was initiated by the same, first-stage species of fungi, and additional species were recruited with host growth, especially within vegetation patches, which was mainly attributed to the accumulation of litter and resultant soil development. Most species which were found only underground were fungi that do not form conspicuous sporocarps (*Cenococcum geophilum*, *Sebacina* sp., species belonging to *Thelephoraceae*). The colonization percentage of these fungi increased with host growth, especially inside large host patches, and was high in unhealthy hosts (Nara et al. 2003b). These species are usually abundant in older forests (Kõljalg et al. 2000).

Production of sporocarps of ECM fungi in early-successional volcanic deserts increased with host size and was positively correlated with the host's photosynthetic rate (Nara et al. 2003a). This rate, in turn, increased linearly with N and P concentrations in leaves, suggesting a bidirectional, positive feedback between both symbionts. An efficient nutrient supply from ECM fungi results in high photosynthetic activity, and, conversely, enough photosynthate can be supplied from hosts to maintain the activity of ECM fungi, also expressed in sporocarp production. In patches of less healthy hosts (based on photosynthetic rate, leaf biomass, and 1-year shoot length), not only lower ECM sporocarp production was observed but also a significant change in the ECM fungal species composition and a decrease in species richness. This might reflect an insufficient carbon allocation to the roots of the hosts and the inability of some ECM fungi (e.g., *S. bovista*) to colonize such roots or sustain growth. However, the colonization of species forming no or inconspicuous sporocarps was supported in these conditions indicating their low carbon demands or ability to use other than host-derived carbon sources.

Nutritional status of the habitat seems to drive ECM species diversity in a variety of plant hosts colonizing volcanic deserts, including *Salix* spp., *Populus* spp., *Betula* spp., *Quercus* spp., and *L. kaempferi* (Yang et al. 1998; Tsuyuzaki et al. 2005). The ECM communities can differ greatly among hosts when the availability of different ECM inoculum is relatively high (Obase et al. 2007), but species of *Laccaria*, *Hebeloma*, *Inocybe*, *Scleroderma*, and *Thelephoraceae* constitute the majority of the ECM community in the roots of early-successional ECM plant species.

10.4.3 Sand Dune Systems

As reviewed by Corkidi and Rincón (1997), sand dunes are successional habitats usually deficient in major nutrients (N, P, K) and organic matter. They may be salinized and are subject to wide fluctuations in soil moisture and temperature, blowing winds moving sand which can cause excessive erosion or accretion and harmful sand blasting.

Coastal sand dunes are a model system in which environmental factors, mainly soil condition gradient, filter for colonization by different groups of plants characterized by different mycorrhizal status. The succession starts with ruderal plant species with low mycorrhizal dependency in nutrient-enriched high-tide zone, then facultatively AM and AM plants prevail in P-limited zone, ECM plants in N-limited zone, and finally plants with ericoid mycorrhizal associations in zone of high organic matter accumulation and limited nutrients bound in acidic organic complexes (Read 1989; Allen and Allen 1992).

ECM host plants usually inhabit stabilized dunes covered with herbaceous vegetation, where soil organic matter has accumulated over time leading to reduction of pH, inhibition of nitrification processes, and deficit of plant available N, and they continue to grow in Ericales-dominated successional stages (Read 1989).

Among sand dune plants, taxa with dual mycorrhizal colonization, AM and ECM, seem to be very important, sometimes dominant in fixed dunes and persisting throughout succession (van der Heijden and Vosatka 1999; Maremmani et al. 2003; Çakan and Karataş 2006). Such an ecological trait may be advantageous in nutrient-limiting dune environment, enabling the plants to have better access to both P and N sources that can fluctuate more within the season than along successional gradient (van der Heijden and Vosatka 1999). The root colonization of *Salix repens* by ECM fungi was high and by AM fungi was low with no difference between successional dune habitats, indicating that ectomycorrhizas on these plants do not increase their importance in later successional stages. However, different habitat preferences of various ECM root associates and AM fungi found in willows suggest that mycorrhizal diversity contributes to the broad ecological amplitude of this plant and may play an important role in facilitation of other plants in this successional series, both AM and ECM dependent (van der Heijden and Vosatka 1999). The huge discrepancy between the high number of ECM species present as sporocarps and relatively low number of ECM morphotypes found on roots of *S. repens* reflects strongly heterogeneous spatial and temporal structure of the ECM community (van der Heijden et al. 1999). A characteristic feature of that community is a high proportion of fungal taxa host specific to *Salicaceae*, especially those of the *Cortinariaceae*, but the mechanisms that underpin this phenomenon or its functional relevance are not clear.

One of the earliest woody colonizers of sand dunes are pines, and they are always associated with their ECM partners, even in the most pioneer and isolated sites (Dominik 1951; Maremmani et al. 2003; Ashkannejhad and Horton 2006). Mechanisms of the establishment of ECM fungi which support primary successional *Pinus contorta* seedlings on coastal dunes of Oregon (USA) involve both long-distance spore dispersal and animal vectors (Ashkannejhad and Horton 2006). The spores, originating from sporocarps occurring in adjacent but relatively remote coniferous forest zones, come to the isolated dune areas either as a long-distance spore rain or with animal feces and create a resistant spore bank in soil. It may comprise propagules of tens of ECM species typical of coniferous forests including Basidiomycetes (*Boletaceae*, *Cortinariaceae*, *Russulaceae*, *Thelephoraceae*, *Tricholomataceae*) and Ascomycetes (*C. geophilum*, *Wilcoxina mikolae*). However, as it was shown in the field and in seedling bioassays, only a subset of these fungi had spores that were resistant to inhospitable, exposed conditions and sustained their ability to infect roots of isolated pine seedlings. These were *C. geophilum*, *W. mikolae*, a few *Thelephoraceae* species, *Laccaria* sp., *I. lacera*, and a whole range of *Pinaceae*-specific species of *Suillus* and *Rhizopogon*. Among these early-successional fungi, suilloid taxa seem uniquely adapted for long-distance dispersal to, and survival in, the isolated areas where mycelial networks are absent. They are the principal ECM group with specific ecological adaptations for establishment of pines in harsh or early-successional habitats (Ashkannejhad and Horton 2006). Also, this study showed a key role of deer in suilloid and other fungi dispersal over the long distances across sparsely vegetated early-successional areas, especially of hypogeous species, as deer feces considerably add to the soil dormant spore bank.

10.4.4 Mine Spoils and Other Soil-Disturbed Sites

Vast areas of stripped land and various kinds of spoils and tailings usually stored in the form of spoil heaps (slag heaps, spoil banks, dumps), resulting from road cuttings, land leveling on construction sites, mining and extraction of different minerals, fuels, etc., lead to the formation of primary successional sites surrounded with the much more successionally advanced landscapes. After removal of the upper layer, the soil is usually low in macronutrients, contains no organic matter or mineralizable nitrogen, and lacks water-stable microaggregates. Spoil heaps frequently have a poor water-holding capacity, low porosity, and unfavorable air–water conditions and absorb solar energy causing high surface temperatures. They tend to get compacted, but are prone to wind and water erosion, and may show high metal(loid) concentrations, salinity, and extreme pH values (Marx 1975; Conesa and Schulin 2010).

Passive reclamation of such sites relies on natural, spontaneous successional processes; active reclamation is supported by human assistance and various amelioration measures, including reinstatement of topsoil layers, fertilization, organic enrichment, seeding grasses and legumes, and tree planting (Hüttl and Weber 2001; Macdonald et al. 2015). However, the establishment of trees and their development is often problematic, as the ECM fungal inoculum potential of strongly disturbed soils and young mine spoils is usually very low (Malajczuk et al. 1994; Lunt and

Hedger 2003; Bois et al. 2005). With time, the amount of the inoculum increases, suggesting air- and animal-vector spore dispersal (Bois et al. 2005). As the succession of the habitat and vegetation proceeds, gradually more niches for diverse ECM fungi appear (Malajczuk et al. 1994) and spore/hyphal network production by developing in-site ECM assemblage increases.

The inoculum potential, ECM root colonization, and host plant growth can be increased significantly by the amelioration measures aiming at soil reconstruction, such as placing the retained topsoil, usually rich in ECM fungal propagules, on the surface of the spoil (Helm and Carling 1993; Lunt and Hedger 2003; Bois et al. 2005), different soil subsurface preparation methods (Bauman et al. 2013), or organic enrichment with leaf litter (Lunt and Hedger 2003).

Spontaneous arrival of ECM pioneer trees (*Salix*, *Populus*, *Betula*, and *Pinus* spp.) onto different types of non-meliorated coal-mine wastes was observed after 10–30 years of their establishment, and after 50 years, a relatively diverse and abundant ECM fungi were found on their roots (Pachlewski 1956, 1958). Similarly, *Pinus halepensis* colonized metal-contaminated tailings after 40 years (Parraga-Aguado et al. 2014), forming isolated "fertility islands" accelerating further vegetation development under their canopies. Thus, an artificial introduction of tree seedlings is an effective way of facilitating succession and forest restoration processes on reclaimed sites (Hüttl and Weber 2001; Macdonald et al. 2015).

The spore bank of row or relatively young reclamation materials is usually rather poor. Greenhouse bioassays proved the ability to germinate and support emerging or transplanted seedlings for *C. geophilum*, *Tuber* sp., *Wilcoxina* sp. (E-strain), *Laccaria* sp., *Rhizopogon*-like, *Suillus* spp., *Thelephora americana*, as well as Sebacinoid and Pezizales species (Danielson et al. 1983; Bois et al. 2005). The indigenous ECM communities, which colonize seedlings and young trees up to 7 years growing on young mine spoils, consist mainly of *C. geophilum* and the species of *Geopora*, *Tuber*, *Wilcoxina*, *Amphinema*, *Hebeloma*, *Inocybe*, *Laccaria*, *Rhizopogon*, *Sebacina*, *Suillus*, *Thelephora*, *Tomentella*, and *Atheliaceae* (Danielson and Visser 1989; Danielson 1991; Lunt and Hedger 2003; Gebhardt et al. 2007; Rincón et al. 2007; Huang et al. 2012; Onwuchekwa et al. 2014). Also, *Pisolithus tinctorius* was reported as a typical early-stage ECM fungus, which is well adapted to low pH and high temperatures of anthracite and hard coal spoils (Marx 1975). This common and widespread multi-host species, known from many highly disturbed sites (Marx 1977), has been widely used for artificial inoculation of tree seedlings prior to outplanting and proved to be effective in increasing the seedlings' survival and growth in post-mining landscapes, especially in warmer and drought-susceptible parts of the world (Marx et al. 2002).

The adverse conditions of young mine spoils, especially in non-meliorated sites, seem to be a very strong limiting factor. Environmental filtering promotes early-successional ECM fungi capable of inhabiting immature soils with low levels of organic matter, low nutrient mineral soils or fertilized, disturbed, polluted, or extreme pH sites, and fluctuations in soil temperature and moisture (Münzenberger et al. 2004; Bois et al. 2005; Staudenrausch et al. 2005; Huang et al. 2012). Some of these taxa are able to withstand high metal(loid) concentrations and can significantly improve the survival and the performance of host trees on metalliferous substrates or smelter wastes (Staudenrausch et al. 2005; Hrynkiewicz et al. 2008; Huang et al. 2012) providing a powerful tool for phytoremediation (Colpaert et al. 2011).

After outplanting, the colonization rate of seedlings roots by indigenous ECM fungi may stay low for a long time. On the roots of *Quercus rubra* seedlings growing on a lignite mine spoil, it reached 15% after 5 years (Gebhardt et al. 2007); on the roots of *Pinus massoniana* seedlings on a Pb–Zn mine tailing, it reached 26% after 7 years (Huang et al. 2012); at a former ore mining area rich in Pb, Zn, Cu, and Cd, the rate reached 3–36% on the roots of 20–30-year-old trees of *Salix caprea* (Hrynkiewicz et al. 2008). The inoculation of seedlings with selected ECM fungi prior to planting can help to overcome the problem of the probable lack of the indigenous ECM inoculum in the sites, to surmount the initial stress of transplantation, and to acclimate to a harsh environment and adverse substrates during the first year(s) after outplanting (Rincón et al. 2007). Despite the fact that they are often poor competitors among field conditions (Danielson and Visser 1989; Xu et al. 2001; Rincón et al. 2007) and most of them may be replaced within 2 years (Rincón et al. 2007), they can help seedlings to develop new short roots susceptible to the colonization of indigenous ECM symbionts (Grossnickle 2005) or to increase the overall percent of ECM root mycorrhization (Onwuchekwa et al. 2014). The best results are expected when inoculated fungi are specifically adapted to the environmental conditions of transplantation sites (Rincón et al. 2007). However, the seedlings usually get colonized by native ECM fungi during their development in nurseries (Leski et al. 2010; Pietras et al. 2013) and intentional inoculation may not be necessary.

After the initial period of low richness and poor diversity of ECM communities in disturbed sites, fungal communities can develop quite rapidly. In *Pinus sylvestris* afforestations on a reclaimed lignite mine spoil heap, the number of ECM species (counted based on sporocarps) reached 5 in the third year after outplanting and 23 in 14 years old stand, coupled with a massive production for some taxa (for *Geopora*, *Hebeloma*, *Helvella*, *Inocybe*, *Rhizopogon*, *Suillus*, and *Tricholoma* spp.) (Kałucka and Jagodziński 2016). In 20–25 years old stands of *Populus tremula* growing in a heavy metal-contaminated site, 54 ECM fungal taxa were found (Krpata et al. 2008).

With time, the number of ECM species and ECM community diversity tend to be more similar between the older stands of spoil heaps and the surrounding forests (Gebhardt et al. 2007; Glen et al. 2008) or other mature forests (Krpata et al. 2008). However, the ECM species composition is different; late-successional fungi, such as members of *Thelephoraceae* or *Russulaceae*, are relatively less represented (Glen et al. 2008; Huang et al. 2012), while the members of the typical early-successional genera, such as *Cenococcum*, *Hebeloma*, *Inocybe*, *Laccaria*, *Pisolithus*, *Scleroderma*, *Tomentella*, *Tuber*, and *Wilcoxina*, tend to be persistent in disturbed habitats (Glen et al. 2008; Hrynkiewicz et al. 2008; Krpata et al. 2008). Variation in successional trajectories of ECM communities on mine spoils may be very high, as they are affected by strong, site-dependent environmental filtering caused by climatic conditions (Glen et al. 2008), host preferences (Huang et al. 2014), and soil properties, such as the content of metalliferous contaminations (Huang et al. 2014), the maturity of soil, and the presence of organic matter (Staudenrausch et al. 2005; Huang et al. 2012, 2014), and they are modified by a variety of restoration and amelioration measures (Kałucka and Jagodziński 2016). Moreover, mine reclamation forests differ from the forests of the same age and grow under better conditions in root biomass (Jagodziński and Kałucka 2010) and allocation of aboveground biomass (Jagodziński et al. 2014), which probably both affect their ECM communities. As a result, although different from natural, the reclamation stands form a unique habitat for many rare and interesting ECM fungal taxa and can significantly contribute to the fungal biodiversity, both at a local and larger scale (Kałucka et al. 2016).

10.5 Secondary Successional Sites

Secondary successional sites are created after natural or human-caused disturbance of different size and severity, where the preexisting vegetation has been destroyed but the soil is developed and contains organic matter and nutrients left by the previous community (Begon et al. 1996). Frequently, such sites show heterogeneous patterns of surviving organisms and their propagules, which can affect early stages of secondary succession and development of new vegetation (Turner et al. 1998). As in case of primary succession, fungal partners are key components of developing mycorrhiza-dependent plant communities. Their establishment in secondary sites is driven by the same rules; however, the propagules may both arrive from "outside" the community (air and animal dispersal) and originate from the previous assemblage of fungi associated with the preexisting plant community (if the disturbance was not severe enough to destroy it completely), enabling regeneration (Jones et al. 2003; Hewitt et al. 2013). Dispersal from adjacent non-disturbed sites, covered with intact vegetation, seems to be strong and very important, especially when connections via common mycorrhizal networks are established (Dickie and Reich 2005; Thiet and Boerner 2007; Teste and Simard 2008). Some ecological filters accounting for secondary successional trajectory are different, especially in terms of enriched soil conditions (Walker and Syers 1976). Among the most distinctive and frequent examples of secondary successional sites are the previously forest-occupied areas affected by natural or man-initiated fire, agriculture, intensive forest management or natural wind-throws, floods, etc. A growing body of mycological research in such environments shows vital role of ectomycorrhizal fungi in the recovery and sustainable continuity of forest ecosystems.

10.5.1 Fire-Affected Sites

Fire is one of the most common and important natural disturbances that influence soils (Bento-Gonçalves et al. 2012), vegetation (Keeley et al. 2011), and microbial communities (Hart et al. 2005; Cairney and Bastias 2007). Human interference with natural fire regimes, both via increasing the frequency of intentional or unintentional fires and via extension of fire intervals and limiting fire extent with suppressing measures, is also of great concern.

As reviewed by Dahlberg et al. (2001) and Dahlberg (2002), succession of ectomycorrhizal fungi following fire depends considerably on the intensity and frequency of fire events. Fennoscandian boreal forests are usually subject to low-intensity fires allowing most of trees to survive; moreover, their mycorrhizal partners escape from combustion in thick organic humus layer and ECM species richness may not be significantly affected (Jonsson et al. 1999). In these forests, no post-fire ectomycorrhizal succession is apparent, unless high-intensity burns cause high tree and ECM mortality. However, at long timescales, fire-driven successional processes may lead to the decline of ECM basidiomycetes and the predominance of ericoid mycorrhizal ascomycetes, potentially resulting in long-term humus accumulation and high C sequestration (Clemmensen et al. 2015).

North American boreal forests, different types of temperate forests such as in temperate North America, southern South America or Mediterranean, and forests in Australia undergo natural fire cycles with much shorter intervals and high intensity (Keeley et al. 2011); prescribed burning is also a common forest management practice to reduce the impact of wildfires (Fernandes and Botelho 2003). More frequently, these fires result in stand replacement and eradication of pre-fire ECM communities.

Post-fire changes in ECM communities may be caused directly by heating (Kipfer et al. 2010), but also by heat-mediated effects on roots including their mortality and limited availability, and chemical and physical changes of soil (loss of organic matter, modified nutrient and moisture levels, changes in pH) (Neary et al. 1999; Hart et al. 2005; Cairney and Bastias 2007). The extent and quality of changes in mycorrhizal fungal assemblages following fire events may vary considerably; the fire legacies, which may persist for decades after fire events (Treseder et al. 2004; Holden et al. 2013), and status of the ECM community after the disturbance cassation are essential for the subsequent course of successional processes.

Tree seedlings regenerating after fire are usually highly mycorrhized with colonization and the number of ECM operational taxonomic units per soil sample increasing with time (Longo et al. 2011; Barker et al. 2013; Holden et al. 2013; Rincón et al. 2014). The species richness and diversity increase significantly during the first 2 years (Rincón and Pueyo 2010; Hernández-Rodríguez et al. 2013). Early stages of ECM fungal succession are often characterized by few dominant and many less frequent taxa (Grogan et al. 2000; Rincón and Pueyo 2010; Rincón et al. 2014). Most of them are the components of the pre-fire community (Horton et al. 1998;

Dahlberg et al. 2001; de Román and de Miguel 2005). The richness of fungi colonizing seedling roots is higher in the areas previously occupied by ECM hosts and on the sites where some of the trees survived the fire, suggesting that viable inocula (ectomycorrhizas, mycelial networks, resident spores, sclerotia) remain in burned sites and readily infect new roots. Wind or water dispersal of spores is probably more important in the previously non-ECM host areas (Horton et al. 1998). Spores, sclerotia, and individual surviving ECM roots form dominant inoculum type for mycorrhizal colonization after severe fires, when most mycelia and mycorrhizae are killed (Baar et al. 1999; Grogan et al. 2000; Bruns et al. 2002). As a consequence, as shown for seedlings of *Pinus muricata* (Grogan et al. 2000) and *Pinus pinaster* (Rincón et al. 2014) established immediately after fire, significantly more ascomycetes and species representing suilloid types, *Rhizopogonaceae*, and *Atheliaceae* were colonizing the root tips than amanitoid, russuloid, cortinarioid, and thelephoroid taxa, which prevailed before fire. Resistant propagule bank, consisting mainly of heat-resistant spores and sclerotia of *Rhizopogon* spp., *C. geophilum*, *Wilcoxina* spp., and also *Hebeloma* spp., *Tuber* sp., and *Tomentella* spp., plays a pivotal role in colonizing emerging tree seedlings and early post-fire succession in the absence of ECM species which are more competitive in non-disturbed sites (Baar et al. 1999; Taylor and Bruns 1999; Izzo et al. 2006; Buscardo et al. 2010; Barker et al. 2013). In case of high severity fires, an increase in the relative abundance of ascomycetous fungi was signaled (Grogan et al. 2000; Cairney and Bastias 2007; Holden et al. 2013; Rincón et al. 2014). Point-source inoculum may still prevail for the regenerating seedlings even 14 years post-fire (Rincón et al. 2014).

Successional changes in above- and belowground ECM fungal communities have been observed with increasing forest-stand age in a few chronosequences following stand-replacing fires. Distinct shifts in species composition and increasing diversity were shown, especially in the first years after disturbance until the canopy closure, in regenerating stands of *Pinus banksiana* (Visser 1995; LeDuc et al. 2013), *P. sylvestris* (Kipfer et al. 2011), as well as *Pseudotsuga menziesii* and, less clearly, *Betula papyrifera* (Twieg et al. 2007). Very few species present in early-successional sites (*Wilcoxina*, *Thelephora terrestris*, *Coltricia perennis*, *Rhizopogon* spp., *Tomentella* spp.) were completely replaced in older sites. Many of them persisted in low abundances over the entire chronosequence, although those dominant in young stages were largely absent after canopy closure. Some species, e.g., *C. geophilum*, were found to persist throughout the lifetime of the stand; in the mature stands, these taxa were joined by the late-successional species of *Clavulina*, *Cortinarius*, *Lactarius*, *Russula*, *Tricholoma*, *Hygrophorus*, and *Piloderma*. The ECM species abundance distribution differed between youngest stage, where few dominant species prevailed, and older stages, characterized by rather lognormal distribution, indicative of a stable, species-rich community (Visser 1995). The overall ECM species diversity increased with the stand age due to both the presence of taxa found in the young sites and the gradual accumulation of new taxa. It reached a plateau right before or at the time of canopy closure, while the community composition stabilized some time later in the mature stands (Visser 1995;

Twieg et al. 2007; LeDuc et al. 2013). Similar trends were observed in the ecosystems that are not adapted to fire, like Central European pine forests (Kipfer et al. 2011). The number of ECM species was strongly reduced after the severe fire event, then significantly increased with time, and reached the level of adjacent undisturbed forests after 15–18 years; however, the community composition did not converge to the pre-fire state, indicating its resilience but much longer time needed for its restoration.

Majority of the early colonizing ECM species in post-fire sites are fruiting species; a rapid increase in the taxonomical diversity of sporophores associated with replanted pine seedlings was observed in the second to fourth year after fire, even after severe stand-replacing fires (Friedrich 2001; Kutorga et al. 2012), with *T. terrestris*, *Laccaria* spp., *C. perennis*, *Paxillus involutus*, *Scleroderma citrinum*, *Amphinema byssoides,* and *Tomentellopsis zygodesmoides* found first. It is possible that the origin of some of these species was from a nursery. However, sporophores of *T. terrestris* and *A. byssoides*, as well as sclerotia of *C. geophilum*, were observed in unmanaged and naturally regenerating burned stand of *Pinus mugo* in the first year after the intensive fire, suggesting the resistant spore bank and aerial deposition to be the main inoculum sources (Motiejūnaitė et al. 2014).

Some of the understory shrubs may exert facilitative effect on the recruitment and development of ECM host trees. *Cistus ladanifer*, pyrophytic shrub typical of Mediterranean early post-fire vegetation, was shown to harbor rich and diverse community of ECM fungi 4 years after fire (Hernández-Rodríguez et al. 2013). Other *Cistus* spp. and *Arbutus unedo* were also shown to share ECM species with trees in fire-affected Mediterranean vegetation (Buscardo et al. 2012). *Arctostaphylos uva-ursi* is another shrub which could play a similar role in alpine forests (Kipfer et al. 2011).

Despite the increasing knowledge about the effects of fire on ECM communities, the understanding of the functional drivers of their post-fire successional traits and role of particular fungal species is still poor. An increase in soluble organic and free amino acid N, pH, organic matter, C/N ratio, and available P and Fe were shown to significantly explain the differences in ECM species assemblages (Twieg et al. 2009; Longo et al. 2011; LeDuc et al. 2013; Rincón et al. 2014), although the relationships are inconsistent and thus difficult to interpret. Apart from soil conditions, multiple biotic and abiotic factors associated with climate, vegetation type and structure, host plant composition and age, topography, fire cycles intervals or time elapsed since last fire event, fire severity, and heterogeneity are proven to drive the recovery of the post-fire ECM fungal communities (Dahlberg et al. 2001; Hart et al. 2005; Cairney and Bastias 2007; Buscardo et al. 2011; LeDuc et al. 2013).

10.5.2 Clear-Cuts and Gap-Forming Disturbances

Timber harvesting is a common practice affecting ECM forests throughout the world. The silvicultural cutting systems may exert impacts of different extent and intensity, from single tree and group selection (selective logging), partial harvesting, and variable retention cutting (e.g., seed tree or green tree retention with single stems, groups, or hectares-size patches of trees left after logging) to variable density thinning, including still widely used clear-cutting (Kuuluvainen et al. 2015). All these activities, to some extent, mimic natural disturbances. Sufficient retention and creating fine-scale heterogeneity in stand structure and composition, promoting development of forests with late-successional stand characteristics including greater structural complexity and biodiversity, are part of the currently recommended ecosystem-based forest restoration and management strategy (Gustafsson et al. 2012; Kuuluvainen et al. 2015).

As reviewed by Jones et al. (2003), sustained harvest practices usually neither lower mycorrhizal colonization rate of regenerating seedlings nor reduce the inoculum potential of ECM fungi to a level that threatens species diversity, while increasing disturbance intensity may cause a reduction in ECM richness and diversity (Lazaruk et al. 2005) (but see Barker et al. 2013 for contrasting results). ECM fungal communities typically benefit from retention practice (Rosenvald and Lõhmus 2008). Their species richness and abundance in retention cuts was shown to be significantly higher than in clear-cuts, and many species typical of mature forests remained present within the rooting zone of single retention trees or their groups (lifeboating effect), providing sources of inocula for regenerating tree seedlings (Kranabetter 1999; Cline et al. 2005; Lazaruk et al. 2005; Rosenvald and Lõhmus 2008).

Although the patch size effect on the species richness may be small or insignificant immediately postharvest (Lazaruk et al. 2005; Jones et al. 2008), a clear effect of habitat fragmentation on the ECM community 10 years after logging was demonstrated by Kranabetter et al. (2013) and attributed to reductions in rooting density, declines in spore dispersal, and alterations in soil conditions. The reductions were apparent especially in the late-successional species of *Clavulina*, *Lactarius*, *Russula*, and *Tylospora* and in the species of *Cortinarius* and *Elaphomyces* forming large, mat-like colonies. Few species, like *Meliniomyces* bicolor (*Hymenoscyphus ericae* aggr.) and *Tomentellopsis submollis*, were more abundant under retained trees.

The size of the retention patches and the proportion of trees cut also affect taxonomical richness and production of ECM epigeous sporophores which seem less resistant to the harvest stress than mycorrhizal root tips. Sharp decrease in ECM sporophore production was found in Douglas-fir stands with 15% retention (Luoma et al. 2004); reduction of fruiting in patches up to 0.12 ha was by 50% on average (Kranabetter et al. 2013). Among the reasons of lower productivity, the authors mention seasonal alterations of temperature and moisture regimes under variable retention systems, especially at the soil surface, soil disturbances, and possible edge effects. Kranabetter et al. (2013) suggest retention patches approximately 0.2 ha in size and culminating in at least 3% of the total cutblock area as having strategic value by capturing much of the spatial heterogeneity and species diversity of the ECM fungal community.

Edges of intact forests and retention trees act as efficient sources of mycorrhizal inoculum for seedlings, creating the potential for including them into the functioning mycorrhizal networks and improving their carbon, nutrient, and water balance (Teste and Simard 2008; Bingham and Simard 2012). However, the beneficial influence of the live trees as an inoculum source disappears gradually within 5–10–25 m into the harvested area from the edge of intact forest (Kranabetter and Wylie 1998; Hagerman et al. 1999a), retention patches (Outerbridge and Trofymow 2004; Jones et al. 2008), and single retention trees (Kranabetter 1999; Cline et al. 2005; Luoma et al. 2006). In most cases, the ECM colonization level and diversity diminishes at a distance of >5 m from the retention trees, and the ECM richness is positively correlated with fine-root-tip density (Luoma et al. 2006). Taking the edge:area ratio of retention patches into account, Jones et al. (2008) suggested that more small patches (but of at least 10 m diameter) would be more effective than a few large patches in supplying ECM inoculum to adjacent harvested areas during the first year after harvest. Although ECM diversity did not differ with retention stand age, seedling root colonization was significantly lower adjacent to second-growth stands than to old-growth (Outerbridge and Trofymow 2004).

Many years following harvest, the richness of ECM fungal community on seedlings emerging in clear-cuts outside the rooting zone of live trees persists on the reduced level or decreases even further (Hagerman et al. 1999b; Kranabetter 1999; Cline et al. 2005), suggesting reduction in the available inoculum potential. Active ECM root tips substantially decline to the level usually not exceeding 5% within 2 years after clear-cutting (Hagerman et al. 1999b; Lazaruk et al. 2005). Naturally regenerated *Tsuga heterophylla* seedlings transplanted from the mature forest into the 6-year-old clear-cut opening, after 2 years, showed reduced morphotype richness, both general and average per seedling, less even species distribution, and increased abundance of pioneer species, suggesting the key role of enough rooting density and hyphal links to mature trees in shaping the ECM community structure (Kranabetter and Friesen 2002).

Clear-cutting puts strong selection pressure on the ECM community, as it cuts off the carbon supply. Often, this results in a substantial reduction of ECM fungal diversity and different ECM species assemblages capable of colonizing the roots of regenerating seedlings (Kranabetter and Wylie 1998; Hagerman et al. 1999b; Cline et al. 2005; Dickie et al. 2009). Among these species, *C. geophilum* and the species of *Hebeloma*, *Phialocephala*, *Rhizopogon*, *Suillus*, *Tuber*, *Wilcoxina*, *Atheliaceae* (e.g., *Amphinema*), *T. terrestris*, and other *Thelephoraceae* are most frequent and abundant (Hagerman et al. 1999b; Jones et al. 2003; Hagerman and Durall 2004; Cline et al. 2005; Barker et al. 2013; Walker and Jones 2013). These are known pioneer and ruderal species, comprising an ECM-resistant propagule bank in disturbed soils. Most of these species are present also in undisturbed forests; however, as the late-successional species dependent upon the presence of mature trees and connections to their living roots decline after clear-cut logging, they become relatively more abundant and readily colonize new seedlings in the absence of competitive interactions (Jones et al. 2003; Cline et al. 2005; Barker et al. 2013). It is possible that the ECM fungal communities on seedling roots in clear-cuts are better adapted to taking up nutrients under the altered conditions (Jones et al. 2003, 2008); functional efficiency of these early-successional ECM assemblages expressed by extracellular enzyme activity may not differ from that of the ECM fungi present in intact forests, at least in the short term (Jones et al. 2010).

Further development of simplified and overall unified ECM fungal communities (Jones et al. 2010; Walker and Jones 2013) in regenerating secondary stands in clear-cuts is influenced by stochastic dispersal limitations and various environmental filters. As summarized by Jones et al. (2003), among the most important interrelated physical, chemical, and biological factors are changes in soil temperature, moisture, insulation, pH, carbon supply, nutrient cycling, soil microbiota, possible ECM fungi introduced from nursery, and host characteristics (host age, species composition of regenerating stand, presence of ECM shrubs) as well as suppressing effects of non-ECM plants. Soil chemistry, especially soil N, C, P, and organic horizon depth, can affect, although inconsistently, the composition of ECM fungal assemblages in clear-cut origin forests, with different species responding differently to soil parameters (Dickie et al. 2009; Twieg et al. 2009). Some forest management practices, besides different retention logging described earlier, were shown to foster reestablishment of ECM symbiotic relationships which accelerate forest regeneration. Although niche partitioning by substrate (mineral soil versus different types of woody debris) was not confirmed to occur among ECM fungal species on very young seedlings planted in clear-cuts, retention of postharvest coarse woody debris was suggested to increase the diversity of the ECM community in the long term (Walker and Jones 2013). Forest floor retention in clear-cuts may result in the richness and diversity of ECM fungi on regenerating seedlings comparable to the undisturbed forest as soon as after 1 year; however, mineral soil inoculum was shown to be sufficient to compensate for losses caused by forest floor disruption in the longer term (Barker et al. 2013). Methods of post-logging site preparation that leave surface organic matter relatively intact may impact indigenous ECM fungi less than those that remove or bury the organic layer (Lazaruk et al. 2008). Sprouts of stumps left after logging (e.g., of *Quercus*) (Dickie et al. 2002), understory trees (e.g., *Salix*, *Populus*), as well as ECM shrubs (e.g., *Arctostaphylos*) (Hagerman and Durall 2004) may serve as retained inoculum sources accelerating colonization of regenerating overstory tree seedlings.

In the chronosequence study, Twieg et al. (2007) demonstrated a significant increase in ECM fungal diversity with host age between two age classes (young, 5 years old, and at canopy closure, 26 years old) in mixed stands of *B. papyrifera* and *P. menziesii* of clear-cut origin. Species representing the genera of *Russula*, *Piloderma*, and *Cortinarius* increased, while *Amphinema*, *Laccaria*, and *Thelephora* decreased their frequency and abundance when stands aged (Smith et al. 2000; Kranabetter et al. 2005; Durall et al. 2006; Twieg et al. 2007).

Changes in the postharvest forest tree composition lead to shifts in the ECM fungal community, however, with the legacies of clear-cut disturbance persisting for decades. In a common garden experiment, in which pure stands of native trees, *P. sylvestris* and *Quercus robur*, were planted directly after 80-year-old *P. sylvestris* stand had been clear-cut (including removal of the root systems), the former, although characterized by lower ECM species richness, maintained more even and more than twice as diverse ECM fungal community than the latter, over 30 years postharvest (Trocha et al. 2012).

10.5.3 Post-Agricultural Sites

Over the last hundred of years, population movements caused by wars, demographic changes (aging of rural population, migration of the youth to the towns), and changes in the agricultural policy of the governments resulted in an intensive and well-documented process of farmland abandonment in many parts of the world. In Europe, post-agricultural sites are expected to be one of the most striking features of the landscape (Faliński 1998; Stoate et al. 2009). Following the abandonment of traditional agricultural land, natural or semi-natural habitats can be reestablished; however, intensive agriculture may lead to increasing number of old fields showing a low recovery of a historic vegetation state (Cramer et al. 2008).

The previous land-use and cultivation practices have persistent effects on most of the physical, chemical, and biological properties of soils, including contents of organic matter, nitrogen and other elements, pH, porosity, bulk density, microbial activity, and the overall rate of soil processes, manifesting in a long-term, 80–100 years long or even longer recovery of forest soils and vegetation (van der Wal et al. 2006; Olszewska and Smal 2008; Smal and Olszewska 2008; von Oheimb et al. 2008). In old fields, the rates of the recovery of a well-structured, mature forest community vary considerably: up to 140 years in European mesophilous pine forests and 350 years in European mixed hardwood forests (Faliński 1986).

In an active agricultural field, ECM inocula are sparse, patchily distributed, and surrounded by large areas with a lack of infectivity (Boerner et al. 1996). However, the infectiveness and richness of ECM community increases significantly with time after disturbance, especially in the first 5–10 years, reaching the pre-disturbance level 25–30 years after agriculture cessation. In this process, a major role of animal faces in ECM dispersal is apparent, as during these years, an increasing number of discrete patches with a high potential of infectiveness were recorded.

With time, diverse ECM propagule bank can be formed in post-agricultural sites. However, as shown by Ding et al. (2011) for a site abandoned for 15 years, it was less rich in species than in the reference forest site and showed a strong dominance of *Wilcoxina micolae*. Also, the ECM colonization rate on the roots of seedlings was relatively high, but, on some hosts, still lower compared to that of the reference site. On the other hand, in a mycorrhization experiment carried out by Menkis et al. (2007), pine and spruce seedlings, pre-inoculated with *C. geophilum*, *Piceirhiza bicolorata*, and *Hebeloma crustuliniforme*, were effectively colonized by a diverse ECM community (19 morphotypes, including inoculated species, *Rhizopogon* sp., *Suillus* spp., *T. terrestris*, *Paxillus* sp., *A. byssoides*, and numerous unidentified taxa) over two growing seasons in post-agricultural soil after 10 years of field abandonment. This indicated a high ECM inoculum potential and a low impact of inoculation on subsequent fungal community development in this type of habitats. Beneficial effects of inoculated *T. terrestris* on the number of mycorrhizal tips of pine seedlings within 2 years of outplantation were shown by Hilszczańska and Sierota (2006). Moreover, 17 morphotypes of ECM fungi were found on the roots of saplings (including the pre-inoculated strain) after 6–8 years, indicating the importance of environmental conditions for the persistent establishment of ECM fungi (Hilszczańska et al. 2011).

Among the drivers of ECM inoculum occurrence and seedling mycorrhization in post-agricultural systems, the distance from established vegetation has proved to be a key element. The abundance of mycorrhizae on the roots of seedlings declined sharply ca. 15 m from the base of mature trees, whereas the mycorrhization level, foliar N concentrations, and survival of seedlings were higher near established trees (Dickie et al. 2002, 2007; Dickie and Reich 2005). The species richness declined with distance from trees with a higher number of infrequent species close to the trees and ascomycete fungi being relatively more represented at greater distances (Dickie and Reich 2005). The facilitation of tree seedlings may be very efficient if they grow near the forest edges and are incorporated into the local common mycorrhizal network (Thiet and Boerner 2007). Moreover, the close vicinity of abandoned fields to well-established woodlands may contribute to a fast restoration of ECM spore bank (Kałucka 2009).

A successful tree establishment in post-agricultural land may be hindered not only by the lack of ECM inocula but the competition with herbs, which can negatively affect seedlings regardless of their distance from mature trees (Dickie et al. 2007). ECM colonization and the survival of tree seedlings is even more suppressed by a management practice of sowing additional herbs onto the sites (Hedlund and Gormsen 2002). The composition and structure of vegetation developing in old fields soon after cultivation cessation may form strong filters for ECM trees or, instead, open "the windows of opportunity" for their establishment (Faliński 1998; van der Putten et al. 2000; Cramer et al. 2008; Tokuoka et al. 2011).

Early-successional old fields proved to be good model systems for the studies of ECM fungal chronosequences. Based on the observations of spatial and temporal occurrence of ECM sporocarps (Mason et al. 1982; Last et al. 1983) and morphotypes (Deacon et al. 1983) in first-generation birch stands planted in abandoned falmland, a model of sequential appearance of "early-" and "late-stage" fungi was proposed (see Sect. 10.2 in this chapter). As first ECM colonizers, species of *Hebeloma*, *Laccaria*, *Thelephora*, *Inocybe*, and *Lactarius* were found, later followed by *Amanita*, *Cortinarius*, and *Leccinum*; then, 10 years after seedlings outplanting, *Russula* appeared. Sporocarps occurred in zones of progressively increasing radius. The new species appeared usually closer to the stem base, which suggested an association with the older roots and relationship of successional changes in ECM community composition with tree age. Similar patterns were observed concerning other tree species, e.g., *P. menziesii* (Chu-Chou and Grace 1981) and *Pinus radiata* (Chu-Chou 1979; Chu-Chou and Grace 1988).

Temporal and spatial sequential appearances of ECM fungi were also observed in longer chronosequences, both in the natural processes of post-agricultural land reforestations and in tree plantations. In the complete series of seral communities representing spontaneous vegetation change from an old-field psammophilous grassland to a 100-year-old mesophilous Scots pine forest, sporophocarps of 98 ECM fungal species were recorded (Kałucka 2009). After 8–15 years of field abandonment, 44 species were found growing already under 6–8 years old pine trees. The first ECM colonizers included the species of *Inocybe*, *Laccaria*, *Suillus*, *Cortinarius*, *Amanita*, and *Hygrophorus* (colonization stage), joined by *Chalciporus*, *Lactarius*, many small-sized *Cortinarius*, and *Tricholoma* species before and at the canopy closure (stabilization stage). Later, a gradual change in ECM species composition, including a partial replacement of species selective for the pre-forest, semi-open habitats and arrival of numerous forest-dwelling species, mainly members of genera *Russula*, *Lactarius*, *Cortinarius*, and *Amanita*, was observed (replacement stage). ECM fungal species diversity continuously increased, while species richness and sporocarp biomass production peaked shortly before the canopy closure (under 18–25 years old pines). Strong relationships among the patterns of ECM species composition, frequency, abundance, and production of sporocarps, and forest development, the age of trees, stand structure, and soil profile formation suggested that ECM fungal succession is best explained by the development of the vegetation and the habitat.

The ECM successional trajectories in spontaneous reforestation systems and in tree plantations in post-agricultural lands seem to be similar (Kałucka and Jagodziński 2016), but in the plantations, canopy closure happens usually earlier and at a younger age of trees. As it was shown in the author's (ILK) study carried out in a chronosequence of 1–42 years old managed Scots pine stands, development of the ECM fungal community can be significantly accelerated. Within 4 years after outplantation, sporocarps of *Hebeloma*, *Laccaria*, *Inocybe*, *Suillus*, *Thelephora*, and *Tomentella* appeared and ECM fungal species richness and diversity continuously increased through the chronosequence of stands examined. However, the peak of sporocarp biomass production connected with the canopy closure was observed already in 7-year-old stands. Moreover, it coincided with an intensive development of fine root biomass (Jagodziński and Kałucka 2011). The canopy closure was then followed by gradual species change and successive arrival of fungi typical of more mature forests. On former agricultural fields, similar pattern of ECM community development was found in the chronosequence of Norway spruce plantations (Gáper and Lizoň 1995; Mihál 1999). Thus, the factors that shape the ECM community include the above- and belowground structure of the stand and both the development of vegetation and habitat, not only the age of trees.

The reciprocal relationship between the development of ECM fungal communities and forest regeneration after agricultural abandonment is well documented at the level of nitrogen biogeochemical cycle. After cultivation cessation, the available N pools decline due to the substantial transfer of mineralized N from soil into plants (Richter et al. 2000). With tightening of N availability, the foliar %N and the total foliar N content of overstory vegetation also decrease over time; as stands aged, foliar $\delta^{15}N$ decreased and became more negative, giving a strong evidence of an increasing participation of ECM fungi in the process of N transfer from soil into the trees via the mineralization–mycorrhizal–plant uptake pathway (Compton et al. 2007; compare Hobbie et al. 2005).

10.6 Conclusions

Succession is one of the most important ecological processes shaping the biosphere. Pioneer or disturbed ecosystems, if not permanently restricted by climatic or edaphic conditions, or if released from disturbance, follow various trajectories of successional processes toward more mature systems. ECM fungi, as obligatory symbionts of the whole range of trees and shrub species, especially in the temperate and boreal zones, are a key organism group which enable and enhance the development of forest communities. On one hand, through their nutritional and protective relationships with host plants and involvement in biogeochemical cycles, ECM fungi interact with vegetation succession and changing soil environment; on the other hand, they are subjected to their own successional dynamics. As it was shown in this chapter, the tree encroachment and successful colonization of open land, both in primary and secondary successional sites, become possible if only the ECM fungal propagules are present in the form of aerially deposited spores from local or remote sources, or spores supplied by animal vectors, spore banks, or mycelial inocula left after the disturbance of preexisting community or functioning ECM networks that can be joined by the roots of arriving tree seedlings. In early-successional sites, ECM fungal inocula and developing ECM communities can speed up or alter the course of forest establishment or restoration. This phenomenon is crucial in both natural and managed systems. Moreover, it is equally important for understanding the mechanisms of woody vegetation succession when making afforestation polices aiming at land reclamation (Rincón et al. 2006; Menkis et al. 2007), in sustainable forest management (Rosenvald and Lõhmus 2008), and in preventing open biocoenoses from being overgrown with trees (Collier and Bidartondo 2009; Dickie et al. 2011). The role of ECM fungi should also be considered in the interpretation of the processes affecting forest development or forest decline as a result of anthropogenic climate change (Milad et al. 2011) and CO_2 enrichment (Higgins and Scheiter 2012). ECM fungal communities may be key players in the intensified forest restoration practices aiming at increased carbon sequestration (van Breugel et al. 2011; Kałucka and Jagodziński 2013; Clemmensen et al. 2013, 2015). Similarly, they strongly influence the processes involving non-native tree invasions (Nuñez et al. 2009; Dickie et al. 2010) and expansions of trees within their native ranges (Collier and Bidartondo 2009). Mechanisms of ECM fungal succession and its interactions with vegetation and habitat development impact ecosystem trajectories after changes in land management practices. This has important implications for biodiversity and sustainability of rare and threatened ecosystems (Thiet and Boerner 2007; Dickie et al. 2011), and thus the role of ECM fungi should be more considered in conservation polices.

Acknowledgements This review was supported by the Ministry of Science and Higher Education, Warsaw, Poland (grant nos. 3 P04G 031 23 and N N305 2996 40) and by General Directorate of State Forests, Warsaw, Poland (research project "Environmental and genetic factors affecting productivity of forest ecosystems in forest and post-industrial habitats"). The authors would like to thank Dr. Douglas Zook (Global Ecology Education Initiative, Boston University, USA) and Dr. Joanna Kazik (University Centre Doncaster, UK) for linguistic revision of the manuscript.

References

Afkhami ME, Rudgers JA, Stachowicz JJ (2014) Multiple mutualist effects: conflict and synergy in multispecies mutualisms. Ecology 95:833–844. doi:10.1890/13-1010.1

Alfredsen G, Høiland K (2001) Succession of terrestrial macrofungi along a deglaciation gradient at Glacier Blaisen, South Norway. Nord J Bot 21:19–37. doi:10.1111/j.1756-1051.2001.tb01335.x

Allen MF, Allen EB (1992) Mycorrhizae and plant community development: mechanisms and patterns. In: Carrol GC, Wicklow DT (eds) The fungal community. Its organization and role in the ecosystem, 2nd edn. Marcel Dekker, Inc, New York, Basel, Hong Kong, pp 455–479

Allen MF, Crisafulli C, Friese CF, Jeakins SL (1992) Re-formation of mycorrhizal symbioses on Mount St Helens, 1980-1990: interactions of rodents and mycorrhizal fungi. Mycol Res 96: 447–453. doi:10.1016/S0953-7562(09)81089-7

Andrews JH (1992) Fungal life-history strategies. In: Carrol GC, Wicklow DT (eds) The fungal community. Its organization and role in the ecosystem, 2nd edn. Marcel Dekker, Inc, New York, Basel, Hong Kong, pp 119–145

Andrews JH, Harris RF (1986) r-selection and K-selection and microbial ecology. Adv Microb Ecol 9:99–147

Ashkannejhad S, Horton TR (2006) Ectomycorrhizal ecology under primary succession on coastal sand dunes: interactions involving *Pinus contorta*, suilloid fungi and deer. New Phytol 169:345–354. doi:10.1111/j.1469-8137.2005.01593.x

Baar J (1996) The ectomycorrhizal flora of primary and secondary stands of *Pinus sylvestris* in relation to soil conditions and ectomycorrhizal succession. J Veg Sci 7:497–504. doi:10.2307/3236298

Baar J, Horton TR, Kretzer AM, Bruns TD (1999) Mycorrhizal colonization of *Pinus muricata* from resistant propagules after a stand-replacing wildfire. New Phytol 143:409–418. doi:10.1046/j.1469-8137.1999.00452.x

Bahram M, Peay KG, Tedersoo L (2015) Local-scale biogeography and spatiotemporal variability in communities of mycorrhizal fungi. New Phytol 205:1454–1463. doi:10.1111/nph.13206

Barker JS, Simard SW, Jones MD, Durall DM (2013) Ectomycorrhizal fungal community assembly on regenerating Douglas-fir after wildfire and clearcut harvesting. Oecologia 172: 1179–1189. doi:10.1007/s00442-012-2562-y

Bauman JM, Keiffer CH, Hiremath S, McCarthy BC (2013) Soil preparation methods promoting ectomycorrhizal colonization and American chestnut *Castanea dentata* establishment in coal mine restoration. J Appl Ecol 50:721–729. doi:10.1111/1365-2664.12070

Begon M, Harper JL, Townsend CR (1996) Ecology. Individuals, populations and communities, 3rd edn. Blackwell Science, Oxford

Bento-Gonçalves A, Vieira A, Ubeda X, Martin D (2012) Fire and soils: key concepts and recent advances. Geoderma 191:3–13. doi:10.1016/j.geoderma.2012.01.004

Berendsen RL, Pieterse CMJ, Bakker PAHM (2012) The rhizosphere microbiome and plant health. Trends Plant Sci 17:478–486. doi:10.1016/j.tplants.2012.04.001

Bergemann SE, Miller SL (2002) Size, distribution, and persistence of genets in local populations of the late-stage ectomycorrhizal basidiomycete, *Russula brevipes*. New Phytol 156:313–320. doi:10.1046/j.1469-8137.2002.00507.x

Bever JD, Dickie IA, Facelli E, Facelli JM, Klironomos J, Moora M, Rillig MC, Stock WD, Tibbett M, Zobel M (2010) Rooting theories of plant community ecology in microbial interactions. Trends Ecol Evol 25:468–478. doi:10.1016/j.tree.2010.05.004

Bingham MA, Simard S (2012) Ectomycorrhizal networks of *Pseudotsuga menziesii* var. *glauca* trees facilitate establishment of conspecific seedlings under drought. Ecosystems 15:188–199. doi:10.1007/s10021-011-9502-2

Blaalid R, Carlsen T, Kumar S, Halvorsen R, Ugland KI, Fontana G, Kauserud H (2012) Changes in the root-associated fungal communities along a primary succession gradient analysed by 454 pyrosequencing. Mol Ecol 21:1897–1908. doi:10.1111/j.1365-294X.2011.05214.x

Blasius D, Oberwinkler F (1989) Succession of mycorrhizae—a matter of tree age or stand age. Ann For Sci 46:758–761. doi:10.1051/forest:198905ART0169

Boerner REJ, DeMars BG, Leicht PN (1996) Spatial patterns of mycorrhizal infectiveness of soils long a successional chronosequence. Mycorrhiza 6:79–90. doi:10.1007/s005720050111

Bois G, Piché Y, Fung MYP, Khasa DP (2005) Mycorrhizal inoculum potentials of pure reclamation materials and revegetated tailing sands from the Canadian oil sand industry. Mycorrhiza 15:149–158. doi:10.1007/s00572-004-0315-4

Brown SP, Jumpponen A (2014) Contrasting primary successional trajectories of fungi and bacteria in retreating glacier soils. Mol Ecol 23:481–497. doi:10.1111/mec.12487

Brundrett MC (2009) Mycorrhizal associations and other means of nutrition of vascular plants: understanding the global diversity of host plants by resolving conflicting information and developing reliable means of diagnosis. Plant Soil 320:37–77. doi:10.1007/s11104-008-9877-9

Bruns TD (1995) Thoughts on the process that maintain local species diversity of ectomycorrhizal fungi. Plant Soil 170:63–73. doi:10.1007/bf02183055

Bruns T, Tan J, Bidartondo M, Szaro T, Redecker D (2002) Survival of *Suillus pungens* and *Amanita francheti* ectomycorrhizal genets was rare or absent after a stand-replacing wildfire. New Phytol 155:517–523. doi:10.1046/j.1469-8137.2002.00468.x

Buscardo E, Rodríguez-Echeverría S, Martín MP, De Angelis P, Pereira JS, Freitas H (2010) Impact of wildfire return interval on the ectomycorrhizal resistant propagules communities of a Mediterranean open forest. Fungal Biol 114:628–636. doi:10.1016/j.funbio.2010.05.004

Buscardo E, Freitas H, Pereira JS, De Angelis P (2011) Common environmental factors explain both ectomycorrhizal species diversity and pine regeneration variability in a post-fire Mediterranean forest. Mycorrhiza 21:549–558. doi:10.1007/s00572-011-0363-5

Buscardo E, Rodríguez-Echeverría S, Barrico L, García MA, Freitas H, Martín MP, De Angelis P, Muller LAH (2012) Is the potential for the formation of common mycorrhizal networks influenced by fire frequency? Soil Biol Biochem 46:136–144. doi:10.1016/j.soilbio.2011.12.007

Buscot F (2015) Implication of evolution and diversity in arbuscular and ectomycorrhizal symbioses. J Plant Physiol 172:55–61. doi:10.1016/j.jplph.2014.08.013

Cairney JWG, Bastias BA (2007) Influences of fire on forest soil fungal communities. Can J For Res 37:207–215. doi:10.1139/x06-190

Çakan H, Karataş C (2006) Interactions between mycorrhizal colonization and plant life forms along the successional gradient of coastal sand dunes in the eastern Mediterranean, Turkey. Ecol Res 21:301–310. doi:10.1007/s11284-005-0134-x

Cázares E, Trappe JM (1994) Spore dispersal of ectomycorrhizal fungi on a glacier forefront by mammal mycophagy. Mycologia 86:507–510. doi:10.2307/3760743

Cázares E, Trappe JM, Jumpponen A (2005) Mycorrhiza-plant colonization patterns on a subalpine glacier forefront as a model system of primary succession. Mycorrhiza 15:405–416. doi:10.1007/s00572-004-0342-1

Chu-Chou M (1979) Mycorrhizal fungi of *Pinus radiata* in New Zealand. Soil Biol Biochem 11: 557–562. doi:10.1016/0038-0717(79)90021-x

Chu-Chou M, Grace LJ (1981) Mycorrhizal fungi of *Pseudotsuga menziesii* in the North Island of New Zealand. Soil Biol Biochem 13:247–249. doi:10.1016/0038-0717(81)90029-8
Chu-Chou M, Grace LJ (1988) Mycorrhizal fungi of radiata pine in different forests of the north and south islands in New Zealand. Soil Biol Biochem 20:883–886. doi:10.1016/0038-0717(88)90098-3
Clements FE (1916) Plant succession. An analysis of the development of vegetation. The Carnegie Institution of Washington, Washington, DC
Clemmensen KE, Bahr A, Ovaskainen O, Dahlberg A, Ekblad A, Wallander H, Stenlid J, Finlay RD, Wardle DA, Lindahl BD (2013) Roots and associated fungi drive long-term carbon sequestration in Boreal Forest. Science 339:1615–1618. doi:10.1126/science.1231923
Clemmensen KE, Finlay RD, Dahlberg A, Stenlid J, Wardle DA, Lindahl BD (2015) Carbon sequestration is related to mycorrhizal fungal community shifts during long-term succession in boreal forests. New Phytol 205:1525–1536. doi:10.1111/nph.13208
Cline ET, Ammirati JF, Edmonds RL (2005) Does proximity to mature trees influence ectomycorrhizal fungus communities of Douglas-fir seedlings? New Phytol 166:993–1009. doi:10.1111/j.1469-8137.2005.01387.x
Collier FA, Bidartondo MI (2009) Waiting for fungi: the ectomycorrhizal invasion of lowland heathlands. J Ecol 97:950–963. doi:10.1111/j.1365-2745.2009.01544.x
Colpaert JV, Wevers JHL, Krznaric E, Adriaensen K (2011) How metal-tolerant ecotypes of ectomycorrhizal fungi protect plants from heavy metal pollution. Ann For Sci 68:17–24. doi:10.1007/s13595-010-0003-9
Compton JE, Hooker TD, Perakis SS (2007) Ecosystem N distribution and delta N-15 during a century of forest regrowth after agricultural abandonment. Ecosystems 10:1197–1208. doi:10.1007/s10021-007-9087-y
Conesa HM, Schulin R (2010) The Cartagena—La Union mining district (SE Spain): a review of environmental problems and emerging phytoremediation solutions after fifteen years research. J Environ Monit 12:1225–1233. doi:10.1039/c000346h
Cooke RC, Rayner ADM (1984) Ecology of Saprotrophic fungi. Longman, London
Corkidi L, Rincón E (1997) Arbuscular mycorrhizae in a tropical sand dune ecosystem on the Gulf of Mexico. 1. Mycorrhizal status and inoculum potential along a successional gradient. Mycorrhiza 7:9–15. doi:10.1007/s005720050157
Cramer VA, Hobbs RJ, Standish RJ (2008) What's new about old fields? Land abandonment and ecosystem assembly. Trends Ecol Evol 23:104–112. doi:10.1016/j.tree.2007.10.005
Dahlberg A (2002) Effects of fire on ectomycorrhizal fungi in fennoscandian boreal forests. Silva Fenn 36:69–80 doi:10.14214/sf.551
Dahlberg A, Schimmel J, Taylor AFS, Johannesson H (2001) Post-fire legacy of ectomycorrhizal fungal communities in the Swedish boreal forest in relation to fire severity and logging intensity. Biol Conserv 100:151–161. doi:10.1016/s0006-3207(00)00230-5
Danielson RM (1991) Temporal changes and effects of amendments on the occurrence of sheathing (ecto-)mycorrhizas of conifers growing in oil sands tailings and coal spoil. Agric Ecosyst Environ 35:261–281. doi:10.1016/0167-8809(91)90054-2
Danielson RM, Visser S (1989) Host response to inoculation and behaviour of introduced and indigenous ectomycorrhizal fungi of jack pine grown on oil-sands tailings. Can J For Res 19:1412–1421. doi:10.1139/x89-216
Danielson RM, Visser S, Parkinson D (1983) Microbial activity and mycorrhizal potential of four overburden types used in the reclamation of extracted oil sands. Can J Soil Sci 63:363–337
de la Bastide PY, Kropp BR, Piche Y (1994) Spatial distribution and temporal persistence of discrete genotypes of the ectomycorrhizal fungus *Laccaria bicolor* (Maire) Orton. New Phytol 127:547–556
de Román M, de Miguel AM (2005) Post-fire, seasonal and annual dynamics of the ectomycorrhizal community in a *Quercus ilex* L. forest over a 3-year period. Mycorrhiza 15:471–482. doi:10.1007/s00572-005-0353-6

Deacon JW, Fleming LV (1992) Interactions of ectomycorrhizal fungi. In: Allen MF (ed) Mycorrhizal functioning: an integrated plant-fungal process. Chapman & Hall, London, pp 249–300

Deacon JW, Donaldson SJ, Last FT (1983) Sequences and interactions of mycorrhizal fungi on birch. Plant Soil 71:257–262. doi:10.1007/bf02182660

Dickie IA, Reich PB (2005) Ectomycorrhizal fungal communities at forest edges. J Ecol 93: 244–255. doi:10.1111/j.1365-2745.2005.00977.x

Dickie IA, Koide RT, Steiner KC (2002) Influences of established trees on mycorrhizas, nutrition, and growth of *Quercus rubra* seedlings. Ecol Monogr 72:505–521. doi:10.1890/0012-9615(2002)072[0505:IOETOM]2.0.CO;2

Dickie IA, Schnitzer SA, Reich PB, Hobbie SE (2007) Is oak establishment in old-fields and savanna openings context dependent? J Ecol 95:309–320. doi:10.1111/j.1365-2745.2006.01202.x

Dickie IA, Richardson SJ, Wiser SK (2009) Ectomycorrhizal fungal communities and soil chemistry in harvested and unharvested temperate *Nothofagus* rainforests. Can J For Res 39: 1069–1079. doi:10.1139/X09-036

Dickie IA, Bolstridge N, Cooper JA, Peltzer DA (2010) Co-invasion by *Pinus* and its mycorrhizal fungi. New Phytol 187:475–484. doi:10.1111/j.1469-8137.2010.03277.x

Dickie IA, Yeates GW, St John MG, Stevenson BA, Scott JT, Rillig MC, Peltzer DA, Orwin KH, Kirschbaum MUF, Hunt JE, Burrows LE, Barbour MM, Aislabie J (2011) Ecosystem service and biodiversity trade-offs in two woody successions. J Appl Ecol 48:926–934. doi:10.1111/j.1365-2664.2011.01980.x

Dickie IA, Martinez-Garcia LB, Koele N, Grelet GA, Tylianakis JM, Peltzer DA, Richardson SJ (2013) Mycorrhizas and mycorrhizal fungal communities throughout ecosystem development. Plant Soil 367:11–39. doi:10.1007/s11104-013-1609-0

Dickie IA, Koele N, Blum JD, Gleason JD, McGlone MS (2014) Mycorrhizas in changing ecosystems. Botany 92:149–160. doi:10.1139/cjb-2013-0091

Dighton J, Mason PA (1985) Mycorrhizal dynamics during forest tree development. In: Moore D, Casselton LA, Wood DA, Frankland JC (eds) Developmental biology of higher fungi. Cambridge University Press, Cambridge, pp 117–139

Dighton J, Poskitt JM, Howard DM (1986) Changes in occurrence of basidiomycete fruit bodies during forest stand development—with specific reference to mycorrhizal species. Trans Br Mycol Soc 87:163–171

Ding Q, Liang Y, Legendre P, He X-h, Pei K-q, Du X-j, Ma K-p (2011) Diversity and composition of ectomycorrhizal community on seedling roots: the role of host preference and soil origin. Mycorrhiza 21:669–680. doi:10.1007/s00572-011-0374-2

Dix NJ, Webster J (1995) Fungal ecology. Chapman and Hall, London, Glasgow, Weinheim, New York, Tokyo, Melbourne, Madras

Dominik T (1951) Badania mykotrofizmu roślinności wydm nadmorskich i śródlądowych. Acta Soc Bot Pol 21:125–164

Douhan GW, Vincenot L, Gryta H, Selosse M-A (2011) Population genetics of ectomycorrhizal fungi: from current knowledge to emerging directions. Fungal Biol 115:569–597. doi:10.1016/j.funbio.2011.03.005

Durall DM, Gamiet S, Simard SW, Kudrna L, Sakakibara SM (2006) Effects of clearcut logging and tree species composition on the diversity and community composition of epigeous fruit bodies formed by ectomycorrhizal fungi. Can J Bot 84:966–980. doi:10.1139/b06-045

Faliński JB (1986) Vegetation dynamics in temperate lowland primeval forests. Ecological studies in Białowieża Forest. Geobotany 8:1–537 (Dr W. Junk Publishers, Dordrecht, Boston, Lancaster)

Faliński JB (1998) Dioecious woody pioneer species (*Juniperus communis*, *Populus tremula*, *Salix* sp. div.) in the secondary succession and regeneration. Phytocoenosis 10 (N.S.) Suppl Cartogr Geobot 8:1–156

Fernandes PM, Botelho HS (2003) A review of prescribed burning effectiveness in fire hazard reduction. Int J Wildland Fire 12:117–128. doi:10.1071/wf02042

Ford ED, Mason PA, Pelham J (1980) Spatial patterns of sporophore distribution around a young birch tree in three successive years. Trans Br Mycol Soc 75:287–296

Frankland JC (1992) Mechanisms in fungal succession. In: Carrol GC, Wicklow DT (eds) The fungal community. Its organization and role in the ecosystem, 2nd edn. Marcel Dekker, Inc, New York, Basel, Hong Kong, pp 383–401

Frankland JC (1998) Fungal succession—unravelling the unpredictable. Mycol Res 102:1–15. doi:10.1017/s0953756297005364

Friedrich S (2001) Macromycetes diversity of pine-tree plantings on a post-fire forest site in Notecka Forest (NW Poland). Acta Mycol 36:127–148

Fujiyoshi M, Yoshitake S, Watanabe K, Murota K, Tsuchiya Y, Uchida M, Nakatsubo T (2011) Successional changes in ectomycorrhizal fungi associated with the polar willow *Salix polaris* in a deglaciated area in the High Arctic, Svalbard. Polar Biol 34:667–673. doi:10.1007/s00300-010-0922-9

Gáper J, Lizoň P (1995) Sporocarp succession of mycorrhizal fungi in the Norway spruce plantations in formerly agricultural land. In: Baluška F, Ciamporova M, Gasparikova O, Barlow PW (eds) Structure and function of roots. Kluver Academic Publishers, Dordrecht, pp 349–352

Gebhardt S, Neubert K, Wöllecke J, Müenzenberger B, Hüttl RF (2007) Ectomycorrhiza communities of red oak (*Quercus rubra* L.) of different age in the Lusatian lignite mining district, East Germany. Mycorrhiza 17:279–290. doi:10.1007/s00572-006-0103-4

Gleason HA (1939) The individualistic concept of the plant association. Am Midl Nat 21:92–110

Glen M, Bougher NL, Colquhoun IJ, Vlahos S, Loneragan WA, O'Brien P, Hardy GESJ (2008) Ectomycorrhizal fungal communities of rehabilitated bauxite mines and adjacent, natural jarrah forest in Western Australia. For Ecol Manage 255:214–225. doi:10.1016/j.foreco.2007.09.007

Glenn-Lewin DC, Peet RK, Veblen TT (1992) Plant succession—theory and prediction. Chapman and Hall, London

Grime JP (1977) Evidence for the existence of three primary strategies in plants and its relevance to ecological and evolutionary theory. Am Nat 111:1169–1194

Grime JP (1979) Plant strategies and vegetation process. Wiley, New York

Grogan P, Baar J, Bruns TD (2000) Below-ground ectomycorrhizal community structure in a recently burned bishop pine forest. J Ecol 88:1051–1062. doi:10.1046/j.1365-2745.2000.00511.x

Grossnickle SC (2005) Importance of root growth in overcoming planting stress. New For 30:273–294. doi:10.1007/s11056-004-8303-2

Gustafsson L, Baker SC, Bauhus J, Beese WJ, Brodie A, Kouki J, Lindenmayer DB, Lohmus A, Pastur GM, Messier C, Neyland M, Palik B, Sverdrup-Thygeson A, Volney WJA, Wayne A, Franklin JF (2012) Retention forestry to maintain multifunctional forests: a world perspective. Bioscience 62:633–645. doi:10.1525/bio.2012.62.7.6

Hagerman SM, Durall DM (2004) Ectomycorrhizal colonization of greenhouse-grown Douglas-fir (*Pseudotsuga menziesii*) seedlings by inoculum associated with the roots of refuge plants sampled from a Douglas-fir forest in the southern interior of British Columbia. Can J Bot 82:742–751. doi:10.1139/b04-047

Hagerman SM, Jones MD, Bradfield GE, Sakakibara SM (1999a) Ectomycorrhizal colonization of *Picea engelmannii* x *Picea glauca* seedlings planted across cut blocks of different sizes. Can J For Res 29:1856–1870

Hagerman SM, Jones MD, Bradfield GE, Gillespie M, Durall DM (1999b) Effects of clear-cut logging on the diversity and persistence of ectomycorrhizae at a subalpine forest. Can J For Res 29:124–134

Hart SC, DeLuca TH, Newman GS, MacKenzie MD, Boyle SI (2005) Post-fire vegetative dynamics as drivers of microbial community structure and function in forest soils. For Ecol Manage 220:166–184. doi:10.1016/j.foreco.2005.08.012

Hedlund K, Gormsen D (2002) Mycorrhizal colonization of plants in set-aside agricultural land. Appl Soil Ecol 19:71–78. doi:10.1016/s0929-1393(01)00174-3

Helm DJ, Carling DE (1993) Use of soil transfer for reforestation on abandoned mined lands in Alaska. II. Effects of soil transfers from different successional stages on growth and mycorrhizal formation by *Populus balsamifera* and *Alnus crispa*. Mycorrhiza 3:107–114. doi:10.1007/bf00208918

Helm DJ, Allen EB, Trappe JM (1996) Mycorrhizal chronosequence near Exit Glacier, Alaska. Can J Bot 74:1496–1506

Helm DJ, Allen EB, Trappe JM (1999) Plant growth and ectomycorrhiza formation by transplants on deglaciated land near Exit Glacier, Alaska. Mycorrhiza 8:297–304. doi:10.1007/s005720050250

Hernández-Rodríguez M, Andres Oria-de-Rueda J, Martín-Pinto P (2013) Post-fire fungal succession in a Mediterranean ecosystem dominated by *Cistus ladanifer* L. For Ecol Manage 289:48–57. doi:10.1016/j.foreco.2012.10.009

Hewitt RE, Bent E, Hollingsworth TN, Chapin FS III, Taylor DL (2013) Resilience of Arctic mycorrhizal fungal communities after wildfire facilitated by resprouting shrubs. Ecoscience 20:296–310. doi:10.2980/20-3-3620

Hibbett DS, Matheny PB (2009) The relative ages of ectomycorrhizal mushrooms and their plant hosts estimated using Bayesian relaxed molecular clock analyses. BMC Biol 7. doi:10.1186/1741-7007-7-13

Higgins SI, Scheiter S (2012) Atmospheric CO_2 forces abrupt vegetation shifts locally, but not globally. Nature 488:209–212. doi:10.1038/nature11238

Hilszczańska D, Sierota Z (2006) Persistence of ectomycorrhizas by *Thelephora terrestris* on outplanted Scots pine seedlings. Acta Mycol 41:313–318

Hilszczańska D, Sierota Z, Małecka M (2011) Ectomycorrhizal status of Scots pine saplings growing in post-agricultural soils. Pol J Environ Stud 20:83–88

Hobbie EA, Jumpponen A, Trappe J (2005) Foliar and fungal ^{15}N: ^{14}N ratios reflect development of mycorrhizae and nitrogen supply during primary succession: testing analytical models. Oecologia 146:258–268. doi:10.1007/s00442-005-0208-z

Hodkinson ID, Webb NR, Coulson SJ (2002) Primary community assembly on land—the missing stages: why are the heterotrophic organisms always there first? J Ecol 90:569–577. doi:10.1046/j.1365-2745.2002.00696.x

Hodkinson ID, Coulson SJ, Webb NR (2003) Community assembly along proglacial chronosequences in the high Arctic: vegetation and soil development in north-west Svalbard. J Ecol 91:651–663. doi:10.1046/j.1365-2745.2003.00786.x

Holden SR, Gutierrez A, Treseder KK (2013) Changes in soil fungal communities, extracellular enzyme activities, and litter decomposition across a fire chronosequence in Alaskan Boreal Forests. Ecosystems 16:34–46. doi:10.1007/s10021-012-9594-3

Horton TR, Cázares E, Bruns TD (1998) Ectomycorrhizal, vesicular-arbuscular and dark septate fungal colonization of bishop pine (*Pinus muricata*) seedlings in the first 5 months of growth after wildfire. Mycorrhiza 8:11–18. doi:10.1007/s005720050205

Hrynkiewicz K, Haug I, Baum C (2008) Ectomycorrhizal community structure under willows at former ore mining sites. Eur J Soil Biol 44:37–44. doi:10.1016/j.ejsobi.2007.10.004

Huang J, Nara K, Lian C, Zong K, Peng K, Xue S, Shen Z (2012) Ectomycorrhizal fungal communities associated with Masson pine (*Pinus massoniana* Lamb.) in Pb-Zn mine sites of central south China. Mycorrhiza 22:589–602. doi:10.1007/s00572-012-0436-0

Huang J, Nara K, Zong K, Wang J, Xue S, Peng K, Shen Z, Lian C (2014) Ectomycorrhizal fungal communities associated with Masson pine (*Pinus massoniana*) and white oak (*Quercus fabri*) in a manganese mining region in Hunan Province, China. Fungal Ecol 9:1–10. doi:10.1016/j.funeco.2014.01.001

Hüttl RF, Weber E (2001) Forest ecosystem development in post-mining landscapes: a case study of the Lusatian lignite district. Naturwissenschaften 88:322–329. doi:10.1007/s001140100241

Ishida TA, Nara K, Hogetsu T (2007) Host effects on ectomycorrhizal fungal communities: insight from eight host species in mixed conifer-broadleaf forests. New Phytol 174:430–440. doi:10.1111/j.1469-8137.2007.02016.x

Ishida TA, Nara K, Tanaka M, Kinoshita A, Hogetsu T (2008) Germination and infectivity of ectomycorrhizal fungal spores in relation to their ecological traits during primary succession. New Phytol 180:491–500. doi:10.1111/j.1469-8137.2008.02572.x

Izzo A, Canright M, Bruns TD (2006) The effects of heat treatments on ectomycorrhizal resistant propagules and their ability to colonize bioassay seedlings. Mycol Res 110:196–202. doi:10.1016/j.mycres.2005.08.010

Jagodziński AM, Kałucka I (2010) Fine roots biomass and morphology in a chronosequence of young *Pinus sylvestris* stands growing on a reclaimed lignite mine spoil heap. Dendrobiology 64:19–30

Jagodziński AM, Kałucka I (2011) Fine root biomass and morphology in an age-sequence of post-agricultural *Pinus sylvestris* L. stands. Dendrobiology 66:71–84

Jagodziński AM, Kałucka I, Horodecki P, Oleksyn J (2014) Aboveground biomass allocation and accumulation in a chronosequence of young *Pinus sylvestris* stands growing on a lignite mine spoil heap. Dendrobiology 72:139–150 doi:10.12657/denbio.072.012

Jansen AE (1991) The mycorrhizal status of Douglas fir in The Netherlands: its relation with stand age, regional factors, atmospheric pollutants and tree vitality. Agric Ecosyst Environ 35: 191–208. doi:10.1016/0167-8809(91)90051-x

Jenny H (1980) The soil resource, origin and behaviour. Springer, New York

Jones MD, Durall DM, Cairney JWG (2003) Ectomycorrhizal fungal communities in young forest stands regenerating after clearcut logging. New Phytol 157:399–422. doi:10.1046/j.1469-8137.2003.00698.x

Jones MD, Twieg BD, Durall DM, Berch SM (2008) Location relative to a retention patch affects the ECM fungal community more than patch size in the first season after timber harvesting on Vancouver Island, British Columbia. For Ecol Manage 255:1342–1352. doi:10.1016/j.foreco.2007.10.042

Jones MD, Twieg BD, Ward V, Barker J, Durall DM, Simard SW (2010) Functional complementarity of Douglas-fir ectomycorrhizas for extracellular enzyme activity after wildfire or clearcut logging. Funct Ecol 24:1139–1151. doi:10.1111/j.1365-2435.2010.01699.x

Jonsson L, Dahlberg A, Nilsson MC, Zackrisson O, Karen O (1999) Ectomycorrhizal fungal communities in late-successional Swedish boreal forests, and their composition following wildfire. Mol Ecol 8:205–215. doi:10.1046/j.1365-294x.1999.00553.x

Jumpponen A (2003) Soil fungal community assembly in a primary successional glacier forefront ecosystem as inferred from rDNA sequence analyses. New Phytol 158:569–578. doi:10.1046/j.1469-8137.2003.00767.x

Jumpponen A, Egerton-Warburton LM (2005) Mycorrhizal fungi in successional environments—a community assembly model incorporating host plant, environmental and biotic filters. In: Dighton J, White JF, Oudemans P (eds) The fungal community. Its organization and role in the ecosystem, 3rd edn. CRC Press, Taylor & Francis Group, Boca Raton, London, New York, Singapore, pp 139–180

Jumpponen A, Trappe JM, Cázares E (1999a) Ectomycorrhizal fungi in Lyman Lake Basin: a comparison between primary and secondary successional sites. Mycologia 91:575–582. doi:10.2307/3761242

Jumpponen A, Vare H, Mattson KG, Ohtonen R, Trappe JM (1999b) Characterization of 'safe sites' for pioneers in primary succession on recently deglaciated terrain. J Ecol 87: 98–105. doi:10.1046/j.1365-2745.1999.00328.x

Jumpponen A, Trappe JM, Cázares E (2002) Occurrence of ectomycorrhizal fungi on the forefront of retreating Lyman Glacier (Washington, USA) in relation to time since deglaciation. Mycorrhiza 12:43–49. doi:10.1007/s00572-001-0152-7

Jumpponen A, Brown SP, Trappe JM, Cázares E, Strommer R (2012) Twenty years of research on fungal-plant interactions on Lyman Glacier forefront—lessons learned and questions yet unanswered. Fungal Ecol 5:430–442. doi:10.1016/j.funeco.2012.01.002

Kałucka I (2009) Macrofungi in the secondary succession on the abandoned farmland near the Białowieża old-growth forest. Monogr Bot 99:1–155

Kałucka I, Jagodziński AM (2013) Grzyby ektomykoryzowe w obiegu węgla w ekosystemach leśnych. Ectomycorrhizal fungi and carbon dynamics in forest ecosystems. Sylwan 157: 817–830

Kałucka IL, Jagodziński AM (2016) Successional traits of ectomycorrhizal fungi in forest reclamation after surface mining and agricultural disturbances: a review. Dendrobiology 76: 91–104. doi:10.12657/denbio.076.009

Kałucka IL, Jagodziński AM, Nowiński M (2016) Biodiversity of ectomycorrhizal fungi in surface mine spoil restoration stands in Poland—first time recorded, rare and red-listed species. Acta Mycol 51(2):1080. doi:10.5586/am.1080

Keeley JE, Pausas JG, Rundel PW, Bond WJ, Bradstock RA (2011) Fire as an evolutionary pressure shaping plant traits. Trends Plant Sci 16:406–411. doi:10.1016/j.tplants.2011.04.002

Keizer PJ, Arnolds E (1994) Succession of ectomycorrhizal fungi in roadside verges planted with common oak (*Quercus robur* L.) in Drenthe, The Netherlands. Mycorrhiza 4:147–159. doi:10.1007/bf00203533

Kipfer T, Egli S, Ghazoul J, Moser B, Wohlgemuth T (2010) Susceptibility of ectomycorrhizal fungi to soil heating. Fungal Biol 114:467–472. doi:10.1016/j.funbio.2010.03.008

Kipfer T, Moser B, Egli S, Wohlgemuth T, Ghazoul J (2011) Ectomycorrhiza succession patterns in *Pinus sylvestris* forests after stand-replacing fire in the Central Alps. Oecologia 167: 219–228. doi:10.1007/s00442-011-1981-5

Klironomos J, Zobel M, Tibbett M, Stock WD, Rillig MC, Parrent JL, Moora M, Koch AM, Facelli JM, Facelli E, Dickie IA, Bever JD (2011) Forces that structure plant communities: quantifying the importance of the mycorrhizal symbiosis. New Phytol 189:366–370. doi:10.1111/j.1469-8137.2010.03550.x

Kõljalg U, Dahlberg A, Taylor AFS, Larsson E, Hallenberg N, Stenlid J, Larsson KH, Fransson PM, Kårén O, Jonsson L (2000) Diversity and abundance of resupinate thelephoroid fungi as ectomycorrhizal symbionts in Swedish boreal forests. Mol Ecol 9:1985–1996. doi:10.1046/j.1365-294X.2000.01105.x

Kranabetter JM (1999) The effect of refuge trees on a paper birch ectomycorrhiza community. Can J Bot 77:1523–1528

Kranabetter JM, Friesen J (2002) Ectomycorrhizal community structure on western hemlock (*Tsuga heterophylla*) seedlings transplanted from forests into openings. Can J Bot 80: 861–868. doi:10.1139/b02-071

Kranabetter JM, Wylie T (1998) Ectomycorrhizal community structure across forest openings on naturally regenerated western hemlock seedlings. Can J Bot 76:189–196

Kranabetter JM, Friesen J, Gamiet S, Kroeger P (2005) Ectomycorrhizal mushroom distribution by stand age in western hemlock—lodgepole pine forests of northwestern British Columbia. Can J For Res 35:1527–1539. doi:10.1139/x05-095

Kranabetter JM, De Montigny L, Ross G (2013) Effectiveness of green-tree retention in the conservation of ectomycorrhizal fungi. Fungal Ecol 6:430–438. doi:10.1016/j.funeco.2013.05.001

Krpata D, Peintner U, Langer I, Fitz WJ, Schweiger P (2008) Ectomycorrhizal communities associated with *Populus tremula* growing on a heavy metal contaminated site. Mycol Res 112: 1069–1079. doi:10.1016/j.mycres.2008.02.004

Kutorga E, Adamonytė G, Iršėnaitė R, Juzėnas S, Kasparavičius J, Markovskaja S, Motiejūnaitė J, Treigienė A (2012) Wildfire and post-fire management effects on early fungal succession in *Pinus mugo* plantations, located in Curonian Spit (Lithuania). Geoderma 191:70–79. doi:10.1016/j.geoderma.2012.02.007

Kuuluvainen T, Bergeron Y, Coates KD (2015) Restoration and ecosystem-based management in the circumboreal forest: background, challenges, and opportunities. In: Stanturf JA (ed) Restoration of boreal and temperate forests, 2nd edn. CRC Press, Taylor & Francis Group, Boca Raton, London, New York, Singapore, pp 251–270

Larsen CPS (1997) Spatial and temporal variations in boreal forest fire frequency in northern Alberta. J Biogeogr 24:663–673. doi:10.1111/j.1365-2699.1997.tb00076.x

Last FT, Mason PA, Wilson J, Deacon JW (1983) Fine roots and sheathing mycorrhizas: their formation, function and dynamics. Plant Soil 71:9–21. doi:10.1007/bf02182637

Last FT, Dighton J, Mason PA (1987) Successions of sheathing mycorrhizal fungi. Trends Ecol Evol 2:157–161. doi:10.1016/0169-5347(87)90066-8

Lazaruk LW, Kernaghan G, Macdonald SE, Khasa D (2005) Effects of partial cutting on the ectomycorrhizae of *Picea glauca* forests in northwestern Alberta. Can J For Res 35: 1442–1454. doi:10.1139/x05-062

Lazaruk LW, Macdonald SE, Kernaghan G (2008) The effect of mechanical site preparation on ectomycorrhizae of planted white spruce seedlings in conifer-dominated boreal mixedwood forest. Can J For Res 38:2072–2079. doi:10.1139/x08-035

LeDuc SD, Lilleskov EA, Horton TR, Rothstein DE (2013) Ectomycorrhizal fungal succession coincides with shifts in organic nitrogen availability and canopy closure in post-wildfire jack pine forests. Oecologia 172:257–269. doi:10.1007/s00442-012-2471-0

Leski T, Pietras M, Rudawska M (2010) Ectomycorrhizal fungal communities of pedunculate and sessile oak seedlings from bare-root forest nurseries. Mycorrhiza 20:179–190. doi:10.1007/s00572-009-0278-6

Longo MS, Urcelay C, Nouhra E (2011) Long term effects of fire on ectomycorrhizas and soil properties in *Nothofagus pumilio* forests in Argentina. For Ecol Manage 262:348–354. doi:10.1016/j.foreco.2011.03.041

Lunt PH, Hedger JN (2003) Effects of organic enrichment of mine spoil on growth and nutrient uptake in oak seedlings inoculated with selected ectomycorrhizal fungi. Restor Ecol 11:125–130. doi:10.1046/j.1526-100X.2003.09968.x

Luoma DL, Eberhart JL, Molina R, Amaranthus MP (2004) Response of ectomycorrhizal fungus sporocarp production to varying levels and patterns of green-tree retention. For Ecol Manage 202:337–354. doi:10.1016/j.foreco.2004.07.041

Luoma DL, Stockdale CA, Molina R, Eberhart JL (2006) The spatial influence of *Pseudotsuga menziesii* retention trees on ectomycorrhiza diversity. Can J For Res 36:2561–2573. doi:10.1139/x06-143

MacArthur RH, Wilson EO (1967) The theory of island biogeography. Princeton University Press, Princeton, NJ

Macdonald SE, Landhäusser SM, Skousen J, Franklin J, Frous J, Hall S, Jacobs DF, Quideau S (2015) Forest restoration following surface mining disturbance: challenges and solutions. New For 46:703–732. doi:10.1007/s11056-015-9506-4

Malajczuk N, Reddell P, Brundrett M (1994) Role of ectomycorrhizal fungi in minesite reclamation. In: Pfleger FL, Linderman RG (eds) Mycorrhizae and plant health, The American phytopathological society symposium series. APS Press, Minnesota, pp 83–100

Maremmani A, Bedini S, Matosevic I, Tomei PE, Giovannetti M (2003) Type of mycorrhizal associations in two coastal nature reserves of the Mediterranean basin. Mycorrhiza 13:33–40. doi:10.1007/s00572-002-0194-5

Marx DH (1975) Mycorrhizae and establishment of trees on strip-mined land. Ohio J Sci 75: 288–297

Marx DH (1977) Tree host range and world distribution of the ectomycorrhizal fungus *Pisolithus tinctorius*. Can J Microbiol 23:217–223

Marx DH, Marrs LF, Cordell CE (2002) Practical use of the mycorrhizal fungal technology in forestry, reclamation, arboriculture, agriculture, and horticulture. Dendrobiology 47:27–40

Mason PA, Last FT, Pelham J, Ingleby K (1982) Ecology of some fungi associated with an ageing stand of birches (*Betula pendula* and *Betula pubescens*). For Ecol Manage 4:19–39. doi:10.1016/0378-1127(82)90026-3

Mason PA, Wilson J, Last FT, Walker C (1983) The concept of succession in relation to the spread of sheathing mycorrhizal fungi on inoculated tree seedlings growing in unsterile soils. Plant Soil 71:247–256. doi:10.1007/bf02182659

Menkis A, Vasiliauskas R, Taylor AFS, Stenlid J, Finlay R (2007) Afforestation of abandoned farmland with conifer seedlings inoculated with three ectomycorrhizal fungi-impact on plant performance and ectomycorrhizal community. Mycorrhiza 17:337–348. doi:10.1007/s00572-007-0110-0

Mihál I (1999) Production of fruiting bodies of ectomycorrhizal fungi in spruce monocultures planted on former arable land. Ekologia (Bratislava) 18:125–133

Milad M, Schaich H, Buergi M, Konold W (2011) Climate change and nature conservation in Central European forests: a review of consequences, concepts and challenges. For Ecol Manage 261:829–843. doi:10.1016/j.foreco.2010.10.038

Montesinos-Navarro A, Segarra-Moragues JG, Valiente-Banuet A, Verdu M (2012) The network structure of plant-arbuscular mycorrhizal fungi. New Phytol 194:536–547. doi:10.1111/j.1469-8137.2011.04045.x

Motiejūnaitė J, Adamonytė G, Iršėnaitė R, Juzėnas S, Kasparavičius J, Kutorga E, Markovskaja S (2014) Early fungal community succession following crown fire in *Pinus mugo* stands and surface fire in *Pinus sylvestris* stands. Eur. J For Res 133:745–756. doi:10.1007/s10342-013-0738-6

Mueller KE, Hobbie SE, Chorover J, Reich PB, Eisenhauer N, Castellano MJ, Chadwick OA, Dobies T, Hale CM, Jagodziński AM, Kałucka I, Kieliszewska-Rokicka B, Modrzyński J, Rożen A, Skorupski M, Sobczyk L, Stasińska M, Trocha LK, Weiner J, Wierzbicka A, Oleksyn J (2015) Effects of litter traits, soil biota, and soil chemistry on soil carbon stocks at a common garden with 14 tree species. Biogeochemistry 123:313–327. doi:10.1007/s10533-015-0083-6

Mühlmann O, Peintner U (2008a) Ectomycorrhiza of *Kobresia myosuroides* at a primary successional glacier forefront. Mycorrhiza 18:355–362. doi:10.1007/s00572-008-0188-z

Mühlmann O, Peintner U (2008b) Mycobionts of *Salix herbacea* on a glacier forefront in the Austrian Alps. Mycorrhiza 18:171–180. doi:10.1007/s00572-008-0169-2

Mühlmann O, Bacher M, Peintner U (2008) *Polygonum viviparum* mycobionts on an alpine primary successional glacier forefront. Mycorrhiza 18:87–95. doi:10.1007/s00572-007-0156-z

Mujic AB, Durall DM, Spatafora JW, Kennedy PG (2016) Competitive avoidance not edaphic specialization drives vertical niche partitioning among sister species of ectomycorrhizal fungi. New Phytol 209:1174–1183. doi:10.1111/nph.13677

Muller LAH, Lambaerts M, Vangronsveld J, Colpaert JV (2004) AFLP-based assessment of the effects of environmental heavy metal pollution on the genetic structure of pioneer populations of *Suillus luteus*. New Phytol 164:297–303. doi:10.1111/j.1469-8137.2004.01190.x

Münzenberger B, Golldack J, Ullrich A, Schmincke B, Hüttl RF (2004) Abundance, diversity, and vitality of mycorrhizae of Scots pine (*Pinus sylvestris* L.) in lignite recultivation sites. Mycorrhiza 14:193–202. doi:10.1007/s00572-003-0257-2

Nara K (2006a) Ectomycorrhizal networks and seedling establishment during early primary succession. New Phytol 169:169–178. doi:10.1111/j.1469-8137.2005.01545.x

Nara K (2006b) Pioneer dwarf willow may facilitate tree succession by providing late colonizers with compatible ectomycorrhizal fungi in a primary successional volcanic desert. New Phytol 171:187–198. doi:10.1111/j.8137.2006.01744.x

Nara K (2009) Spores of ectomycorrhizal fungi: ecological strategies for germination and dormancy. New Phytol 181:245–248. doi:10.1111/j.1469-8137.2008.02691.x

Nara K, Hogetsu T (2004) Ectomycorrhizal fungi on established shrubs facilitate subsequent seedling establishment of successional plant species. Ecology 85:1700–1707. doi:10.1890/03-0373

Nara K, Nakaya H, Hogetsu T (2003a) Ectomycorrhizal sporocarp succession and production during early primary succession on Mount Fuji. New Phytol 158:193–206. doi:10.1046/j.1469-8137.2003.00724.x

Nara K, Nakaya H, Wu BY, Zhou ZH, Hogetsu T (2003b) Underground primary succession of ectomycorrhizal fungi in a volcanic desert on Mount Fuji. New Phytol 159:743–756. doi:10.1046/j.1469-8137.2003.00844.x

Neary DG, Klopatek CC, DeBano LF, Ffolliott PF (1999) Fire effects on belowground sustainability: a review and synthesis. For Ecol Manage 122:51–71. doi:10.1016/s0378-1127(99)00032-8

Nuñez MA, Horton TR, Simberloff D (2009) Lack of belowground mutualisms hinders Pinaceae invasions. Ecology 90:2352–2359. doi:10.1890/08-2139.1

Obase K, Tamai Y, Miyamoto T, Yajima T (2005) Macrofungal flora on the volcano Usu, deforested by 2000 eruptions. Eurasian J For Res 8:65–70

Obase K, Tamai Y, Yajima T, Miyamoto T (2007) Mycorrhizal associations in woody plant species at the Mt. Usu volcano, Japan. Mycorrhiza 17:209–215. doi:10.1007/s00572-006-0097-y

Obase K, Tamai Y, Yajima T, Miyamoto T (2008) Mycorrhizal colonization status of plant species established in an exposed area following the 2000 eruption of Mt. Usu, Hokkaido, Japan. Landsc Ecol Eng 4:57–61. doi:10.1007/s11355-008-0035-6

Olszewska M, Smal H (2008) The effect of afforestation with Scots pine (*Pinus silvestris* L.) of sandy post-arable soils on their selected properties. I. Physical and sorptive properties. Plant Soil 305:157–169. doi:10.1007/s11104-008-9537-0

Onwuchekwa NE, Zwiazek JJ, Quoreshi A, Khasa DP (2014) Growth of mycorrhizal jack pine (*Pinus banksiana*) and white spruce (*Picea glauca*) seedlings planted in oil sands reclaimed areas. Mycorrhiza 24:431–441. doi:10.1007/s00572-014-0555-x

Outerbridge RA, Trofymow JA (2004) Diversity of ectomycorrhizae on experimentally planted Douglas-fir seedlings in variable retention forestry sites on southern Vancouver Island. Can J Bot 82:1671–1681. doi:10.1139/b04-134

Pachlewski R (1956) Badania mikotrofizmu naturalnych zespołów roślinnych na hałdach żużlowo-łupkowych w Wałbrzychu. Untersuchungen uber den Mykotrophismus der naturlichen Pflanzengesellschaften auf Schlacken-Schieferhalden in Wałbrzych. Rocz Nauk Lesn 14:267–292

Pachlewski R (1958) Badania mikotrofizmu naturalnych zespołów roślinnych na hałdach górniczych w Knurowie i Gliwicach na Górnym Śląsku. Forschungen uber den Mykotrophismus naturlicher Pflanzengesellschaften auf Berghalden in Knurów und Gliwice (Oberschlesien). Pr Inst Badaw Lesn 182:173–209

Park D (1968) The ecology of terrestrial fungi. In: Ainsworth GC, Sussman AS (eds) The fungi, vol 3. Academic Press, New York, pp 5–39

Parraga-Aguado I, Querejeta J-I, González-Alcaraz M-N, Jiménez-Cárceles FJ, Conesa HM (2014) Usefulness of pioneer vegetation for the phytomanagement of metal(loid)s enriched tailings: grasses vs. shrubs vs. trees. J Environ Manage 133:51–58. doi:10.1016/j.jenvman.2013.12.001

Peña-Ramírez MV, Vazquez-Selem L, Siebe C (2009) Soil organic carbon stocks and forest productivity in volcanic ash soils of different age (1835-30,500 years BP) in Mexico. Geoderma 149:224–234. doi:10.1016/j.geoderma.2008.11.038

Pickett STA (1989) Space for time substitutions as an alternative to long-term studies. In: Likens GE (ed) Long-term studies in ecology. Springer, New York, pp 110–135

Pickett STA, Kolasa J, Armesto JJ, Collins SL (1989) The ecological concept of disturbance and its expression at various hierarchical levels. Oikos 54:129–136. doi:10.2307/3565258

Pietras M, Rudawska M, Leski T, Karliński L (2013) Diversity of ectomycorrhizal fungus assemblages on nursery grown European beech seedlings. Ann For Sci 70:115–121. doi:10.1007/s13595-012-0243-y

Pugh GJF, Boddy L (1988) A view of disturbance and life strategies in fungi. In: Boddy L, Watling R, Lyon AJE (eds) Fungi and ecological disturbance. Proc R Soc Edinburgh 94:3–11

Rayner ADM, Todd NK (1979) Population and community structure and dynamics of fungi in decaying wood. Adv Bot Res 7:333–420

Read DJ (1989) Mycorrhizas and nutrient cycling in sand dune ecosystems. Proc R Soc Edinb B 96:89–110

Read DJ, Perez-Moreno J (2003) Mycorrhizas and nutrient cycling in ecosystems—a journey towards relevance? New Phytol 157:475–492. doi:10.1046/j.1469-8137.2003.00704.x

Richard F, Selosse M-A, Gardes M (2009) Facilitated establishment of *Quercus ilex* in shrub-dominated communities within a Mediterranean ecosystem: do mycorrhizal partners matter? FEMS Microbiol Ecol 68:14–24. doi:10.1111/j.1574-6941.2009.00646.x

Richter DD, Markewitz D, Heine PR, Jin V, Raikes J, Tian K, Wells CG (2000) Legacies of agriculture and forest regrowth in the nitrogen of old-field soils. For Ecol Manage 138: 233–248. doi:10.1016/s0378-1127(00)00399-6

Rinaldi AC, Comandini O, Kuyper TW (2008) Ectomycorrhizal fungal diversity: separating the wheat from the chaff. Fungal Divers 33:1–45

Rincón A, Pueyo JJ (2010) Effect of fire severity and site slope on diversity and structure of the ectomycorrhizal fungal community associated with post-fire regenerated *Pinus pinaster* Ait. seedlings. For Ecol Manage 260:361–369. doi:10.1016/j.foreco.2010.04.028

Rincón A, Ruíz-Díez B, Fernández-Pascual M, Probanza A, Pozuelo JM, de Felipe MR (2006) Afforestation of degraded soils with *Pinus halepensis* Mill.: effects of inoculation with selected microorganisms and soil amendment on plant growth, rhizospheric microbial activity and ectomycorrhizal formation. Appl Soil Ecol 34:42–51. doi:10.1016/j.apsoil.2005.12.004

Rincón A, de Felipe MR, Fernández-Pascual M (2007) Inoculation of *Pinus halepensis* Mill. with selected ectomycorrhizal fungi improves seedling establishment 2 years after planting in a degraded gypsum soil. Mycorrhiza 18:23–32. doi:10.1007/s00572-007-0149-y

Rincón A, Santamaría BP, Ocaña L, Verdú M (2014) Structure and phylogenetic diversity of post-fire ectomycorrhizal communities of maritime pine. Mycorrhiza 24:131–141. doi:10.1007/s00572-013-0520-0

Rosenvald R, Lõhmus A (2008) For what, when, and where is green-tree retention better than clear-cutting? A review of the biodiversity aspects. For Ecol Manage 255:1–15. doi:10.1016/j.foreco.2007.09.016

Sawyer NA, Chambers SM, Cairney JWG (1999) Molecular investigation of genet distribution and genetic variation of *Cortinarius rotundisporus* in eastern Australian sclerophyll forests. New Phytol 142:561–568. doi:10.1046/j.1469-8137.1999.00417.x

Simard SW, Beiler KJ, Bingham MA, Deslippe JR, Philip LJ, Teste FP (2012) Mycorrhizal networks: mechanisms, ecology and modelling. Fungal Biol Rev 26:39–60. doi:10.1016/j.fbr.2012.01.001

Smal H, Olszewska M (2008) The effect of afforestation with Scots pine (*Pinus silvestris* L.) of sandy post-arable soils on their selected properties. II. Reaction, carbon, nitrogen and phosphorus. Plant Soil 305:171–187. doi:10.1007/s11104-008-9538-z

Smith SE, Read DJ (2008) Mycorrhizal symbiosis, 3rd edn. Academic Press, Amsterdam, Boston, Heidelberg, London, New York, Oxford, Paris, San Diego, San Francisco, Singapore, Sydney, Tokyo

Smith JE, Molina R, Huso MMP, Larsen MJ (2000) Occurrence of *Piloderma fallax* in young, rotation-age, and old-growth stands of Douglas-fir (*Pseudotsuga menziesii*) in the Cascade Range of Oregon, USA. Can J Bot 78:995–1001

Staudenrausch S, Kaldorf M, Renker C, Luis P, Buscot F (2005) Diversity of the ectomycorrhiza community at a uranium mining heap. Biol Fertil Soils 41:439–446. doi:10.1007/s00374-005-0849-4

Stoate C, Báldi A, Beja P, Boatman ND, Herzon I, van Doorn A, de Snoo GR, Rakosy L, Ramwell C (2009) Ecological impacts of early 21st century agricultural change in Europe—a review. J Environ Manage 91:22–46. doi:10.1016/j.jenvman.2009.07.005

Taylor AFS, Alexander I (2005) The ectomycorrhizal symbiosis: life in the real world. Mycologist 19:102–112

Taylor DL, Bruns TD (1999) Community structure of ectomycorrhizal fungi in a *Pinus muricata* forest: minimal overlap between the mature forest and resistant propagule communities. Mol Ecol 8:1837–1850. doi:10.1046/j.1365-294x.1999.00773.x

Tedersoo L, Smith ME (2013) Lineages of ectomycorrhizal fungi revisited: foraging strategies and novel lineages revealed by sequences from belowground. Fungal Biol Rev 27:83–99. doi:10.1016/j.fbr.2013.09.001

Tedersoo L, Jairus T, Horton BM, Abarenkov K, Suvi T, Saar I, Koljalg U (2008) Strong host preference of ectomycorrhizal fungi in a Tasmanian wet sclerophyll forest as revealed by DNA barcoding and taxon-specific primers. New Phytol 180:479–490. doi:10.1111/j.1469-8137.2008.02561.x

Tedersoo L, May TW, Smith ME (2010) Ectomycorrhizal lifestyle in fungi: global diversity, distribution, and evolution of phylogenetic lineages. Mycorrhiza 20:217–263. doi:10.1007/s00572-009-0274-x

Tedersoo L, Bahram M, Ryberg M, Otsing E, Koljalg U, Abarenkov K (2014) Global biogeography of the ectomycorrhizal/sebacina lineage (Fungi, Sebacinales) as revealed from comparative phylogenetic analyses. Mol Ecol 23:4168–4183. doi:10.1111/mec.12849

Termorshuizen AJ (1991) Succession of mycorrhizal fungi in stands of *Pinus sylvestris* in the Netherlands. J Veg Sci 2:555–564. doi:10.2307/3236038

Teste FP, Simard SW (2008) Mycorrhizal networks and distance from mature trees alter patterns of competition and facilitation in dry Douglas-fir forests. Oecologia 158:193–203. doi:10.1007/s00442-008-1136-5

Theodorou C, Bowen GD (1987) Germination of basidiospores of mycorrhizal fungi in the rhizosphere of *Pinus radiata* D. Don. New Phytol 106:217–223. doi:10.1111/j.1469-8137.1987.tb00137.x

Thiet RK, Boerner RE (2007) Spatial patterns of ectomycorrhizal fungal inoculum in arbuscular mycorrhizal barrens communities: implications for controlling invasion by *Pinus virginiana*. Mycorrhiza 17:507–517. doi:10.1007/s00572-007-0123-8

Tokuoka Y, Ohigashi K, Nakagoshi N (2011) Limitations on tree seedling establishment across ecotones between abandoned fields and adjacent broad-leaved forests in eastern Japan. Plant Ecol 212:923–944. doi:10.1007/s11258-010-9868-9

Trappe JM, Luoma DL (1992) The ties that bind: fungi in ecosystems. In: Carrol GC, Wicklow DT (eds) The fungal community. Its organization and role in the ecosystem, 2nd edn. Marcel Dekker, Inc, New York, Basel, Hong Kong, pp 17–27

Treseder KK, Mack MC, Cross A (2004) Relationships among fires, fungi, and soil dynamics in Alaskan Boreal Forests. Ecol Appl 14:1826–1838. doi:10.1890/03-5133

Treseder KK, Torn MS, Masiello CA (2006) An ecosystem-scale radiocarbon tracer to test use of litter carbon by ectomycorrhizal fungi. Soil Biol Biochem 38:1077–1082. doi:10.1016/j.soilbio.2005.09.006

Trocha LK, Kałucka I, Stasińska M, Nowak W, Dabert M, Leski T, Rudawska M, Oleksyn J (2012) Ectomycorrhizal fungal communities of native and non-native *Pinus* and *Quercus* species in a common garden of 35-year-old trees. Mycorrhiza 22:121–134. doi:10.1007/s00572-011-0387-x

Trowbridge J, Jumpponen A (2004) Fungal colonization of shrub willow roots at the forefront of a receding glacier. Mycorrhiza 14:283–293. doi:10.1007/s00572-003-0264-3

Tsuyuzaki S, Hase A, Niinuma H (2005) Distribution of different mycorrhizal classes on Mount Koma, northern Japan. Mycorrhiza 15:93–100. doi:10.1007/s00572-004-0304-7

Turner MG, Baker WL, Peterson CJ, Peet RK (1998) Factors influencing succession: lessons from large, infrequent natural disturbances. Ecosystems 1:511–523. doi:10.1007/s100219900047

Twieg BD, Durall DM, Simard SW (2007) Ectomycorrhizal fungal succession in mixed temperate forests. New Phytol 176:437–447. doi:10.1111/j.1469-8137.2007.02173.x

Twieg BD, Durall DM, Simard SW, Jones MD (2009) Influence of soil nutrients on ectomycorrhizal communities in a chronosequence of mixed temperate forests. Mycorrhiza 19: 305–316. doi:10.1007/s00572-009-0232-7

van Breugel M, Ransijn J, Craven D, Bongers F, Hall JS (2011) Estimating carbon stock in secondary forests: decisions and uncertainties associated with allometric biomass models. For Ecol Manage 262:1648–1657. doi:10.1016/j.foreco.2011.07.018

van der Heijden MGA, Horton TR (2009) Socialism in soil? The importance of mycorrhizal fungal networks for facilitation in natural ecosystems. J Ecol 97:1139–1150. doi:10.1111/j.1365-2745.2009.01570.x

van der Heijden EW, Vosatka M (1999) Mycorrhizal associations of *Salix repens* L. communities in succession of dune ecosystems. II. Mycorrhizal dynamics and interactions of ectomycorrhizal and arbuscular mycorrhizal fungi. Can J Bot 77:1833–1841

van der Heijden MGA, Klironomos JN, Ursic M, Moutoglis P, Streitwolf-Engel R, Boller T, Wiemken A, Sanders IR (1998) Mycorrhizal fungal diversity determines plant biodiversity, ecosystem variability and productivity. Nature 396:69–72

van der Heijden EW, Vries FW, Kuyper TW (1999) Mycorrhizal associations of *Salix repens* L. communities in succession of dune ecosystems. I. Above-ground and below-ground views of ectomycorrhizal fungi in relation to soil chemistry. Can J Bot 77:1821–1832

van der Maarel E (1993) Some remarks on disturbance and its relations to diversity and stability. J Veg Sci 3:733–736

van der Putten WH, Mortimer SR, Hedlund K, Van Dijk C, Brown VK, Lepš J, Rodriguez-Barrueco C, Roy J, Len TAD, Gormsen D, Korthals GW, Lavorel S, Santa Regina I, Smilauer P (2000) Plant species diversity as a driver of early succession in abandoned fields: a multi-site approach. Oecologia 124:91–99. doi:10.1007/s004420050028

van der Wal A, van Veen JA, Smant W, Boschker HTS, Bloem J, Kardol P, van der Putten WH, de Boer W (2006) Fungal biomass development in a chronosequence of land abandonment. Soil Biol Biochem 38:51–60. doi:10.1016/j.soilbio.2005.04.017

Visser S (1995) Ectomycorrhizal fungal succession in jack pine stands following wildfire. New Phytol 129:389–401. doi:10.1111/j.1469-8137.1995.tb04309.x

von Oheimb G, Härdtle W, Naumann PS, Westphal C, Assmann T, Meyer H (2008) Long-term effects of historical heathland farming on soil properties of forest ecosystems. For Ecol Manage 255:1984–1993. doi:10.1016/j.foreco.2007.12.021

Walker JKM, Jones MD (2013) Little evidence for niche partitioning among ectomycorrhizal fungi on spruce seedlings planted in decayed wood versus mineral soil microsites. Oecologia 173:1499–1511. doi:10.1007/s00442-013-2713-9

Walker TW, Syers JK (1976) The fate of phosphorus during pedogenesis. Geoderma 15:1–ll. doi:10.1016/0016-7061(76)90066-5

Walker LR, Wardle DA, Bardgett RD, Clarkson BD (2010) The use of chronosequences in studies of ecological succession and soil development. J Ecol 98:725–736. doi:10.1111/j.1365-2745.2010.01664.x

Wardle DA, Walker LR, Bardgett RD (2004) Ecosystem properties and forest decline in contrasting long-term chronosequences. Science 305:509–513. doi:10.1126/science.1098778

Wardle DA, Jonsson M, Bansal S, Bardgett RD, Gundale MJ, Metcalfe DB (2012) Linking vegetation change, carbon sequestration and biodiversity: insights from island ecosystems in a long-term natural experiment. J Ecol 100:16–30. doi:10.1111/j.1365-2745.2011.01907.x

White PS, Pickett STA (1985) Natural disturbance and patch dynamics: an introduction. In: Picket STA, White PS (eds) The ecology of natural disturbances and patch dynamics. Academic Press, New York, pp 3–13

Xu D, Dell B, Malajczuk N, Gong M (2001) Effects of P fertilisation and ectomycorrhizal fungal inoculation on early growth of eucalypt plantations in southern China. Plant Soil 233:47–57. doi:10.1023/a:1010355620452

Yang GT, Cha JY, Shibuya M, Yajima T, Takahashi K (1998) The occurrence and diversity of ectomycorrhizas of *Larix kaempferi* seedlings on a volcanic mountain in Japan. Mycol Res 102:1503–1508. doi:10.1017/s0953756298006480

Zak JC (1992) Response of soil fungal communities to disturbance. In: Carrol GC, Wicklow DT (eds) The fungal community. Its organization and role in the ecosystem, 2nd edn. Marcel Dekker, Inc, New York, Basel, Hong Kong, pp 403–425

Chapter 11
Truffle Ecology: Genetic Diversity, Soil Interactions and Functioning

Antonietta Mello, Elisa Zampieri, and Alessandra Zambonelli

Abstract Truffles are fungi producing hypogeous fruiting bodies belonging to at least 13 phylogenetically distant orders. The most studied are "true truffles" belonging to the genus *Tuber*, which is the most economically important group. Truffle fruiting bodies are colonized by bacteria, yeasts, guest filamentous fungi and viruses that, all together, constitute the truffle microbiota. Research on the role of this community has demonstrated that bacteria contribute to truffle aroma. From the ecological point of view, truffle aroma attracts mycophagous animals, which in turn disperse and diffuse truffle spores in the soil, and mediates interactions with microorganisms and plant roots. Truffles have a heterothallic organization, whereby for truffle reproduction it is necessary that strains of opposite mating type meet. Regarding truffle development, the truffle ascocarps use carbon coming from the host plant and not from dead host tissues or soil organic matter as believed so far. In addition to form ectomycorrhizae with a wide diversity of host plants, some truffle species are able to form also arbutoid and orchid mycorrhizas.

Knowledge of truffle diversity, traditionally relied on the survey and molecular identification of fruiting bodies, moved over the years towards the survey of mycorrhizas and, recently, on the distribution in soil of the mycelium, with the determination of genets and mating types. The possibility of studying (micro) organisms directly in the field (metagenomics or environmental genomics) and the introduction of high-throughput sequencing techniques (454 pyrosequencing)

A. Mello (✉)
Istituto per la Protezione Sostenibile delle Piante, SS Torino - CNR, Viale Mattioli 25, 10125 Torino, Italy
e-mail: antonietta.mello@ipsp.cnr.it

E. Zampieri
Dipartimento di Scienze Agrarie, Forestali e Alimentari (DISAFA), Università di Torino, Largo Paolo Braccini 2, 10095 Grugliasco, TO, Italy

Present Address: Rice Research Unit, CREA, Council for Agricultural Research and Economics, S.S. 11 to Torino, Km 2,5, 13100 Vercelli, Italy
e-mail: elisa.zampieri@unito.it

A. Zambonelli
Dipartimento di Scienze Agrarie (DipSA), Alma Mater Studiorum Università di Bologna, Via Fanin 46, 40127 Bologna, Italy
e-mail: alessandr.zambonelli@unibo.it

A. Varma et al. (eds.), *Mycorrhiza - Function, Diversity, State of the Art*,
DOI 10.1007/978-3-319-53064-2_11

have given a strong impulse to the study of the microbial communities interacting with truffles and their habitat.

11.1 Introduction: Truffles and Their Life Cycle

Truffles are generically considered the fungi producing hypogeous fruiting bodies (Trappe and Claridge 2010). They have generally globose, subglobose fruiting bodies, which protect their spores that are packed inside their internal tissue. Because truffles are sequestrate fungi, their spores are dispersed by mycophagous animals, which are attracted by their aroma (Piattoni et al. 2016; Urban 2016). Truffle volatiles not only attract mammals and insects but also, diffusing in the soil, mediate interactions with microorganisms and plant roots (Splivallo et al. 2011).

Fungi that form truffle sporomata belong to at least 13 orders that represent phylogenetically distant fungal lineages comprised in *Zygomycota*, *Ascomycetes*, *Basidiomycetes* and *Glomeromycota* (Bonito and Smith 2016). However, Italian and French researchers restrict the term of "true truffles" at only the *Ascomycetes* belonging to the genus *Tuber*, which is the most economically important group of hypogeous fungi (Jeandroz et al. 2008). The genus is estimated to comprise around 200 species (Bonito et al. 2010a, 2013) worldwide distributed. However, only few *Tuber* spp. have fruiting bodies, which are considered gastronomic delicacy for their unique aroma. The most worldwide appreciated species grow naturally in Europe and include *T. magnatum* Pico (the Italian white truffle), *T. melanosporum* Vittad. (the Perigord black truffle or black truffle of Norcia and Spoleto), *T. aestivum* Vittad. (the burgundy truffle or summer truffle) and *T. borchii* Vittad. (the bianchetto or pine truffle). By cultivation, these species have been introduced in non-native areas not only in Europe but also in numerous American, African and Oceanian countries (Hall and Haslam 2012; Reyna and García Barreda 2014). Recently also some species of American and Chinese truffles began to be appreciated in the cuisine even if their cultivation is still in its infancy.

True truffles form ectomycorrhizae with a wide diversity of host plants including several species of angiosperms (forest trees and shrubs) and Pinaceae (Hall et al. 2007; Bonito et al. 2013). Recently, it has been found that some species are able to form also different types of mycorrhizal association such as arbutoid mycorrhizas and orchid mycorrhizas (Lancellotti et al. 2014; Selosse et al. 2004). Detailed studies on their natural growth habitat have mainly focused on the commercially important species. In general, they are found in calcareous, slightly to moderately alkaline soils in different climatic conditions depending on the *Tuber* species. Truffle fruiting bodies are colonized by a dense microbial community made of bacteria, yeasts, guest filamentous fungi and viruses that, all together, constitute the truffle microbiota. According to Vahdatzadeh et al. (2015), bacteria, which are the dominant group in the truffle's microbiome, are the most important contributors to *T. borchii* truffle aroma, and this might also be the case for *T. magnatum*, *T. aestivum* and *T. melanosporum*. Mycoviruses have been found to infect *T. excavatum* (Stielow et al. 2012) and *T. aestivum* (Stielow and Menzel 2010;

Stielow et al. 2011a, b) symptomless fruiting bodies and recently were found in *T. magnatum* fruiting bodies showing suberized consistency and brown spots irregularly distributed on the peridium (Ratti et al. 2016).

Regarding the life cycle, truffles have been considered for long-time self-fertile (Bertault et al. 1998). This opinion could not be tested in absence of an experimental system, based on spore germination, and therefore of the classical breeding of the resulting mycelia (Mello et al. 2005). Only after the detection of heterozygosis in *T. melanosporum* and in *T. magnatum* it has been possible to discover that truffles outcross (Paolocci et al. 2006; Riccioni et al. 2008) and, thanks to the *T. melanosporum* genome sequencing, that it has an heterothallic organization (Martin et al. 2010; Rubini et al. 2011a). Heterothallic organization with a MAT locus structured in two idiomorphs harboured by different strains was also found in other truffles: *T. borchii* and *T. indicum* (Belfiori et al. 2013, 2016). That means that for truffle reproduction it is necessary that strains of opposite mating type meet. These discoveries have stimulated researchers to follow the spatiotemporal distribution of these strains in soil by exploiting the new approaches of environmental microbiology and molecular ecology. New important advances were reached in truffle soil biology and ecology as will be discussed in the following paragraphs. However, many aspects on their reproduction are still unravelled, for example, the fungal structures involved in the fertilization are still unknown (Le Tacon et al. 2016). In the *Pezizales* fertilization is accomplished by ascogonial fusion with an antheridium, more rarely by spermatization or by somatogamy (Pfister and Kimbrough 2001). Somatogamy seems to be excluded in truffles because anastomosis between genetically different strains of *T. melanosporum* was never observed in laboratory trials (Iotti et al. 2012c). Moreover in the field competitive exclusion between strains of different mating type occurs (Selosse et al. 2013). Although in the past structures attributable to ascogonia were reported in immature fruiting bodies, antheridia were never observed (Callot 1999; Le Tacon et al. 2016). Recently it was found that many *Tuber* spp. are able to produce mitotic (asexual) spores (Urban et al. 2004; Healy et al. 2013). These spores are hypothesized to function in reproduction acting as spermatia for sexual outcrossing (Healy et al. 2013).

Regarding truffle development, for long time it was believed that after the formation of the primordium the ascoma was able to use dead host tissues or soil organic matter as main carbon and nitrogen sources (Callot 1999). In an in situ $^{13}CO2$ pulse-labelling experiment performed on a 20-year-old hazel tree in a truffle orchard established in the northeast of France, Le Tacon et al. (2013) could establish that almost all of the carbon allocated to the truffle ascomata come from the host and not via saprotrophic pathways. The development of truffles requires that carbon is stored in the host plant for several weeks/months because *Tuber* ascomata take at least 6 months to grow between the production of the primordia and full ascoma development.

Moving from truffle life cycle, this chapter will provide insights on the distribution of the most studied species in Europe, their diversity and interactions in soil. The discovery of new species in America and the cultivation in Australia, as well as the interest for Chinese truffles, will also be discussed. Finally, recent findings on the combination of metagenomics and metaproteomics to unravel the habitat of the brûlé and its functioning will be presented.

11.2 Genetic Diversity of the Most Studied Species and Their Soil Community Composition

Knowledge of truffles diversity has traditionally relied on the survey and molecular identification of fruiting bodies, as done for other ECM fungi. Given that fruiting bodies are temporary organs, the search of truffle diversity moved over the years towards the survey of mycorrhizas, through the molecular typing of their morphotypes. However, as mycorrhizas mostly depend on the host's physiological activity, later when the technologies have made possible DNA extraction and amplification from the soil, the research focused on the distribution in the soil of the mycelium, which is expected to be present year-round in the soil. The possibility of studying (micro)organisms directly in the field (metagenomics or environmental genomics) and the introduction of high-throughput sequencing techniques (i.e. 454 pyrosequencing) have given a strong impulse to the study of the microbial communities interacting with truffles and their habitat.

Hereafter we report these studies on *T. magnatum*, *T. borchii*, *T. aestivum* and *T. melanosporum*.

11.2.1 *Tuber magnatum*

Tuber magnatum is the most precious truffle because of its aroma, a very limited distribution area (Italy, Switzerland-Ticino, Romania, Hungary and some parts of the Balkans), and the absence of reliable methods for its cultivation (Hall et al. 1998; Murat et al. 2005; Iotti et al. 2012a; Salerni et al. 2014b). In Italy, *T. magnatum* grows mainly in Piedmont, Marche, Tuscany, Emilia-Romagna, Abruzzo and Molise (Zambonelli et al. 2012c). It was also found in Liguria, Umbria, Lazio, Veneto, Campania and Basilicata (Zambonelli et al. 2012c) and recently in Greece (Christopoulos et al. 2013), Sicilia (Vasquez et al. 2014) and France (http://lci.tf1.fr/jt-13h/videos/2011/des-truffes-blanches-trouvees-dans-la-drome-une-revolution-6848528.html). For these reasons, several studies were carried out to describe the natural habitat and the soil-associated microorganisms (Murat et al. 2005; Bertini et al. 2006; Mello et al. 2010; Leonardi et al. 2013; Orgiazzi et al. 2013; Lalli et al. 2015). Lalli et al. (2015) have recently proposed a double check of the truffle-grounds, basing on root tip and fruiting body harvesting. Although the ECM community characterized by fruiting body sampling was not specular to that found by root tip collections, this type of analysis has allowed to have a more comprehensive record of the ectomycorrhizal fungal species sharing the same environment of *T. magnatum* (Lalli et al. 2015). The presence of some fruiting bodies (*Amanita stenospora*, *Cortinarius aprinus*, *Hebeloma quercetorum* and *Hygrophorus arbustivus* var. *quercetorum*) could be considered as bioindicators of habitats typical of the precious white truffle (Lalli et al. 2015). The surveys in *T. magnatum* truffle-ground put in evidence as the ectomycorrhizae of this truffle were very rare (Murat et al. 2005; Bertini et al. 2006; Leonardi et al. 2013; Zambonelli et al. 2012a), even if in controlled conditions the white truffle is able to form them (Iotti et al. 2012b). Its mycelium is more widespread (Zampieri et al. 2010) than can be inferred from the distribution of fruiting bodies (Mello et al. 2005), and it is moreover positively correlated to the ascoma production (Iotti et al. 2012a), soil tillage (Salerni et al. 2014b) and the seasonal fluctuations (Iotti et al. 2014). The phenomenon of non-correspondence between the production of fruiting bodies and presence of ectomycorrhizae in the soil has not found an explanation yet, but it has been supposed that its ectomycorrhizae are camouflaged by the copresence of other ectomycorrhizal fungi present on the same root or that *T. magnatum* may form other types of symbiosis (Leonardi et al. 2013). In general, in Italy and Istria (Croatia), the most productive hosts are *Quercus* spp., *Populus* spp., *Salix* spp. and *Tilia* spp. in presence of soils with an extremely soft and porous texture, moderately alkaline pH, a high amount of calcium carbonate, sufficient not stagnant water and a climate with abundant annual precipitation with very short dry periods (Bencivenga and Urbani 1992; Lulli et al. 1993; Hall et al. 1998; Bragato et al. 2010; Salerni et al. 2014b). On the contrary, the habitats of *T. magnatum* in Serbia show features different from the Italian ones (Marjanović et al. 2015). The soils have no traces of $CaCO_3$, with 50% clay and lower pH (6.8–7.5) (Marjanović et al. 2010). The vegetation is, moreover, characterized by *Q. robur*, *Populus* sp. and *Fraxinus angustifolia* (Marjanović et al. 2015).

11.2.2 *Tuber borchii*

Among the white truffles, *T. borchii* is appreciated in the market and used both as fresh product and as conserved products (Hall et al. 2007). It is found widely from Finland to Sicily and from Ireland to Hungary and Poland both in calcareous soils and in acidic soils, in association with a wide range of broad leaf trees (oak, hazel, poplar, linden, chestnut and alder) and of coniferous species such as pine and cedar (Gardin 2005; Hall et al. 2007; Shamekh et al. 2009). Recently, it was found in association with strawberry tree (*Arbutus unedo*) with which it forms arbutoid mycorrhizas (Lancellotti et al. 2014). Recently *T. borchii* cultivation has been introduced in several European countries and in New Zealand, Australia and the USA (Zambonelli et al. 2015).

The ectomycorrhizal fungal communities of natural *T. borchii* truffle-grounds have been examined by Iotti et al. (2010) who have found that the presence of *T. borchii* did not have an effect on the richness of other ectomycorrhizal fungi. Regarding to the genetic diversity of *T. borchii*, the presence of two cryptic species undistinguishable by morphological analysis was shown by Bonuso et al. (2010). The life cycle of the whitish truffle has just been clarified, demonstrating its heterothallic nature, with homokaryotic hyphae in the ascomata and ectomycorrhizae (Belfiori et al. 2016). Recently, Iotti et al. (2016) have demonstrated how it is possible to produce *T. borchii* fruiting bodies starting from the mycorrhization of plants with mycelial pure culture and then the establishment of the truffle-ground. Both mating types in the environment have to be present in different blocks of trees in order to limit their competition and to ensure their presence and their fertilization. This modern approach addresses towards a more technological trufficulture based on management strategies of a balanced presence of the mycelia of the two opposite mating partners in an orchard, as suggested by Rubini et al. (2014). A major problem in the cultivation of mycorrhized plants can be the replacement of the truffles by aggressive competing fungi (Zambonelli and Iotti 2001). On the contrary, Zambonelli et al. (2000) have demonstrated that *T. borchii* is able to replace other mycorrhizal fungi such as *Laccaria bicolor* when both fungi are inoculated on the *P. pinea* roots. This result could be explained in two ways: *T. borchii* and *P. pinea* had a stronger association than that of the other ectomycorrhizal fungi and the pine, or the other ectomycorrhizal fungi were less adapt to the environment in which the plants were transplanted (Zambonelli et al. 2000). In vitro experiments showed also that the bacterial soil composition may influence the competitive relationships between *T. borchii* and other ectomycorrhizal fungi (Zambonelli et al. 2009). The VOCs produced by a ubiquitous bacterium, *Staphylococcus pasteuri*, are able to inhibit the mycelial growth of *T. borchii* as well as the ectomycorrhizal basidiomycete *Boletus luridus*, but do not affect the mycelial growth of *H. radicosum* (Barbieri et al. 2005).

11.2.3 *Tuber aestivum*

The black truffle *T. aestivum* (syn. *T. uncinatum*) is reported from Spain to China across East Europe and the UK and from Gotland (Sweden) to North Africa (Morocco) (Wedén et al. 2004; Song et al. 2005; Hilszczańska et al. 2008; Jeandroz et al. 2008; Stobbe et al. 2012). It is cultivated in New Zealand and Australia (Hall et al. 2007), but also in Canada in regions of British Columbia (Berch and Bonito 2014), in Finland (Shamekh et al. 2014) and in USA-Missouri using *Q. bicolor* × *Q. robur* hybrid (Pruett et al. 2009). The fungus prefers a soil pH of 6.8–8, low phosphorus concentrations (0.30–0.59 g × kg^{-1}), and soils that are poor in the readily degradable nitrogen (Wedén et al. 2004; Hilszczańska et al. 2008), and it is adapted in semi-continental region as Burgundy in France (Chevalier et al. 1979). It reaches higher altitudes in the Southern habitats with a warm climate, and it is located near the sea level in the Northern areas (Stobbe et al. 2013a). Following the presence of *T. aestivum* in a field in the Czech Republic, it was demonstrated that warm temperatures help the fungus to colonize new spaces (Gryndler et al. 2015). Its presence is characterized by the formation of a burnt area (brûlé), around the host tree, with a scant vegetation, probably due to the production of allelopathic compounds (Streiblová et al. 2012; Gryndler et al. 2014). Molinier et al. (2013) also demonstrated a *T. aestivum* competitive power and expandability in environmental conditions that are appropriate in terms of soil, climate, exposure and plant cover.

In Israel, *T. aestivum* similarly demonstrated its capacity to adapt to severe conditions and to replace *T. melanosporum* in orchards (Turgeman et al. 2012).

The fungus can enter in association with more than 20 species of host plants, including species of *Abies*, *Betula*, *Carpinus*, *Carya*, *Castanea*, *Cistus*, *Corylus*, *Fagus*, *Ostrya*, *Tilia*, *Picea*, *Pinus*, *Populus*, *Quercus* and *Ulmus* (Wedén et al. 2009; Benucci et al. 2012; Stobbe et al. 2013a, b; Gryndler et al. 2014). It was demonstrated that in natural conditions, it can form brûlé with *Pinus nigra* subsp. *salzmannii* and *P. sylvestris* and *Quercus ilex* subsp. *ballota* and *Q. faginea* hosts (García-Montero et al. 2014). The brûlé with the conifers was smaller than that of hardwood, but the fruiting body production was greater (García-Montero et al. 2014). Roots of non-host plants (i.e. the plants usually not forming fully functional mycorrhizae with a specific fungal symbiont) showed the presence of *T. aestivum* mycelium biomass (Gryndler et al. 2014). The mycelium was localized in the decomposing root cell layers on the root surface, suggesting a functional interaction with the non-host plants (Gryndler et al. 2013, 2014). Its quantification by qPCR has allowed to suggest that the mycelium of *T. aestivum* is relatively dense and that reaches higher biomass densities than other ectomycorrhizal fungi previously comparably quantified (Gryndler et al. 2013). The production of allelopathic compounds could have different effects in relation to the type of non-host plants (Gryndler et al. 2014). It is known that *T. aestivum* can reduce the growth of *Vicia faba* (Lanza et al. 2004) or that can cause the formation of the brûlé due to an allelopathic activity on non-host plants (Streiblová et al. 2012). Gryndler et al. (2014) highlighted the capacity of *T. aestivum* to change vegetation cover and therefore to have an impact on carbon and nutrient cycles in the ecosystem.

The ectomycorrhizal community of a *T. aestivum* truffle-ground close Spoleto (Italy) showed a dominance of *T. aestivum* ectomycorrhizae and a diversity of ectomycorrhizal species mediated by the host plants (hornbeam trees and hazel), but not host specific (Benucci et al. 2011). Salerni et al. (2014a) analysed in a truffle-ground the fungal communities above and below ground, by means of observations of fruiting bodies and ectomycorrhizal morphotypes associated with *Quercus cerris*, *Q. pubescens* and *P. nigra*. The data were put in relation with the production of *T. aestivum*, expressed in both the number and weight of fruiting bodies, demonstrating that mycorrhizal fungi and *T. aestivum* production are not in competition. The orchard showed high species richness with fungal species that changed marginally in relation to host tree. In detail, the most frequent fungi were *Tricholoma* spp. and *Tomentella* spp., above and below, respectively (Salerni et al. 2014a). In order to understand the effects of indigenous mycorrhizal fungi on *T. aestivum* colonization in plants, the ectomycorrhizal community of an artificial truffle-ground was monitored for 2 years after the transplantation of mycorrhized seedlings (Pruett et al. 2008). Pruett et al. (2008) found that *T. aestivum* colonization and indigenous ectomycorrhizal fungi increased in terms of richness and abundance, demonstrating that *T. aestivum* was not replaced by native species in a short term. On the contrary, *T. aestivum* affects soil microarthropod community: *Symphyla* and *Pauropoda* are negatively affected, whereas *Folsomia* spp. are positively affected by *T. aestivum* (Menta et al. 2014).

A recent study based on a large-scale population genetic analysis suggests that genetically distinct populations and likely ecotypes are present within *T. aestivum* (Molinier et al. 2016). In the light of the previous finding that the aroma of *T. aestivum* is influenced by the identity of single clones/genets (Molinier et al. 2015), the future research will be aimed at verifying if ecotypes adapted to particular environments exist within this species and if they differ for their aroma.

11.2.4 *Tuber melanosporum* and the *Brûlé* Functioning

This truffle species grows naturally in France, Italy and Spain where its cultivation is an agroforestry alternative for rural areas. In the last 20 years, attempts have been made to grow mycorrhized oaks and hazel with *T. melanosporum*, in areas and countries outside its natural growth habitats. The first black truffles to be harvested outside of Europe came from a northern Californian truffière in 1991; 2 years later, in 1993, the first truffles were harvested also in New Zealand (Hall et al. 1998), and in the same year, a cultivation project was initiated in Israel (Pinkas et al. 2000). However in Israel after the *T. melanosporum* fruiting body production (Kagan Zur et al. 2000), this species was replaced with the locally more competitive *T. aestivum* (Turgeman et al. 2012). In Australia, the first black truffles were harvested in Tasmania in 1999, and truffle production increased considerably in the following years. Three tons of black truffles were produced in 2011 (Hall and Haslam 2012), and Australian truffle production could soon overtake French truffle production at least in a poor fruiting year (Hall and Zambonelli 2012).

Cultivated black truffles have also been produced in Canada, Morocco, Chile, China and South Africa (Reyna and García Barreda 2014; Berch and Bonito 2014; Zambonelli et al. 2015; Wang 2012). In Chile a local plant, *Nothofagus obliqua*, was also mycorrhized with black truffle under greenhouse conditions, with the aim of cultivating this truffle as a secondary crop during reforestation (Pérez et al. 2007).

Genetic diversity and mating type distribution of *T. melanosporum* have been investigated in artificially planted truffiéres in Australia to increase ascoma production (Linde and Selmes 2012), highlighting that the lacking of production of certain truffle orchard is due to the scarce quality of nursery-inoculated seedlings which can be contaminated with *T. brumale*, rather than to the absence of both mating types in the field.

The distribution of mating type genes of *T. melanosporum* has been also investigated in *T. melanosporum* orchards, located in central Italy (Rubini et al. 2011b). Contrary to what is expected, strains with opposite mating types were never present on the same root apparatus, while both mating types were detected in the soil of the truffle-ground (Rubini et al. 2011b). Experiments made by the same authors and based on inoculation of host plants in controlled conditions showed that the coexistence of both mating types on the roots of the same host plant can happen but lasts until their competition excludes one of the two mating types. The genetic basis of the competition between strains belonging to different mating types is still unknown. It seems related to a self-/non-self-recognition system acting before hyphal contact rather than the presence of a heterokaryon incompatibility (HI) system which leads to the death of the heterokaryotic cells in incompatible reactions (Iotti et al. 2012c). In fact, although orthologs of the genes which control HI in other filamentous ascomycetes are present also in the *T. melanosporum* genome, they lack the key functional domains involved in the HI process (Rubini et al. 2014).

Anyway, according to Zampieri et al. (2012), the detection of the two mating types in soil extracts can be a promising predictor of *T. melanosporum* production, given that mating type genes for *T. melanosporum* were detected under productive and formally productive trees and, generally, not under unproductive trees. However, in all the three situations, the mating type genes were detected only when more than 0.3 ng of *T. melanosporum* DNA was present in soil. These observations clearly suggest that quantification of *T. melanosporum* DNA coupled to detection of its mating type genes can be a suitable tool to predict the fertility of truffle orchards. However, the production of sporomata depends not only from the encounter of the two mating types, but from several abiotic and biotic factors. In any case, the discovery of competition between the two mating types on the host roots has practical implications in trufficulture. In fact, inoculation techniques should be developed for production of seedlings of opposite mating type to potentially improve truffle productivity (Rubini et al. 2014).

The spatial genetic structure of *T. melanosporum* populations at a small scale has been investigated in two productive *T. melanosporum* orchards, one located in the northern France and the other in central Italy (Murat et al. 2013). The analysis of the genetic profiles of ECMs using both SSR markers and mating type genes and the monitoring of the distribution of *T. melanosporum* mycelia of the two mating types in the soil allowed the authors to demonstrate a pronounced spatial genetic structure of *T. melanosporum*, characterized by non-random distribution of small genets. Several small *T. melanosporum* genets that shared the same mating types can be found on the same host plant, suggesting that the genet distributional pattern is related to the allelic configuration of the MAT locus.

Following the distribution of ectomycorrhizae and mycelium of *T. melanosporum* in natural ecosystems dominated by four ectomycorrhizal host species (*Q. ilex*, *Q. coccifera*, *A. unedo*, *C. albidus*), Taschen et al. (2015) documented a spatial heterogeneous pattern of *T. melanosporum* distribution in soils, and, in contrast to most multi-host ECM fungal species, a marked affinity for *Q. ilex*. This work underlines that the frequency of ECM tips and the distribution patterns of soil mycelia do not mirror fruiting distribution within brûlés, which cannot be then predicted, on this basis.

The development of *T. melanosporum* is associated with the production of a "burnt" area (commonly referred to the French word brûlés), around its symbiotic plant, characterized by scanty vegetation (Pacioni 1991). Bragato (2014) observed in an experimental truffle-producing area in Italy that the disappearance of grasses in the brûlé induces modifications of soil aggregation that are determined by freeze-thaw cycles occurring in the winter season. These changes play an important function in the exchange of water and air between soil and the atmosphere helping the growth of *T. melanosporum*. This observation can have practical applications for truffle farming that could increase aggregate breakdown inside the brûlé by watering in the frost days of late winter, thus limiting the use of more expensive tillage practices. What is known on the origin of the brûlés? Splivallo et al. (2007) and Splivallo et al. (2011) have indicated that fungal volatiles can mediate fungal-plant interactions and that the production of ethylene and indole-3-acetic acid (IAA) could in large quantities act as herbicides, thus suggesting their role in the origin of the brûlés. According to Streiblová et al. (2012), truffles adopt an efficient survival strategy by diffusing their metabolites, which are regarded as having allelopathic effects on the herbaceous plants and the microorganisms in the rhizosphere. In order to investigate the potential effects of *T. melanosporum* growth and metabolites on the soil communities with which interacts, two studies used metagenomics to compare the fungal composition inside and outside the brûlé in French truffle-ground soils at Cahors (Napoli et al. 2010; Mello et al. 2011). Clear differences were shown between the fungal communities, together with a lower fungal biodiversity inside the brûlé in which the ectomycorrhizal fungus *T. melanosporum* was the dominant fungus. By contrast, *Basidiomycota*, which are mostly ECM fungi, showed decreased abundance in the brûlé, suggesting a competition with *T. melanosporum*. In this way *T. melanosporum* ensures its growth at the expense of fungi having the same trophic strategy. Mello et al. (2013) also showed differences in the bacterial community composition between the interior and exterior zones of the same brûlé through DGGE profiles and microarrays of 16S rRNA gene fragments. *Firmicutes* (e.g. *Bacillus*), several genera of *Actinobacteria* and a few *Cyanobacteria* were found more frequently inside the brûlé than outside, whereas *Pseudomonas* and several genera from the *Flavobacteriaceae* family (e.g. *Flavobacterium*, Fig. 11.1) were more abundant outside the brûlé. As most herbaceous plants form symbioses with arbuscular mycorrhizal fungi (AMF), Mello et al. (2015) wandered whether the scant plant coverage in the brûlé is mycorrhized as the plant coverage in the area outside the brûlé. The results showed that the patchy herbaceous plants in the brûlé were extensively colonized by AMF, as were the plants outside the brûlé, and AMF richness on the roots of the herbaceous plants inside the brûlé was not affected. By contrast, reduced species richness of AMF was observed in the soil inside the brûlé compared with the soil outside the brûlé. However, members of *Diversispora*, *Acaulospora* and *Archaeospora* were only found in the brûlé, in roots or soil, suggesting that this habitat specifically affected some taxa. Taking together, all these metagenomic studies have identified the microorganisms present inside and outside the brûlé of a *T. melanosporum* truffle-ground, but the molecular mechanisms that occur in this ecological niche have not been faced, thus remaining elusive. To elucidate the metabolic pathways present in the brûlé, Zampieri et al. (2016) constructed a database, incorporating the metagenomic data for the organism previously identified in the brûlé, and conducted a metaproteomic analysis on the same soil cross-referencing the resulting proteins with the database. As a result,

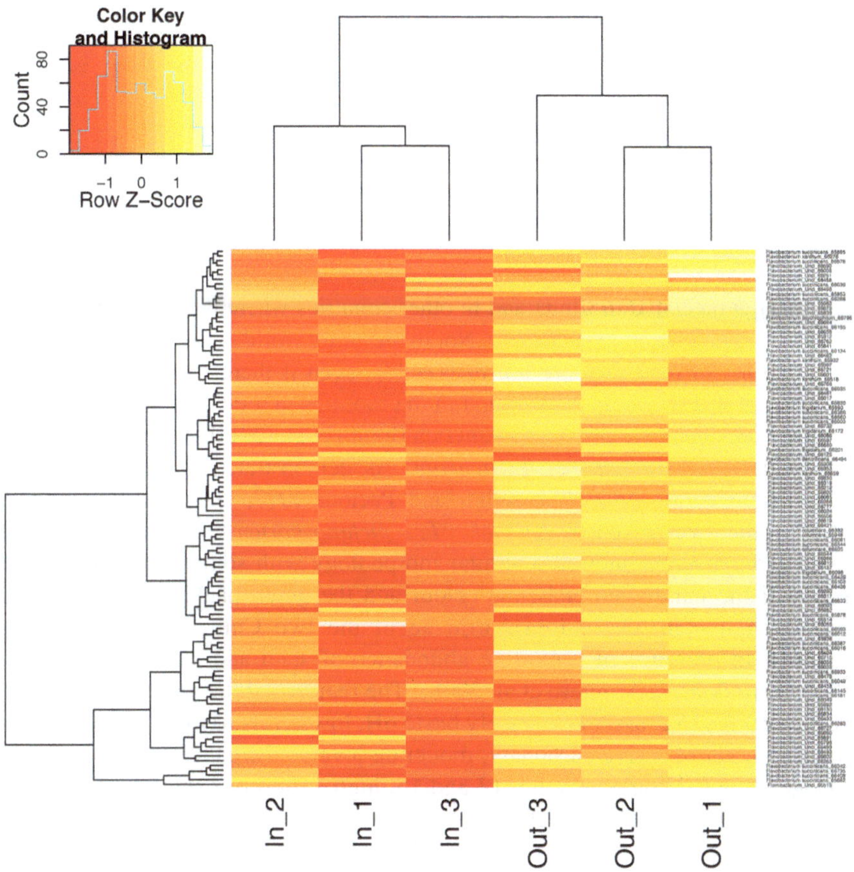

Fig. 11.1 Heat map of the operational taxonomic units (OTUs) that were both significantly different and had nearly a twofold difference in average intensity between inside and outside the brûlé for *Flavobacterium*. In_1, In_2, In_3 and Out_1, Out_2, Out_3, respectively, were pools from *inside* and *outside* the brûlé and were used as replicate samples. Supplementary material from Mello et al. (2013), PLoS One 8(4): e61945. doi:10.1371/journal.pone.0061945

the soil inside the brûlé revealed to contain a larger number of proteins compared with the soil outside the brûlé. Fisher's Exact Tests detected more biological processes inside the brûlé, especially processes related to responses to multiple types of stress. Thus, surprisingly, the organisms living in the brûlé show strong metabolic activity, despite this niche being characterized by a reduced diversity of plant and microbial species. Since the category "response to stress" principally consisted of proteins identified in herbaceous plants, the authors hypothesized that those few plants living in the brûlé experience stress conditions. From these results, it is possibly hypothesized that truffle metabolites, such as volatile organic compounds, may directly or indirectly elicit stress and defence responses in fungi and bacteria, but mostly in the surrounding herbaceous plants. In conclusion, the combination of metagenomics and metaproteomics has provided a powerful tool to reveal the functioning of a complex soil niche as the brûlé.

11.2.5 Chinese Truffles

Chinese truffles can be considered as a good complement to *T. melanosporum* in the market due to their lower prices and more availability (García-Montero et al. 2010). The most appreciated in China and in other countries where they are exported are *T. indicum* and *T. pseudoexcavatum*. Not in the entire world these fungi can be sold; for example, in Italy they are not marketable, even if there was evidence of the presence of *T. indicum* in the soil and in the root tips in a Piedmont truffle-ground (Murat et al. 2008). Also in North America Bonito et al. (2011b) inferred about two independent introductions of *T. indicum*. As demonstrated by several authors (Zambonelli et al. 1997; Comandini and Pacioni 1997; García-Montero et al. 2008), the ectomycorrhizae of *T. melanosporum*, *T. indicum* and *T. pseudoexcavatum* showed a similar morphology with some exceptions as the length of the unramified ends, the colour and the outer mantle pseudocells and cystidia diameter. The ectomycorrhizae of the two Chinese truffles can be obtained in presence of a relative abundance of sand and clay and a modest level of silt (García-Montero et al. 2008). The features of the soil, where the two truffle species live, are a moderately basic pH, low levels of total carbonates, an elevated level of active carbonate, moderate levels of organic carbon and total nitrogen, the C/N ratio between 13 and 15, high values of exchangeable cation complex, 100% of the degree of saturation of exchangeable cations, a good proportion of exchangeable Ca^{2+} and Mg^{2+}, K^+ in relatively large concentrations and scarce Na^+ (García-Montero et al. 2008). In general, the two Chinese truffles prefer calcareous soils, even if Riousset et al. (2001) found *T. indicum* in substrates devoid of calcium carbonate, and rich in organic matter. In China, the natural truffle habitats are often damaged during the random digging without pigs and dogs and out-of-season harvesting, causing a decline in the local production (Wang et al. 2008, Wang 2012). Some fungal species have been already considered "endangered fungi taxa" because of the excessive human collections (Liu et al. 2003). It becomes therefore crucial both to take care of the truffle habitats and to regulate the commercial harvesting and conservation of truffles in China (Wang et al. 2008, Wang 2012).

Regarding host plants, in general Chinese truffle species are mostly associated to conifers, broad-leaved trees and mixed forests of pines and broadleaved trees with the following plants: *Pinus tabulaeformis* var. *yunnanensis*, *P. armandii*, *Quercus acutissima*, *Q. pannosa*, *Viburnum cylindricum*, *Alnus cremastogyme*, *Coriaria nepalensis*, *Camptotheca acuminate*, *Prunus mume*, *Pyrus pashia* and *Berberis poiretii* (Yang 2001). However, it was demonstrated that *T. pseudoexcavatum* and *T. indicum* showed a high capacity of mycorrhization with *Q. ilex* subsp. *ballota* (García-Montero et al. 2008). Moreover, *T. indicum* is also able to interact with

Q. ilex subsp. *ilex*, *Q. pubescens* and *Q. cerris* that are the same host plants of *T. melanosporum* (Comandini and Pacioni 1997; Zambonelli et al. 1997; Di Massimo et al. 1996) and with North American angiosperm and gymnosperm hosts (Bonito et al. 2011b).

Seen in the common host range and soil preference of *T. pseudoexcavatum* and *T. indicum* and the Perigord black truffle, it could be strongly hypothesized that both Chinese truffle species have a potential ability to go through numerous *T. melanosporum* plantations and Mediterranean ecosystems (García-Montero et al. 2008). Another Chinese truffle, which may threaten European truffles, is the new Chinese species *T. sinoaestivum* (Zhang et al. 2012) which is sold as *T. aestivum* on the international market. *T. sinoaestivum* has already been found mixed to *T. aestivum* in Italy (Zambonelli et al. 2012b). This truffle could be unintentionally used as inoculum for the production of mycorrhizal plants for truffle cultivation and potentially contaminate European truffle areas.

Due to the increasing interest in import truffles from China and the fraudulent commercial practices related to these truffles (sold as *T. melanosporum* in Europe and in Japan), researchers have tried to characterize 16 Chinese truffles and to show a comprehensive summary of their taxonomy, ecology, mycorrhizae, genetics, biochemistry and cultivation in order to distinguish them from European black truffles (García-Montero et al. 2010). It is acknowledged that the morphological classification among truffle species is controversial (Ceruti et al. 2003) and that molecular tools are necessary for quality control both of commercial plants mycorrhized with *Tuber* and of ascomata sold in the market (García-Montero et al. 2008, 2010). Several phylogenetic studies were performed in order to describe the *Tuber* genus (Chen and Liu 2007; Wang et al. 2007; Jeandroz et al. 2008; Huang et al. 2009; Bonito et al. 2010a, 2013; Kinoshita et al. 2011). Thanks to these studies, new species were highlighted, for example, *T. latisporum*, which was separated from other white truffles (Chen and Liu 2007); *T. huidongense*, *T. liaotongense* and *T. taiyuanense*, which are different species or subspecies inside the *Puberulum* group (Wang et al. 2007); *T. pseudosphaerosporum*, which is related to *T. borchii* and *T. gibbosum* (Fan and Yue 2013); *T. turmericum* sp. nov., which belongs to the *T. turmericum* group (=*japonicum* group *sensu* Kinoshita et al. 2011) (Fan et al. 2015); or *T. hubeiense* and *T. wumengense*, resembling *T. borchii* (Fan et al. 2016b). Recently, it has been demonstrated that species of *T. californicum* in China were misidentified and they were *T. xuanhuaense*, *T. jinshajiangense*, *T. caoi* and *T. parvomurphium* (Fan et al. 2016a). Other species were on the contrary discovered as synonyms: *T. indicum*, *T. himalayense*, *T. sinense* and *T. pseudohimalayense* are one species, *T. indicum* complex (Wang et al. 2006a, b).

While several studies have been carried out to clarify the Chinese *Tuber* taxonomy, few focused on other aspects such as Chinese *Tuber* ecology (García-Montero et al. 2010). Further studies are therefore requested to better characterize these valuable products, which in addition to be eaten for their taste are considered as source of natural antioxidants for use in functional foods or medicine (Luo et al. 2011; Zhao et al. 2012).

Recently, the life cycle of *T. indicum* has been clarified, demonstrating its heterothallic nature with a *MAT* locus organized in two idiomorphs harboured by different strains (Belfiori et al. 2013). Given that *T. melanosporum* and *T. indicum* are very similar, a breeding between them cannot be excluded, thus leading to have potentially detrimental effects related to erosion of their biodiversity and specificity (Belfiori et al. 2013).

11.2.6 American Truffles

Truffles are dominant in ectomycorrhizal communities of commercial pecan orchards in North America (Bonito et al. 2011a). *T. lyonii*, native of North America, is sold in limited quantities because its aroma is not so similar to that of European truffles; its production fluctuates from year to year, limiting its supply; and the harvesting is performed by raking, allowing to collect both mature and immature ascomata (Bonito et al. 2011a). Several attempts are in progress to obtain pecan seedlings mycorrhized with European truffle species (e.g. *T. borchii* and *T. aestivum*) in order to have nuts, truffles and woods (Benucci et al. 2012). Although Benucci et al. (2012) obtained mycorrhizae of *T. borchii* and *T. aestivum* and not of *T. macrosporum* with pecan, further studies are needed to assess whether these relationships are maintained after planting in the field and whether truffle production can be supported by this host species.

Among the American truffles, there is *T. anniae* complex. Interestingly, ascomata of *T. anniae* complex were found under *P. sylvestris* in Finland, demonstrating a dispersal of truffle species between America and Europe (Wang et al. 2013). The *T. anniae* complex seems to be able to live in a wide range of soil types and pHs, also characterized by extreme conditions as the presence of heavy metals (Staudenrausch et al. 2005; Krpata et al. 2008). Liming of the forest may help to increase the incidence of *T. anniae* ectomycorrhizae, while the ascomata were collected only in sites with renewed substrates; this last finding leads to hypothesize that fresh organic matter and aerobic soil conditions are important to ascoma production (Wang et al. 2013). In recent time the *T. anniae* complex has been found in British Columbia in association both with *C. avellana* in truffle orchards and with *Pinus contorta*, *Pseudotsuga menziesii* and *Betula papyrifera* in forestry sites (Berch and Bonito 2016).

Recently seven new species (*T. beyerlei*, *T. castilloi*, *T. guevarai*, *T. lauryi*, *T. mexiusanum*, *T. miquihuanense* and *T. walker*) were described and collected in the USA and in Mexico (Guevara et al. 2013). They belong to *Maculatum* group, and they are associated with angiosperm hosts, including monocot species of *Epipactis* and woody dicot species of *Quercus*, *Populus*, *Salix*, *Carya* and *Notholithocarpus* with the exception of *T. beyerlei*, which was found under *P. menziesii* (Guevara et al. 2013). Moreover, the *Tuber gibbosum* complex was better characterized by phylogenetic analysis, showing the presence of four distinct species (Bonito et al. 2010b). The species in this group are associated exclusively with Pinaceae hosts, particularly with Douglas-fir (*P. menziesii*), but also occasionally with *Pinus. T. gibbosum* and *T. oregonense* are two of the most important species of this group (Bonito and Smith 2016). These truffles have not been cultivated yet, but in the Pacific Northwest of the United States are wild-harvested during winter and spring (Lefevre 2013).

Ascomata of a previously unknown species described by Healy et al. (2016) as *T. arnoldianum* have been found in both native and non-native tree roots at the Arnold Arboretum in Massachusetts (USA). From data showing its abundance, *T. arnoldianum* appears to be a strong competitor when exposed to other native ECM fungi in disturbed environments, and this suggests that it could be used in forestry and restoration.

11.3 Conclusions

In the last years, the knowledge of the ecology of truffles has greatly increased, thanks to many new scientific insights and technologies as the sequencing of *T. melanosporum* genome, the new approaches of environmental microbiology and the high-throughput sequencing. The discovery of the heterothallic nature of truffles, the correlation between the genet distributional pattern of *T. melanosporum* and the allelic configuration of the MAT locus, the successful production of plants inoculated with mycelial pure cultures and the likely presence of ecotypes within *T. aestivum* will move the researchers towards the optimization and selection of the inoculum to be used for the set-up of new orchards. At the same time, multi-marker phylogenetic analyses have allowed the discovery of new species increasing the knowledge of the *Tuber* genus.

Recently, the combination of metagenomics and metaproteomics is starting to reveal the proteins expressed by the organisms within an ecosystem at a specific time and, consequently, to describe metabolic processes active in these ecosystems. The application of this approach to the *T. melanosporum* brûlé has been the first step towards the knowledge of its functioning.

Altogether, the recent advent of "omics" technologies has greatly increased our understanding of truffle ecology that is relevant to answer applied questions of importance to the management of the truffle habitat.

References

Barbieri E, Gioacchini AM, Zambonelli A et al (2005) Determination of microbial volatile organic compounds from *Staphylococcus pasteuri* against *Tuber borchii* using solid-phase microextraction and gas chromatography/ion trap mass spectrometry. Rapid Commun Mass Spectrom 19:3411–3415

Belfiori B, Riccioni C, Paolocci F et al (2013) Mating type locus of Chinese black truffles reveals heterothallism and the presence of cryptic species within the *Tuber indicum* species complex. PLoS One 8:e82353

Belfiori B, Riccioni C, Paolocci F et al (2016) Characterization of the reproductive mode and life cycle of the whitish truffle *T. borchii*. Mycorrhiza 26:515–527

Bencivenga M, Urbani G (1992) Note sull'ecologia e sulla coltivazione del tartufo bianco (*Tuber magnatum* Pico). L'inf Agrar 48:64–69

Benucci GM, Raggi L, Albertini E et al (2011) Ectomycorrhizal communities in a productive *Tuber aestivum* Vittad. orchard: composition, host influence and species replacement. FEMS Microbiol Ecol 76:170–184

Benucci GMN, Bonito G, Baciarelli LF et al (2012) Mycorrhization of Pecan trees (*Carya illinoinensis*) with commercial truffle species: *Tuber aestivum* Vittad. and *Tuber borchii* Vittad. Mycorrhiza 22:383–392

Berch SM, Bonito G (2014) Cultivation of Mediterranean species of *Tuber* (Tuberaceae) in British Columbia, Canada. Mycorrhiza 24:473–479

Berch SM, Bonito G (2016) Truffle diversity (*Tuber*, Tuberaceae) in British Columbia. Mycorrhiza 26:587–594

Bertault G, Raymond M, Berthomieu A et al (1998) Trifling variation in truffles. Nature 394:734–734

Bertini L, Rossi I, Zambonelli A et al (2006) Molecular identification of *Tuber magnatum* ectomycorrhizas in the field. Microbiol Res 161:59–64

Bonito G, Smith ME (2016) General systematic position of the truffles: evolutionary theories. In: Zambonelli A, Iotti M, Murat C (eds) True truffle (*Tuber* spp.) in the World. Soil Biol 47:3–18 (in press)

Bonito G, Gryganskyi A, Vilgalys R et al (2010a) A global metaanalysis of *Tuber* ITS rDNA sequences: species diversity, host specificity, and long-distance dispersal. Mol Ecol 19:4994–5008

Bonito G, Trappe JM, Rawlinson P et al (2010b) Improved resolution of major clades within *Tuber* and taxonomy of species within the *Tuber gibbosum* complex. Mycologia 102:1042–1057

Bonito G, Brenneman T, Vilgalys R (2011a) Ectomycorrhizal fungal diversity in orchards of cultivated pecan (*Carya illinoinensis*; Juglandaceae). Mycorrhiza 21(7):601–612

Bonito G, Trappe JM, Donovan S et al (2011b) The Asian black truffle *Tuber indicum* can form ectomycorrhizas with North American host plants and complete its life cycle in non-native soils. Fungal Ecol 4:83–93

Bonito G, Smith ME, Nowak M et al (2013) Historical biogeography and diversification of truffles in the Tuberaceae and their newly identified southern hemisphere sister lineage. PLoS One 8: e52765

Bonuso E, Zambonelli A, Bergemann SE et al (2010) Multilocus phylogenetic and coalescent analyses identify two cryptic species in the Italian bianchetto truffle, *Tuber borchii* Vittad. Conserv Genet 11:1453–1466

Bragato G (2014) Is truffle brûlé a case of perturbational niche construction? For Syst 23:349–356

Bragato G, Vignozzi N, Pellegrini S et al (2010) Physical characteristics of the soil environment suitable for *Tuber magnatum* production in fluvial landscapes. Plant Soil 329:51–63

Callot G (1999) La truffe, la terre, la vie. INRA, Quae, p 210

Ceruti A, Fontana A, Nosenzo C (2003) Le Specie Europee del Genere *Tuber*, Una revisione storica. Museo regionale di scienze Naturali, monografie XXXVII, Torino

Chen J, Liu PG (2007) *Tuber latisporum* sp. nov. and related taxa, based on morphology and DNA sequence data. Mycologia 99:475–481

Chevalier G, Desmas C, Frochot H et al (1979) L'espéce *Tuber aestivum* Vitt: I. Dèfinition. Mushroom Sci X (Part 1) 10:957–975

Christopoulos V, Psoma P, Diamandis S (2013) Site characteristics of *Tuber magnatum* in Greece. Acta Mycol 48:27–32

Comandini O, Pacioni G (1997) Mycorrhizae of Asian black truffles, *Tuber himalayense* and *T. indicum*. Mycotaxon 63:77–86

Di Massimo G, García-Montero LG, Bencivenga M et al (1996) *Tuber indicum* Cooke et Massee, un tartufo orientale simile a *Tuber melanosporum* Vitt. Micol Veg Mediterr 11:107–114

Fan L, Yue SF (2013) Phylogenetic divergence of three morphologically similar truffles: *Tuber sphaerosporum, T. sinosphaerosporum,* and T. *pseudosphaerosporum* sp. nov. Mycotaxon 125:283–288

Fan L, Liu X, Cao J (2015) *Tuber turmericum* sp. nov., a Chinese truffle species based on morphological and molecular data. Mycol Prog 14:1–6

Fan L, Han L, Zhang PR et al (2016a) Molecular analysis of Chinese truffles resembling *Tuber californicum* in morphology reveals a rich pattern of species diversity with emphasis on four new species. Mycologia 108:344–353

Fan L, Zhang PR, Yan XY et al (2016b) Phylogenetic analyses of Chinese *Tuber* species that resemble *T. borchii* reveal the existence of the new species *T. hubeiense* and *T. wumengense*. Mycologia 108:354–362

García-Montero LG, Di Massimo G, Manjón JL et al (2008) New data on ectomycorrhizae and soils of the Chinese truffles *Tuber pseudoexcavatum* and *Tuber indicum*, and their impact on truffle cultivation. Mycorrhiza 19:7–14

García-Montero LG, Díaz P, Di Massimo G et al (2010) A review of research on Chinese *Tuber* species. Mycol Prog 9:315–335

García-Montero LG, Moreno D, Monleon VJ et al (2014) Natural production of *Tuber aestivum* in central Spain: *Pinus* spp. *versus Quercus* spp. brûles. For Syst 23:394–399

Gardin L (2005) I tartufi minori in Toscana. Gli ambienti di crescita dei tartufi marzuolo e scorzone. Quad ARSIA 1:1–56

Gryndler M, Trilčová J, Hršelová H et al (2013) *Tuber aestivum* Vittad. mycelium quantified: advantages and limitations of a qPCR approach. Mycorrhiza 23:341–348

Gryndler M, Cerná L, Bukovská P et al (2014) *Tuber aestivum* association with non-host roots. Mycorrhiza 24:603–610

Gryndler M, Beskid O, Hršelová H et al (2015) Mutabilis in mutabili: spatiotemporal dynamics of a truffle colony in soil. Soil Biol Biochem 90:62–70

Guevara G, Bonito G, Trappe JM et al (2013) New North American truffles (*Tuber* spp.) and their ectomycorrhizal associations. Mycologia 105:194–209

Hall IR, Haslam W (2012) Truffle cultivation in southern hemisphere. In: Zambonelli A, Bonito GM (eds) Edible ectomycorrhizal mushrooms. Soil Biol 34:191–208

Hall IR, Zambonelli A (2012) The cultivation of mycorrhizal mushrooms—still the next frontier! In: Zhang J, Wang H, Chen M (eds) Mushroom science XVIII. China Agricultural Press, Beijing, pp 16–27

Hall IR, Zambonelli A, Primavera F (1998) Ectomycorrhizal fungi with edible fruiting bodies 3. *Tuber magnatum*, Tuberaceae. Econ Bot 52:192–200

Hall I, Brown G, Zambonelli A (2007) Taming the truffle. The history, lore, and science of the ultimate mushroom. Timber Press, Portland

Healy RA, Smith ME, Bonito GM et al (2013) High diversity and wide spread occurrence of mitotic spore mats in ectomycorrhizal Pezizales. Mol Ecol 22:1717–1732

Healy RA, Zurier H, Bonito G et al (2016) Mycorrhizal detection of native and non-native truffles in a historic arboretum and the discovery of a new North American species, *Tuber arnoldianum* sp. nov. Mycorrhiza 26:781–792

Hilszczańska D, Sierota Z, Palenzona M (2008) New *Tuber* species found in Poland. Mycorrhiza 18:223–226

Huang JY, Hu HT, Shen WC (2009) Phylogenetic study of two truffles, *Tuber formosanum* and *Tuber furfuraceum*, identified from Taiwan. FEMS Microbiol Lett 294:157–171

Iotti M, Lancellotti E, Hall I et al (2010) The ectomycorrhizal community in natural *Tuber borchii* grounds. FEMS Microbiol Ecol 72:250–260

Iotti M, Leonardi M, Oddis M et al (2012a) Development and validation of a real time PCR assay for detection and quantification of *Tuber magnatum* in soil. BMC Microbiol 12:93

Iotti M, Piattoni F, Zambonelli A (2012b) Techniques for host plant inoculation with truffles and other edible ectomycorrhizal mushrooms. In: Zambonelli A, Bonito G (eds) Edible ectomycorrhizal mushrooms. Soil Biol 34:145–161

Iotti M, Rubini A, Tisserant E et al (2012c) Self/nonself recognition in *Tuber melanosporum* is not mediated by a heterokaryon incompatibility system. Fungal Biol 116:261–275

Iotti M, Leonardi M, Lancellotti E et al (2014) Spatio-temporal dynamic of *Tuber magnatum* mycelium in natural truffle grounds. PLoS One 9:e115921

Iotti M, Piattoni F, Leonardi P et al (2016) First evidence for truffle production from plants inoculated with mycelial pure cultures. Mycorrhiza 26:793–798

Jeandroz S, Murat C, Wang Y et al (2008) Molecular phylogeny and historical biogeography of the genus *Tuber*, the "true truffles". J Biogeogr 35:815–829

Kagan Zur V, Freeman S, Luzzati Y et al (2000) Emergence of the first black Périgord truffle (*Tuber melanosporum*) in Israel. Mycol Veg Mediterr 15:187–192

Kinoshita A, Sasaki H, Nara K (2011) Phylogeny and diversity of Japanese truffles (*Tuber* spp.) inferred from sequences of four nuclear loci. Mycologia 103:779–794

Krpata D, Peintner U, Langer I et al (2008) Ectomycorrhizal communities associated with *Populus tremula* growing on a heavy metal contaminated site. Mycol Res 112:1069–1079

Lalli G, Leonardi M, Oddis M et al (2015) Expanding the understanding of a forest ectomycorrhizal community by combining root tips and fruiting bodies: a case study of *Tuber magnatum* stands. Turk J Bot 39:527–534

Lancellotti E, Iotti M, Zambonelli A et al (2014) Characterization of *Tuber borchii* and *Arbutus unedo* mycorrhizas. Mycorrhiza 24:481–486

Lanza B, Owczarek M, DeMarco A et al (2004) Evaluation of phytotoxicity and genotoxicity of substances produced by *Tuber aestivum* and distributed in the soil using *Vicia faba* root micronucleus test. Fresen Environ Bull 13:1410–1414

Le Tacon F, Zeller B, Plain C et al (2013) Carbon transfer from the host to *Tuber melanosporum* mycorrhizas and ascocarps followed using a13C pulse-labeling technique. PLoS One 8:e64626

Le Tacon F, Rubini R, Murat C et al (2016) Certainties and uncertainties about the life cycle of the perigord black truffle (*Tuber melanosporum* Vittad.) Ann For Sci 73:105–117

Lefevre C (2013) Native and cultivated truffles of North America. In: Zambonelli A, Bonito G (eds) Edible ectomycorrhizal mushrooms. Soil Biol 34:209–226

Leonardi M, Iotti M, Oddis M et al (2013) Assessment of ectomycorrhizal fungal communities in the natural habitats of *Tuber magnatum* (Ascomycota, Pezizales). Mycorrhiza 23:349–358

Linde CC, Selmes H (2012) Genetic diversity and mating type distribution of *Tuber melanosporum* and their significance to truffle cultivation in artificially planted truffiéres in Australia. Appl Environ Microbiol 78:6534–6539

Liu PG, Wang XH, Yu FQ et al (2003) Key taxa of larger members in higher fungi of biodiversity from China. Acta Bot Yunnan 25:285–296

Lulli L, Pagliai M, Bragato G et al (1993) La combinazione dei caratteri che determinano il pedoambiente favorevole alla crescita del *Tuber magnatum* Pico nei suoli dei depositi marnosi dello Schlier in Acqualagna (Marche). Quad Sci Suolo 5:143–155

Luo Q, Zhang J, Yan L et al (2011) Composition and antioxidant activity of water-soluble polysaccharides from *Tuber indicum*. J Med Food 14:1609–1616

Marjanović Ž, Grebenc T, Marković M et al (2010) Ecological specificities and molecular diversity of truffles (genus *Tuber*) originating from mid-west of the Balkan Peninsula. Sydowia 62:67–87

Marjanović Ž, Glišić A, Mutavdžić D et al (2015) Ecosystems supporting *Tuber magnatum* Pico production in Serbia experience specific soil environment seasonality that may facilitate truffle lifecycle completion. Appl Soil Ecol 95:179–190

Martin F, Kohler A, Murat C et al (2010) Périgord black truffle genome uncovers evolutionary origins and mechanisms of symbiosis. Nature 464:1033–1038

Mello A, Murat C, Vizzini A et al (2005) *Tuber magnatum* Pico, a species of limited geographical distribution: its genetic diversity inside and outside a truffleground. Environ Microbiol 7:55–65
Mello A, Miozzi L, Vizzini A et al (2010) Bacterial and fungal communities associated with *Tuber magnatum*-productive niches. Plant Biosyst 144:323–332
Mello A, Napoli C, Murat C et al (2011) ITS-1 versus ITS-2 pyrosequencing: a comparison of fungal populations in truffle-grounds. Mycologia 103:1184–1193
Mello A, Ding GC, Piceno YM et al (2013) Truffle brûlés have an impact on the diversity of soil bacterial communities. PLoS One 8:e61945
Mello A, Lumini E, Napoli C et al (2015) Arbuscular mycorrhizal fungal diversity in the *Tuber melanosporum* brûlé. Fungal Biol 19:518–527
Menta C, García-Montero LG, Pinto S et al (2014) Does the natural "microcosm" created by *Tuber aestivum* affect soil microarthropods? A new hypothesis based on Collembola in truffle culture. Appl Soil Ecol 84:31–37
Molinier V, Bouffaud ML, Castel T et al (2013) Monitoring the fate of a 30-year-old truffle orchard in Burgundy: from *Tuber melanosporum* to *Tuber aestivum*. Agrofor Syst 87:1439–1449
Molinier V, Murat C, Frochot H et al (2015) Fine-scale spatial genetic structure analysis of the black truffle *Tuber aestivum* and its link to aroma variability. Environ Microbiol 17:3039–3050
Molinier V, Murat C, Peter M et al (2016) SSR-based identification of genetic groups within European populations of *Tuber aestivum* Vittad. Mycorrhiza 26:99–110
Murat C, Vizzini A, Bonfante P, Mello A (2005) Morphological and molecular typing of the below-ground fungal community in a natural *Tuber magnatum* truffle-ground. FEMS Microbiol Lett 245:307–313
Murat C, Zampieri E, Vizzini A et al (2008) Is the Perigord black truffle threatened by an invasive species? We dreaded it and it has happened! New Phytol 178:699–702
Murat C, Rubini A, Riccioni C et al (2013) Fine-scale spatial genetic structure of the black truffle (*Tuber melanosporum*) investigated with neutral microsatellites and functional mating type genes. New Phytol 199:176–187
Napoli C, Mello A, Borra A et al (2010) *Tuber melanosporum*, when dominant, affects fungal dynamics in truffle grounds. New Phytol 85:237–247
Orgiazzi A, Bianciotto V, Bonfante P et al (2013) 454 pyrosequencing analysis of fungal assemblages from geographically distant, disparate soils reveals spatial patterning and a core mycobiome. Diversity 51:73–98
Pacioni G (1991) Effects of *Tuber* metabolites on the rhizospheric environment. Mycol Res 95:1355–1358
Paolocci F, Rubini A, Riccioni C et al (2006) Reevaluation of the life cycle of *Tuber magnatum*. Appl Environ Microbiol 72:2390–2393
Pérez F, Palfner G, Brunel N et al (2007) Synthesis and establishment of *Tuber melanosporum* Vitt. ectomycorrhizae on two *Nothofagus* species in Chile. Mycorrhiza 17:627–632
Pfister DH, Kimbrough JW (2001) Discomycetes. In: McLaughlin DJ, McLaughlin EG, Lemke PA (eds) The mycota VII Part A. Systematics and evolution. Springer, New York, Berlin, Heidelberg, pp 257–281
Piattoni F, Ori F, Amicucci A et al (2016) Interrelationships between wild boars (*Sus scrofa*) and truffles. In Zambonelli A, Iotti M, Murat C (eds) True truffle (*Tuber* spp.) in the world. Soil Biol 47:375–389 (in press)
Pinkas Y, Maimon M, Maimon E et al (2000) Inoculation, isolation and identification of *Tuber melanosporum* from old and new oak hosts in Israel. Mycol Res 104:472–477
Pruett G, Bruhn J, Mihail J (2008) Temporal dynamics of ectomycorrhizal community composition on root systems of oak seedlings infected with Burgundy truffle. Mycol Res 112:1344–1354
Pruett GE, Bruhn JN, Mihail JD (2009) Greenhouse production of Burgundy truffle mycorrhizae on oak roots. New For 37:43–52

Ratti C, Iotti M, Zambonelli A et al (2016) Mycoviruses infecting true truffles. In: Zambonelli A, Iotti M, Murat C (eds) True truffle (*Tuber* spp.) in the world. Soil Biol 47:333–349
Reyna S, García Barreda S (2014) Black truffle cultivation: a global reality. For Syst 23:317–328
Riccioni C, Belfiori B, Rubini A et al (2008) *Tuber melanosporum* outcrosses: analysis of the genetic diversity within and among its natural populations under this new scenario. New Phytol 180:466–478
Riousset L, Riousset G, Chevalier G et al (2001) Truffes d'Europe et de Chine. Institut National de la Recherche Agronomique INRA, Paris
Rubini A, Belfiori B, Riccioni C et al (2011a) Isolation and characterization of MAT genes in the symbiotic ascomycete *Tuber melanosporum*. New Phytol 189:710–722
Rubini A, Belfiori B, Riccioni C et al (2011b) *Tuber melanosporum*: mating type distribution in a natural plantation and dynamics of strains of different mating types on the roots of nursery-inoculated host plants. New Phytol 189:723–735
Rubini A, Riccioni C, Belfiori B et al (2014) Impact of the competition between mating types on the cultivation of *Tuber melanosporum*: Romeo and Juliet and the matter of space and time. Mycorrhiza 24:19–27
Salerni E, D'Aguanno M, Leonardi P et al (2014a) Ectomycorrhizal communities above and below ground and truffle productivity in a *Tuber aestivum* orchard. For Syst 23:329–338
Salerni E, Iotti M, Leonardi P et al (2014b) Effects of soil tillage on *Tuber magnatum* development in natural truffières. Mycorrhiza 24(Suppl 1):79–87
Selosse M-A, Faccio A, Scappaticci P et al (2004) Chlorophyllous and achlorophyllous specimens of *Epipactis microphylla* (Neottieae, Orchidaceae) are associated with ectomycorrhizal septomycetes, including truffles. Microbiol Ecol 47:416–426
Selosse M-A, Taschen E, Giraud T (2013) Do black truffles avoid sexual harassment by linking mating type and vegetative incompatibility? New Phytol 199:10–13
Shamekh S, Donnini D, Zambonelli A et al (2009) Wild Finnish truffles. Yunnan Zhiwu Yanjiu 16:69–71
Shamekh S, Grebenc T, Leisola M et al (2014) The cultivation of oak seedlings inoculated with *Tuber aestivum* Vittad. in the boreal region of Finland. Mycol Prog 13:373–380
Song MS, Cao JZ, Yao YJ (2005) Occurrence of *Tuber aestivum* in China. Mycotaxon 91:75–80
Splivallo R, Novero M, Bertea CM et al (2007) Truffle volatiles inhibit growth and induce an oxidative burst in *Arabidopsis thaliana*. New Phytol 175:3417–3424
Splivallo R, Ottonello S, Mello A et al (2011) Truffle volatiles: from chemical ecology to aroma biosynthesis. New Phytol 189:688–699
Staudenrausch S, Kaldorf M, Renker C (2005) Diversity of the ectomycorrhiza community at a uranium mining heap. Biol Fertil Soils 41:439–446
Stielow B, Menzel W (2010) Complete nucleotide sequence of TaV1, a novel totivirus isolated from a black truffle ascocarp (*Tuber aestivum* Vittad.). Arch Virol 155:2075–2078
Stielow JB, Klenk HP, Menzel W (2011a) Complete genome sequence of the first endornavirus from the ascocarp of the ectomycorrhizal fungus *Tuber aestivum* Vittad. Arch Virol 156:343–345
Stielow B, Klenk HP, Winter S et al (2011b) A novel *Tuber aestivum* (Vittad.) mitovirus. Arch Virol 156:1107–1110
Stielow B, Bratek Z, Klenk HP et al (2012) A novel mitovirus from the hypogeous ectomycorrhizal fungus *Tuber excavatum*. Arch Virol 157:787–790
Stobbe U, Buntgen U, Sproll L et al (2012) Spatial distribution and ecological variation of re-discovered German truffle habitats. Fungal Ecol 5:591–599
Stobbe U, Egli S, Tegel W et al (2013a) Potential and limitations of Burgundy truffle cultivation. Mini-review. Appl Microbiol Biotechnol 97:5215–5224
Stobbe U, Stobbe A, Sproll L et al (2013b) New evidence for the symbiosis between *Tuber aestivum* and *Picea abies*. Mycorrhiza 23:669–673
Streiblová E, Gryndlerová H, Gryndler M (2012) Truffle brûlé: an efficient fungal life strategy. FEMS Microbiol Ecol 80:1–8

Taschen E, Sauve M, Taudiere A et al (2015) Whose truffle is this? Distribution patterns of ectomycorrhizal fungal diversity in *Tuber melanosporum* brûlés developed in multi-host Mediterranean plant communities. Environ Microbiol 17:2747–2761

Trappe JM, Claridge AW (2010) The hidden life of truffles. Sci Am 302:78–84

Turgeman T, Sitrit Y, Danai O et al (2012) Introduced *Tuber aestivum* replacing introduced *Tuber melanosporum*: a case study. Agrofor Syst 84:337–343

Urban A (2016) Truffles and small mammals. In: Zambonelli A, Iotti M, Murat C (eds) True truffle (*Tuber* spp.) in the world. Soil Biol 47:353–374

Urban A, Neuner-Plattner I, Krisai-Greilhuber I et al (2004) Molecular studies on terricolous microfungi reveal novel anamorphs of two *Tuber* species. Mycol Res 108:749–758

Vahdatzadeh M, Deveau, Splivallo R (2015) The role of the microbiome of truffles in aroma formation: a meta-analysis approach. Appl Environ Microbiol 81:6946–6952

Vasquez G, Gargano ML, Zambonelli A et al (2014) New distributive and ecological data on *Tuber magnatum* (Tuberaceae) in Italy. FL Mediterr 24:239–245

Wang XH (2012) Truffle cultivation in China. In: Zambonelli A, Bonito G (eds) Edible ectomycorrhizal mushrooms. Soil Biol 34:227–240

Wang Y, Tan ZM, Zhang DG et al (2006a) Phylogenetic and populational study of the *Tuber indicum* complex. Mycol Res 110:1034–1045

Wang Y, Tan ZM, Zhang DC et al (2006b) Phylogenetic relationships between *Tuber pseudoexcavatum*, a Chinese truffle, and other *Tuber* species based on parsimony and distance analysis of four different gene sequences. FEMS Microbiol Lett 259:269–281

Wang Y, Tan ZM, Murat C et al (2007) Molecular taxonomy of Chinese truffles belonging to the *Tuber rufum* and *Tuber puberulum* groups. Fungal Divers 24:301–328

Wang Y, Peigui L, Chen J et al (2008) China–a newly emerging truffle-producing nation. In: Reynal B (ed) Actes du colloques la culture de la truffe dans le monde. Imprimerie Georges Lachaise, Brive-La-Gaillarde, pp 35–44

Wang XH, Benucci GMN, Xie XD et al (2013) Morphological, mycorrhizal and molecular characterization of Finnish truffles belonging to the *Tuber anniae* species-complex. Fungal Ecol 6:269–280

Wedén K, Danell E, Camacho FJ et al (2004) The population of the hypogeous fungus *Tuber aestivum* syn. *T. uncinatum* on the island of Gotland. Mycorrhiza 14:19–23

Wedén C, Pettersson L, Danell E (2009) Truffle cultivation in Sweden: results from *Quercus robur* and *Corylus avellana* field trials on the island of Gotland. Scand J For Res 24:37–53

Yang MC (2001) Truffles in Southwest China. In: Corvoisier M, Olivier JM, Chevalier G (eds) Federation Française des trufficulteurs. Proceedings du Ve Congrès International Science et Culture de la Truffe, Aix-en-Provence, 4–6 Mars 1999. Paris, pp 248–249

Zambonelli A, Iotti M (2001) Effects of fungicides on *Tuber borchii* and *Hebeloma sinapizans* ectomycorrhizas. Mycol Res 105:611–614

Zambonelli A, Tibiletti E, Pisi A (1997) Caratterizzazione anatomo-morfologica delle micorriza di *Tuber indicum* Cooke and Massee su *Pinus pinea* L. and *Quercus cerris* L. Micol Ital 1:29–36

Zambonelli A, Iotti M, Rossi I et al (2000) Interactions between *Tuber borchii* and other ectomycorrhizal fungi in a field plantation. Mycol Res 104:698–702

Zambonelli A, Iotti M, Barbieri E et al (2009) The microbial communities and fruiting of edible ectomycorrhizal mushrooms. Yunnan Zhiwu Yanjiu 16:81–85

Zambonelli A, Iotti M, Boutahir S et al (2012a) Ectomycorrhizal fungal communities of edible ectomycorrhizal mushrooms. In: Zambonelli A, Bonito G (eds) Edible ectomycorrhizal mushrooms, Soil Biol 34:105–124

Zambonelli A, Iotti M, Piattoni F (2012b) Chinese *Tuber aestivum* sensu lato in Europe. Open Mycol J 6:22–26

Zambonelli A, Perini C, Pacioni G (2012c) Progetto MAGNATUM. Monitoraggio delle Attività di Gestione delle Tartufaie Naturali di *Tuber magnatum*. Risultati e consigli pratici. Alimat Edizioni Cesena

Zambonelli A, Iotti M, Hall I (2015) Current status of truffle cultivation: recent results and future perspectives. Micol Ital 44:31–40
Zampieri E, Murat C, Cagnasso M et al (2010) Soil analysis reveals the presence of an extended mycelial network in a *Tuber magnatum* truffle-ground. FEMS Microbiol Ecol 71:43–49
Zampieri E, Rizzello R, Bonfante P et al (2012) The detection of mating type genes of *Tuber melanosporum* in productive and non productive soils. Appl Soil Ecol 57:9–15
Zampieri E, Chiapello M, Daghino S et al (2016) Soil metaproteomics reveals an inter-kingdom stress response to the presence of black truffles. Sci Rep 6:25773
Zhang JP, Liu PG, Chen J (2012) *Tuber sinoaestivum* sp. nov. an edible truffle from southwestern China. Mycotaxon 122:73–82
Zhao L, Wang K, Yang R et al (2012) Antioxidant activity of the water-soluble and alkali-soluble polysaccharides from Chinese Truffle *Tuber sinense*. J Anim Vet Adv 11:1753–1756

Chapter 12
Inter- and Intraspecific Fungal Diversity in the Arbuscular Mycorrhizal Symbiosis

Brandon Monier, Vincent Peta, Jerry Mensah, and Heike Bücking

Abstract The 450-million-year-old arbuscular mycorrhizal (AM) symbiosis plays a critical role for the nutrient uptake and abiotic (drought, salinity, and heavy metals) and biotic stress resistance of the majority of land plants. The fungal extraradical mycelium takes up nutrients, such as phosphate and nitrogen, and delivers them to the intraradical mycelium, where the fungus exchanges these nutrients against carbon from the host. It is known for decades that AM fungi can improve the nutrient acquisition of many important crops under low input conditions and are able to increase plant productivity in stressful environments. However, despite their application potential as biofertilizers and bioprotectors, AM fungi have so far not been widely adopted. This is mainly due to the high variability and context dependency of mycorrhizal growth and nutrient uptake responses that make benefits by AM fungal communities difficult to predict. In this review, we summarize our current understanding of interspecific and intraspecific fungal diversity in mycorrhizal growth benefits and discuss the role of fungal genetic variability and host and fungal compatibility in this functional diversity. A better understanding of these processes is key to exploit the whole potential of AM fungi for agricultural applications and to increase the nutrient acquisition efficiency and productivity of economically important crop species.

12.1 Introduction

Plants from practically all environments can form symbiotic relationships with arbuscular mycorrhizal (AM) fungi, all comprised within the phylum, Glomeromycota. AM fungi were previously placed in the Zygomycota but were later grouped into their own phylum, because molecular data confirmed that this group of fungi is unique and has no obvious affinity to other major phylogenetic groups in the fungal kingdom (Schüßler and Walker 2010). AM associations are formed by approximately 65% of all terrestrial plant species including, but not

B. Monier • V. Peta • J. Mensah • H. Bücking (✉)
Biology and Microbiology Department, South Dakota State University, Brookings, SD 57007, USA
e-mail: heike.bucking@sdstate.edu

A. Varma et al. (eds.), *Mycorrhiza - Function, Diversity, State of the Art*,
DOI 10.1007/978-3-319-53064-2_12

limited to, a gamut of economically important crops (e.g., corn, rice, soybean, wheat) and even bryophytes including hornworts and liverworts (Smith and Smith 2011; Pressel et al. 2010; Wang and Qiu 2006). The wide distribution of these interactions within the plant kingdom and fossil records suggest that this symbiosis evolved ~450 million years ago and played a key role for the evolution of land plants (Taylor et al. 1995).

AM interactions are formed by a large number of different plant species ($n > 200,000$), but the number of fungal species is relatively low and has been estimated as < 350 (Öpik et al. 2013; Brundrett 2009). A high beta diversity among different sampling sites, however, indicates that the global species richness of AM fungi is possibly higher than these estimates (Kivlin et al. 2011). However, the exact number of AM fungal species is difficult to determine, because some species were placed into genera based on older relatively vague descriptions that cannot longer be verified (Rosendahl 2008).

Plants are also able to form other mycorrhizal associations, such as ectomycorrhizal (ECM), ericoid, or orchid mycorrhizas, with fungi from the phyla Ascomycota and Basidiomycota. However, these relationships are not as prevalent as AM fungal associations and entail a relatively small proportion of the overall mycorrhizal interactions between plants and fungi (Moore et al. 2011; Brundrett 2009). ECM interactions are formed predominantly by woody perennials from cool temperate, boreal, montane to tropical ecosystems (Brundrett 2009). The number of plant species that develop ECM interactions is relatively small with 6000–8000 species, but ECM fungi exhibit a wide taxonomic range with 20,000–25,000 species (Tedersoo et al. 2010; Rinaldi et al. 2008). The ECM fungal diversity can be very high in ecosystems and can consist of hundreds of different ECM fungal species (Henkel et al. 2012; Buée et al. 2011; Newton and Haigh 1998).

Both, AM and ECM symbioses play a key role for the nutrient uptake of their host plant and improve the uptake of phosphate and nitrogen, but also of trace elements such as copper and zinc. In addition, the symbiosis increases the resistance of plants against abiotic (drought, heavy metals, salinity) and biotic (pathogen) stresses (Smith and Read 2008). But the symbiosis is also costly for the plant, and plants transfer up to 20% of their assimilated carbon to their fungal partner (Wright et al. 1998). ECM fungi have also saprophytic capabilities, but AM fungi are obligate biotrophs that completely rely on their host plant for their carbon supply and are unable to complete their life cycle without the symbiosis to the host.

AM fungi are not equally beneficial for the host, and mycorrhizal benefits have been described as a mutualism to parasitism continuum (Smith and Smith 2013; Johnson and Graham 2013; Johnson et al. 1997). However, the mechanisms responsible for these functional differences and the more or less beneficial outcomes for the host are currently unknown. We will discuss here factors that may contribute to the observed high interspecific and intraspecific fungal diversity and will particularly focus on the AM symbiosis because a better understanding of these processes is critical for a useful application of these fungi in efforts to increase crop production and food security in the future (Rodriguez and Sanders 2015).

12.2 Genetic Diversity of Arbuscular Mycorrhizal Fungi

12.2.1 The Arbuscular Mycorrhizal Fungus: An Asexual Symbiont?

In the past the standard for "species" identification of AM fungi was through the determination of morphological traits found among resting spore types (Mosse and Bowen 1968; Morton and Benny 1990; Schüßler and Walker 2010). The validity of this methodology is rather limited due to similarities in the spore morphology of different fungal species and significant differences in spore size and color within one fungal species (Redecker et al. 2013; Merryweather and Fitter 1998; Bentivenga et al. 1997; Morton 1985). Another limitation is that the composition of AM fungal communities in colonized roots cannot be accurately identified. Characteristics of fungal structures (arbuscules, vesicles, intercellular hyphae) within colonized roots are not species specific, and the correlation between the presence of resting spores in the soil and the AM fungal communities in roots is not reliable, because not all identifiable spores may really contribute to the AM community composition of the root. Based on differences in spore morphology, about 250 AM fungal species have been validly described.

Progress in sequencing technologies allow now to classify AM fungal species by morphological data in combination with sequence information of ribosomal RNA genes (SSU or LSU). The availability of these sequence data led to substantial changes in the AM fungal taxonomy and the establishment of several new genera and families within the Glomeromycota (Schüßler and Walker 2010). The new sequencing technologies also revealed that the AM fungal diversity in ecosystems is larger than previously been expected. However, our current understanding of the AM community composition is still limited by the availability of reliable sequence data for all species within the Glomeromycota and the lack of a universal standard for the identification of operational taxanomic units (OTUs) of AM fungi. When different OTU delineation techniques are compared, one based on the evolutionary origin of monophyletic clades and the other based on sequence similarities with published sequences, the latter generally leads to a significantly higher number of OTUs and a change in absolute OTU richness (Lekberg et al. 2014). Depending on target gene and sequence similarity cutoff, the number of virtual AM fungal taxa (taxa without morphological analogues) ranges from 300 to 700 in different environmental samples (Öpik et al. 2013; Kivlin et al. 2011).

The biological species concept, however, is difficult to apply to Glomeromycota (Sanders 1999, 2002). The biological species concept defines species as groups of actually or potentially interbreeding natural populations that occupy a specific niche in nature and is not solely based on morphological concepts (Mayr 1942, 2000). However, all fungi within the phylum Glomeromycota (in contrast to fungi within the Ascomycota or Basidiomycota) lack any obvious sexual structures, and the low morphological diversity within this group of fungi led to the overall assumption that AM fungi are ancient asexuals. According to evolutionary theory, sexual reproduction is advantageous because the recombination of genes leads to genetic variations and allows the elimination of deleterious mutations and unfavorable traits. The conservation of an asexual lifestyle in AM fungi over such a long coevolution with plants (~450 million years) therefore represents a paradox (Sanders 1999, 2011).

Earlier studies of AM fungi from pot cultures or field-collected spores provided no evidence for gene recombination in AM fungi (Rosendahl 2008; Stukenbrock and Rosendahl 2005). But over the past decade, the question on whether AM fungi are ancient asexuals without an opportunity for genetic recombination is more controversially discussed. Recent studies revealed that the genomes of several AM fungal species contain genes that are in other organisms involved in sexual reproduction processes. In the transcriptome of *Rhizophagus irregularis* (previously *Glomus intraradices*) (Stockinger et al. 2009), for example, several meiosis-specific genes [homologous-pairing protein 2 (*HOP2*) and meiotic nuclear division protein 1 (*MND1*)] were identified, which are conserved among eukaryotes and are only known to function in eukaryotic meiosis (Tisserant et al. 2012). More than 85% of the core meiotic genes that are involved in the meiosis of *Saccharomyces cerevisiae* can be identified in the AM fungal genome, indicating that AM fungi may be able to undergo a conventional meiosis (Halary et al. 2011).

Recent genomic and transcriptomic surveys also demonstrated the presence of mating-type gene homologues and putative sex pheromone-sensing mitogen-activated protein (MAP) kinases in several AM fungal species. In the genomes of *Rhizophagus* spp. and *Glomus cerebriforme*, orthologues of the sex pheromone-sensing pathway of *S. cerevisiae* were identified, which is highly conserved in Asco- and Basidiomycota and involved in the signal transduction pathway between pheromone receptors at the hyphal surface and the transcription factors that regulate mating in these fungi (Halary et al. 2013). However, as long as the exact function of these genes in AM fungi is unknown, their existence is not conclusive evidence for a sort of cryptic sexuality in AM fungi (Corradi and Bonfante 2012). Nevertheless, the identification of these sex-related genes in AM fungi opens up the possibility that the previous view of AM fungi as ancient asexuals and as evolutionary aberration is oversimplified and that cryptic sexuality could be an important pathway in this ecologically important group of fungi (Halary et al. 2013; Corradi and Bonfante 2012).

12.2.2 Arbuscular Mycorrhizal Fungi Have a Diverse Set of Nuclei

AM fungi are unique, because their spores and hyphae are coenocytic and contain multiple nuclei in a common cytoplasm. The number of nuclei in spores can be as high as several hundred or even thousand nuclei per spore, and in the coenocytic mycelium of the fungus, up to 100 nuclei can be found per 100 μm of hyphae (Marleau et al. 2011).

Genetic diversity in, e.g., ribosomal gene sequences of AM fungi can not only be caused by genetic variation among fungal individuals but also by the heterogeneity found within one individual. It has been hypothesized that in the absence of sexual recombination (see above), evolution should favor individuals with highly divergent genetically different nuclei (Kuhn et al. 2001; Sanders 1999), and indeed individual spores of AM fungi contain a population of genetically divergent nuclei (Kuhn et al. 2001; Sanders 1999; Hijri et al. 1999). It has been hypothesized that AM fungi evolved to be multi-genomic and that this multi-genomic life style could explain the fitness and the long-term evolutionary persistence of this group of fungi (Hijri and Sanders 2005; Pawlowska and Taylor 2004). Genetic divergence of spores cannot only be found in ribosomal genes but also in protein-coding genes, and these genetic variants are passed on from generation to generation through spores (Hijri and Sanders 2005). Kuhn and co-workers assumed (2001) that the genetic diversity is the result of multiple mutations in an otherwise clonal genome and that recombination events cannot explain the majority of mutations in the genome sequences. Genome polyploidization has also been discussed as a potential origin of the spore divergence in AM fungi (Pawlowska and Taylor 2004), but this view has been questioned by other authors, who reported that even species with a very large nuclear DNA content are haploid (Hijri and Sanders 2005).

12.2.3 The Role of Hyphal Fusions in Fungal Diversity

Anastomosis, the fusion between encountering AM fungal hyphae could also explain the high nuclei divergence in AM fungi, and there is increasing evidence that these fusion events can contribute to genetic exchange and diversification in AM fungi. Genetically distinct AM fungi can exchange nuclei through anastomosis, and it has been demonstrated that genetic markers from each parent are transmitted to the progeny of this hyphal fusion (Croll et al. 2009). However, AM fungi differ in their frequency with which they anastomose, and it has been shown that in *Funneliformis mosseae*, the likelihood that hyphal contacts lead to hyphal fusions is more than seven times higher than in *F. coronatus* (Pepe et al. 2016). However, even in pairings in which the anastomosis frequency is relatively low, a genetic exchange between the hyphae can be observed (Croll et al. 2009).

Fungal compatibility plays a role in the frequency with which fungal isolates anastomose. While, for example, high anastomosis frequency and high compatibility were found between isolates of *R. irregularis* that were isolated from a single site (Croll et al. 2009), no anastomosis was observed between geographically distant isolates of *Funneliformis mosseae*, but all these isolates were capable of self-anastomosing (Giovannetti et al. 2003). It has been suggested that similar environments and proximity are important factors for the vegetative compatibility among AM fungi. Successful anastomosis only occurs when the isolates are either genetically similar or from the same habitat (Purin and Morton 2013).

Interestingly it has also been demonstrated that the symbiotic growth phase plays a role for successful anastomosis (Purin and Morton 2013). Before the symbiosis with the host is established and host root and fungus enter the symbiotic growth phase, the fungus undergoes a presymbiotic growth phase that is characterized by spore germination, the exchange of signal molecules between both partners [root exudates (e.g., strigolactones) and the so-called "myc factors" (lipochitooligosaccharides)] (Maillet et al. 2011; Akiyama and Hayashi 2006) and extensive hyphal branching. While in the presymbiotic growth phase, anastomosis was relatively unconstraint between hyphae from either genetically identical or different isolates from the same habitat, hyphal anastomosis was suppressed during the symbiotic growth phase (Purin and Morton 2013). This suggests that hyphal anastomosis may fulfill different functions during the presymbiotic or symbiotic growth phase. A potential explanation could be that during the presymbiotic growth phase, fungal anastomosis allows to redistribute water and nutrients within the growing hyphal network, while during the symbiotic growth phase, anastomosis could cause a significant slowdown in the water and nutrient transport to the host and a dilution of the carbon transport from the source (mycorrhizal interface within the root cortical cells) to the sink (growing hyphal tips and developing spores).

This high genetic diversity among nuclei within one fungal individual may explain the high intraspecific diversity found in AM fungi and the high functional differences and context dependency of mycorrhizal growth responses. If nuclei with different genetic potential are randomly distributed during spore formation, the offspring of this fungal individual will carry a different composition of nucleotypes compared to the parent or the siblings and may also differ from the parent or the siblings in its effect on plant growth. Angelard and co-workers (2010) tested this hypothesis and examined the growth response of *Plantago lanceolata* and *Oryza sativa* after inoculation with parental, crossed, and offspring lines of the AM fungus *R. irregularis* (previously *G. intraradices*) and found that the growth of both plants was reduced by an inoculation with crossed lines, compared to the parental lines. Some offspring lines differed also from the other lines in their effect on plant growth. While offspring lines reduced the plant growth of *P. lanceolata* compared to the crossed lines, the growth of rice significantly increased by the colonization with certain offspring lines. The offspring lines had also a different effect on plant gene expression than the crossed lines (Angelard et al. 2010) and expressed a different fungal phenotype and colonization pattern compared to their respective crossed lines (Angelard and Sanders 2011). This considerable genetic and phenotypic diversity among different single spore lines that share the same parent is also stable over multiple single spore generations (Ehinger et al. 2012).

However, it has also been demonstrated that while there is a high genetic and phenotypic variation among different single spore lines, the differences among subcultured replicates of these single spore lines are small. This suggests that while the genetic potential of each spore is randomly selected during spore development, the phenotypes of these cultures are still relatively stable (Ehinger et al. 2012; Koch et al. 2004). Recently, it was shown that while in crossed isolates, the nuclei are inherited by both parents; mitochondria seem to be inherited only by one parent. Based on putative orthologs in the genome of the AM fungus *R. irregularis* to the set of genes involved in the mitochondrial segregation in *S. cerevisiae*, the authors assume that mitochondrial segregation processes are independent from nuclear segregation processes (Daubois et al. 2016).

12.2.4 Is There an Effect of Endobacteria in Fungal Diversity?

Recently, it has been demonstrated that endobacteria are widely distributed across the whole phylogenetic range of AM fungi. These mycoplasma-related endobacteria (MRE) are related to the recently discovered bacterial lineage of Mollicutes and live in the fungal cytoplasm. There are indications that this fungal-bacterial symbiosis evolved ~400 million years ago (Mondo et al. 2012) and therefore close to the evolution of the AM symbiosis. The bacterial symbiont depends on its host for carbon, phosphate, and nitrogen supply, while the dependence of the fungal partner from these endobacteria has been suggested to be relatively low compared to the dependence from the plant partner.

The analysis of the genome of some of these endobacteria revealed typical determinants of symbiotic, pathogenic, and free-living bacteria that are integrated in an otherwise reduced genome (Ghignone et al. 2012; Naito and Pawlowska 2016). The endobacterium *Candidatus Glomeribacter gigasporarum*, for example, is unable to synthesize essential amino acids indicating a strong metabolic dependence of this endobacterium from its fungal partner and an obligate biotrophic lifestyle (Naito et al. 2015). The bacterial genome also contains a substantial proportion of genes that were potentially acquired horizontally from their fungal host. One potential example for a horizontally acquired gene is a SUMO protease that may allow these endobacteria to change the SUMOylation level of fungal proteins (Naito et al. 2015).

The role of MRE in the biology of their fungal hosts is largely unknown (Toomer et al. 2015), but there are indications that these endobacteria may have a functioning pathway for the synthesis of folate and of cobalamin (vitamin B12) and contain the genes for a type III secretion system that are used by other pathogenic and symbiotic Gram-negative bacteria to release effector molecules into their host cell (Ghignone et al. 2012). Recent fungal transcriptome and proteome studies demonstrated that endobacteria play an important role during the fungal presymbiotic growth phase (Vannini et al. 2016; Salvioli et al. 2010, 2016). The endosymbiosis has an influence on fungal growth and calcium signaling and enhances the bioenergetic capacity during the presymbiotic growth phase and plays thereby an important role for the successful establishment of the AM symbiosis with the host plant.

Germinating spores that are colonized by endobacteria accumulate proteins involved in DNA replication, transcription, and protein synthesis and have higher transcript levels of a Rho-GDP-dissociation inhibitor (Vannini et al. 2016) than control spores. This dissociation inhibitor regulates Rho-GTPases, which are involved in cytoskeletal organization, vesicle trafficking, and bud site selection and are all important processes during fungal growth. Several genes that are involved in oxidative phosphorylation are upregulated, and ATP biosynthesis and fungal respiration are increased in germinating spores with endobacteria, indicating that the colonization with endobacteria increases the bioenergetic potential and the ecological fitness of the fungal host during the critical presymbiotic growth phase (Salvioli et al. 2016; Vannini et al. 2016).

In contrast, the fungal phenotype in the symbiotic growth phase does not seem to be affected by the colonization with the endobacterium *Candidatus G. gigasporarum* (Salvioli et al. 2016). However, under consideration that each AM fungal species harbors a distinct group of MRE (Naito et al. 2015) and that there is also a considerable MRE diversity across AM fungal individuals (Toomer et al. 2015; Agnolucci et al. 2015), more research is necessary to evaluate whether bacterial endophytes can also contribute to the functional diversity of AM fungi during the symbiotic growth phase. Based on the currently available evidence, it can be assumed that endobacteria at least play a significant role for the successful establishment of the symbiosis and may have an effect on the AM community composition of the host plant.

In addition to MREs, spores of different AM fungal species and fungal isolates have been shown to be associated with diverse bacterial communities, and several of these spore-associated bacteria exhibit plant growth-promoting capabilities (Battini et al. 2016; Agnolucci et al. 2015). Several bacterial isolates showed, for example, the capability to produce plant growth hormones and are able to solubilize phosphate from mineral phosphate and phytate. It can be assumed that these bacterial capabilities can also contribute to the mycorrhizal benefits for the host plant, but the composition and effects of these bacterial communities are largely unexplored.

12.3 Host Specificity in the Arbuscular Mycorrhizal Symbiosis

Since the AM fungus is an obligate biotroph that is unable to complete its life cycle without the symbiosis to its host, AM fungal species were seen as generalists with a low host specificity, and are able to colonize a wide range of host plants (Ehinger et al. 2009). In fact, the low fungus to host species ratio (350 fungal species to 200,000 plant species, see above) has led to the overall assumption that there is a high functional redundancy among fungal species and that the role of inter- and intraspecific fungal diversity does not play an important role for ecosystem functioning (Klironomos 2000).

AM interactions are many-to-many interactions, and each individual host plant is colonized simultaneously with multiple fungal species, and each fungal individual is associated with multiple host plants of the same or of different plant species. These host plants share a common mycorrhizal network (CMN), and it has been demonstrated that AM fungi allocate nutrient resources preferentially to specific host plants within these CMNs (Bücking et al. 2016; Fellbaum et al. 2014; Walder et al. 2012). It has been estimated that in any community between 30 and 50 different AM fungal species could exist (Fitter 2005). For example, in a boreonemoral forest, up to 47 fungal taxa were identified (Öpik et al. 2008, 2009). If AM fungi are not host specific, all these species could potentially contribute to the AM fungal community composition of a single host plant.

However, new sequencing technologies provide now much more evidence for a host specificity or at least host preference of AM fungi. When fungal communities in the roots of forest plant species were compared to the roots of generalist plant species, the fungal taxon richness was significantly higher for forest than for generalist plant species (28.8–13.0 fungal taxa), and the AM fungal community composition differed significantly among these two plant groups (Öpik et al. 2009). Almost half of the fungal virtual taxa that were identified colonized exclusively forest plant species, while only one fungal taxa colonized specifically generalist plant species, and these differences in these fungal communities were unrelated to plant community spatial structure or environmental conditions (Öpik et al. 2009).

Distinct AM fungal communities among different host plant species were also found in a semiarid prairie ecosystem and temperate grasslands (Valyi et al. 2015; Torrecillas et al. 2012). Perennial plant species harbored a lower AM fungal diversity than annual plant species, and half of the AM fungal species that were identified were specific for one plant species (Torrecillas et al. 2012). These data suggest that the host specificity of AM fungi is higher than previously assumed, and this has also implications for the success and survival of introduced AM fungi and the establishment of designed AM fungal communities in agricultural applications for enhanced crop productivity.

12.4 Functional Diversity in the Arbuscular Mycorrhizal Symbiosis

The impact of different AM fungi on plant growth can range from highly mutualistic to antagonistic (Klironomos 2003), and mycorrhizal growth responses have been described as a mutualism to parasitism continuum (Johnson and Graham 2013; Smith and Smith 2013; Johnson et al. 1997). Mycorrhizal growth responses are highly context dependent, and it has been suggested that particularly, the nutrient availability in the soil determines the position of AM fungi along this mutualism to parasitism continuum (Johnson and Graham 2013). High phosphate availabilities in the soil in general reduce mycorrhizal colonization and mycorrhizal growth benefits for the plant, and negative mycorrhizal growth responses have been discussed as a consequence of the high carbon costs of the symbiosis for the plant that are not counterbalanced by a net gain in phosphate (Peng et al. 1993). However, it has also been suggested that negative mycorrhizal growth responses could be the result of the suppression of the phosphate uptake via the plant pathway (via the epidermis and root hairs) which is not compensated for by an increase in the phosphate uptake via the mycorrhizal uptake pathway (via the extraradical mycelium and the mycorrhizal interface) (Smith et al. 2011). AM fungal species differ in the efficiency with which they suppress the plant uptake pathway (Grunwald et al. 2009), and this suppression could lead to an overall reduction in total phosphate uptake and even phosphate deficiency of the plant (Smith et al. 2011).

However, a meta-analysis of about 2000 field and laboratory studies suggest that functional differences not only depend on soil fertility but also on functional characteristics of the host plants and the complexity of the soil microbial community, which includes AM fungi and non-mycorrhizal microbial species (Hoeksema et al. 2010). Below we discuss different factors that may contribute to the functional diversity in mycorrhizal growth responses.

12.4.1 Fungal Identity

Genetic and functional diversity (see also above) have been observed at all levels of biological organization in AM fungi (Antunes et al. 2011; Powell et al. 2009; Koch et al. 2006; Munkvold et al. 2004; Hart and Reader 2002). However, the reasons for the high functional variability among AM fungi are largely unknown. It has been suggested that fungal growth traits are conserved within one phylogenetic group. For example, Hart and Reader (2002), who screened different phylogenetic groups within the Glomeromycota for their colonization strategies, found that members of the Gigasporaceae tend to extensively colonize the soil, while the colonization of the roots is limited. In contrast, members of the Glomeraceae exhibit a different colonization strategy and extensively colonize the host roots but show only a relatively low hyphal exploration into the soil. Based on these fungal growth traits, the authors assumed that the phylogenetically determined variability in colonization strategies could also lead to differences in the mechanisms by which these fungi promote host plant growth. The extensive colonization of the root system of the Glomeraceae could suppress the colonization of the root system with root pathogens and thereby contribute to a higher biotic stress resistance of the host, while the better exploration of the soil by hyphae of the Gigasporaceae could have a stronger effect on the nutrient and water uptake of the host. Evidence that fungal growth traits such as levels of root colonization, spore production, and extraradical hyphal extension are phylogenetically conserved within the Glomeromycota has also been described by other authors (Antunes et al. 2011; Powell et al. 2009).

Some studies have shown that mycorrhizal growth responses, such as shoot biomass and phosphate and nitrogen contents, are positively correlated to fungal growth traits, such as hyphal length, area covered by ERM, hyphal density, or hyphal length per mm of colonized root length (Avio et al. 2006). However, fungal growth traits are not necessarily correlated to mycorrhizal growth benefits or the capability of the AM fungi to increase the phosphate or nitrogen uptake of the plant. In a study, in which the effect of 31 different AM fungal isolates from 10 AM species on plant biomass (*Medicago sativa*) and phosphate and nitrogen uptake was examined, no correlation between fungal growth traits and mycorrhizal benefits was observed. The authors reported that the capability of AM fungi to increase the growth and nutrient uptake of *Medicago* is not related to the fungal phylogeny and is relatively widely distributed in the phylum Glomeromycota (Koch et al. 2017; Mensah et al. 2015). This is consistent with the results of de Novais et al. (2014), who reported that the ability to promote plant growth is unrelated to the taxonomic classification of AM fungal isolates. This asymmetry between phylogenetically conserved fungal growth traits and evolutionary not conserved host plant effects indicates that other processes such as more efficient nutrient uptake and/or higher nutrient transport rates to the host contribute to the observed functional diversity among AM fungi.

However, there is not only a high interspecific but also intraspecific functional diversity among different isolates of one fungal species (Koch et al. 2017; Börstler et al. 2008, 2010; Koch et al. 2004). Mensah and co-workers (2015), who tested three different isolates of ten fungal species, found in all fungal species a high intraspecific variability in the effects on host plant biomass and phosphate and nitrogen uptake. High within-species diversity among different isolates of the same fungal species has been reported in several studies and in symbiosis with different host plant species (Koch et al. 2017; de Novais et al. 2014; Campagnac and Khasa 2014; Koch et al. 2006; Avio et al. 2006; Munkvold et al. 2004). The high intraspecific functional diversity can likely be explained by the high genetic variability among different isolates (see also above). In *F. mosseae* (previously *Glomus mosseae*), for example, a genetic diversity of more than 50% was found among different geographical isolates (Avio et al. 2009). However, similarly high genetic and phenotypic differences can also exist among individuals from one AM fungal population. Fivefold differences in hyphal length were observed among isolates of *R. irregularis* (previously *G. intraradices*) that were isolated from one population (Koch et al. 2004). Hyphal length has previously been used as an important criterion to explain differences in the phosphate uptake by mycorrhizal plants (Jakobsen et al. 1992). Koch et al. (2017) found that host plant performance was not related to the AM fungal morphology or growth traits, and they hypothesized that differences in plant growth benefits among isolates within an AM fungal species could be the result of a co-evolution between co-existing fungal and plant populations. The reason for this high functional diversity among different isolates of one AM species is largely unexplored and should be more strongly considered, when fungal gene expression and function is studied.

12.4.2 *Fungal-Host Compatibility*

Before the symbiosis to the host plant can be established, AM fungi undergo a presymbiotic growth phase and respond to their potential host plants with enhanced hyphal branching of germinating spores and a more target-oriented growth of their hyphae (Bücking et al. 2008; Buée et al. 2000). To attract AM fungi, host plants release root exudates that contain several active molecules, e.g., strigolactones, and there are indications that host plants change the composition of their root exudates to attract AM fungi particularly under stressful conditions (Tripathi et al. 2016). Strigolactones, for example, stimulate hyphal branching and fungal metabolic activity during the presymbiotic growth phase of the fungus (Bücking et al. 2008; Tamasloukht et al. 2007; Besserer et al. 2006; Akiyama and Hayashi 2006; Akiyama et al. 2005; Tamasloukht et al. 2003). However, there is evidence that the plant genotype plays a critical role in the microbial community composition and that these differences could be the result of quantitative or qualitative changes in the root exudate composition among different plant genotypes (Aira et al. 2010). Consistently, plant genotypes have been shown to differ in their responsiveness to mycorrhizal fungi (Wang and Bücking 2015; Aira et al. 2010).

Branched fungal hyphae on the other hand also secrete a diffusible signal (lipochitooligosaccharides) to the roots, also referred to as "myc factor" that induces a symbiosis program in the roots and prepares the roots for colonization. In response to myc factors, specific cells in the roots support the formation of the prepenetration apparatus that provides the fungus with a pathway through the epidermis to the inner cortex where the fungus forms arbuscules (Parniske 2008; Genre et al. 2005). These intracellular highly branched structures are involved in the nutrient to carbon exchange processes between both partners and are characterized by the expression of mycorrhiza-inducible plant phosphate and nitrogen transporters in the periarbuscular membrane and carbohydrate transporters in the fungal membrane (Breuillin-Sessoms et al. 2015; Helber et al. 2011; Guether et al. 2009; Gomez et al. 2009; Javot et al. 2007). The successful colonization of the root depends on a common symbiosis signaling pathway that is highly evolutionary conserved in mycorrhizal plants. Plants with mutations in this pathway are unable to form a successful symbiosis (Parniske 2008; Gherbi et al. 2008; Kistner et al. 2005). The perception of myc factors leads to a transcriptional reprogramming of host gene expression (e.g., transcription factors) (Czaja et al. 2012), but whether AM fungi differ in their myc factor composition and lead to different changes in host plant gene expression is currently unknown.

The mycorrhizal colonization percentage is a common metric to describe the abundance of AM fungal structures in roots, and it is generally assumed that mycorrhizal colonization is positively correlated to host plant benefit. Accordingly, Treseder (2013) found in her meta-analysis an increase in plant biomass and host plant phosphate content with higher mycorrhizal colonization rates. However, differences in the mycorrhizal colonization are only in part responsible for the variability in host plant responses, and AM fungi differ greatly in the benefit that they provide per root length colonized (Mensah et al. 2015; Treseder 2013).

Mycorrhizal nutrient transport per root length colonized, however, depends on an effective interplay between resource release (carbon from the host plant and nutrients from the AM fungus) into the mycorrhizal interface and the efficient uptake of resources by both partners from the interface (nutrients by the host plant and carbon by the fungal partner). If an essential component in these processes is interrupted, a successful symbiosis will not be established. For example, if *MtPt4*, the mycorrhiza-inducible phosphate transporter of *Medicago truncatula* is not expressed, the plant is unable to take up phosphate from the mycorrhizal interface, and arbuscules are prematurely degenerated (Javot et al. 2007). AM fungi can escape this premature degeneration when they are able to transfer nitrogen to their host (Javot et al. 2011). Similarly, if the transcript levels of *MST2*, a high-affinity monosaccharide transporter2 of the AM fungus, are reduced, arbuscules are malformed, and the expression of *MtPt4* is reduced (Helber et al. 2011). This indicates that the exchange processes of carbon for nutrients are linked and that both processes are critical for an efficient AM symbiosis. However, AM fungal species differ in their effect on *MtPt4* expression, and there are indications that the expression of this transporter is correlated to the fungal phosphate transport to the host (Fellbaum et al. 2014).

It has been suggested that a reciprocal reward system, in which carbon or nutrients are preferentially allocated to more beneficial partners, contributed to the evolutionary stability of the AM symbiosis (Kiers et al. 2011). Plants are able to distinguish between high-quality and low-quality AM fungi and allocate more carbon to fungi that provide more benefit (Kiers et al. 2011). Similarly, AM fungi transfer more phosphate or nitrogen to plants that are able to provide more carbon benefit (Fellbaum et al. 2012, 2014; Hammer et al. 2011; Bücking and Shachar-Hill 2005). However, resource exchange to multiple partners in the AM symbiosis is not an all-or-nothing process, and fungi still provide nutrients to low-quality hosts, and plants still invest carbon into fungal structures of low-quality fungal partners (Fellbaum et al. 2014; Kiers et al. 2011). This indicates that resource exchange in the AM symbiosis is controlled by biological market dynamics, and there are indications that the cost to nutrient benefit ratio varies among different host plant species (Walder et al. 2012).

The carbon transport from the host is an important trigger for phosphate and nitrogen transport and leads to changes in fungal gene expression and in the polyphosphate metabolism of the AM fungus (Fellbaum et al. 2012; Hammer et al. 2011; Kiers et al. 2011; Bücking and Shachar-Hill 2005). Polyphosphates are linear polymers of inorganic phosphate residues linked by phosphoanhydride bonds that play a role for the phosphate and nitrogen transport through the fungal hyphae to the host (Kikuchi et al. 2014; Cruz et al. 2007). It has been suggested that AM fungi control the nutrient release into the mycorrhizal interface by regulating polyphosphate formation and/or remobilization in the intraradical mycelium (Bücking and Shachar-Hill 2005; Takanishi et al. 2009; Ohtomo and Saito 2005). However, the mechanisms that control the resource exchange between partners are only poorly understood, and more research is needed to understand whether and how the processes in the mycorrhizal interface contribute to the functional diversity in the AM symbiosis.

12.4.3 *Effects of Microbial Communities on Functional Diversity*

Our current understanding of functional diversity in the AM symbiosis is primarily based on laboratory experiments and single plant/single fungus interactions. However, plant responses are substantially lower when the plant is colonized with one fungal partner compared with inoculations with multiple fungal species or a whole-soil microbial inoculum (with multiple AM fungal species and non-AM microorganisms) (Hoeksema et al. 2010). The higher plant responses after inoculations with multiple AM species could be the result of (1) a complementarity effect, in which different members of the AM community provide different benefits to the host (Hart and Reader 2002), (2) an establishment of a more beneficial AM fungal community, or (3) a competition effect, in which the competition among fungi for host plant carbon changes the cost to benefit ratio in favor of the host (Bücking et al. 2016). However, there are also reports in which negative effects of multi-fungal communities on host plant growth were observed. Violi et al. (2007), for example, demonstrated that the inoculation with multiple fungi reduced host plant growth and nutrient uptake compared to host plants that were inoculated only with one fungus. Gosling et al. (2016) also reported that an increase in AM fungal diversity does not lead to higher plant growth benefits. It has been suggested that functional complementarity or redundancy among different fungal species controls whether fungal communities act synergistically or antagonistically (Jansa et al. 2008; Maherali and Klironomos 2007). However, it has also been shown that the relatedness of coexisting AM fungi in a fungal community affects plant growth benefit. Communities of more closely related AM fungi were more likely to coexist, and also led to higher plant growth responses compared to communities of more distantly related AM fungi (Roger et al. 2013). This indicates that the general belief that host plant benefits will be higher with more diverse AM fungal communities is not necessarily applicable to all host plants and that more research is needed to better understand how AM fungal communities (in comparison to single inoculations) affect host plant growth and nutrient uptake.

12.5 Conclusions

There is an increasing interest to apply AM fungi in environmentally sustainable agriculture, but the application of AM fungi is still hindered by the high functional diversity in the AM symbiosis that makes host plant responses and/or benefits difficult to predict. Our current understanding of mycorrhizal host plant benefits is mainly based on observations with single AM fungal inoculations that provide only a limited insight into the application potential of specific fungi in certain environments and conditions, and/or for different host plants. In order to identify AM fungi that can provide specific benefits for their host plant, it is critical to better understand the interspecific genetic diversity, but also the intraspecific genetic diversity within AM fungal species and its effect on host plant benefit. In addition, more research is needed to identify AM fungal communities of specific host plants and under different environmental conditions and to characterize the contributions of individual AM fungi alone and in the community to host plant benefit. For the commercial application of AM fungi or AM fungal communities, it is also necessary to examine how specific communities can be established and whether introduced AM fungi are able to survive and to colonize host plants in the presence of an already existing AM fungal community (Rodriguez and Sanders 2015).

Acknowledgments The authors wish to acknowledge the financial support of the National Science Foundation (Award: 1051397), the South Dakota Soybean Research and Promotion Council, the USDA NIFA Hatch project SD00H423-12, and the South Dakota Oilseed Initiative.

References

Agnolucci M, Battini F, Cristani C, Giovannetti M (2015) Diverse bacterial communities are recruited on spores of different arbuscular mycorrhizal fungal isolates. Biol Fertil Soils 51:379–389. doi:10.1007/s00374-014-0989-5

Aira M, Gomez-Brandon M, Lazcano C, Baath E, Dominguez J (2010) Plant genotype strongly modifies the structure and growth of maize rhizosphere microbial communities. Soil Biol Biochem 42:2276–2281. doi:10.1016/j.soilbio.2010.08.029

Akiyama K, Hayashi H (2006) Strigolactones: chemical signals for fungal symbionts and parasitic weeds in plant roots. Ann Bot 97:925–931. doi:10.1093/aob/mcl063

Akiyama K, Matsuzaki KI, Hayashi H (2005) Plant sesquiterpenes induce hyphal branching in arbuscular mycorrhizal fungi. Nature 435:824–827. doi:10.1038/nature03608

Angelard C, Sanders IR (2011) Effect of segregation and genetic exchange on arbuscular mycorrhizal fungi in colonization of roots. New Physician 189:652–657. doi:10.1111/j.1469-8137.2010.03602.x

Angelard C, Colard A, Niculita-Hirzel H, Croll D, Sanders IR (2010) Segregation in a mycorrhizal fungus alters rice growth and symbiosis-specific gene transcription. Curr Biol 20:1216–1221. doi:10.1016/j.cub.2010.05.031

Antunes PM, Koch AM, Morton JB, Rillig MC, Klironomos JN (2011) Evidence for functional divergence in arbuscular mycorrhizal fungi from contrasting climatic origins. New Physician 189:507–514. doi:10.1111/j.1469-8137.2010.03480.x

Avio L, Pellegrino E, Bonari E, Giovannetti M (2006) Functional diversity of arbuscular mycorrhizal fungal isolates in relation to extraradical mycelial networks. New Physician 172:347–357. doi:10.1111/j.1469-8137.2006.01839.x

Avio L, Cristani C, Strani P, Giovannetti M (2009) Genetic and phenotypic diversity of geographically different isolates of *Glomus mosseae*. Can J Microbiol 55:242–253. doi:10.1139/w08-129

Battini F, Cristani C, Giovannetti M, Agnolucci M (2016) Multifunctionality and diversity of culturable bacterial communities strictly associated with spores of the plant beneficial symbiont *Rhizophagus intraradices*. Microbiol Res 183:68–79. doi:10.1016/j.micres.2015.11.012

Bentivenga SP, Bever JD, Morton JB (1997) Genetic variation of morphological characters within a single isolate of the endomycorrhizal fungus *Glomus clarum*. Am J Bot 84:1211–1216

Besserer A, Puech-Pagès V, Kiefer P, Gomez-Roldan V, Jauneau A, Roy S, Portais JC, Roux C, Bécard G, Séjalon-Delmas N (2006) Strigolactones stimulate arbuscular mycorrhizal fungi by activating mitochondria. PLoS Biol 4:1239–1247. doi:10.1371/journal.pbio.0040226

Börstler B, Raab PA, Thiery O, Morton JB, Redecker D (2008) Genetic diversity of the arbuscular mycorrhizal fungus *Glomus intraradices* as determined by mitochondrial large subunit rRNA gene sequences is considerably higher than previously expected. New Physician 180:452-465 doi:10.1111/j.1469-8137.2008.02574.x

Börstler B, Thiéry O, Sýkorova Z, Berner A, Redecker D (2010) Diversity of mitochondrial large subunit rDNA haplotypes of *Glomus intraradices* in two agricultural field experiments and two semi-natural grasslands. Mol Ecol 19:1497–1511. doi:10.1111/j.1365-294X.2010.04590.x

Breuillin-Sessoms F, Floss DS, Gomez SK, Pumplin N, Ding Y, Levesque-Tremblay V, Noar RD, Daniels DA, Bravo A, Eaglesham JB, Benedito VA, Udvardi MK, Harrison MJ (2015) Suppression of arbuscule degeneration in *Medicago truncatula* phosphate transporter 4 mutants is dependent on the ammonium transporter 2 family protein AMT2;3. Plant Cell. doi:10.1105/tpc.114.131144. pii:tpc.114.131144

Brundrett MC (2009) Mycorrhizal associations and other means of nutrition of vascular plants: understanding the global diversity of host plants by resolving conflicting information and developing reliable means of diagnosis. Plant Soil 320:37–77. doi:10.1007/s11104-008-9877-9

Bücking H, Shachar-Hill Y (2005) Phosphate uptake, transport and transfer by the arbuscular mycorrhizal fungus *Glomus intraradices* is stimulated by increased carbohydrate availability. New Physician 165:899–912. doi:10.1111/j.1469-8137.2004.01274.x

Bücking H, Abubaker J, Govindarajulu M, Tala M, Pfeffer PE, Nagahashi G, Lammers PJ, Shachar-Hill Y (2008) Root exudates stimulate the uptake and metabolism of organic carbon in germinating spores of *Glomus intraradices*. New Physician 180:684–695. doi:10.1111/j.1469-8137.2008.02590.x

Bücking H, Mensah JA, Fellbaum CR (2016) Common mycorrhizal networks and their effect on the bargaining power of the fungal partner in the arbuscular mycorrhizal symbiosis. Commun Integr Biol 9(1):e1107684. doi:10.1080/19420889.2015.1107684

Buée M, Rossignol M, Jauneau A, Ranjeva R, Bécard G (2000) The pre-symbiotic growth of arbuscular mycorrhizal fungi is induced by a branching factor partially purified from plant root exudates. Mol Plant Microbe Interact 13:693–698. doi:10.1094/MPMI.2000.13.6.693

Buée M, Maurice J-P, Zeller B, Andrianarisoa S, Ranger J, Courtecuisse R, Marçais B, Le Tacon F (2011) Influence of tree species on richness and diversity of epigeous fungal communities in a French temperate forest stand. Fungal Ecol 4:22–31. doi:10.1016/j.funeco.2010.07.003

Campagnac E, Khasa DP (2014) Relationship between genetic variability in *Rhizophagus irregularis* and tolerance to saline conditions. Mycorrhiza 24:121–129. doi:10.1007/s00572-013-0517-8

Corradi N, Bonfante P (2012) The arbuscular mycorrhizal symbiosis: origin and evolution of a beneficial plant infection. PLoS Pathog 8:3. doi:10.1371/journal.ppat.1002600

Croll D, Giovannetti M, Koch AM, Sbrana C, Ehinger M, Lammers PJ, Sanders IR (2009) Nonself vegetative fusion and genetic exchange in the arbuscular mycorrhizal fungus *Glomus intraradices*. New Physician 181:924–937. doi:10.1111/j.1469-8137.2008.02726.x

Cruz C, Egsgaard H, Trujillo C, Ambus P, Requena N, Martins-Loucao MA, Jakobsen I (2007) Enzymatic evidence for the key role of arginine in nitrogen translocation by arbuscular mycorrhizal fungi. Plant Physiol 144:782–792. doi:10.1104/pp.106.090522

Czaja LF, Hogekamp C, Lamm P, Maillet F, Martinez EA, Samain E, Dénarié J, Küster H, Hohnjec N (2012) Transcriptional responses toward diffusible signals from symbiotic microbes reveal *MtNFP*- and *MtDMI3*-dependent reprogramming of host gene expression by arbuscular mycorrhizal fungal lipochitooligosaccharides. Plant Physiol 159:1671–1685. doi:10.1104/pp.112.195990

Daubois L, Beaudet D, Hijri M, de la Providencia I (2016) Independent mitochondrial and nuclear exchanges arising in *Rhizophagus irregularis* crossed-isolates support the presence of a mitochondrial segregation mechanism. BMC Microbiol 16:12. doi:10.1186/s12866-016-0627-5 de Novais CB, Borges WL, Jesus ED, Saggin OJ, Siqueira JO (2014) Inter- and intraspecific functional variability of tropical arbuscular mycorrhizal fungi isolates colonizing corn plants. Appl Soil Ecol 76:78–86. doi:10.1016/j.apsoil.2013.12.010

Ehinger M, Koch AM, Sanders IR (2009) Changes in arbuscular mycorrhizal fungal phenotypes and genotypes in response to plant species identity and phosphorus concentration. New Physician 184:412–423. doi:10.1111/j.1469-8137.2009.02983.x

Ehinger MO, Croll D, Koch AM, Sanders IR (2012) Significant genetic and phenotypic changes arising from clonal growth of a single spore of an arbuscular mycorrhizal fungus over multiple generations. New Physician 196:853–861. doi:10.1111/j.1469-8137.2012.04278.x

Fellbaum CR, Gachomo EW, Beesetty Y, Choudhari S, Strahan GD, Pfeffer PE, Kiers ET, Bücking H (2012) Carbon availability triggers fungal nitrogen uptake and transport in the arbuscular mycorrhizal symbiosis. Proc Natl Acad Sci USA 109:2666–2671. doi:10.1073/pnas.1118650109

Fellbaum CR, Mensah JA, Cloos AJ, Strahan GD, Pfeffer PE, Kiers ET, Bücking H (2014) Fungal nutrient allocation in common mycelia networks is regulated by the carbon source strength of individual host plants. New Phytol 203:645–656. doi:10.1111/nph.12827

Fitter AH (2005) Darkness visible: reflections on underground ecology. J Ecol 93:231–243. doi:10.1111/j.0022-0477.2005.00990.x

Genre A, Chabaud M, Timmers T, Bonfante P, Barker DG (2005) Arbuscular mycorrhizal fungi elicit a novel intracellular apparatus in *Medicago truncatula* root epidermal cells before infection. Mol Plant Microbe Interact 17:3489–3499. doi:10.1105/tpc.105.035410

Gherbi H, Markmann K, Svistoonoff S, Estevan J, Autran D, Giczey G, Auguy F, Péret B, Laplaze L, Franche C, Parniske M, Bogusz D (2008) SymRK defines a common genetic basis for plant root endosymbioses with arbuscular mycorrhizal fungi, rhizobia, and Frankia bacteria. Proc Natl Acad Sci USA 105:4928–4932. doi:10.1073/pnas.0710618105

Ghignone S, Salvioli A, Anca I, Lumini E, Ortu G, Petiti L, Cruveiller S, Bianciotto V, Piffanelli P, Lanfranco L, Bonfante P (2012) The genome of the obligate endobacterium of an AM fungus reveals an interphylum network of nutritional interactions. ISME J 6:136–145. doi:10.1038/ismej.2011.110

Giovannetti M, Sbrana C, Strani P, Agnolucci M, Rinaudo V, Avio L (2003) Genetic diversity of isolates of *Glomus mosseae* from different geographic areas detected by vegetative compatibility testing and biochemical and molecular analysis. Appl Environ Microbiol 69:616–624. doi:10.1128/aem.69.1.616-624.2003

Gomez SK, Javot H, Deewatthanawong P, Torres-Jerez I, Tang Y, Blancaflor EB, Udvardi MK, Harrison MJ (2009) *Medicago truncatula* and *Glomus intraradices* gene expression in cortical cells harboring arbuscules in the arbuscular mycorrhizal symbiosis. BMC Plant Biol 9(10). doi:10.1186/1471-2229-9-10

Gosling P, Jones J, Bending GD (2016) Evidence for functional redundancy in arbuscular mycorrhizal fungi and implications for agroecosystem management. Mycorrhiza 26:77–83. doi:10.1007/s00572-015-0651-6

Grunwald U, Guo WB, Fischer K, Isayenkov S, Ludwig-Müller J, Hause B, Yan XL, Küster H, Franken P (2009) Overlapping expression patterns and differential transcript levels of phosphate transporter genes in arbuscular mycorrhizal, P_i-fertilised and phytohormone-treated *Medicago truncatula* roots. Planta 229:1023–1034. doi:10.1007/s00425-008-0877-z

Guether M, Neuhauser B, Balestrini R, Dynowski M, Ludewig U, Bonfante P (2009) A mycorrhizal-specific ammonium transporter from *Lotus japonicus* acquires nitrogen released by arbuscular mycorrhizal fungi. Plant Physiol 150:73–83. doi:10.1104/pp.109.136390

Halary S, Malik S-B, Lildhar L, Slamovits CH, Hijri M, Corradi N (2011) Conserved meiotic machinery in *Glomus* spp., a putatively ancient asexual fungal lineage. Genome Biol Evol 3:950–958. doi:10.1093/gbe/evr089

Halary S, Daubois L, Terrat Y, Ellenberger S, Wostemeyer J, Hijri M (2013) Mating type gene homologues and putative sex pheromone-sensing pathway in arbuscular mycorrhizal fungi, a presumably asexual plant root symbiont. PLoS One 8(11):12. doi:10.1371/journal.pone.0080729

Hammer EC, Pallon J, Wallander H, Olsson PA (2011) Tit for tat? A mycorrhizal fungus accumulates phosphorus under low plant carbon availability. FEMS Microbiol Ecol 76:236–244. doi:10.1111/j.1574-6941.2011.01043.x

Hart MM, Reader RJ (2002) Taxonomic basis for variation in the colonization strategy of arbuscular mycorrhizal fungi. New Physician 153:335–344. doi:10.1046/j.0028-646X.2001.00312.x

Helber N, Wippel K, Sauer N, Schaarschmidt S, Hause B, Requena N (2011) A versatile monosaccharide transporter that operates in the arbuscular mycorrhizal fungus *Glomus* sp. is crucial for the symbiotic relationship with plants. Plant Cell 23:3812–3823. doi:10.1105/tpc.111.089813

Henkel TW, Aime MC, Chin MML, Miller SL, Vilgalys R, Smith ME (2012) Ectomycorrhizal fungal sporocarp diversity and discovery of new taxa in Dicymbe monodominant forests of the Guiana Shield. Biodivers Conserv 21:2195–2220. doi:10.1007/s10531-011-0166-1

Hijri M, Sanders IR (2005) Low gene copy number shows that arbuscular mycorrhizal fungi inherit genetically different nuclei. Nature 433:160–163. doi:10.1038/nature03069

Hijri M, Hosny M, van Tuinen D, Dulieu H (1999) Intraspecific ITS polymorphism in *Scutellospora castanea* (Glomales, Zygomycota) is structured within multinucleate spores. Fungal Genet Biol 26:141–151. doi:10.1006/fgbi.1998.1112

Hoeksema JD, Chaudhary VB, Gehring CA, Johnson NC, Karst J, Koide RT, Pringle A, Zabinski C, Bever JD, Moore JC, Wilson GWT, Klironomos JN, Umbanhowar J (2010) A meta-analysis of context-dependency in plant response to inoculation with mycorrhizal fungi. Ecol Lett 13:394–407. doi:10.1111/j.1461-0248.2009.01430.x

Jakobsen I, Abbott LK, Robson AD (1992) External hyphae of vesicular-arbuscular mycorrhizal fungi associated with *Trifolium subterraneum* L. I. Spread of hyphae and phosphorus inflow into roots. New Physician 120:371–380. doi:10.1111/j.1469-8137.1992.tb01077.x

Jansa J, Smith FA, Smith SE (2008) Are there benefits of simultaneous root colonization by different arbuscular mycorrhizal fungi? New Phytol 177:779–789. doi:10.1111/j.1469-8137.2007.02294.x

Javot H, Penmetsa RV, Terzaghi N, Cook DR, Harrison MJ (2007) A *Medicago truncatula* phosphate transporter indispensable for the arbuscular mycorrhizal symbiosis. Proc Natl Acad Sci USA 104:1720–1725. doi:10.1073/pnas.0608136104

Javot H, Penmetsa RV, Breuillin F, Bhattarai KK, Noar RD, Gomez SK, Zhang Q, Cook DR, Harrison MJ (2011) *Medicago truncatula Mtpt4* mutants reveal a role for nitrogen in the regulation of arbuscule degeneration in arbuscular mycorrhizal symbiosis. Plant J 68:954–965

Johnson NC, Graham JH (2013) The continuum concept remains a useful framework for studying mycorrhizal functioning. Plant Soil 363:411–419. doi:10.1007/s11104-012-1406-1

Johnson NC, Graham JH, Smith FA (1997) Functioning of mycorrhizal associations along the mutualism-parasitism continuum. New Physician 135:575–585. doi:10.1046/j.1469-8137.1997.00729.x

Kiers ET, Duhamel M, Beesetty Y, Mensah JA, Franken O, Verbruggen E, Fellbaum CR, Kowalchuk GA, Hart MM, Bago A, Palmer TM, West SA, Vandenkoornhuyse P, Jansa J, Bücking H (2011) Reciprocal rewards stabilize cooperation in the mycorrhizal symbiosis. Science 333:880–882. doi:10.1126/science.1208473

Kikuchi Y, Hijikata N, Yokoyama K, Ohtomo R, Handa Y, Kawaguchi M, Saito K, Ezawa T (2014) Polyphosphate accumulation is driven by transcriptome alterations that lead to near-synchronous and near-equivalent uptake of inorganic cations in an arbuscular mycorrhizal fungus. New Physician 204:638–649. doi:10.1111/nph.12937

Kistner C, Winzer T, Pitzschke A, Mulder L, Sato S, Kaneko T, Tabata S, Sandal N, Stougaard J, Webb KJ, Szczyglowski K, Parniske M (2005) Seven *Lotus japonicus* genes required for transcriptional reprogramming of the root during fungal and bacterial symbiosis. Plant Cell 17:2217–2229. doi:10.1105/tpc.105.032714

Kivlin SN, Hawkes CV, Treseder KK (2011) Global diversity and distribution of arbuscular mycorrhizal fungi. Soil Biol Biochem 43:2294–2303. doi:10.1016/j.soilbio.2011.07.012
Klironomos JN (2000) Host-specificity and functional diversity among arbuscular mycorrhizal fungi. In: Bell CR, Brylinsky M, Johnson-Green P (eds) Microbial biosystems: new frontiers. Proceedings of the 8th International Symposium on microbial ecology. Atlantic Canada Society for Microbial Ecology, Halifax, Canada, pp 845–851
Klironomos JN (2003) Variation in plant response to native and exotic arbuscular mycorrhizal fungi. Ecology 84:2292–2301. doi:10.1890/02-0413
Koch AM, Antunes PM, Maherali H, Hart MM, Klironomos JN (2017) Evolutionary assymmetry in the arbuscular mycorrhizal symbiosis: conservatism in fungal morphology does not pretict host plant growth. New Phytol. doi:10.1111/nph.1446
Koch AM, Kuhn G, Fontanillas P, Fumagalli L, Goudet J, Sanders IR (2004) High genetic variability and low local diversity in a population of arbuscular mycorrhizal fungi. Proc Natl Acad Sci USA 101:2369–2374. doi:10.1073/pnas.0306441101
Koch AM, Croll D, Sanders IR (2006) Genetic variability in a population of arbuscular mycorrhizal fungi causes variation in plant growth. Ecol Lett 9:103–110. doi:10.1111/j.1461-0248.2005.00853.x
Kuhn G, Hijri M, Sanders IR (2001) Evidence for the evolution of multiple genomes in arbuscular mycorrhizal fungi. Nature 414:745–748. doi:10.1038/414745a
Lekberg Y, Gibbons SM, Rosendahl S (2014) Will different OTU delineation methods change interpretation of arbuscular mycorrhizal fungal community patterns? New Physician 202:1101–1104. doi:10.1111/nph.12758
Maherali H, Klironomos JN (2007) Influence of phylogeny on fungal community assembly and ecosystem functioning. Science 316(5832):1746–1748. doi:10.1126/science.1143082
Maillet F, Poinsot V, André O, Puech-Pagès V, Haouy A, Gueunier M, Cromer L, Giraudet D, Formey D, Niebel A, Martinez EA, Driguez H, Bécard G, Dénarié J (2011) Fungal lipochitooligosaccharide symbiotic signals in arbuscular mycorrhiza. Nature 469:58–63
Marleau J, Dalpé Y, St-Arnaud M, Hijri M (2011) Spore development and nuclear inheritance in arbuscular mycorrhizal fungi. BMC Evol Biol 11:1–11. doi:10.1186/1471-2148-11-51
Mayr E (1942) Systematics and the origin of species, from the viewpoint of a zoologist. Harvard University Press, London
Mayr E (2000) The biological species concept. In: Wheeler QD, Meier R (eds) Species concepts and phylogenetic theory: a debate. Columbia University Press, New York, pp 17–29
Mensah JA, Koch AM, Antunes PM, Hart MM, Kiers ET, Bücking H (2015) High functional diversity within arbuscular mycorrhizal fungal species is associated with differences in phosphate and nitrogen uptake and fungal phosphate metabolism. Mycorrhiza 7:533–546. doi:10.1007/s00572-015-0631-x
Merryweather J, Fitter A (1998) The arbuscular mycorrhizal fungi of hyacinthoides non-scripta. II. Seasonal and spatial patterns of fungal populations. New Phytol 138:131–142
Mondo SJ, Toomer KH, Morton JB, Lekberg Y, Pawlowska TE (2012) Evolutionary stability in a 400-million-year-old heritable facultative mutualism. Evolution 66:2564–2576. doi:10.1111/j.1558-5646.2012.01611.x
Moore D, Robson GD, Trinci AP (2011) 21st century guidebook to fungi. Cambridge University Press, Cambridge
Morton JB (1985) Variation in mycorrhizal and spore morphology of *Glomus occultum* and *Glomus diaphanum* as influenced by plant host and soil environment. Mycologia 77:192–204. doi:10.2307/3793068
Morton JB, Benny GL (1990) Revised classification of arbuscular mycorrhizal fungi (Zygomycetes): a new order, Glomales, two new suborders, Glomineae and Gigasporineae, and two new families, Acaulosporaceae and Gigasporaceae, with an emendation of Glomaceae. Mycotaxon 37:471–491

Mosse B, Bowen G (1968) The distribution of Endogone spores in some Australian and New Zealand soils, and in an experimental field soil at Rothamsted. Trans Br Mycol Soc 51:485–492. doi:10.1016/S0007-1536(68)80015-4

Munkvold L, Kjoller R, Vestberg M, Rosendahl S, Jakobsen I (2004) High functional diversity within species of arbuscular mycorrhizal fungi. New Physician 164:357–364. doi:10.1111/j.1469-8137.2004.01169.x

Naito M, Pawlowska TE (2016) Defying Muller's ratchet: ancient heritable endobacteria escape extinction through retention of recombination and genome plasticity. mBIO 7(3):e02057–e02015. doi:10.1128/mBio.02057-15

Naito M, Morton JB, Pawlowska TE (2015) Minimal genomes of mycoplasma-related endobacteria are plastic and contain host-derived genes for sustained life within Glomeromycota. Proc Natl Acad Sci USA 112:7791–7796. doi:10.1073/pnas.1501676112

Newton A, Haigh J (1998) Diversity of ectomycorrhizal fungi in Britain: a test of the species–area relationship, and the role of host specificity. New Physician 138:619–627. doi:10.1046/j.1469-8137.1998.00143.x

Ohtomo R, Saito M (2005) Polyphosphate dynamics in mycorrhizal roots during colonization of an arbuscular mycorrhizal fungus. New Physician 167:571–578. doi:10.1111/j.1469-8137.2005.01425.x

Öpik M, Moora M, Zobel M, Saks U, Wheatley R, Wright F, Daniell T (2008) High diversity of arbuscular mycorrhizal fungi in a boreal herb-rich coniferous forest. New Physician 179:867–876. doi:10.1111/j.1469-8137.2008.02515.x

Öpik M, Metsis M, Daniell TJ, Zobel M, Moora M (2009) Large-scale parallel 454 sequencing reveals host ecological group specificity of arbuscular mycorrhizal fungi in a boreonemoral forest. New Physician 184:424–437. doi:10.1111/j.1469-8137.2009.02920.x

Öpik M, Zobel M, Cantero JJ, Davison J, Facelli JM, Hiiesalu I, Jairus T, Kalwij JM, Koorem K, Leal ME (2013) Global sampling of plant roots expands the described molecular diversity of arbuscular mycorrhizal fungi. Mycorrhiza 23:411–430. doi:10.1007/s00572-013-0482-2

Parniske M (2008) Arbuscular mycorrhiza: the mother of plant root endosymbioses. Nat Rev Microbiol 6:763–775. doi:10.1038/nrmicro1987

Pawlowska TE, Taylor JW (2004) Organization of genetic variation in individuals of arbuscular mycorrhizal fungi. Nature 427:733–737. doi:10.1038/nature02290

Peng S, Eissenstat DM, Graham JH, Williams K, Hodge NC (1993) Growth depression in mycorrhizal citrus at high-phosphorus supply. Plant Physiol 101:1063–1071

Pepe A, Giovannetti M, Sbrana C (2016) Different levels of hyphal self-incompatibility modulate interconnectedness of mycorrhizal networks in three arbuscular mycorrhizal fungi within the Glomeraceae. Mycorrhiza 26:325–332. doi:10.1007/s00572-015-0671-2

Powell JR, Parrent JL, Hart MM, Klironomos JN, Rillig MC, Maherali H (2009) Phylogenetic trait conservatism and the evolution of functional trade-offs in arbuscular mycorrhizal fungi. Proc R Soc Biol Sci 276:4237–4245. doi:10.1098/rspb.2009.1015

Pressel S, Bidartondo MI, Ligrone R, Duckett JG (2010) Fungal symbioses in bryophytes: new insights in the twenty first century. Phytotaxa 9:238-253 doi: 10.11646/phytotaxa.9.1.13

Purin S, Morton J (2013) Anastomosis behavior differs between asymbiotic and symbiotic hyphae of *Rhizophagus clarus*. Mycologia:12-135. doi: 10.3852/12-135

Redecker D, Schüßler A, Stockinger H, Stürmer SL, Morton JB, Walker C (2013) An evidence-based consensus for the classification of arbuscular mycorrhizal fungi (Glomeromycota). Mycorrhiza 23:515–531. doi:10.1007/s00572-013-0486-y

Rinaldi AC, Comandini O, Kuyper TW (2008) Ectomycorrhizal fungal diversity: separating the wheat from the chaff. Fungal Divers 33:1–45

Rodriguez A, Sanders IR (2015) The role of community and population ecology in applying mycorrhizal fungi for improved food security. ISME J 9:1053–1061. doi:10.1038/ismej.2014.207

Roger A, Colard A, Angelard C, Sanders IR (2013) Relatedness among arbuscular mycorrhizal fungi drives plant growth and intraspecific fungal coexistence. ISME J 7:2137–2146. doi:10.1038/ismej.2013.112

Rosendahl S (2008) Communities, populations and individuals of arbuscular mycorrhizal fungi. New Physician 178:253–266. doi:10.1111/j.1469-8137.2008.02378.x

Salvioli A, Chiapello M, Fontaine J, Hadj-Sahraoui AL, Grandmougin-Ferjani A, Lanfranco L, Bonfante P (2010) Endobacteria affect the metabolic profile of their host *Gigaspora margarita*, an arbuscular mycorrhizal fungus. Environ Microbiol 12:2083–2095. doi:10.1111/j.1462-2920.2010.02246.x

Salvioli A, Ghignone S, Novero M, Navazio L, Venice F, Bagnaresi P, Bonfante P (2016) Symbiosis with an endobacterium increases the fitness of a mycorrhizal fungus, raising its bioenergetic potential. ISME J 10:130–144. doi:10.1038/ismej.2015.91

Sanders IR (1999) No sex please, we're fungi. Nature 399:737–739. doi:10.1038/21544

Sanders IR (2002) Ecology and evolution of multigenomic arbuscular mycorrhizal fungi. Am Nat 160:128–141. doi:10.1086/342085

Sanders IR (2011) Fungal sex: meiosis machinery in ancient symbiotic fungi. Curr Biol 21:896–897. doi:10.1016/j.cub.2011.09.021

Schüßler A, Walker C (2010) The Glomeromycota. A species list with new families and new genera. In: Libraries at the Royal Botanic Garden Edinburgh, The Royal Botanic Garden Kew, Botanische Staatssammlung Munich, and Oregon State University, Gloucester

Smith SE, Read D (2008) Mycorrhizal symbiosis, 3rd edn. Academic Press, New York

Smith SE, Smith FA (2011) Roles of arbuscular mycorrhizas in plant nutrition and growth: new paradigms from cellular to ecosystem scales. Annu Rev Plant Biol 62:227–250. doi:10.1146/annurev-arplant-042110-103846

Smith FA, Smith SE (2013) How useful is the mutualism-parasitism continuum of arbuscular mycorrhizal functioning? Plant Soil 363:7–18. doi:10.1007/s11104-012-1583-y

Smith SE, Jakobsen I, Grønlund M, Smith FA (2011) Roles of arbuscular mycorrhizas in plant phosphorus nutrition: interactions between pathways of phosphorus uptake in arbuscular mycorrhizal roots have important implications for understanding and manipulating plant phosphorus acquisition. Plant Physiol 156:1050–1057. doi:10.1104/pp.111.174581

Stockinger H, Walker C, Schussler A (2009) '*Glomus intraradices* DAOM197198', a model fungus in arbuscular mycorrhiza research, is not *Glomus intraradices*. New Physician 183 (4):1176–1187. doi:10.1111/j.1469-8137.2009.02874.x

Stukenbrock EH, Rosendahl S (2005) Clonal diversity and population genetic structure of arbuscular mycorrhizal fungi (*Glomus* spp.) studied by multilocus genotyping of single spores. Mol Ecol 14:743–752. doi:10.1111/j.1365-294X.2005.02453.x

Takanishi I, Ohtomo R, Hayatsu M, Saito M (2009) Short-chain polyphosphate in arbuscular mycorrhizal roots colonized by *Glomus* spp.: a possible phosphate pool for host plants. Soil Biol Biochem 41:1571–1573. doi:10.1016/j.soilbio.2009.04.002

Tamasloukht M, Séjalon-Delmas N, Kluever A, Jauneau A, Roux C, Bécard G, Franken P (2003) Root factors induce mitochondrial related gene expression and fungal respiration during developmental switch from asymbiosis to presymbiosis in the arbuscular mycorrhizal fungus *Gigaspora rosea*. Plant Physiol 131:1468–1478. doi:10.1104/pp.012898

Tamasloukht M, Waschke A, Franken P (2007) Root exudate-stimulated RNA accumulation in the arbuscular mycorrhizal fungus *Gigaspora rosea*. Soil Biol Biochem 39:1824–1827. doi:10.1016/j.soilbio.2007.01.031

Taylor TN, Remy W, Hass H, Kerp H (1995) Fossil arbuscular mycorrhizae from the early devonian. Mycologia 87:560–573. doi:10.2307/3760776

Tedersoo L, May TW, Smith ME (2010) Ectomycorrhizal lifestyle in fungi: global diversity, distribution, and evolution of phylogenetic lineages. Mycorrhiza 20:217–263. doi:10.1007/s00572-009-0274-x

Tisserant E, Kohler A, Dozolme-Seddas P, Balestrini R, Benabdellah K, Colard A, Croll D, Da Silva C, Gomez SK, Koul R, Ferrol N, Fiorilli V, Formey D, Franken P, Helber N, Hijri M,

Lanfranco L, Lindquist E, Liu Y, Malbreil M, Morin E, Poulain J, Shapiro H, van Tuinen D, Waschke A, Azcón-Aguilar C, Bécard G, Bonfante P, Harrison MJ, Küster H, Lammers P, Paszkowski U, Requena N, Rensing SA, Roux C, Sanders IR, Shachar-Hill Y, Tuskan G, Young JPW, Gianinazzi-Pearson V, Martin F (2012) The transcriptome of the arbuscular mycorrhizal fungus *Glomus intraradices* (DAOM 197198) reveals functional tradeoffs in an obligate symbiont. New Physician 193:755–769. doi:10.1111/j.1469-8137.2011.03948.x
Toomer KH, Chen XH, Naito M, Mondo SJ, den Bakker HC, VanKuren NW, Lekberg Y, Morton JB, Pawlowska TE (2015) Molecular evolution patterns reveal life history features of mycoplasma-related endobacteria associated with arbuscular mycorrhizal fungi. Mol Ecol 24:3485–3500. doi:10.1111/mec.13250
Torrecillas E, Alguacil MM, Roldán A (2012) Host preferences of arbuscular mycorrhizal fungi colonizing annual herbaceous plant species in semiarid Mediterranean Prairies. Appl Environ Microbiol 78:6180–6186. doi:10.1128/AEM.01287-12
Treseder KK (2013) The extent of mycorrhizal colonization of roots and its influence on plant growth and phosphorus content. Plant Soil 371:1–13. doi:10.1007/s11104-013-1681-5
Tripathi P, Rabara RC, Reese RN, Miller MA, Rohila JS, Subramanian S, Shen QJ, Morandi D, Bücking H, Shulaev V, Rushton PJ (2016) A toolbox of genes, proteins, metabolites and promoters for improving drought tolerance in soybean includes the metabolite coumestrol and stomatal development genes. BMC Genomics 17:1–22. doi:10.1186/s12864-016-2420-0
Valyi K, Rillig MC, Hempel S (2015) Land-use intensity and host plant identity interactively shape communities of arbuscular mycorrhizal fungi in roots of grassland plants. New Physician 205:1577–1586. doi:10.1111/nph.13236
Vannini C, Carpentieri A, Salvioli A, Novero M, Marsoni M, Testa L, de Pinto MC, Amoresano A, Ortolani F, Bracale M, Bonfante P (2016) An interdomain network: the endobacterium of a mycorrhizal fungus promotes antioxidative responses in both fungal and plant hosts. New Phytol 211:265–275. doi:10.1111/nph.13895
Violi HA, Treseder KK, Menge JA, Wright SF, Lovatt CJ (2007) Density dependence and interspecific interactions between arbuscular mycorrhizal fungi mediated plant growth, glomalin production, and sporulation. Can J Bot 85:63–75. doi:10.1139/b06-151
Walder F, Niemann H, Natarajan M, Lehmann MF, Boller T, Wiemken A (2012) Mycorrhizal networks: common goods of plants shared under unequal terms of trade. Plant Physiol 159:789–797. doi:10.1104/pp.112.195727
Wang X, Bücking H (2015) Arbuscular mycorrhizal growth and phosphate uptake responses are fungal specific but do not differ between a soybean genotype with a high and a low phosphate acquisition efficiency. Ann Bot. doi:10.1093/aob/mcw074. pii:mcw074
Wang B, Qiu Y-L (2006) Phylogenetic distribution and evolution of mycorrhizae in land plants. Mycorrhiza 16:2. doi:10.1007/s00572-005-0033-6 99-363
Wright DP, Read DJ, Scholes JD (1998) Mycorrhizal sink strength influences whole plant carbon balance of *Trifolium repens* L. Plant Cell Environ 21:881–891. doi:10.1046/j.1365-3040.1998.00351.x

Chapter 13
Arbuscular Mycorrhizal Fungi and Dark Septate Endophytes in Grapevine: The Potential for Sustainable Viticulture?

M. Likar and M. Regvar

Abstract Viticulture is an important agronomic sector that has the potential to greatly benefit by improvements in our understanding of grapevine cultivation. Although conventional viticulture relies to a great extent on pesticide and fertilizer application, more sustainable approaches involve management practices that favor plant–fungus interactions that have positive effects on the nutritional quality of the grapes and reduce production costs (i.e., of pesticides and fertilizers) and thus reduce the negative effects on the environment. Fungal endophytes that colonize grapevines belong to different taxa, with the majority of reports focusing on fungi that form arbuscular mycorrhizal associations. These fungal endophytes have been demonstrated to confer beneficial growth and nutrition effects to their plant hosts via improved exploitation of the substrate and improved tolerance of the grapevine to abiotic and biotic stresses. Here, we review current knowledge on the importance and potential of these diverse fungal groups for grapevine production and expose the gaps in our understanding of possible functions of fungal groups that are currently little studied. In addition, we underline the effects of sustainable agricultural practices on fungal communities, to boost the progress in different viticultural techniques on the interactions between fungal endophytes and grapevines.

13.1 Introduction

The majority of terrestrial ecosystems are dominated by plants that form associations with mutualistic fungi, such as arbuscular mycorrhizal fungi (AMF) (Smith and Read 2008) and dark septate endophytes (DSEs). There is an abundance of literature devoted to AMF colonization of grapevine roots (Karagiannidis and Nikolaou 1999; Likar et al. 2013; Radić et al. 2012, 2014; Schreiner and Mihara 2009; Schubert and Cravero 1985). Grapevine roots are also colonized by diverse

M. Likar (✉) • M. Regvar
Biotechnical Faculty, Department of Biology, University of Ljubljana, Večna pot 111, 1000 Ljubljana, Slovenia
e-mail: matevz.likar@bf.uni-lj.si

A. Varma et al. (eds.), *Mycorrhiza - Function, Diversity, State of the Art*,
DOI 10.1007/978-3-319-53064-2_13

groups of ascomycetes, basidiomycetes, and chytridiomycetes (Likar et al. 2013). The status of colonization of grapevine roots with DSEs remains elusive.

13.2 Arbuscular Mycorrhizal Fungi of Grapevines

The AMF in vineyard soils and in grapevine roots are dominated by *Glomus* s. l. spp.: *Rhizophagus intraradices/Rhizophagus irregularis*[1], *Diversispora epigaea*, *Septoglomus viscosum*, and other unidentified representatives of *Glomus* s. l. (Balestrini et al. 2010; Cheng and Baumgartner 2004; Likar et al. 2013; Lumini et al. 2010; Menge et al. 1983; Nappi et al. 1985; Radić et al. 2012, 2014; Schreiner and Mihara 2009; Schubert and Cravero 1985). In addition, the *Paraglonimus/ Archaeospora* and *Scutellospora* spp. have been reported as colonizers of grapevine roots (Schreiner and Mihara 2009). A large-scale study of vineyards in Burgundy (France) showed that six molecular taxa of AMF (*Claroideoglomus* sp.; *Glomeraceae* sp. 1, 4, and 5; *Glomerales* sp.; and *Rhizophagus irregularis*) represented 77% of all of the sequences obtained, which indicated relatively low diversity (Bouffaud et al. 2016). A screening study in production vineyards along an even longer stretch along the east Adriatic coast (*ca.* 500 km) reported *Sclerocystis sinuosa* (= *Glomus sinuosum*) as the dominant AMF species observed on grapevine roots (Likar et al. 2013). This is an AMF species that is typically found in undisturbed habitats (Borstler et al. 2006; Oehl et al. 2005; Rosendahl and Stukenbrock 2004) and rarely in low-input agroecosystems (Li et al. 2007). In addition, Likar et al. (2013) reported *Glomus indicum* as a colonizer of grapevine roots. *G. indicum* was described recently by Blaszkowski et al. (2010) and is known to colonize various hosts, including *Vitis berlandieri*. As such it might be a common colonizer of *Vitis* rootstock based on *V. berlandieri* (including SO4 and 5BB). Similarly, several species from the genera *Claroideoglomus*, *Funneliformis*, *Glomus* s. str., and other AMF genera have also been identified on roots of the wild grapevine *V. vinifera* ssp. *sylvestris* (Ocete et al. 2015). Studies on wild grapevine have indicated high diversity of AMF communities on their roots, thus stressing the important quality and ecological value of AMF colonization, and, as a consequence, also the potential for improvement of cultivation techniques used in production vineyards.

However, the picture of the diversity of AMF that colonizes grapevine and the factors that affect the structure and changes in AMF communities in different parts of the world is still far from complete. Likar et al. (2013) showed that the AMF community along a 500-km-long stretch of karst slowly changed from north to south (Fig. 13.1). In addition to communities that were similar along the entire

[1]In the past, the name *G. intraradices* was in the majority of cases misapplied to a present species *R. irregularis*, and not to *R. intraradices*, which has a very limited distribution (Stockinger et al. 2009).

transect, some southern locations showed no resemblance to the northern locations, which suggested different influences of the more temperate and more Mediterranean parts of the range. This finding, however, contrasts with the reported stability of fungal diversity in vineyards (Schreiner and Mihara 2009), although it is in line with the dependency on high soil moisture reported by the same authors. One of the reasons might be the geographical scales of these studies, which for Likar et al. (2013) was along the Adriatic coast. This study was sufficient to demonstrate that although AMF from grapevine roots are dominated by Glomerales, there is a shift to members of Diversisporales in sandy vineyard soils, which is in line with the findings of Balestrini et al. (2010). Some of the changes in the composition of fungal endophyte communities, however, have been attributed to soil phosphorous (P) concentrations, again confirming that AMF communities are highly influenced by soil characteristics (Balestrini et al. 2010; Lin et al. 2012; Liu et al. 2016; Schreiner and Mihara 2009).

13.3 Dark Septate Endophytes and Other Non-arbuscular Mycorrhizal Endophytes of Grapevine

As well as AMF strains, fungi that form septate hyphae with melanized cell walls are common colonizers of plant roots in natural environments. These fungi are referred to as DSEs, and they represent a heterogeneous group of ascomycetes that occur across a wide range of terrestrial ecosystems, although they are most numerous in extreme habitats (Mandyam and Jumpponen 2005). Among the DSEs, members of the order Helotiales are commonly observed, although frequently these isolates are not fully identified or they are grouped with the anamorphic genera *Phialocephala*, *Rhynchosporium*, and *Phialophora*. This last not only includes many genera within Helotiales (Hennebert and Bellemere 1979), but also the pyrenomycetous genera *Coniochaeta* (Coniochaetaceae) (Schol-Schwarz 1970), *Lasiosphaeria* (Lasiosphaeriaceae, Gams and Holubova-Jechova 1976), and *Gaeumannomyces* (Diaporthaceae, Walker 1980). In a study performed by Knapp et al. (2012) in semiarid sandy grasslands with wood steppe patches on the Great Hungarian Plain, they identified members belonging to the additional orders of Pleosporales, Hypocreales, Eurotiales, and Xylariales.

The presence of DSEs on grapevine is still not fully confirmed, and the majority of studies have reported the presence of pathogens (e.g., *Pestalopsis*), the genera *Fusarium* and *Alternaria*, and different yeast (e.g., *Aureobasidium pullulans*). Despite the increasing number of studies on DSEs, the bulk of our knowledge on DSE diversity and taxonomy is still restricted to the colder and temperate zones of our planet. Furthermore, although DSEs are cosmopolitan (for a review on their specific habitats, see Mandyam and Jumpponen 2005), and some DSE taxa appear to reoccur across unrelated biomes, studies performed in Mediterranean regions have shown that the DSE taxa that dominate the colder regions are not present

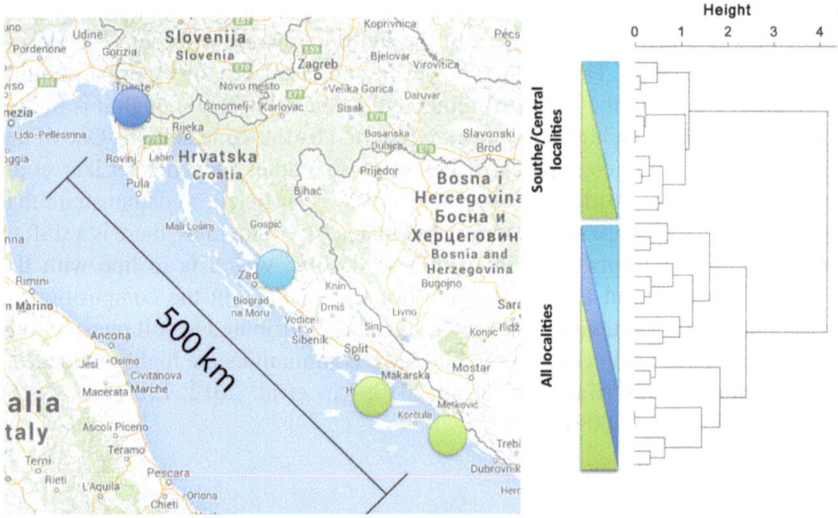

Fig. 13.1 Communities of arbuscular mycorrhizal fungi in production vineyards along a 500-km stretch of the Adriatic coast. Color-code: wine regions in western Slovenia (*blue*), middle Dalmatia (*cyan*), and south Dalmatia (*green*). *Right*: Cluster analysis based on Bray–Curtis dissimilarity index and created using unweighted pair-group analysis with arithmetic average linking (Likar et al. 2013)

(Girlanda et al. 2006). These findings suggest that there is at least some environmental selection in the distribution of DSEs. This might result in completely different DSE assemblages in vineyards, which are by nature based in warmer climates. To our knowledge, there have also been no reports on root colonization of grapevine with hyaline hyphal endophytes, which are otherwise frequent plant colonizers (Varma et al. 1999, 2012). These might be particularly interesting for the Mediterranean basin, as aquatic hyphomycetes have been reported to be particularly frequent root colonizers of plants in riparian and coastal ecosystems (Porras-Alfaro and Bayman 2011).

13.4 Impact of Fungal Endophytes on the Physiology of Grapevine

Arbuscular mycorrhizal fungi have an increasingly important role in vineyard production systems, as many vineyards receive little water and are planted on not very fertile soils (Schreiner 2005). Inoculation of grapevine with AMF has been associated with increased growth (Linderman and Davis 2001; Schubert et al. 1988) and improved drought tolerance (Nikolaou et al. 2003; Schreiner 2003) and nutrient uptake (Schreiner 2007), in comparison with non-inoculated grapevine. As such,

AMF represent an integral and important component of these vineyard ecosystems and might have significant applications for sustainable agricultural ecosystems.

13.4.1 Improved Mineral Nutrition of Grapevine

Studies under controlled conditions have shown that AMF-inoculated grapevine have higher shoot and root weights in P-sufficient (Biricolti et al. 1997; Schubert et al. 1988) and P-limiting (Linderman and Davis 2001) soils, higher tissue P concentrations in P-sufficient soil (Biricolti et al. 1997), and more compact, highly branched, roots than non-inoculated grapevines (Schellenbaum et al. 1991). In addition to P and nitrogen (N), other chemical elements are also essential for grapevine growth, such as potassium (K), magnesium (Mg), zinc (Zn), and boron (B). In particular, K and Fe deficiency can occur in sensitive rootstock grown on calcareous soils (Havlin et al. 1990). K-deficient grapevines are more sensitive to drought (Reynolds 2010), which can be exacerbated by the low water capacity of the calcareous soils that develop in karst regions.

A study on the mineral composition of grapevine in production vineyards along the east Adriatic coast revealed that differences in the macronutrients K and P and the micronutrient Mn explained two-thirds of the total variance in the mineral composition of the grapevine leaves (Likar et al. 2015). Twenty-six to thirty-four percent of the variance of these three elements was explained by the abiotic and biotic soil parameters, with soil concentrations of K, Fe, and Cu, organic matter content, and vesicular colonization having the strongest effects on the mineral composition of the grapevines. This study confirmed that AMF are one of the important factors that affect the mineral composition of grapevine. Due to their specific root architecture, with low root density and coarse roots (Smart et al. 2006), grapevines can be strongly dependent on beneficial fungal endophytes (Smith and Read 2008). AMF mycelia increase the exploitable soil volume due to the small diameter of the hyphae, which allows access to soil pores that would otherwise not be explorable (Smith and Read 2008), and this improves the mineral nutrition of the grapevines. Also, Maherali (2014) observed no correlations between root architecture and mycorrhizal dependency in a meta-analysis of 12 studies across 196 experimental trials on AMF–plant interactions. In line with these results, it appears that grapevines still retain strong control over their colonization levels, as reduced colonization of grapevines was observed with high soil fertility and/or high nutrient status (reviewed by Schreiner 2005).

The involvement of AMF in plant nutrition has been studied in depth; however, the knowledge of DSEs is very limited. The first reports on DSE isolates showed that DSEs can promote the uptake of N and P into plants (Haselwandter and Read 1980; Jumpponen and Trappe 1998), although their effects on the overall plant biomass appeared to be dependent on host–symbiont association and soil nutrient status (Fernando and Currah 1996; Jumpponen and Trappe 1998). With further studies, a wider range of effects have been reported, from negative or negligible to positive (Jumpponen 2001). In contrast, a more recent meta-analysis failed to confirm any negative effects of DSEs on plant performance (Newsham 2011) and confirmed DSE-dependent improvements in shoot and root biomass, and shoot N and P content, with up to 138% increase in the biomass. In addition, Mullen et al. (1998) proposed that N uptake can occur through DSEs early in the season, before any root growth or AMF colonization takes place.

13.4.2 Non-nutritional Effects on the Physiology of Grapevine

Beneficial effects of fungal endophytes on plant hosts can also result from altered metabolic pathways. Secondary phenolic metabolites are an important part of plant defense mechanisms and most are also known to be beneficial to human health. Increases in their concentrations might therefore improve the resistance of grapevine and at the same time the quality of the grapes produced. AMF can influence the levels of phenolics in plant roots and shoots (Ulrichs et al. 2008; Regvar et al. 2012), including grapevine (Eftekhari et al. 2012). Eftekhari et al. (2012) reported that inoculation with *Funneliformis mosseae* (= *G. mosseae*) increased leaf quercetin content two- to sixfold, which might improve the resistance of grapevine to different plant pathogens. In addition, AMF were shown to affect induced plant defenses—systemic acquired resistance (salicylic acid dependent; Gallou et al. 2011) and induced systemic resistance (jasmonic acid dependent; Li et al. 2010; Hao et al. 2012). These protective effects appear to be active in grapevine through the latter pathway that is mediated by jasmonic acid, and it was shown to be active against the fungi *Armillaria melea* (Nogales et al. 2009) and *Cylindrocarpon macrodidymum* (Petit and Gubler 2006) and also against the ectoparasitic nematode *Xiphinema index* (Hao et al. 2012).

Fungal endophytes themselves can also produce secondary metabolites with beneficial effects. Among the non-AMF endophytes identified on grapevine roots in different studies (Pancher et al. 2012; Radić et al. 2014), *Fusarium* and *Penicillium* isolates are known to produce secondary metabolites that are antagonistic to soilborne phytopathogenic fungi (Nunez-Trujillo et al. 2012; De Stefano et al. 1999). Furthermore, *Cladosporium* sp., which can produce the secondary metabolite with antimicrobial potential known as brefeldin A (Wang et al. 2007), and *Aureobasidium pullulans*, which was confirmed to inhibit various grapevine pathogens (de Felice et al. 2008; Schmid et al. 2011), were also observed.

Bioprotection of plants by fungal endophytes might therefore provide an innovative protection strategy that is worth considering as part of overall vineyard management.

13.5 Endophyte-Related Stress Alleviation in Grapevine

In addition to biotic stress, fungal endophytes can help to alleviate abiotic stress. AMF have been reported to counterbalance adverse effects of salinity stress and to thereby increase plant growth (Giri et al. 2003; Giri and Mukerji 2004). In a pilot experiment on grapevine, *Rhizophagus fasciculatus* (= *Glomus fasciculatum*) increased the growth of grape rootstock, further improving the tolerance of the rootstock itself (Belew et al. 2010). AMF are believed to improve the supply of mineral nutrients to plants in salt-stressed soil (e.g., enhanced acquisition of P, N, Mg, Ca) and to promote maintenance of the K^+/Na^+ ratio and changes in biochemical composition (e.g., accumulation of proline), plant physiology, gene expression, and ultrastructure (Evelin et al. 2009).

Also, even if vineyard soils are not polluted by heavy metals to an extent that would push grapevine plants toward the limit of tolerance, fungal endophytes can still improve the overall quality of the grapes and their product, by limiting the uptake of heavy metals (Jiang et al. 2016; Likar and Regvar 2013). Again, these beneficial effects are a combination of maintained, or even increased, uptake of essential elements and changes in host physiology that limit the uptake of toxic elements, such as through induced secretion of organic acids that reduce the bioavailability of these metals (Hildebrandt et al. 2007; Klugh and Cumming 2007).

Arbuscular mycorrhizal fungi colonization appears to decrease transplantation shock in young grapevines. In an experiment under controlled conditions, there was enhanced photosynthetic performance of inoculated plants (van Rooyen et al. 2004). These increased photosynthesis rates of the inoculated plants were related to improved water relations. These results indicate that AMF inoculation can influence the water relations of transplanted grapevine rootstock, thereby increasing potential survival during the initial growth stages of the grapevine.

13.5.1 Indirect Effects Through Mycorrhizal Networks

The beneficial effects of fungal endophytes on plant hosts can be exerted not only through direct interactions between fungal symbionts and plants but also through indirect interactions through common mycelial networks. As fungal endophytes tend to be nonspecific in their choice of hosts, many plants can be linked through fungal hyphae in a common mycelial network. These networks can be enormous, with around 200 m of AMF hyphae present in a single gram of typical forest soil (Dickie 2006). The flow of nutrients between plants and fungal endophytes, and the resulting redistribution of nutrients throughout a community, can affect the fitness of the individual plants (Fitter et al. 1998).

Radić et al. (2012) compared AMF communities on grapevine roots and the most frequent herbaceous perennials growing in vineyards. They observed that several specialist AMF groups were exclusive to either the grapevines or the herbaceous

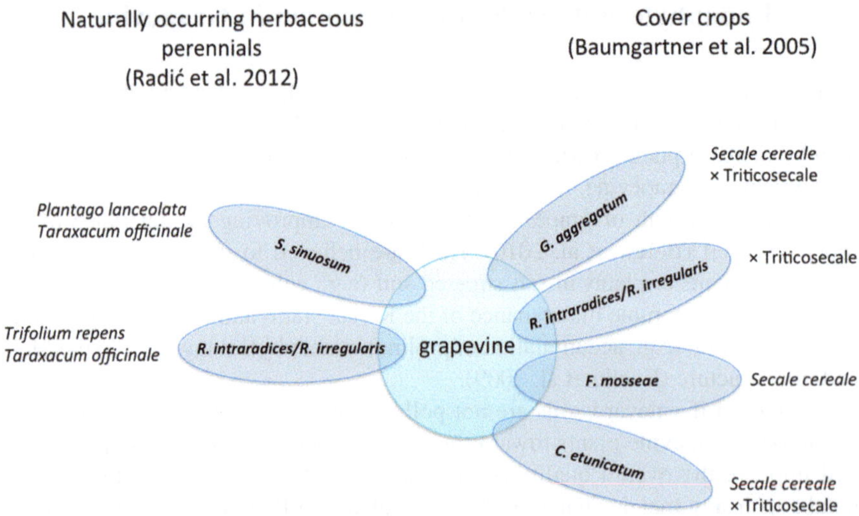

Fig. 13.2 Arbuscular mycorrhizal species shared between the dominant herbaceous perennials that occur naturally in vineyards (Radić et al. 2012), the cover crops (Baumgartner et al. 2005), and the grapevines

perennials and also that generalist AMF species were shared between both plant groups. The most frequent AMF colonizer of grapevine, *Sclerocystis sinuosa*, was shared between grapevine, *Plantago lanceolate*, and *Taraxacum officinale*, whereas *Rhizophagus intraradices*/*Rhizophagus irregularis* colonized grapevine, *Trifolium repens*, and *T. officinale* (Fig. 13.2). Similarly, Baumgartner et al. (2005) observed four AMF species (i.e., *Glomus aggregatum*, *Claroideoglomus etunicatum* = *G. etunicatum*, *Funneliformis mosseae*, *Rhizophagus intraradices*/*Rhizophagus irregularis*) that were shared between grapevine and cover crops. These results confirm the possibility of beneficial effects for grapevine promoted through the common mycelial networks. As such, cover crop management practices might enhance the microbiological function in vineyard ecosystems (Steenwerth and Belina 2008), as they have been shown to affect AMF colonization of grapevine roots through these shared common mycelial networks (Baumgartner et al. 2005).

13.6 Effects of Agricultural Practices on Fungal Endophyte–Grapevine Interactions

In viticulture, the concept of terroir relates the sensory attributes of wines to the environmental conditions of the grapes and is therefore an important descriptor of the connection between a wine and its origin. This encompasses both the natural factors of soil, climate, and topography and the human role in vineyard management (Van Leeuwen and Seguin 2006).

Among the terroir determinants, vineyard management in particular can drastically change, and due to increased awareness of consumers, increasing areas of viticultural land are being devoted to ecological soil management. The new ecological approaches to viticulture emphasize ecologically sound grape production and recognize vineyards as a complex agroecosystem where many organisms coexist and interact. In particular, ecological approaches recognize the importance of interactions between the microbe communities and the plants (Likar et al. 2015; Regvar et al. 2012), as these influence the growth, physiology, and yield of the crop.

Vineyards infected with soil pathogens such as phylloxera often require fumigation treatments. However, the fumigant clears the soil of both desired and undesired soil microbes, including AMF (Menge et al. 1983; Linderman and Davis 2001). With this in mind, soil management practices (especially in ecological agriculture) are focused on preserving a more natural diversity of plants, which can result in the formation of different soil microbiomes (compared to conventional agriculture). It has been shown that ecological agriculture increases the abundance of AMF spores in different agricultural systems (Oehl et al. 2004; Verbruggen et al. 2012), including vineyards (de Oliveira Freitas et al. 2011; Radić et al. 2014). Increased AMF spore abundance results in potentially increased AMF colonization. In agreement with this, Radić et al. (2014) observed not only greater total colonization and spore abundance in ecological vineyards but also greater abundance of arbuscules and vesicles, in comparison to conventional vineyards. Despite the increase in overall spore abundance in ecological vineyards, the fractions of different spores were changed. Furthermore, changes in number of observed and estimated AMF species were increased in ecological vineyards, which suggests changes in the whole AMF community structure. Similarly, Pancher et al. (2012) observed that fungal endophyte communities in grapevine from organically managed farms were different from those from farms that used integrated pest management. Differences in soil management can alter the physicochemical characteristics of the soil and thus affect the AMF spore community through, e.g., N levels or soil temperature (Zhang et al. 2016). Although grapevine cultivar and cultivar-dependent plant physiology can have roles in the shaping of endophytic communities of grapevine, they have lesser effects than crop management techniques (Pancher et al. 2012).

Vineyard soil management practices, and specifically weed control and cover cropping, can impact upon AMF colonization of grapevine roots, as the composition and diversity of the plant community can influence the structure of the AMF community (Burrows and Pfleger 2002; Hausmann and Hawkes 2010; Johnson et al. 2004). In general, plant richness is known to increase AMF diversity (Alguacil et al. 2011), which in the case of vineyards might increase the number of possible AMF colonizers of grapevine and result in improved physiological status of the plants. Also, in addition to plant–host specificity and environmental factors, AMF community composition is influenced by the neighboring plants, which can have beneficial, neutral, or negative effects (Mummey et al. 2005). In line with this, an examination of AMF colonization of grapevine when grown alone or with different

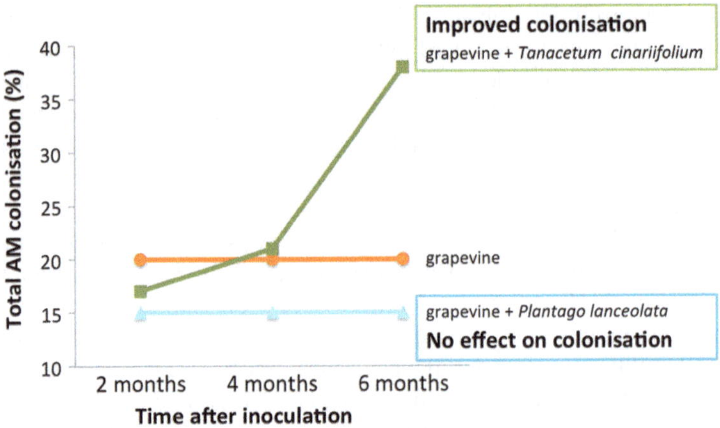

Fig. 13.3 Arbuscular mycorrhizal fungi colonization of grapevine when grown alone or in combination with *Plantago lanceolata* or *Tanacetum cinariifolium* (Radić et al. 2012)

companion plants (Fig. 13.3) showed that some of the weeds increased the colonization levels of grapevine, whereas others had negative effects (Radić et al. 2012).

Furthermore, the appropriate cover crops effectively enhance the soil organic matter content, with effects that include reduced run-off water and improved soil water content, which is of particular importance in Mediterranean and semi-arid agroecoystems (Steenwerth and Belina 2008).

13.7 Outlook

This chapter has focused on the recent evidence of great diversity of fungal endophytes in grapevine plants and the potential that their presence has for viticulture. It appears that despite high selective pressures in agricultural ecosystems, fungal endophyte communities are well developed and can provide benefits for grapevine, which is by default highly dependent on interactions with beneficial fungi. Despite the commercial allure of a more extensive approach, promotion of well-formed fungal endophyte communities can provide long-lasting benefits for sustainable grape production. In the long-term, this can provide commercially sound production, due to lowered costs with reduced need for fumigation and/or fertilization of vineyards.

In particular, the focus needs to be on the benefits that fungal endophytes can provide for grape quality and to relate these to the grapevines. A wider spectrum of interactions needs to be considered (e.g., Piccolo et al. 2016). In more depth, improved knowledge that relates interactions between fungal endophytes and grapevine, and the positive effects that these will bring, will ultimately lead to improvements in viticulture techniques and more sustainably oriented production of grapevine products.

References

Alguacil MM, Torres MP, Torrecillas E et al (2011) Plant type differently promote the arbuscular mycorrhizal fungi biodiversity in the rhizosphere after revegetation of a degraded, semiarid land. Soil Biol Biochem 43:167–173. doi:10.1016/j.soilbio.2010.09.029

Balestrini R, Magurno F, Walker C et al (2010) Cohorts of arbuscular mycorrhizal fungi (AMF) in *Vitis vinifera*, a typical Mediterranean fruit crop. Environ Microbiol Rep 2:594–604. doi:10.1111/j.1758-2229.2010.00160.x

Baumgartner K, Smith RF, Bettiga L (2005) Weed control and cover crop management affect mycorrhizal colonization of grapevine roots and arbuscular mycorrhizal fungal spore populations in a California vineyard. Mycorrhiza 15:111–119. doi:10.1007/s00572-004-0309-2

Belew D, Astatkie T, Mokashi MN et al (2010) Effects of salinity and mycorrhizal inoculation (*Glomus fasciculatum*) on growth responses of grape rootstocks (*Vitis* spp.) S Afr J Enol Vitic 31:82–88

Biricolti S, Ferrini F, Rinaldeli E et al (1997) VAM fungi and soil lime content influence rootstock growth and nutrient content. Am J Enol Vitic 48:93–99

Błaszkowski J, Wubet T, Harikumar VSS et al (2010) *Glomus indicum*, a new arbuscular mycorrhizal fungus. Botany 88:132–143. doi:10.1139/B09-104

Borstler B, Renker C, Kahmen A, Buscot F (2006) Species composition of arbuscular mycorrhizal fungi in two mountain meadows with differing management types and levels of plant biodiversity. Biol Fertil Soils 42:286–298. doi:10.1007/s00374-005-0026-9

Bouffaud M, Bernaud E, Colombet A et al (2016) Regional-scale analysis of arbuscular mycorrhizal fungi: the case of Burgundy vineyards. J Int des Sci la Vigne du Vin 50:1–8

Burrows RL, Pfleger FL (2002) Arbuscular mycorrhizal fungi respond to increasing plant diversity. Can J Bot 80:120–130. doi:10.1139/b01-138

Cheng X, Baumgartner K (2004) Arbuscular mycorrhizal fungi-mediated nitrogen transfer from vineyard cover crops to grapevines. Biol Fertil Soils 40:406–412. doi:10.1007/s00374-004-0797-4

de Felice DV, Solfrizzo M, De Curtis F et al (2008) Strains of *Aureobasidium pullulans* can lower ochratoxin A contamination in wine grapes. Phytopathology 98:1261–1270. doi:10.1094/PHYTO-98-12-1261

de Oliveira Freitas N, Yano-Melo AM, Barbosa da Silva FS, de Melo NF, Costa Maia L (2011) Soil biochemistry and microbial activity in vineyards under conventional and organic management at Northeast Brazil. Sci Agric 68:223–229

De Stefano S, Nicoletti R, Milone A, Zambardino S (1999) 3-o-Methylfunicone, a fungitoxic metabolite produced by the fungus *Penicillium pinophilum*. Phytochemistry 52:1399–1401. doi:10.1016/S0031-9422(99)00320-9

Dickie IA (2006) Mycorrhiza of forest ecosystems. In: Lal E (ed) Encyclopedia of soil science. Taylor and Francis, New York, pp 1111–1114

Eftekhari M, Alizadeh M, Ebrahimi P (2012) Evaluation of the total phenolics and quercetin content of foliage in mycorrhizal grape (*Vitis vinifera* L.) varieties and effect of postharvest drying on quercetin yield. Ind Crop Prod 38:160–165. doi:10.1016/j.indcrop.2012.01.022

Evelin H, Kapoor R, Giri B (2009) Arbuscular mycorrhizal fungi in alleviation of salt stress: a review. Ann Bot 104:1263–1280

Fernando AA, Currah RS (1996) A comparative study of the effects of the root endophytes *Leptodontidium orchidicola* and *Phialocephala fortinii* (Fungi Imperfecti) on the growth of some subalpine plants in culture. Can J Bot 74:1071–1078. doi:10.1139/b96-131

Fitter AH, Graves JD, Watkins NK et al (1998) Carbon transfer between plants and its control in networks of arbuscular mycorrhizas. Funct Ecol 12:406–412

Gallou A, Mosquera HPL, Cranenbrouck S et al (2011) Mycorrhiza-induced resistance in potato plantlets challenged by *Phytophthora infestans*. Physiol Mol Plant Pathol 76:20–26. doi:10.1016/j.pmpp.2011.06.005

Gams W, Holubova-Jechova V (1976) Chloridium and some other dematiaceous Hyphomycetes growing on decaying wood. Stud Mycol 13:1–99
Giri B, Mukerji KG (2004) Mycorrhizal inoculant alleviates salt stress in *Sesbania aegyptiaca* and *Sesbania grandiflora* under field conditions: evidence for reduced sodium and improved magnesium uptake. Mycorrhiza 14:307–312. doi:10.1007/s00572-003-0274-1
Giri B, Kapoor R, Mukerji KG (2003) Influence of arbuscular mycorrhizal fungi and salinity on growth, biomass, and mineral nutrition of *Acacia auriculiformis*. Biol Fertil Soils 38:170–175. doi:10.1007/s00374-003-0636-z
Girlanda M, Perotto S, Luppi AM (2006) Molecular diversity and ecological roles of mycorrhiza-associated sterile fungal endophytes in mediterranean ecosystems. In: Schulz B, Boyle C, Sieber T (eds) Soil biology. Springer, Berlin, pp 207–226
Hao Z, Fayolle L, van Tuinen D et al (2012) Local and systemic mycorrhiza-induced protection against the ectoparasitic nematode Xiphinema index involves priming of defence gene responses in grapevine. J Exp Bot 63:3657–3672. doi:10.1093/jxb/ers046
Haselwandter K, Read D (1980) Fungal associations of roots of dominant and sub-dominant plants in high-alpine vegetation systems with special reference to mycorrhiza. Oecologia 45:57–62. doi:10.1007/BF00346707
Hausmann NT, Hawkes CV (2010) Order of plant host establishment alters the composition of arbuscular mycorrhizal communities. Ecology 91:2333–2343. doi:10.1890/09-0924.1
Havlin JL, Kissel DE, Maddux LD et al (1990) Crop rotation and tillage effects on soil organic carbon and nitrogen. Soil Sci Soc Am J 54:448–452. doi:10.2136/sssaj1990.03615995005400020026x
Hennebert G, Bellemere A (1979) Les formes conidiennes des Discomycetes. Essai taxonomique. Rev Mycol 43:259–352
Hildebrandt U, Regvar M, Bothe H (2007) Arbuscular mycorrhiza and heavy metal tolerance. Phytochemistry 68:139–146. doi:10.1016/j.phytochem
Jiang Q, Zhuo F, Long S et al (2016) Can arbuscular mycorrhizal fungi reduce Cd uptake and alleviate Cd toxicity of *Lonicera japonica* grown in Cd-added soils? Sci Rep 6:21805. doi:10.1038/srep21805
Johnson D, Vandenkoornhuyse PJ, Leake JR et al (2004) Plant communities affect arbuscular mycorrhizal fungal diversity and community composition in grassland microcosms. New Phytol 161:503–515. doi:10.1046/j.1469-8137.2003.00938.x
Jumpponen A (2001) Dark septate endophytes—are they mycorrhizal ? Mycorrhiza 11:207–211. doi:10.1007/s005720100112
Jumpponen A, Trappe JM (1998) Dark septate endophytes: a review of facultative biotrophic root-colonizing fungi. New Phytol 140:295–310. doi:10.1046/j.1469-8137.1998.00265.x
Karagiannidis N, Nikolaou N (1999) Arbuscular mycorrhizal root infection as an important factor of grapevine nutrition status. Multivariate analysis application for evaluation and characterization of the soil and leaf parameters. Agrochimica 43:151–165
Klugh KR, Cumming JR (2007) Variations in organic acid exudation and aluminum resistance among arbuscular mycorrhizal species colonizing *Liriodendron tulipifera*. Tree Physiol 27:1103–1112
Knapp DG, Pintye A, Kovacs GM (2012) The dark side is not fastidious? Dark septate endophytic fungi of native and invasive plants of semiarid sandy areas. PLoS One 7:1–8. doi:10.1371/journal.pone.0032570
Li LF, Li T, Zhao ZW (2007) Differences of arbuscular mycorrhizal fungal diversity and community between a cultivated land, an old field, and a never-cultivated field in a hot and arid ecosystem of southwest China. Mycorrhiza 17:655–665
Li L-F, Li T, Zhang Y, Zhao Z-W (2010) Molecular diversity of arbuscular mycorrhizal fungi and their distribution patterns related to host-plants and habitats in a hot and arid ecosystem, southwest China. FEMS Microbiol Ecol 71:418–427. doi:10.1111/j.1574-6941.2009.00815.x
Likar M, Regvar M (2013) Isolates of dark septate endophytes reduce metal uptake and improve physiology of *Salix caprea* L. Plant Soil 370:593–604. doi:10.1007/s11104-013-1656-6

Likar M, Hančević K, Radić T, Regvar M (2013) Distribution and diversity of arbuscular mycorrhizal fungi in grapevines from production vineyards along the eastern Adriatic coast. Mycorrhiza 23:209–219. doi:10.1007/s00572-012-0463-x

Likar M, Vogel-Mikuš K, Potisek M et al (2015) Importance of soil and vineyard management in the determination of grapevine mineral composition. Sci Total Environ 505:724–731. doi:10.1016/j.scitotenv.2014.10.057

Lin X, Feng Y, Zhang H et al (2012) Long-term balanced fertilization decreases arbuscular mycorrhizal fungal diversity in an arable soil in North China revealed by 454 pyrosequencing. Environ Sci Technol 46:5764–5771. doi:10.1021/es3001695

Linderman RG, Davis EA (2001) Comparative response of selected grapevine rootstocks and cultivars to inoculation with different mycorrhizal fungi. Am J Enol Vitic 52:8–11

Liu W, Zhang Y, Jiang S et al (2016) Arbuscular mycorrhizal fungi in soil and roots respond differently to phosphorus inputs in an intensively managed calcareous agricultural soil. Sci Rep 6:24902. doi:10.1038/srep24902

Lumini E, Orgiazzi A, Borriello R et al (2010) Disclosing arbuscular mycorrhizal fungal biodiversity in soil through a land-use gradient using a pyrosequencing approach. Environ Microbiol 12:2165–2179. doi:10.1111/j.1462-2920.2009.02099.x

Maherali H (2014) Is there an association between root architecture and mycorrhizal growth response? New Phytol 204:192–200. doi:10.1111/nph.12927

Mandyam K, Jumpponen A (2005) Seeking the elusive function of the root-colonising dark septate endophytic fungi. Stud Mycol 53:173–189. doi:10.3114/sim.53.1.173

Menge J, Raski D, Lider L (1983) Interactions between mycorrhizal fungi, soil fumigation, and growth of grapes in California. Am J Enol Vitic 34:117–121

Mullen RB, Schmidt SK, Jaeger CH (1998) Nitrogen uptake during snowmelt by the snow buttercup, *Ranunculus adoneus*. Arct Alp Res 30:121–125. doi:10.2307/1552126

Mummey DL, Rillig MC, Holben WE (2005) Neighboring plant influences on arbuscular mycorrhizal fungal community composition as assessed by T-RFLP analysis. Plant Soil 271:83–90. doi:10.1007/s11104-004-2066-6

Nappi P, Jodice R, Luzzati A, Corino L (1985) Grapevine root system and VA mycorrhizae in some soils of Piedmont (Italy). Plant Soil 85:205–210. doi:10.1007/BF02139624

Newsham KK (2011) A meta-analysis of plant responses to dark septate root endophytes. New Phytol 190:783–793. doi:10.1111/j.1469-8137.2010.03611.x

Nikolaou N, Angelopoulos K, Karagiannidis N (2003) Effects of drought stress on mycorrhizal and non-mycorrhizal cabernet sauvignon grapevine, grafted onto various rootstocks. Exp Agric 39:241–252. doi:10.1017/S001447970300125X

Nogales A, Luque J, Estaún V et al (2009) Differential growth of mycorrhizal field-inoculated grapevine rootstocks in two replant soils. Am J Enol Vitic 60:484–489

Nunez-Trujillo G, Cabrera R, Burgos-Reyes RL, Da Silva E, Gimenez C, Cosoveanu A, Brito N (2012) Endophytic fungi from *Vitis vinifera* L. isolated in Canary Islands and Azores as potential biocontrol agents of *Botrytis cinerea* Pers.:Fr. J Hortic Forestry Biotechnol 16:1–6

Ocete R, Armendariz I, Cantos M et al (2015) Ecological characterization of wild grapevine habitats focused on arbuscular mycorrhizal symbiosis. Vitis 54:207–211

Oehl F, Sieverding E, Mäder P et al (2004) Impact of long-term conventional and organic farming on the diversity of arbuscular mycorrhizal fungi. Oecologia 138:574–583. doi:10.1007/s00442-003-1458-2

Oehl F, Sieverding E, Ineichen K et al (2005) Community structure of arbuscular mycorrhizal fungi at different soil depths in extensively and intensively managed agroecosystems. New Phytol 165:273–283. doi:10.1111/j.1469-8137.2004.01235.x

Pancher M, Ceol M, Corneo PE et al (2012) Fungal endophytic communities in grapevines (*Vitis vinifera* L.) Respond to crop management. Appl Environ Microbiol 78:4308–4317. doi:10.1128/AEM.07655-11

Petit E, Gubler WD (2006) Influence of *Glomus intraradices* on black foot disease caused by *Cylindrocarpon macrodidymum* on *Vitis rupestris* under controlled conditions. Plant Dis 90:1481–1484. doi:10.1094/PD-90-1481

Piccolo S, Alfonzo A, Burruano S, Moschetti G (2016) Detection of bacterial endophytes in *Vitis vinifera* L. and antibiotic activity against grapevine fungal pathogens. In: Compant S, Methieu F (eds) Biocontrol of major grapevine diseases: leading research. CABI, Boston, pp 182–190

Porras-Alfaro A, Bayman P (2011) Hidden fungi, emergent properties: endophytes and microbiomes. Annu Rev Phytopathol 49:291–315. doi:10.1146/annurev-phyto-080508-081831

Radić T, Hančević K, Likar M, Bogdanović I (2012) Neighbouring weeds influence the formation of arbuscular mycorrhiza in grapevine. Symbiosis 56:111–120. doi:10.1007/s13199-012-0165-3

Radić T, Likar M, Hančević K et al (2014) Occurrence of root endophytic fungi in organic versus conventional vineyards on the Croatian coast. Agric Ecosyst Environ 192:115–121. doi:10.1016/j.agee.2014.04.008

Regvar M, Bukovnik U, Likar M, Kreft I (2012) UV-B radiation affects flavonoids and fungal colonisation in *Fagopyrum esculentum* and *F. tataricum*. Cent Eur J Biol 7:275–283. doi:10.2478/s11535-012-0017-4

Reynolds AG (ed) (2010) Front matter. In: Managing wine quality. Woodhead Publishing, Oxford, pp 1–3

Rosendahl S, Stukenbrock EH (2004) Community structure of arbuscular mycorrhizal fungi in undisturbed vegetation revealed by analyses of LSU rDNA sequences. Mol Ecol 13:3179–3186. doi:10.1111/j.1365-294X.2004.02295.x

Schellenbaum L, Berta G, Ravolanirina F, Tisserant B, Gianinazzi S, Fitter AH (1991) Influence of endomycorrhizal infection on root morphology in a micropropagated woody plant species (*Vitis vinifera* L.) Ann Bot 68:135–141

Schmid F, Moser G, Müller H, Berg G (2011) Functional and structural microbial diversity in organic and conventional viticulture: organic farming benefits natural biocontrol agents. Appl Environ Microbiol 77:2188–2191. doi:10.1128/AEM.02187-10

Schol-Schwarz M (1970) Revision of the genus *Phialophora* (Moniliales). Persoonia 6:59–94

Schreiner RP (2003) Mycorrhizal colonization of grapevine rootstocks under field conditions. J Enol Vitic 3:143–149

Schreiner RP (2005) Mycorrhizas and mineral acquisition in grapevines. In: Christensen LP, Smart DR (eds) Proceedings of the soil environment and vine mineral nutrition symposium. pp 49–60

Schreiner PR (2007) Effects of native and nonnative arbuscular mycorrhizal fungi on growth and nutrient uptake of "Pinot noir" (*Vitis vinifera* L.) in two soils with contrasting levels of phosphorus. Appl Soil Ecol 36:205–215. doi:10.1016/j.apsoil.2007.03.002

Schreiner RP, Mihara KL (2009) The diversity of arbuscular mycorrhizal fungi amplified from grapevine roots (*Vitis vinifera* L.) in Oregon vineyards is seasonally stable and influenced by soil and vine age. Mycologia 101:599–611. doi:10.3852/08-169

Schubert A, Cravero MC (1985) Occurrence and infectivity of vesicular-arbuscular mycorrhizal fungi in north-western Italy vineyards. Vitis 24:129–138

Schubert A, Cammarata S, Eynard I (1988) Growth and root colonization of grapevines inoculated with different mycorrhizal endophytes. Hortic Sci 23:302–303

Smart DR, Schwass E, Lakso A, Morano L (2006) Grapevine rooting patterns: a comprehensive Analysis and a review. Am J Enol Vitic 57:89–104

Smith SE, Read D (2008) Mycorrhizal symbiosis, 3rd edn. Academic Press, London

Steenwerth K, Belina KM (2008) Cover crops enhance soil organic matter, carbon dynamics and microbiological function in a vineyard agroecosystem. Appl Soil Ecol 40:359–369. doi:10.1016/j.apsoil.2008.06.006

Stockinger H, Walker C, Schüssler A (2009) Glomus intraradices DAOM197198, a model fungus in arbuscular mycorrhiza research, is not Glomus intraradices. New Phytol 183: 1176–1187

Ulrichs C, Fischer G, Büttner C, Mewis I (2008) Comparison of lycopene, b -carotene and phenolic contents of tomato using conventional and ecological horticultural practices, and arbuscular mycorrhizal fungi (AMF). Agron Colomb 26:40–46

van Leeuwen C, Seguin G (2006) The concept of terroir in viticulture. J Wine Res 17:1–10. doi:10.1080/09571260600633135

van Rooyen M, Valentine AJ, Archer E (2004) Arbuscular mycorrhizal colonisation modifies the water relations of young transplanted grapevines (*Vitis*). S Afr J Enol Vitic 25:37–42

Varma A, Verma S, Sudha et al (1999) *Piriformospora indica*, a cultivable plant-growth-promoting root endophyte. Appl Environ Microbiol 65:2741–2744

Varma A, Bakshi M, Lou B et al (2012) *Piriformospora indica*: a novel plant growth-promoting mycorrhizal fungus. Agric Res 1:117–131. doi:10.1007/s40003-012-0019-5

Verbruggen E, Van Der Heijden MGA, Weedon JT et al (2012) Community assembly, species richness and nestedness of arbuscular mycorrhizal fungi in agricultural soils. Mol Ecol 21:2341–2353. doi:10.1111/j.1365-294X.2012.05534.x

Walker J (1980) *Gaeumannomyces, Linocarpon, Ophiobolus* and several other genera of scolecospored Ascomycetes and Phialophora conidial states, with a note on hyphopodia. Mycotaxon 11:1–129

Wang FW, Jiao RH, Cheng AB et al (2007) Antimicrobial potentials of endophytic fungi residing in *Quercus variabilis* and brefeldin A obtained from *Cladosporium* sp. World J Microbiol Biotechnol 23:79–83. doi:10.1007/s11274-006-9195-4

Zhang T, Yang X, Guo R, Guo J (2016) Response of AM fungi spore population to elevated temperature and nitrogen addition and their influence on the plant community composition and productivity. Sci Rep 6:24749. doi:10.1038/srep24749

Chapter 14
What Have We Learnt from Studying Mycorrhizal Colonisation of Wetland Plant Species?

Alenka Gaberščik, Nataša Dolinar, Nina Šraj, and Marjana Regvar

Abstract Wetlands are ecosystems where the water regime is the main factor that shapes the physical, chemical and biological characteristics. Wetland plants are rooted in water-saturated soils that are frequently anoxic. In spite of this, the rhizosphere can be oxygenated due to the aerenchyma of the wetland plants, which enable active ventilation of roots, rhizomes and the nearby rhizosphere. Some wetland species have an amphibious character, whereby they can thrive both in water and on dry land, with the development of structurally different aquatic and terrestrial forms. Studies of fungal colonisation in wetlands have revealed the presence of fungal endophytes and mycorrhizal fungi. These colonisers are affected by the hydrological regime of the specific wetland. The availability of oxygen also alters the morphology and density of the individual fungal structures. It has been shown that occurrence of arbuscular mycorrhiza is negatively correlated with water depth and duration of flooding. In wetlands, the availability of nutrients depends on a variety of factors, which can mask the role of these fungi. This is particularly the case for phosphorus, which is the main plant benefit from mycorrhizal symbiosis. The same holds true for the potentially positive role of aerenchyma, as the conditions that induce their development inhibit colonisation by arbuscular mycorrhiza. Studies carried out in an intermittent lake, Lake Cerknica, have revealed relatively high arbuscular mycorrhizal colonisation of amphibious species. This appears to be due to the low organic matter content and the low level of plant-available phosphorus in the rhizosphere. At the same time, the frequency of colonisation is lower in aquatic specimens. The impact of water level fluctuations and season on fungal root colonisation of the common reed *Phragmites australis* is reflected in an altered frequency and intensity of fungal colonisation. The structures of dark septate endophytes that might have a similar role in plants as arbuscular mycorrhiza under stress conditions are relatively frequent in this species.

A. Gaberščik • N. Dolinar • N. Šraj • M. Regvar (✉)
Biotechnical Faculty, Department of Biology, University of Ljubljana, Jamnikarjeva 101, SI1000 Ljubljana, Slovenia
e-mail: marjana.regvar@bf.uni-lj.si

A. Varma et al. (eds.), *Mycorrhiza - Function, Diversity, State of the Art*,
DOI 10.1007/978-3-319-53064-2_14

14.1 Wetlands as Habitats for Plant Species

Wetlands are ecosystems where the water regime is the main factor that shapes the physical, chemical and biological characteristics. Wetland plants are rooted in water-saturated, or even flooded, soils, which frequently results in anoxic conditions and therefore the unpredictable availability of nutrients, especially phosphorus (Baldwin et al. 2000). The element cycles are altered due to anoxic conditions, which results in decreased availability of metabolic energy for the plants and in the presence of a variety of reduced elemental forms that can interfere with biotic processes, such as Mn^{2+}, Fe^{2+}, S^{2-}, HS^{-} and H_2S (Mitsch and Gosselink 2007). However, flooded soils are usually hypoxic rather than completely anoxic, as the topsoil can remain oxygenated. Oxygen can also be present in the root rhizosphere, because of the special tissue known as aerenchyma. These aerenchyma have abundant intercellular spaces that develop in leaves, petioles, stems and roots of wetland plant species, and they provide active ventilation of the roots and rhizomes that are anchored in oxygen-deficient soils, thus maintaining favourable oxygen conditions (Cronk and Fennessy 2001; Dickopp et al. 2011). As a consequence, oxygen is also released radially from the roots into the surrounding rhizosphere, with this oxygenation supporting aerobic processes and preventing production of reduced forms of different substances that can have toxic effects on the biota, including fungi (Sasikala et al. 2009). The conditions in the rhizosphere are more favourable in wetlands with intermittent flooding, where the exchange of wet and dry periods enables soil aeration. This thus has positive effects on the biogeochemical cycles and biological processes, such as primary production and decomposition (Dolinar et al. 2015), as well as the diversity of the microbial communities (Boulton and Brock 1999).

14.2 Fungal Colonisation of Plant Roots in Wetlands

Plants in wetlands can be colonised by fungal endophytes (Bärlocher 2006; Weishampel and Bedford 2006; Rodriguez et al. 2009) and mycorrhizal fungi (Bohrer et al. 2004; Šraj-Kržič et al. 2006; Ipsilantis and Sylvia 2007; Dolinar et al. 2010a, 2015; Dolinar and Gaberščik 2010), although most of the studies carried out to date have focussed on arbuscular mycorrhiza (AM). Wigand et al. (1998) reported that AM have a similar role in the aquatic environment as seen for terrestrial habitats. The presence of AM increases nutrient uptake (Wigand and Stevenson 1997; Miller and Sharitz 2000; Jayachandran and Shetty 2003), has positive effects on plant shoot and root biomass and alters the composition of the plant communities (Wolfe et al. 2006).

Fungal colonisation of plant roots in wetlands has kindled little research interest in the past (Thormann et al. 1999), and some wetland species have even been recognised as non-mycorrhizal, such as sedges (Muthukumar et al. 2004). However, recent studies have shown that AM are relatively common in wetlands. In some wetlands, the majority of the plant species examined have roots colonised by fungi (e.g. Dolinar et al. 2010a; Kandalepas et al. 2010; Stevens et al. 2010), which is also the case for some submerged plant species (Šraj-Kržič et al. 2009) and even for plants in saline wetlands (Welsh et al. 2010). Roots colonised by AM have been found up to 100 cm deep in the soil (Taniguchi et al. 2012). Also, Wang et al. (2010) reported the presence of orthologous copies of mycorrhizal genes in the common ancestors of land plants, which provides evolutionary and functional molecular evidence of the mycorrhiza-assisted transition of plants onto land. This means that these fungi had to acquire some tolerance to anoxic conditions beforehand.

Among endophytes, dark septate endophytes (DSEs) are found in plant roots, where they produce structures known as microsclerotia—clusters of round cells with thick walls (Fig. 14.1) that have storage or vegetative propagation functions (Mandyam and Jumpponen 2008). DSEs are not recognised as pathogens, because they are usually found in healthy roots. They are more abundant in extreme environments, although they have been reported for a variety of ecosystems (Rodriguez et al. 2009). Studies focusing on DSEs and their effects on plant growth are rare. In the review of Mandyam and Jumpponen (2005), they concluded that DSEs are functionally similar to AM and have an important role in the acquisition of nutrients, and so they might complement AM in the foraging for organic nutrient sources. DSEs are especially important for plants that are subjected to stress (Jumpponen and Trappe 1998; Jumpponen 2001). DSEs have been reported for aquatic environments only in a few studies (e.g. Miller et al. 1999; Fuchs and Haselwandter 2004; Kai and Zhiwei 2006; Dolinar and Gaberščik 2010; Dolinar et al. 2015) (Fig. 14.1).

14.3 The Impact of Flooding on Fungal Colonisation

Flooding affects the processes of fungal root colonisation. The survival of mycorrhizal fungi under the anoxic conditions is possible according to two main scenarios: some fungi can tolerate hypoxic conditions, while others concentrate near the roots, which can provide the fungi with sufficient oxygen (Miller and Bever 1999). If the roots have already been colonised by fungi before there is flooding, their abundance can be maintained and, in some cases, even increased (Neto et al. 2006). This is also the case for the amounts of spores from mycorrhizal fungi in flooded soil (Miller and Sharitz 2000). Nevertheless, this cannot be generalised, as reductions in fungal colonisation due to flooding have also been reported (García and Mendoza 2008; Dolinar and Gaberščik 2010; Dolinar et al. 2015).

The occurrence of AM shows negative correlation with water depth and duration of flooding (Stevens and Peterson 1996; Miller 2000; Jayachandran and Shetty 2003; Mendoza et al. 2005). It has been shown that the activity of mycorrhizal fungi decreases with increased soil moisture, due to the lack of oxygen (Turner and Friese 1998). AM colonisation of wetland plants in lakes is less frequent in comparison

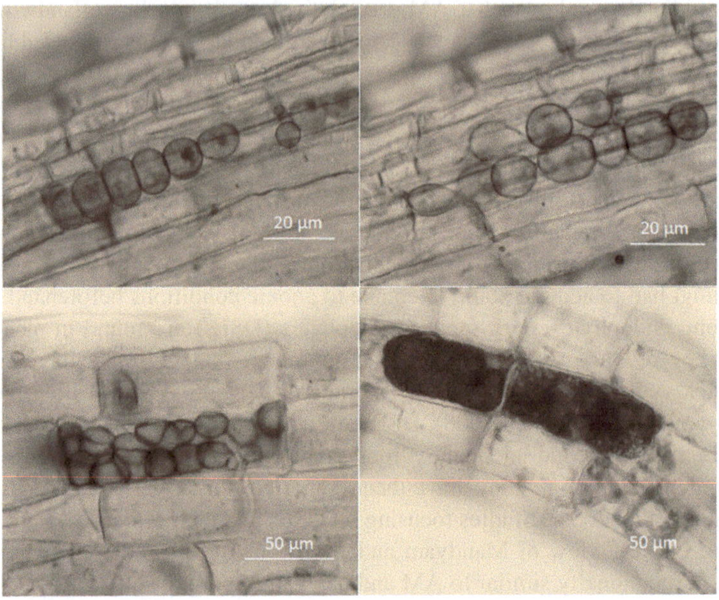

Fig. 14.1 Different types of microsclerotia, as the DSE structures found in *P. australis* roots

with that seen for wetland plants that thrive along rivers, because aeration of the sediments is more efficient for the latter (Kai and Zhiwei 2006). The potential for fungal colonisation of wetland plants can be increased with their development of aerenchyma, which can supply the rhizosphere with oxygen (Tanner and Clayton 1985; Mendoza et al. 2005; Voesenek et al. 2006; Dickopp et al. 2011), and also sometimes by occasional draining of wetlands. AM can also be negatively affected by increased soil salinity (Carvalho et al. 2003) and high phosphorus content or high organic matter content of the soil (Wigand et al. 1998). The rhizosphere soil characteristics affect AM, and AM fungi in nutrient poor soils can increase the nutrient uptake of the plants (Wigand and Stevenson 1997; Andersen and Andersen 2006).

The availability of oxygen alters the morphology and density of the individual structures that are characteristic of fungal colonisation. Ipsilantis and Sylvia (2007) reported a reduction in the frequency of coils, arbuscules and vesicles in the wetland plants *Panicum hemitomon* and *Typha latifolia* during flooding, where these structures practically disappeared. In contrast, Ray and Inouye (2006) only reported a reduction in the colonisation of *T. latifolia* due to flooding. Plants of the genus *Typha* have extensive aerenchyma that are very effective for aeration of the rhizosphere and which promote fungal colonisation also during flooding. These large intercellular spaces enable the development of arbuscules and has a negative effect on the production of coils (Nielsen et al. 2004). Similarly, Smith and Smith (1997) reported that increased oxygen availability supports the development of arbuscules, while hypoxic conditions result in the development of coils.

An increased supply of phosphorus for plants is one of the most important ecological roles of mycorrhiza. However, in wetlands, the availability of nutrients, and particularly of phosphorus, depends on other factors that can mask the direct benefits of mycorrhizal symbiosis, such as the oxygen conditions. In terrestrial ecosystems, mycorrhizal colonisation is often lower in phosphorus-rich soils (Smith and Read 1997), which has also been confirmed for wetland plants (Stevens and Peterson 2007). Johnson (2010) proposed a model for the availability of nutrients that defined the relationships between the fungi and the plant that ranged from parasitism to mutualism. This model assumes that the role of AM depends on the interactions between the availability of nitrogen and phosphorus and the demand of the fungi for carbon. According to this model, mutualistic relations are not likely to be established under phosphorus-rich conditions. For example, the wetland species *Lythrum salicaria* showed significantly higher AM colonisation when grown in phosphorus-poor soil (White and Charvat 1999; Stevens et al. 2002a, b). The relationship between phosphorus and mycorrhizal colonisation becomes more complex in wetlands, where the changing water level results in changes in soil oxygenation and in mobility and solubility of phosphorus (Baldwin et al. 2000). This relationship can also be affected by the timing of the phosphorus availability and the need of the plants for phosphorus (Carvalho et al. 2001).

14.4 Case Study

14.4.1 Fungal Colonisation of Plant Roots in a Wetland with Intermittent Flooding

Studies of the fungal colonisation of roots have been performed in a wetland with intermittent flooding (Lake Cerknica) in different habitats along a hydrological gradient (i.e. from water to dry land). This area shows zonation of the vegetation due to differences in the frequency of water level fluctuations, the timing and extent of flooding and the substantial differences in the characteristics of the rhizosphere soil (Šraj-Kržič et al. 2006, 2009; Dolinar and Gaberščik 2010). This last might also critically affect fungal colonisation of the plant roots (Smith and Read 1997; Andersen and Andersen 2006).

14.4.2 Fungal Colonisation of Amphibious Species

Fungal colonisation of wetland species with an amphibious character is poorly understood. Only a handful of studies are currently available that have included comparisons of submerged and terrestrial plant species (Beck-Nielsen and Madsen 2001; Miller 2000). Submerged and terrestrial forms of amphibious plants differ significantly in their structural, biochemical and functional characteristics (Kržič et al. 2004; Šraj-Kržič and Gaberščik 2005; Šraj-Kržič et al. 2006, 2009; Dolinar et al. 2010b, c; Klančnik et al. 2012, 2014), and they provide a perfect model system for studying the levels of fungal root colonisation in the same species in contrasting hydrological environments. In submerged specimens of amphibious plants, the presence of root colonisation was confirmed by Šraj-Kržič et al. (2006, 2009); however, the levels of root colonisation were significantly lower in the aquatic specimens of the same species in comparison with their terrestrial counterparts (Fig. 14.2). A redundancy analysis of these data revealed clustering of the surveyed species into two groups, namely, the terrestrial and aquatic forms. The presence of aerenchyma was negatively related to the frequency of fungal colonisation. In spite of this, AM colonisation in aquatic specimens appears to be the result of efficient aeration of the soil due to the intermittent flooding (Mendoza et al. 2005). These kinds of studies are, however, hampered by several difficulties. First, effective aeration within a root system is not easy to establish. In addition, some fungi can be more tolerant to anoxic conditions than others, thus blurring the correlation between aerenchyma formation and AM. As a rule, the conditions that induce development of aerenchyma also inhibit AM colonisation, i.e. flooding and oxygen deficiency (Turner and Friese 1998; Braendle and Crawford 1999; Rascio 2002; Visser and Voesenek 2004). Cornwell et al. (2001) reported higher AM colonisation for plants with fewer aerenchyma that aerate the rhizosphere less efficiently, possibly due to a lower availability of nutrients. In the case of a limited availability of nutrients (e.g. under the influence of anoxia) or nutrient deficiencies, the production of anthocyanins presents plant strategy to increase the protection against stress conditions (Chalker-Scott 1999; Steyn et al. 2002). Our data analysis (Fig. 14.2) indicates that anthocyanins are an important factor explaining the frequency of mycorrhiza and contributes to the separation of these plants into these terrestrial and aquatic groups; here, the aquatic plants are richer in anthocyanin content per dry mass.

A study of Šraj-Kržič et al. (2009) in Lake Cerknica revealed relatively high AM colonisation in amphibious species, which appeared to be due to low organic matter content ($<10\%$) in the rhizosphere soils of the plants studied. In addition, the levels of plant-available phosphorus in the rhizosphere were also low (<5.3 mg L^{-1}). In organic soils, oxygen can be consumed relatively rapidly and the phosphorus content increases, which in turn has a negative effect on mycorrhizal colonisation (Wigand et al. 1998). AM are likely to be an advantage for plants in habitats with intermittent flooding, as they can increase the efficiency of nutrient uptake from the soil, which is especially important during the flooding. As a rule, aquatic specimens of amphibious plants cannot take up nutrients *via* the leaf surfaces, and their nutrient uptake from the soil is slow, due to the absence of the transpiration flow (Nilsen and Orcutt 1996; Rascio 2002). Studies in Lake Cerknica revealed a significant positive correlation between the soil content of plant-available phosphorus and the AM parameters. We believe that greater AM colonisation of the terrestrial plant forms is a consequence of better aerated soil, rather than having a direct association with the soil content of plant-available phosphorus. Similarly,

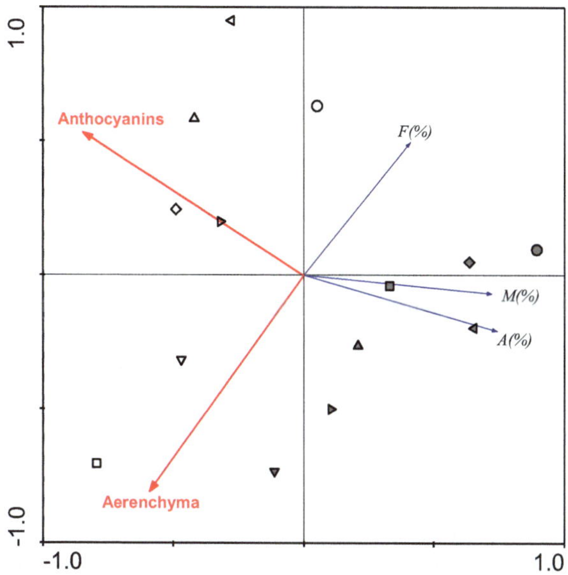

Fig. 14.2 Redundancy analysis plot showing the relationships between leaf properties (anthocyanins, aerenchyma) and mycorrhizal frequency (*F%*), mycorrhizal intensity (*M%*) and density of arbuscules (*A%*). Anthocyanins explained 32% and aerenchyma 19% of the variability of the samples. Diamond, *Gratiola officinalis*; square, *Glyceria fluitans*; triangle-up, *Ranunculus lingua*; triangle-left, *Teucrium scordium*; circle, *Sium latifolium*; triangle-down, *Sparganium emersum*; triangle-right, *Veronica anagallis-aquatica*. Filled symbols represent terrestrial specimens, and empty symbols represent aquatic specimens

Miller (2000) noted that with the same availability of phosphorus, AM increase with decreasing soil moisture. We believe that in wetlands with intermittent flooding, the processes in the rhizosphere benefit from fungal colonisation due to the changes between flooding and drought and diminish the effects of the availability of nutrients. The positive correlation between the AM parameters and the photochemical efficiency of PSII in some plants additionally highlights the importance of mycorrhiza for amphibious species (Šraj-Kržič et al. 2006, 2009).

14.4.3 Fungal Colonisation of Selected Helophytes and Mire Species

The roots of wetland plant species like the grass *Phalaris arundinacea*, the bog rushes *Schoenus nigricans* and *Schoenus ferrugineus*, the plantain *Plantago altissima* and the summer snowflake *Leucojum aestivum* are all abundantly colonised by fungi (Dolinar et al. 2010a). An exception was seen for the sedge *Carex elata*, which showed very low frequency of fungal colonisation and only a rare presence of coils and microsclerotia. In *C. elata*, unicellular root hairs were observed, which are reported to be the consequence of anoxic conditions in some species of the genus *Carex* and to also be associated with a lack of fungal colonisation (Miller et al. 1999). However, the presence of unicellular root hairs and the level of fungal colonisation are not necessarily exclusive (Muthukumar et al. 2004). Mejstrik (1972) focussed on fungal colonisation of the roots of different plant species in wet meadow community of *Molinietum caeruleae*. There was no fungal colonisation in only nine of the 55 species studied here. In a meta-analysis of mycorrhizal colonisation of the British flora, Harley and Harley (1987) reported that there are many species, like *P. arundinacea*, *S. nigricans*, *S. ferrugineus* and *L. aestivum*, where different sources have come to conflicting conclusions to define these species as mycorrhizal or non-mycorrhizal. Among these others, *C. elata* was indicated as non-mycorrhizal. Furthermore, for *P. arundinacea*, Bauer et al. (2003) and Bohrer et al. (2004) reported mycorrhizal colonisation, as was also the case for Lake Cerknica (our unpublished data). Here it also needs to be noted that in some plant species that show low mycorrhizal frequencies, root fungal colonisation with arbuscules (the most characteristic structure of AM colonisation) can be limited to certain periods of the plant ontogenesis and in particular to flowering and seeding (Regvar et al. 2006). This appears to be due to the larger demands for essential nutrients under otherwise sparing conditions of available elements in the soil. Similarly, in aquatic and intermittently flooded habitats, the formation of arbuscules can be confined to the periods of high nutrient demands in the plants, thus leading to the conflicting results in the literature.

14.4.4 Fungal Colonisation on *Phragmites australis* Roots

The common reed, *Phragmites australis*, is the prevailing species in Lake Cerknica, where it colonises habitats with different hydrological conditions (Dolinar et al. 2010a). It usually develops according to the water level fluctuations (Fig. 14.3). Fungal colonisation of the *P. australis* roots was observed in sites with different water regimes, which did not depend on the present or past water conditions (Dolinar and Gaberščik 2010). Pronounced differences, however, occurred in terms of the degree of colonisation in three locations (Fig. 14.4) and the density of the different fungal structures observed within the root systems (data not shown). AM fungi in the roots of *P. australis* have also been confirmed by Cooke and Lefor (1998) and Oliveira et al. (2001), although during flooding, AM were not always present (Wirsel 2004; Dolinar et al. 2015). *P. australis* has been associated with more than 300 fungal taxa, although only a few species were found in abundance (Neubert et al. 2006). Furthermore, the roots of *P. australis* can also contain hyphae and structures of DSEs (Wu et al. 2009), as was also seen for Lake Cerknica (Dolinar and Gaberščik 2010). The impact of the water level fluctuations and the seasonal changes on the fungal root colonisation of *P. australis* was related to the frequency and intensity of the fungal colonisation (Dolinar et al. 2015). This appears to be a consequence of the sudden rises in the water level, during which

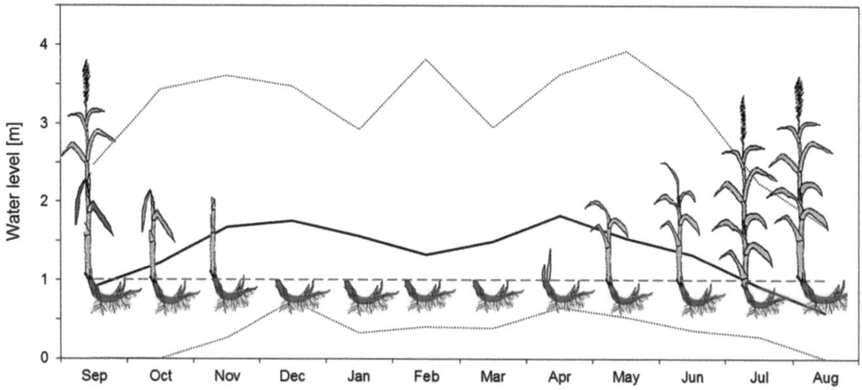

Fig. 14.3 Development of *P. australis* in Lake Cerknica and the average and extreme water levels at the measuring location Dolenje Jezero. The measurements were obtained from the Environmental Agency of Slovenia (ARSO)

the fungi failed to become established within new root hairs, due to the unfavourable conditions, as was also seen by Ray and Inouye (2006).

In wetlands with intermittent flooding, it is difficult to distinguish between the influence of environmental conditions and plant phenology on the fungal colonisation (Koide 1991). This was also reported in a study of the seasonal changes of fungal colonisation in a saline wetland (Carvalho et al. 2001). In Lake Cerknica, the fungal colonisation was monitored for three different locations, namely, in a dry zone (Dujice transitional mire), a riparian zone (near the village of Gorenje Jezero) and lake littoral zone (Zadnji kraj—the area of extent reed stands) (Fig. 14.4). The greatest intensity of fungal colonisation for *P. australis* in the driest location was at the end of the growing season, in late August and September, although this was still relatively low regarding the whole root system (Fig. 14.4). Structures typical of AM were seen for all of these three sites, namely, arbuscules, vesicles and coils, the abundance of which increased towards the end of the growing season (Dolinar and Gaberščik 2010). This study in Lake Cerknica also confirmed that the rise in the water level had a negative effect on the proportion of roots colonised by fungi, as revealed for *P. australis* from Zadnji kraj, although the densities of the arbuscules and vesicles were higher for the colonised roots (our unpublished data). As storage structures, the increased density of the vesicles under high water level conditions indicated the altered activity of the fungi as a consequence of acclimatisation to the changed conditions (Bonfante and Genre 2010).

Studies of *P. australis* have revealed negative correlation between frequency of fungal colonisation and water level (Dolinar and Gaberščik 2010; Dolinar et al. 2015) and indicated that DSEs can have similar roles in plants as AM under stress conditions, as was also suggested by Jumpponen (2001). Some studies have revealed that roots that are suddenly exposed to water-saturated soils show either reduced levels of fungal colonisation (Miller 2000; Ray and Inouye 2006; Stevens et al. 2011) or changes in the form of root fungal colonisation, as well as in the

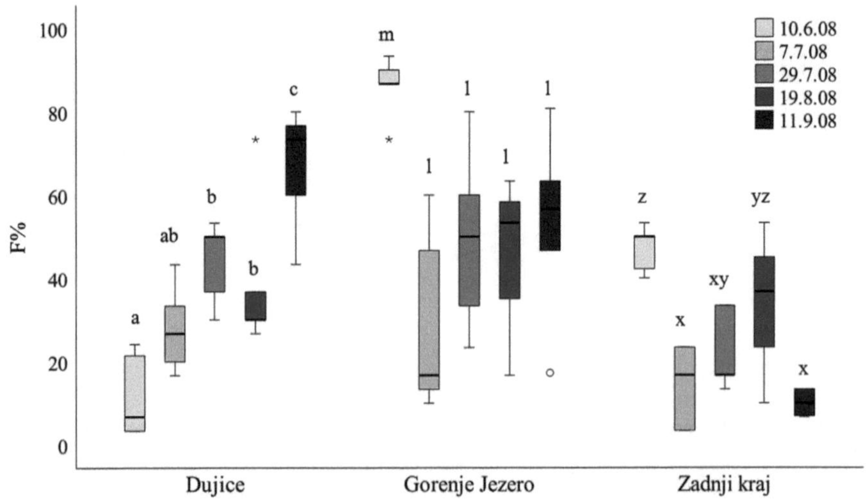

Fig. 14.4 Seasonal changes in fungal root colonisation frequency (*F%*) for *P. australis* roots in Lake Cerknica at locations Dujice, Gorenje Jezero and Zadnji kraj. Data ($n = 5$) with different letters are significantly different ($P < 0.05$). *a*, *b*, *c*, Dujice; *m*, *l*, Gorenje Jezero; *x*, *y*, *z*, Zadnji kraj. *Open circle* outliers; *asterisk* extreme outliers

fungal species composition (Wang et al. 2011), compared to non-saturated conditions (Bauer et al. 2003). However, it has not been clarified yet whether colonisation of new root hairs occurs in flooded soils or that the structures that develop in non-saturated soil are maintained (Miller and Sharitz 2000).

Koide (1991) showed that fungal colonisation increases when plants have greater requirements for nutrients. This is usually during the time of flowering and fruiting (Bohrer et al. 2004; Regvar et al. 2006; Likar et al. 2009), which was also observed in a survey of fungal colonisation of *P. arundinacea* in Lake Cerknica (our unpublished data); for *P. australis*, increased frequencies and intensities of root colonisation were also seen during the storage and senescence processes in autumn (Fig. 14.3, Dujice), which is in line with the study of Escudero and Mendoza (2005). Some studies have reported the most abundant root fungal colonisation during spring, in the period of the most intensive vegetative plant growth (Bajwa et al. 2001; Welsh et al. 2010). This is, however, in contrast to the findings of Miller (2000), who noted that at the time of the most intensive growth there can also be a reduction in the root mycorrhizal colonisation as a consequence of the trade-off of between cost and benefit for the plants. This confirms that fungal colonisation of roots in wetlands is a result of many exogenous and endogenous factors, the most important being the presence of nutrients and water (Gaur and Kaushik 2011).

DSE colonisation of *P. australis* roots is relatively abundant, with a variety of different forms of microsclerotia seen (Fig. 14.1). Two peaks were seen at two sampling plots according to the water level changes: one at the beginning of the growing season and one at the end (Dolinar and Gaberščik 2010). The question however remains in terms of what the cost and benefit are with DSE colonisation in *P. australis* roots in wetlands and habitats with intermittent flooding.

Acknowledgements This study was supported by Slovenian Research Agency, through projects "Young researchers" grants no. 3311-03-831202 (2003-2007) and 1000-06310153 (2006-2013) and Plant Biology programme (P1-0212). This support is gratefully acknowledged.

References

Andersen FØ, Andersen T (2006) Effects of arbuscular mycorrhiza on biomass and nutrients in the aquatic plant *Littorella uniflora*. Freshw Biol 51:1623–1633

Bajwa R, Yaqoob A, Javaid A (2001) Seasonal variation in VAM in wetland plants. Pak J Biol Sci 4:464–470

Baldwin DS, Mitchell AM, Rees GN (2000) The effects of in situ drying on sediment-phosphate interactions in sediments from an old wetland. Hydrobiologia 431:3–12

Bärlocher F (2006) Fungal endophytes in submerged roots. In: Schulz B, Boyle C, Sieber TN (eds) Microbial root endophytes. Soil biology, vol 9. Springer, Berlin, pp 179–190

Bauer CR, Kellogg CH, Bridgham SD, Lamberti GA (2003) Mycorrhizal colonisation across hydrologic gradients in restored and reference freshwater wetlands. Wetlands 23:961–968

Beck-Nielsen D, Madsen TV (2001) Occurrence of vesicular-arbuscular mycorrhiza in aquatic macrophytes from lakes and streams. Aquat Bot 71:141–148

Bohrer KE, Friese CF, Amon JP (2004) Seasonal dynamics of arbuscular mycorrhizal fungi in differing wetland habitats. Mycorrhiza 14:329–337

Bonfante P, Genre A (2010) Mechanisms underlying beneficial plant-fungus interactions in mycorrhizal symbiosis. Nat Commun 48:1–11

Boulton AJ, Brock MA (1999) Australian freshwater ecology. Processes and management. Gleneagles, Glen Osmond, pp 149–167

Braendle R, Crawford RMM (1999) Plants as amphibians. Perspect Plant Ecol Evol Syst 2:56–78

Carvalho LM, Caçador I, Martins-Loução MA (2001) Temporal and spatial variation of arbuscular mycorrhizas in salt marsh plants of the Tagus estuary (Portugal). Mycorrhiza 11:303–309

Carvalho LM, Correia PM, Caçador I, Martins-Loução MA (2003) Effects of salitnity and flooding on the infectivity of salt marsh arbuscular mycorrhizal fungi in *Aster tripolium* L. Biol Fertil Soils 38:137–143

Chalker-Scott L (1999) Environmental significance of anthocyanins in plant stress responses. Photochem Photobiol 70:1–9

Cooke JC, Lefor MW (1998) The mycorrhizal status of selected plant species from Connecticut wetlands and transition zones. Restor Ecol 6:214–222

Cornwell WK, Bedford BL, Chapin CT (2001) Occurrence of arbuscular mycorrhizal fungi in a phosphorus-poor wetland and mycorrhizal response to phosphorus fertilization. Am J Bot 88:1824–1829

Cronk JK, Fennessy MS (2001) Wetland plants—biology and ecology. Lewis Publisher, London

Dickopp J, Kazda M, Čížková H (2011) Differences in rhizome aeration of *Phragmites australis* in a constructed wetland. Ecol Eng 37:1647–1653

Dolinar N, Gaberščik A (2010) Mycorrhizal colonization and growth of *Phragmites australis* in an intermittent wetland. Aquat Bot 93:93–98

Dolinar N, Šraj-Kržič N, Pongrac P, Regvar M, Gaberščik A (2010a) The presence of mycorrhiza in different habitats of an intermittent aquatic ecosystem. In: Vymazal J (ed) Water and nutrient management in natural and constructed wetlands. Springer, Dordrecht, pp 299–308

Dolinar N, Šraj-Kržič N, Gaberščik A (2010b) Water regime changes and the function of an intermittent wetland. In: Vymazal J (ed) Water and nutrient management in natural and constructed wetlands. Springer, Dordrecht, pp 251–262
Dolinar N, Rudolf M, Šraj N, Gaberščik A (2010c) Environmental changes affect ecosystem services of the intermittent Lake Cerknica. Ecol Complex 7:403–409
Dolinar N, Regvar M, Abram D, Gaberščik A (2015) Water-level fluctuations as a driver of *Phragmites australis* primary productivity, litter decomposition, and fungal root colonisation in an intermittent wetland. Hydrobiologia 774:69. doi:10.1007/s10750-015-2492-x
Escudero V, Mendoza R (2005) Seasonal variation of arbuscular mycorrhizal fungi in temperate grasslands along a wide hydrological gradient. Mycorrhiza 15:291–299
Fuchs B, Haselwandter K (2004) Red list plants: colonization by arbuscular mycorrhizal fungi and dark septate endophytes. Mycorrhiza 14:277–281
García IV, Mendoza RE (2008) Relationships among soil properties, plant nutrition and arbuscular mycorrhizal fungi-plant symbioses in a temperate grassland along hydrologic, saline and sodic gradients. FEMS Microbiol Ecol 63:359–371
Gaur S, Kaushik P (2011) Influence of edaphic factors on distribution of mycorrhiza associated with medicinal plants in Indian central Himalayas. J Biol Sci 11:349–358
Harley JL, Harley EL (1987) A check list of mycorrhiza in the British flora. New Phytol (Suppl) 105:1–102
Ipsilantis I, Sylvia DM (2007) Interactions of assemblages of mycorrhizal fungi with two Florida wetland plants. Appl Soil Ecol 35:261–271
Jayachandran K, Shetty KG (2003) Growth response and phosphorous uptake by arbuscular mycorrhizae of wet prairie sawgrass. Aquat Bot 76:281–290
Johnson NC (2010) Resource stoichiometry elucidates the structure and function of arbuscular mycorrhizas across scales. New Phytol 185:631–647
Jumpponen A (2001) Dark septate endophytes—are they mycorrhizal? Mycorrhiza 11:207–211
Jumpponen A, Trappe JM (1998) Dark septate endophytes: a review of facultative biotrophic root-colonizing fungi. New Phytol 140:295–310
Kai W, Zhiwei Z (2006) Occurrence of arbuscular mycorrhizas and dark septate endophytes in hydrophytes from lakes and streams in Southwest China. Int Rev Hydrobiol 91:29–37
Kandalepas D, Stevens KJ, Shaffer GP, Platt WJ (2010) How abundant are root-colonizing fungi in southeastern Louisiana's degraded marshes? Wetlands 30:189–199
Klančnik K, Mlinar M, Gaberščik A (2012) Heterophylly results in a variety of "spectral signatures" in aquatic plant species. Aquat Bot 98:20–26
Klančnik K, Pančić M, Gaberščik A (2014) Leaf optical properties in amphibious plant species are affected by multiple leaf traits. Hydrobiologia 737:121–130
Koide RT (1991) Tansley review no. 29. Nutrient supply, nutrient demand and plant response to mycorrhizal infection. New Phytol 117:365–386
Kržič N, Gaberščik A, Germ M (2004) The phenotypic plasticity of *Glyceria fluitans* growing over the enviromnetal gradient. Acta Biol Slov 47:65–73
Likar M, Regvar M, Mandic-Mulec I, Stres B, Bothe H (2009) Diversity and seasonal variations of mycorrhiza and rhizosphere bacteria in three common plant species at the Slovenian Ljubljana marsh. Biol Fertil Soils 45:573–583
Mandyam K, Jumpponen A (2005) Seeking the elusive function of the root-colonising dark septate endophytic fungi. Stud Mycol 53:173–189
Mandyam K, Jumpponen A (2008) Seasonal and temporal dynamics of arbuscular mycorrhizal and dark septate endophytic fungi in a tallgrass prairie ecosystem are minimally affected by nitrogen enrichment. Mycorrhiza 18:145–155
Mejstrik VK (1972) Vesicular-arbuscular mycorrhizas of the species of a *Molinietum coeruleae* L. I. association: the ecology. New Phytol 71:883–890
Mendoza R, Escudero V, García I (2005) Plant growth, nutrient acquisition and mycorrhizal symbioses of a waterlogging tolerant legume (*Lotus glaber* Mill.) in a saline sodic soil. Plant Soil 275:305–315

Miller SP (2000) Arbuscular mycorrhizal colonisation of semi-aquatic grasses along a wide hydrologic gradient. New Phytol 145:145–155
Miller SP, Bever JD (1999) Distribution of arbuscular mycorrhizal fungi in stands of the wetland grass *Panicum hemitomon* along wide hydrologic gradient. Oecologia 119:586–592
Miller SP, Sharitz RR (2000) Manipulation of flooding and arbuscular mycorrhiza formation influences growth and nutrition of two semiaquatic grass species. Funct Ecol 14:738–748
Miller RM, Smith CI, Jastrow JD, Bever JD (1999) Mycorrhizal status of the genus *Carex* (Cyperaceae). Am J Bot 86:547–553
Mitsch WJ, Gosselink JG (2007) Wetlands, 4th edn. Wiley, Hoboken, 582 p
Muthukumar T, Udaiyan K, Shanmughavel P (2004) Mycorrhiza in sedges—an overview. Mycorrhiza 14:65–77
Neto D, Carvalho LM, Cruz C, Martins-Loução MA (2006) How do mycorrhizas affect C and N relationships in flooded *Aster tripolium* plants? Plant Soil 279:51–63
Neubert K, Mendgen K, Brinkmann H, Wirsel SGR (2006) Only a few fungal species dominate highly diverse mycofloras associated with the common reed. Appl Environ Microbiol 72:1118–1128
Nielsen KB, Kjøller R, Olsson PA, Schweiger PF, Andersen FO, Rosendahl S (2004) Colonisation and molecular diversity of arbuscular mycorrhizal fungi in the aquatic plants *Littorella uniflora* and *Lobelia dortmanna* in southern Sweden. Mycol Res 108:616–625
Nilsen ET, Orcutt DM (1996) The physiology of plants under stress. Abiotic factors. Willey, New York, pp 362–400
Oliveira RS, Dodd JC, Castro PML (2001) The mycorrhizal status of *Phragmites australis* in several polluted soils and sediments of an industrialised region of Northern Portugal. Mycorrhiza 10:241–247
Rascio N (2002) The underwater life of secondarily aquatic plants: some problems and solutions. Crit Rev Plant Sci 21:401–427
Ray AM, Inouye RS (2006) Effects of water-level fluctuations on the arbuscular mycorrhizal colonization of *Typha latifolia* L. Aquat Bot 84:210–216
Regvar M, Vogel-Mikuš K, Kugonič N, Turk B, Batič F (2006) Vegetational and mycorrhizal successions at a metal polluted site: indications for the direction of phytostabilisation? Environ Pollut (Barking, Essex: 1987) 144:976–984
Rodriguez RJ, White JF Jr, Arnold AE, Redman RS (2009) Fungal endophytes: diversity and functional roles. New Phytol 182:314–330
Sasikala S, Tanaka N, Wah Wah HSY, Jinadasa KBSN (2009) Effects of water level fluctuation on radial oxygen loss, root porosity, and nitrogen removal in subsurface vertical flow wetland mesocosms. Ecol Eng 35:410–417
Smith SE, Read DJ (1997) Mycorrhizal symbiosis, 2nd edn. Academic Press, San Diego
Smith FA, Smith SE (1997) Structural diversity in (vesicular)-arbuscular mycorrhizal symbioses. New Phytol 137:373–388
Stevens KJ, Peterson RL (1996) The effect of a water gradient on the vesicular-arbuscular mycorrhizal status of *Lythrum salicaria* L. (purple loosestrife). Mycorrhiza 6:99–104
Stevens KJ, Peterson RL (2007) Relationships among Three pathways for resource acquisition and their contribution to plant performance in the emergent aquatic plant *Lythrum salicaria* (L.) Plant Biol 9:758–765
Stevens KJ, Peterson RL, Reader RJ (2002a) The aerenchymatous phellem of *Lythrum salicaria* (L.): a pathway for gas transport and its role in flood tolerance. Ann Bot 89:621–625
Stevens KJ, Spender SW, Petersen RL (2002b) Phosphorous, arbuscular mycorrhizal fungi and performance of the wetland plant *Lythrum salicaria* L. under inundated conditions. Mycorrhiza 12:277–283
Stevens KJ, Wellner MR, Acevedo MF (2010) Dark septate endophyte and arbuscular mycorrhizal status of vegetation colonizing a bottomland hardwood forest after a 100 year flood. Aquat Bot 92:105–111

Stevens KJ, Wall CB, Janssen JA (2011) Effects of arbuscular mycorrhizal fungi on seedling growth and development of two wetland plants, *Bidens frondosa* L., and *Eclipta prostrata* (L.) L., grown under three levels of water availability. Mycorrhiza 21:279–288

Steyn WJ, Wand SJE, Holcroft DM, Jacobs G (2002) Anthocyanins in vegetative tissues: a proposed unified function in photoprotection. New Phytol 155:349–361

Šraj-Kržič N, Gaberščik A (2005) Photochemical efficiency of amphibious plants in an intermittent lake. Aquat Bot 83:281–288

Šraj-Kržič N, Pongrac P, Klemenc M, Kladnik A, Regvar M, Gaberščik A (2006) Mycorrhizal colonisation in plants from intermittent aquatic habitats. Aquat Bot 85:331–336

Šraj-Kržič N, Pongrac P, Regvar M, Gaberščik A (2009) Photon-harvesting efficiency and arbuscular mycorrhiza in amphibious plants. Photosynthetica 47:61–67

Taniguchi T, Usuki H, Kikuchi J, Hirobe M, Miki N, Fukuda K, Zhang G, Wang L, Yoshikawa K, Yamanaka N (2012) Colonization and community structure of root-associated microorganisms of Sabina vulgaris with soil depth in a semiarid desert ecosystem with shallow groundwater. Mycorrhiza 22:419–428

Tanner CC, Clayton JS (1985) Effects of vesicular-arbuscular mycorrhizas on growth and nutrition of submerged aquatic plants. Aquat Bot 22:377–386

Thormann MN, Currah RS, Bayley SE (1999) The mycorrhizal status of the dominant vegetation along a peatland gradient in southern boreal Alberta, Canada. Wetlands 19:438–450

Turner SD, Friese CF (1998) Plant-mycorrhizal community dynamics associated with a moisture gradient within a rehabilitated prairie fen. Restor Ecol 6:44–51

Visser EWJ, Voesenek LACJ (2004) Acclimation to soil flooding—sensing and signal transduction. Plant Soil 254:197–214

Voesenek LACJ, Colmer TD, Pierik R, Millenaar FF, Peeters AJM (2006) How plants cope with complete submergence. New Phytol 170:213–226

Wang B, Yeun LH, Xue JY, Liu Y, Ané JM, Qiu YL (2010) Presence of three mycorrhizal genes in the common ancestor of land plants suggests a key role of mycorrhizas in the colonization of land by plants. New Phytol 186:514–525

Wang Y, Huang Y, Qiu Q, Xin G, Yang Z, Shi S (2011) Flooding greatly affects the diversity of arbuscular mycorrhizal fungi communities in the roots of wetland plants. PLoS One 6:e24512

Weishampel PA, Bedford BL (2006) Wetland dicots and monocots differ in colonization by arbuscular mycorrhizal fungi and dark septate endophytes. Mycorrhiza 16:495–502

Welsh AK, Burke DJ, Hamerlynck EP, Hahn D (2010) Seasonal analyses of arbuscular mycorrhizae, nitrogen-fixing bacteria and growth performance of the salt marsh grass *Spartina patens*. Plant Soil 330:251–266

White JA, Charvat I (1999) The mycorrhizal status of an emergent aquatic, *Lythrum salicaria* L., at different levels of phosphorous availability. Mycorriza 9:191–197

Wigand C, Stevenson JC (1997) Facilitation of phosphate assimilation by aquatic mycorrhizae of *Vallisneria americana* Michx. Hydrobiologia 342:35–41

Wigand C, Andersen FØ, Christensen KK, Holmer M, Jensen HS (1998) Endomycorrhizae of isoetids along a biogeochemical gradient. Limnol Oceanogr 43:508–515

Wirsel SGR (2004) Homogenous stands of a wetland grass harbour diverse consortia of arbuscular mycorrhizal fungi. FEMS Microbiol Ecol 48:129–138

Wolfe BE, Weishampel PA, Klironomos JN (2006) Arbuscular mycorrhizal fungi and water table affect wetland plant community composition. J Ecol 94:905–914

Wu Y, Liu T, He X (2009) Mycorrhizal and dark septate endophytic fungi under the canopies of desert plants in Mu Us Sandy Land of China. Front Agric China 3:164–170. doi:10.1007/s11703-009-0026-x

Chapter 15
Response of Arbuscular Mycorrhizal Fungi to Global Climate Change and Their Role in Terrestrial Ecosystem C and N Cycling

Bhoopander Giri and Bhawna Saxena

Abstract The global climate change presents a serious threat to nature and has been predicted to largely impact the life of human beings in the twenty-first century. The Intergovernmental Panel on Climate Change predicted that human-induced climate change is a major threat and also emphasized to develop global plans for mitigation and adaptation to climate change. Taking into consideration the existing feedbacks between carbon cycle and climate change, understanding whether terrestrial ecosystems will respond to elevated atmospheric carbon dioxide concentration (eCO_2) or up to what extent is of utmost significance. In the global ecosystems, CO_2 is largely used by plants in the process of photosynthesis (net primary production). On the other hand, microbes contribute directly, to a great extent, to net carbon exchange through decomposition and respiration and indirectly by developing symbiotic associations with plants. One of the most common symbiotic associations established between plants and fungi is known as arbuscular mycorrhizal fungi (AMF). This association facilitates the host plants for the better acquisition of water and nutrients and seems to sequester soil organic carbon. AMF could play a vital role in the global carbon cycle, as they can utilize a large proportion of the carbon fixed by the plants, deposit slow-cycling organic compounds (glomalin), and protect organic matter from microbial attack by promoting soil aggregation. In view of the importance of AM symbiosis in the terrestrial ecosystems, this chapter highlights whether the arbuscular mycorrhizal fungi contribute to soil carbon sequestration or influence soil carbon decomposition.

15.1 Introduction

Predicting the impacts of human activities on Earth's climate, the Intergovernmental Panel on Climate Change has decisively advocated that human-induced environmental change is a major area of concern. In the recent UN climate conference

B. Giri (✉) • B. Saxena
Department of Botany, Swami Shraddhanand College, University of Delhi, 110036 Delhi, India
e-mail: bhoopg@yahoo.com; bhawnasaxena5@gmail.com

A. Varma et al. (eds.), *Mycorrhiza - Function, Diversity, State of the Art*,
DOI 10.1007/978-3-319-53064-2_15

held in Paris, scientists, bureaucrats, policy makers, and economists discussed global plans for mitigating and adapting to climate change issues as predicted to occur in the twenty-first century (Treseder 2016). These projections of global climate change are framed from Earth system models, which are the large-scale integrated models, predicting about greenhouse gas emissions in the atmosphere (Treseder 2016). Researchers have received attention to identify natural sinks of atmospheric carbon dioxide (CO_2) and other greenhouse gases to minimize the risks of global climate change. Nonetheless, in view of the existing feedbacks between global carbon (C) cycle and climate change, it is of paramount importance to understand whether terrestrial ecosystems will respond to elevated atmospheric carbon dioxide concentration (CO_2) or up to what extent (Selsted et al. 2012).

The marine cyanobacteria (also known as blue-green algae) produced first molecules of oxygen about three and a half billion years ago; since then microbial processes have been considered the crucial drivers of climate change (Schopf and Packer 1987; Singh et al. 2010). It is widely accepted that microorganisms play an important role in determining the atmospheric concentrations of greenhouse gasses (CO_2, CH_4, and N_2O) right through the Earth's history (Bardgett et al. 2008). As the twenty-first century projected to experience some of the most rapid climatic changes in our planet's history (Pachauri and Reisinger 2007), one of the most important debates could be how do we exploit and improve our understanding of microbial processes to manage or mitigate the emerging problem of climate change (Singh et al. 2010).

The current level of atmospheric CO_2 depends largely on the balance between photosynthesis and respiration. In the ocean ecosystems, photosynthesis is primarily carried out by the phytoplankton and autotrophic and heterotrophic respiration returns much of the carbon assimilated during photosynthesis to the dissolved inorganic carbon pool (Del Giorgio and Duarte 2002; Arrigo 2005). In terrestrial ecosystems, atmospheric CO_2 is largely utilized by higher plants by the process of photosynthesis for the net primary production, but microorganisms contribute greatly to net carbon exchange through the processes of decomposition and heterotrophic respiration, as well as indirectly, through their role as plant symbionts like arbuscular mycorrhizal fungi, which improve nutrient and water availability to host plant in soil (Van der Heijden et al. 2006, 2008).

Arbuscular mycorrhizal fungi (AMF) are important components of terrestrial ecosystems as they provide a crucial link for nutrient exchange between plants and soil. The photosynthate produced by the host plant is required by AMF to fulfill their C requirement for growth and reproduction; thus, it flows from plant to the fungi; in reverse, mycorrhizal fungi facilitate plant by providing mineral nutrients and other benefits (Smith and Read 1997). AMF are expected to modulate plant response to CO_2, by means of improving nutrient status and increasing plant tolerance against a range of environmental stresses (Smith and Read 1997). AMF are likely to play a critical role in the global carbon cycle, as they can utilize a large proportion of the carbon fixed by the plants (up to 20% or approximately 5 billion tons of carbon per year) under ambient atmospheric CO_2 (aCO_2) (Jakobsen and Rosendahl 1990; Bago et al. 2000; Drigo et al. 2010), deposit slow-cycling organic compounds (Smith and Read 2008), and protect organic matter from microbial attack by promoting soil aggregation (Wilson et al. 2009). Treseder (2016) suggested that AMF play an important role in soil C sequestration, particularly under elevated CO_2 (eCO_2) concentration. The organic substance like chitin has been reported to constitute about 60% of fungal cell walls. The compound is not readily decomposed, thus storing C for a long period in the soil. Moreover, AMF are the sole producers of recalcitrant glue-like glycoprotein, the glomalin, and its concentration in soil is positively correlated with the water stability of soil aggregation (Wright et al. 1996; Wright and Upadhyaya 1996, 1999). After the death of fungal hyphae, glomalin enters into the soil and becomes a source of soil organic carbon sink. As AMF represent a major link between atmospheric and soil-retained carbon (C), thus in order to understand the fate of atmospheric CO_2 under the projected increase in near future, it is of utmost significance to understand how plants and AMF interaction could manage atmospheric CO_2 rise through their effects on C pooling or soil C decomposition (Verbruggen et al. 2013).

15.2 Mycorrhiza: A Fungal–Plant Symbiotic Association

The term mycorrhiza (*myco* (fungus) + *rhiza* (roots)), meaning fungus root, was used for the first time in the year 1885 by A. B. Frank. Among the myriad of soil microorganisms, the fungi are distinctive because of their ubiquity in soils around the globe and their ability to form symbiotic associations with plant species. They are known to colonize all types of land plants, including vascular and nonvascular plants. They develop bridges connecting the plant roots with surrounding soil particles improving nutrient acquisition, water absorption, and soil structure and enabling the plant to combat environmental stresses and therefore may have a profound impact on terrestrial ecosystems (Hooker and Black 1995).

So far, seven different categories of mycorrhizal symbiosis have been distinguished on the basis of their morphological characteristics and the fungal and plant species involved, namely:

(1) Arbuscular mycorrhiza—the most ancient and widespread symbiotic relationship (discussed in detail in the next section).
(2) Ectomycorrhiza—the second largest group of fungi associated with woody trees and shrubs and some herbaceous taxa.
(3) Ericoid mycorrhiza—this group of fungi colonizes a taxonomically narrow group of plant families (Ericaceae, Empetraceae, and Epacridaceae (order Ericales)) and *Basidiomycetes*, *Sebacinales* are also common ericoid mycorrhizal fungi. *Ascomycetes* show affinity to this association.
(4) Orchid mycorrhiza—restricted to family Orchidaceae.
(5) Monotropoid mycorrhiza—achlorophyllous monotropoid plants are completely dependent on the fungi for reduced carbon and soil nutrients. Structurally these fungi are like ectomycorrhizas.

(6) Arbutoid mycorrhiza—the association formed between fungi that are normally ectomycorrhizal and plants in the genera *Arbutus* and *Arctostaphylos* and the family Pyrolaceae, where intracellular fungal penetration occurs.
(7) Ectendomycorrhiza—exhibits characters of arbuscular mycorrhiza or ectomycorrhiza, here sheath is reduced or may be absent, Hertig's net is well developed, and intracellular penetration is also found (Srivastava et al. 1996; Smith and Read 1997).

15.3 Arbuscular Mycorrhiza: The Most Ancient Plant–Fungal Association

Arbuscular mycorrhizal fungi (AMF) are one of the most ancient associations having a long evolutionary history (Schüβler and Walker 2011). On the basis of fossil records and phylogenetic analysis, it has been suggested that in the early Devonian era, about 450 million years ago, plants have established an intimate association with certain filamentous fungi, which are now known as mycorrhizal fungi (Srivastava et al. 1996; Torres-Cortes et al. 2015). It is also believed that the progression of plants from water to land would not have been possible without mycorrhizal association. Many types of mycorrhizal associations are recognized by the organisms involved in the formation of symbiosis and the development and structure of mycorrhiza (Smith and Smith 1997). The fungal associates range from largely aseptate endophytes belonging to the order *Glomerales* of phylum *Glomeromycota* to septate fungi belonging to *Ascomycetes* to *Basidiomycetes*. Mycorrhizal fungi are classified on the basis of nature of fungal hyphae, extent of penetration of root, production of external fungal sheath, and various inter- and intracellular structures produced by fungal partners on association with root (Morton and Benny 1990; Walker 1992; Smith and Read 1997) (Fig. 15.1).

AMF are ubiquitous both in terms of their wide geographical distribution and their association with land plants, as they are associated with the roots of more than 90% of all plant species (Bonfante and Genre 2010). They extend their hyphae (100 m of hyphae per cubic centimeter of soil) in soil to explore soil nutrients and thereby facilitate the host plant for acquiring more mineral nutrients from soil (Miller et al. 1995). Though AMF are found in different types of soils, these fungi occur in their active phase of life in the rhizosphere, where they are associated with plant root to reproduce their progeny (asexually). The AMF symbionts are considered intractable organisms by researchers because they are obligate biotrophs, aseptate, multinucleate, and produced only asexually and cannot be cultured on an artificial medium under controlled conditions like other microbes (Rosendahl 2008).

Rosewarne et al. (1997) pointed out that the development of a functional AM symbiosis is the most complex process, which is dependent on molecular

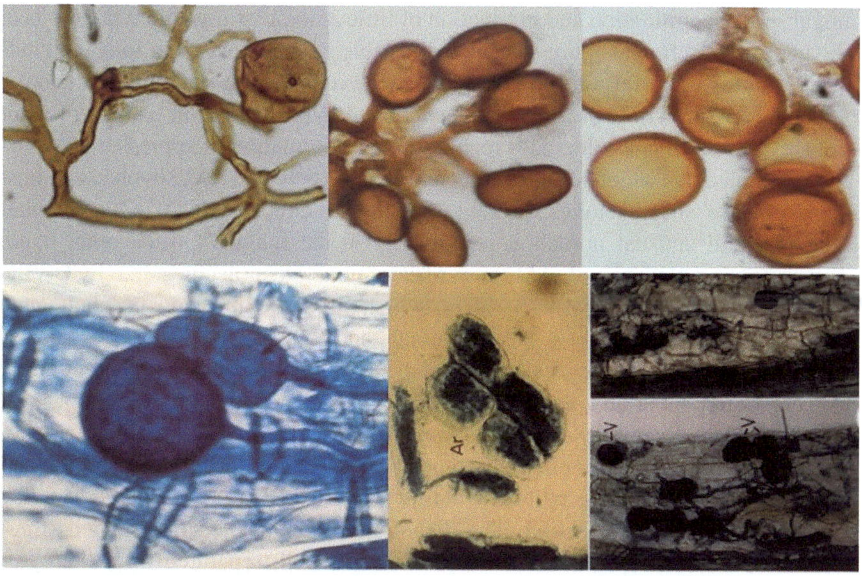

Fig. 15.1 The figure depicts formation of AMF structures in host plant root. The first row shows AM fungal spores formation in soil. The second row shows the dichotomously branched fingerlike projections arising from the branches of intraradical hyphae, the arbuscules, and balloon-like vesicles, produced intercalary from intraradical hyphae in the cortical cells of root; arbuscules serve as the site for nutrient exchange between AMF and host plants, while vesicles act as the storage organ of the fungus

communications between plant and fungal symbionts. Till late 1990s, the molecular mechanisms that govern signaling and recognition between AMF and their host plants were poorly understood. However, with the beginning of the twenty-first century, pioneer works carried out by Kosuta et al. (2003), Akiyama et al. (2005), Navazio et al. (2007), Gutjahr et al. (2008), and Chabaud et al. (2011) identified molecular mechanisms and bioactive compounds involved in the establishment of functional AM symbiosis. During host recognition stage and before hyphopodia (or appressoria) formation at root surface, AMF hyphae go through extensive branching in the vicinity of host plant roots. The host plant roots release low-molecular-weight signaling molecules that trigger hyphal branching and play central roles in inter-organism communication. However, the chemical nature and mode of action of such molecules remained unknown till late 1990s. In a pioneer work, Akiyama et al. (2005) isolated a branching factor from the root exudates of *Lotus japonicus* and identified it as a strigolactone (5-deoxy-strigol). The strigolactone (a group of sesquiterpene lactones) isolated from *L. japonicus* was reported to induce extensive hyphal branching in germinating spores of the AM fungus *Gigaspora margarita* at very low concentration (Akiyama et al. 2005).

The establishment and development of AM symbiosis in the plant roots may involve three phases: (1) asymbiotic phase, (2) pre-symbiotic phase, and (3) symbiotic phase (Bonfante and Anca 2010). In the asymbiotic phase, induction of AM fungal spores germination and production of little amount of mycelium takes place. The germinating fungal spores produce diffusible Myc factors—the fungal signaling molecules—and respond to host plant root exudates by switching to an active pre-symbiotic growth phase. In this phase fungal hyphae come in contact of plant root surfaces. This phase includes recognition and formation of appressoria events. The key molecules which initiate molecular dialogs between AMF hyphae and host plant root are strigolactones, present in the root exudates, and induce hyphal branching and unusual mitochondrial activity in the fungus. On the other hand, after perceiving plant signal, AM fungi release bioactive molecules, like LCOs, the Myc factor (lipochitooligosaccharides), which induce expression of plant gene *ENOD11* and *SYMRK* and *CCaMK* (Bonfante and Requena 2011).

The symbiotic phase includes hyphal penetration into roots, organization of pre-penetration apparatus, and release of cell-wall degrading hydrolytic enzymes, membrane depolarization, and Ca spiking, which consequently establish a functional symbiosis. The pre-penetration apparatus (PPA), a novel cytoskeletal organization, develops in epidermal cells before infection. PPA is considered as a key cellular factor in AM infection. The dichotomously branched tree-shaped structures—the arbuscules—are formed in the cortical cell, separated from the cytoplasm with the help of periarbuscular membrane (PAM). It is also reported that a type of apoplastic interface is formed between the plant and fungus by PAM (Bonfante and Requena 2011).

15.4 Systematic Position of Arbuscular Mycorrhiza

In the last few decades, the classification of AMF has undergone considerable transformations (Fig. 15.2). Till the early 1990s, AMF were considered the members of *Zygomycota* (Gerdemann and Trappe 1974). By the time, their taxonomy was exclusively based on phenotypic characters like spore morphology and the structure and development of spore wall (Morton and Benny 1990; Walker 1992). In 1990, Morton and Benny transferred genera of AMF from the order *Endogonales* to a new order *Glomerales*.

On the basis of cladistic analysis of genetic and phyletic characters (nuclear-encoded rRNA gene markers, sequence variation of the SSU rDNA), AMF have been classified in the phylum *Glomeromycota*, a monophyletic group that diverged from the same common ancestors as *Ascomycota* and *Basidiomycota* (Redecker and Raab 2006; Redecker et al. 2013; Schüβler et al. 2001, 2006; Schüßler and Walker 2010). The phylum *Glomeromycota* is currently divided into four orders, namely, *Diversisporales*, *Glomerales*, *Archaesporales*, and *Paraglomerales* (Schüβler et al. 2001). Three families *Acaulosporaceae*, *Diversisporaceae*, and *Gigasporaceae* have been described within the order *Diversisporales*. Two families, the *Acaulosporaceae* and *Gigasporaceae*, contain the greatest number of described species. Order *Glomerales* consists of monotypic family, the *Glomeraceae* with genus *Glomus*, although species have been recognized in two distinct clades

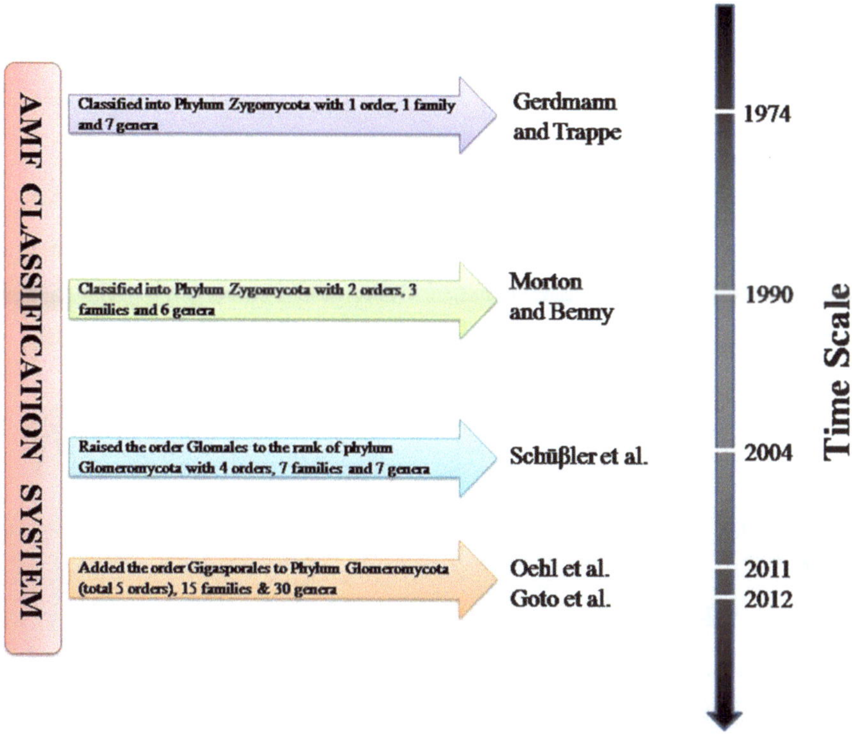

Fig. 15.2 Evolutionary transformation of AMF classification

(Schwarzott et al. 2001; Powell et al. 2009). Oehl et al. (2008, 2011) erected a few new families and genera in the new classification based on the combination of both genetic and phenotypic characters. Recently, Redecker et al. (2013) compiled all information from the different classifications and presented a recent classification of *Glomeromycota* which is based on a consensus of regions spanning ribosomal RNA genes: 18S (SSU), ITS1-5.8S-ITS2 (ITS), and/or 28S (LSU).

15.5 Relevance of Arbuscular Mycorrhizal Fungi to Global Climate Change

Over the last 20 years, considerable attention has been received to understand the role of soil microorganisms in carbon (C) and nitrogen (N) cycling, particularly under elevated carbon dioxide, which is predicted to be augmented in coming decades, if the current state of affairs prevails. Several models have been proposed to understand a link between terrestrial ecosystem nutrients cycling, in respect to global climate change. The role of soil microorganisms has been considered central to it, as they could convert soil organic matter in atmospheric carbon dioxide and mineral nutrients

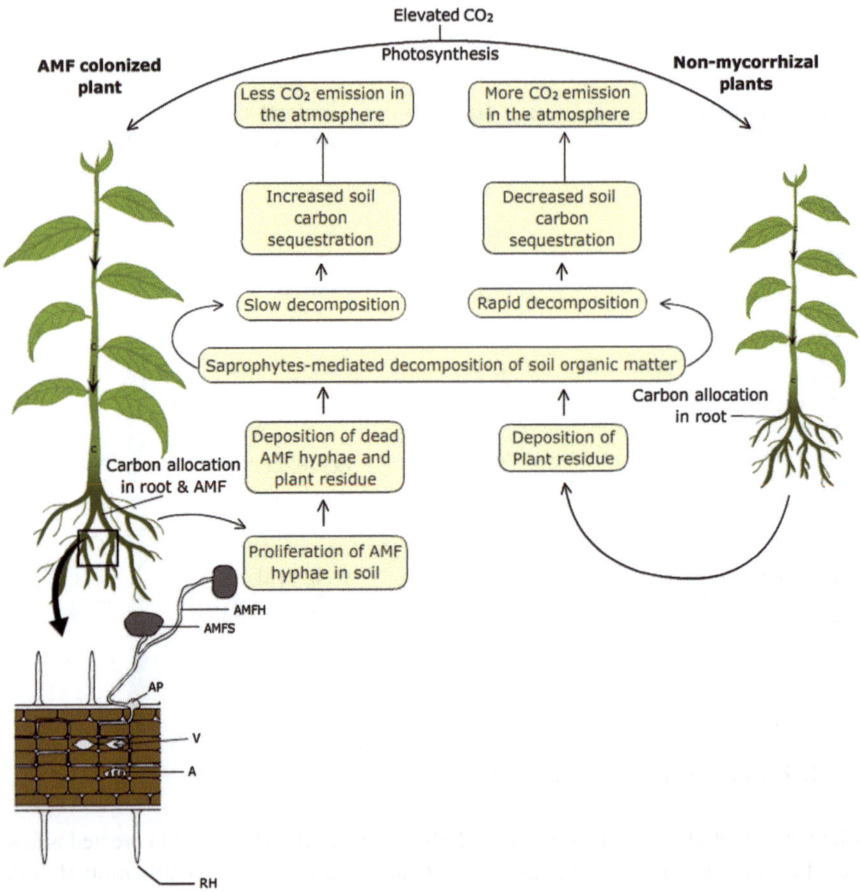

Fig. 15.3 The model framework shows influence of elevated atmospheric CO_2 on plant growth and AMF. The elevated CO_2 concentration in the atmosphere results in increased allocation of C to root and AMF, which indeed influence root growth, and health and proliferation of AMF extraradical hyphae in soil. AMF hyphae release recalcitrant substance in soil, which are difficult to be decomposed by saprophytes and may remain in soil up to several decades; thus AMF help in increased soil C sequestration and mitigate C emission in the atmosphere, while non-mycorrhizal plants deposit only plant residue in soil, which is rapidly decomposed by the activity of saprophytes, thus enhancing emission of C in the atmosphere

(saprophytic fungi) or into a pool of extremely slow-decaying recalcitrant organic matters in the soil (Talbot and Treseder 2011). Indeed, a diverse group of microorganisms occurs in the soils, but they differ in their metabolic abilities to decay organic sources like plant litter, animal residues, humus, and microbial biomass, consequently differentially contributing to C emission or sequestration (Fig. 15.3).

Fungi play a crucial role in circulating atmospheric CO_2 in the terrestrial ecosystems. A majority of soil fungi are saprophytes, which act on dead organic matter, carry out the process of mineralization, decompose the complex organic compounds in to simple compounds, and by the way facilitate release of mineral nutrients (including C) in soil. In turn, other group behaves as obligate biotrophs—the arbuscular mycorrhizal fungi (AMF)—which lack saprophytic abilities. The impacts of AMF on soil carbon cycling are important because these fungi develop a large and complex hyphal network in the soil and constitute about 20–30% of total soil microbial biomass, thus playing a critical role in soil C storage (Rillig et al. 2001a, b, 2003; Treseder and Turner 2007; Treseder 2016; Verbruggen et al. 2013, 2016). Their non-saprophytic nature basically promotes soil C sequestration. AMF extraradical hyphae facilitate the formation of soil aggregates by entangling, enmeshing, and binding soil particles with the help of glomalin-related soil protein, thus making soil organic carbon resistant to be decomposed by other soil microbes (Zhang et al. 2016). In contrary, Cheng et al. (2012) recently suggested that AMF could exert a priming effect on the rhizosphere microbes, therefore participating indirectly in decomposition of organic matter. In this chapter, on the basis of previous and recently published research papers and review articles, we attempt to highlight both aspects, whether the AMF fungi help in soil C sequestration or influence soil C decomposition.

15.6 AMF Synthesis of Glomalin: The Soil Glue

Glomalin is a glycoprotein produced in abundance by the extraradical hyphae (within hyphal walls) of arbuscular mycorrhizal fungi (Wright and Upadhyaya 1996). Glomalin protein accounts for a large amount of organic carbon (30–40%) in undisturbed soils (Rillig et al. 2003). In addition to containing iron, which imparts red color to this compound, it appears to also contain N-linked oligosaccharides (Wright et al. 1999, 2006; Wright 2000). Glomalin is insoluble in water but soluble at high temperature (121 °C). In its natural state, it seems to be hydrophobic (Wright et al. 1996; Rillig and Steinberg 2002). Glomalin helps the extraradical hyphae to keep water and nutrient supply intact during their absorption and movement in the plant. Glomalin is deposited in soils as the fungal hyphae die, and remain accumulates until it represents about 5% of soil carbon and nitrogen (Rillig 2004a, b; Lovelock et al. 2004a, b).

Glomalin was thought to be exuded by the member of other fungal groups; however, no fungal group other than *Glomeromycota* produces this recalcitrant glutinous glycoprotein in copious amount (Wright and Upadhyaya 1996). Glomalin is extremely tough and resistant to microbial decay and hence may last 7–42 years in soil environment, depending on conditions (Nichols et al. 2002). Nevertheless, it is a good protector of fungal hyphae. Glomalin is considered to be distributed ubiquitously as it is found in agriculture, grasslands, desert forests, and noncultivated soils (Wright and Upadhyaya 1996; Rillig et al. 2003; Nichols and Wright 2004; Antibus et al. 2006; Bai et al. 2009). Glomalin acts as glue with hydrophobic properties; however, its direct biochemical evidences remain unknown. Driver et al. (2005) demonstrated that instead of being secreted in the liquid growth medium, glomalin is strongly associated with fungal mycelium, which enters the soil after mycelia death and decomposition (Treseder and Allen 2000).

15.7 Extraction Methods of Glomalin

From soil, glomalin can be extracted with sodium citrate (50 mM) buffer with high temperature (121 °C). A soil containing glomalin appears rich brown colored, which gets transformed into a mineral gray color after removal of glomalin. Four common methods have been proposed to determine glomalin (Rillig 2004a, b; Rosier et al. 2006):

1. Bradford-reactive soil protein (BRSP) analysis
2. Easily extractable BRSP (EE-BRSP) analysis
3. Immunoreactive soil protein (IRSP) analysis
4. Easily extractable IRSP (EE-IRSP) analysis

In most soils, total glomalin is extracted within 1 h with heat. Though a fraction of glomalin can be extracted in the first extraction cycle within 30 min at 121 °C using 20 mm sodium citrate at neutral pH (it is termed as easily extractable), additional extraction cycles with a long extraction time and more concentrated sodium citrate solution are required to remove all the glomalin from a soil (Wright and Upadhyaya 1996). With the help of IRSP and EE-IRSP using a specific monoclonal antibody, the concentrations of immunoreactive glomalin can be estimated, while the BRSP and EE-BRSP analysis can be undertaken for the general assay of protein content (Wright and Upadhyaya 1996). Researchers considered IRSP and EE-IRSP (antibody-based determination) analysis of glomalin more specific than the BRSP and EE-BRSP analysis of glomalin protein (Treseder and Turner 2007). Glomalin concentrations range from 2 to 15 mm/g of soil in temperate climates and over 60 mg cm^{-3} in a chronosequence of Hawaiian soils (Wright and Upadhyaya 1998; Rillig et al. 2001a, b; Lovelock et al. 2004a).

15.8 The Role of AMF in Soil Particle Aggregation

Soil structure is important for facilitating water infiltration, biogeochemical cycling processes, resistance against erosional loss, and soil carbon storage (Oades 1984; Elliott and Coleman 1988; Hartge and Stewart 1995; Jastrow and Miller 1997). The process of soil aggregation is complex. In this process, numerous organisms and binding agents are involved (Tisdall and Oades 1982; Miller and Jastrow 2000). Since their widespread distribution, broad host range and abundant members of the soil biota, AMF are considered an important biotic factor in the terrestrial ecosystems. Their extraradical hyphae have been found to substantially contribute in the formation and maintenance of soil aggregates (Tisdall and Oades 1982; Six et al.

2004; Rillig and Mummey 2006) (Fig. 15.4). The hyphal growth of AM fungi offers itself to stabilize structures, and the relative persistence of hyphae and their products make AMF important in longer-term aggregate stabilization (Miller and Jastrow 2000; Rillig et al. 2001a; Leifheit et al. 2014). Wright and Upadhyaya (1998) found a strong relationship of AMF hyphal product "glomalin" with soil aggregate water stability. Furthermore, Jastrow et al. (1998) also found that AMF hyphae provided the most important direct effect on soil aggregation of all soil factors. Using a path analysis study, Rillig et al. (2002) observed that glomalin produced by the hyphae of AMF significantly contributes to soil aggregate water stability. A positive correlation was observed between soil aggregate water stability and GRSP concentrations in different soils under different cropping systems and management practices by Wright and Anderson (2000). Moreover, glomalin was found to play an important role in soil aeration and drainage, plant nutrient uptake, and productivity (Nichols and Wright 2004). Owing to the importance of AMF, it is recently advocated to improve their representation in Earth system models, as well as into ecosystem-scale models (Treseder 2016). Integrating AMF dynamic into large-scale models may improve projections of soil C sequestration as they are globally abundant (Treseder 2016).

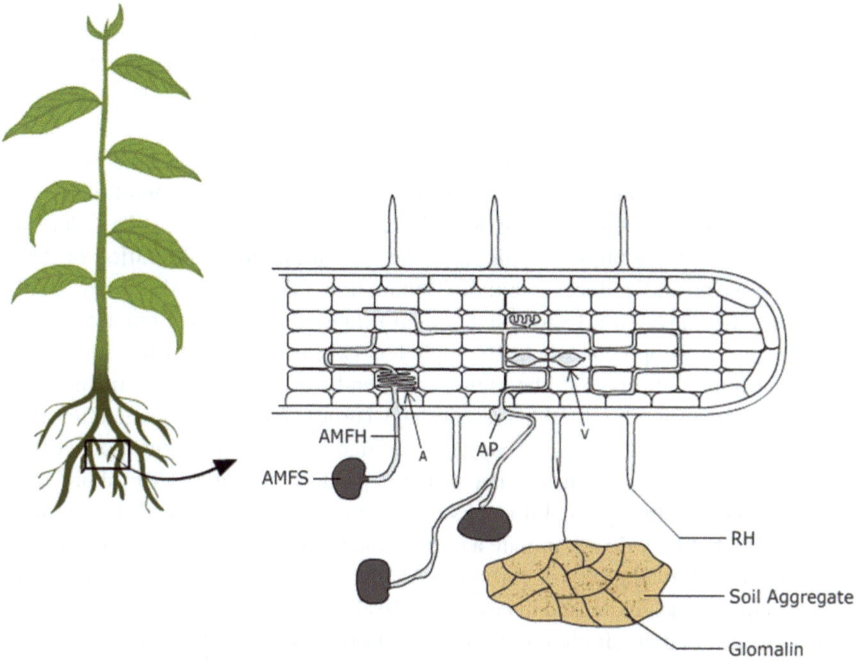

Fig. 15.4 The picture depicts formation of soil aggregates by AMF

15.9 Factors Affecting Glomalin Stocks in Soil

The deposition of glomalin in a soil is directly proportional to the growth of AMF extraradical hyphae. Glomalin is not exuded by AMF hyphae; instead, it is present within hyphal walls. Therefore, the rate of glomalin deposition in a soil is determined by the availability of hyphal standing stocks, hyphal glomalin content, and hyphal turnover rate (Treseder and Turner 2007). Glomalin stocks in soils may be indirectly influenced by a number of factors, which control both plant and AMF growth. One of the most common factors is rate of photosynthesis, which could determine absolute amounts of carbon available for mycorrhizal fungi (Johnson et al. 2002; Lovelock et al. 2004a). Photosynthetic rates could control both the growth and abundance of AMF. Plants often allocate more carbon to AMF under nutrient-deficient conditions where plant growth is limited, while excessive nutrition or fertilization of soil with P and N often reduces AMF growth and may be due to alleviation of nutrient limitation (Treseder 2004). On the other hand, excessive amount of atmospheric CO_2 has been found to increase both AMF growth and abundance and therefore could enhance the production of glomalin. Increased growth and abundance of AMF under eCO_2 may be as plants are more N or P limited or plant carbohydrate is more readily available or both (Read 1991). The availability of inorganic resources (mineral, water, etc.), composition of the plant community, and soil texture could also influence glomalin yield (Rillig and Steinberg 2002; Rillig et al. 2002). Wright and Anderson (2000) observed that lower AM root colonization rate in sunflower decreases glomalin production in soil, which was higher in the case of corn or proso millet crops. The crop rotation, cropping systems, and land management practices also alter accumulation of glomalin in soils (Wright and Anderson 2000; Wright et al. 2007). Further, water availability also affects glomalin yield. It was found that AMF are more abundant if the availability of water is low, which could be due to more investment of C by plant to AMF for efficient utilization of water under water-limited conditions (Augé 2001; Al-Karaki and Clark 1999; Ruiz-Lozano and Azcon 1995; Treseder and Turner 2007). Tillage physically disrupts the formation and proliferation of AMF hyphal network and reduces glomalin production (Treseder and Turner 2007). Therefore, the conversion from conventional tillage to zero-tillage practices could enhance the development of AMF hyphal network. Consequently, the rate of glomalin accumulation would be higher in an undisturbed forest as compared to a regularly plow agricultural field. Treseder and Turner (2007) conducted a literature survey of glomalin from 22 ecosystems to understand a relationship between standing stocks of glomalin and net primary productivity and AMF abundance. They summarized that glomalin stocks are positively correlated with net primary productivity, but they do not initiate a correlation with AMF abundance. It was also summarized that availability of carbon to AMF could influence glomalin dynamics and glomalin stocks, increased in case of AMF abundance and efficient AMF hosts, which narrates that an efficient host supplies more C to AMF, leading to enhanced AMF health, and glomalin synthesis by fungal hyphae. They also found that CO_2

shows a positive impact on the growth of AMF and glomalin concentration in soil; however, the effect of land use change did not show a consistent effect on the synthesis of glomalin stocks, but the glomalin dynamics appear to be linked to C dynamics.

15.10 The Role of AMF in Nitrogen Cycling

Until the 1980, AMF were thought to uptake inorganic forms (NH^{4+}, NO_3^-) of nitrogen (N). Conversely, results of recent studies stress upon the ability of AMF to acquire nitrogen from organic compounds; however, the extent to which they can obtain N from organic materials remains unclear. The inorganic forms particularly ammonium (NH^{4+}) and nitrate (NO_3^-) of N are more mobile and therefore diffuse rapidly, especially NO_3^- through soil matrix (Tinker and Nye 2000). In fact, in a majority of natural ecosystems, most of N is in complex organic forms and distributed hetérogeneously in the soil matrix. Complex organic forms of N present in soil necessitate depolymerization with the help of extracellular enzymes secreted by soil microbes, and then it could transport to plant roots (Schimel and Bennett 2004). Otherwise, these forms of N are generally beyond the approach of plant roots. Unlike to free-living soil fungi, which commence decomposition and mineralization of N from organic compounds, particularly from recalcitrant organic materials like lignin, there is no evidence of direct participation of AMF in such activities, obviously due to their lack of saprophytic capabilities. Recent investigation carried out by Leigh et al. (2009) provides evidence that extraradical hyphae of AMF colonize soil extremely, thus increasing attempts to encounter a nutrient-rich stock of organic material undergoing decomposition. They tested whether AMF (*Glomus hoi* and *Glomus intraradices*) could capture N from organic compounds and transfer it back to their host plant or not. The study demonstrated that two species differ in their capacity to take up N and transfer it to host plant back. Comparing both the *Glomus* species, *Glomus intraradices* was found to be more efficient than *G. hoi*, likely as *Glomus intraradices* produced large amount of external hyphae. Experiments conducted by Hodge and Fitter (2010) to understand the role of AMF for acquisition of N from organic materials also suggested that AMF can obtain substantial amount of N from decomposed organic residue patches. Fellbaum and his co-workers recently observed interactions between C and N transfer in AM symbiosis. They used AMF root organ cultures and manipulated the C flux to the host plant and AMF. Manipulating C supply of the host plant elicits the acquisition and transport of N in AMF symbiosis that could happen due to changes in the fungal gene expression. Further, transport of N gets stimulated if the C is delivered by the host across the AMF interface, instead of its direct supply to the extraradical hyphae of fungi in the form of acetate (Fellbaum et al. 2012). In this study, AMF culture was of artificial nature; therefore, it was suggested that further studies should be conducted to better understand interactions between C and N transport in host plants under natural conditions.

Recently, Cheng et al. (2012) advocated the role of AMF in acquiring N from organic matter under elevated CO_2 (eCO_2) condition by conducting an elegant microcosm experiment. These researchers demonstrated that in the presence of AMF, at elevated CO_2, fresh aboveground dead organic material decomposes faster, increasing N status of host plant. They recommended that it could happen due to the active participation of AMF in accelerating decomposition of organic compounds. Countering the report Verbruggen et al. (2013) pointed out that these experiments were conducted for a short term; therefore, such results should not be directly correlated with long-term studies (decade-scale studies). They appealed that short-term experiments do not account for potential accumulation of plant or microbial-based organic matter, triggered by increased decomposition; therefore, it is difficult to conclude. Moreover, the long-term effects of AMF, particularly on the decadal scale, could be qualitatively different from the short-term effects.

15.11 The Role of AMF in Soil Carbon Cycling

On the basis of the presence of organic matter content, soils can be characterized as mineral or organic. A mineral soil forms most of the world's cultivated land. Such soils may contain up to 30% of organic matter. Indeed organic soils are rich in organic matter and contain more than 30% of organic matter. Soil organic matter (SOM) is predominantly produced by living organisms (plants and animals) by the process of decomposition, which is returned back to the soil. In addition to 60–90% of moisture, plant biomass consists of carbon, oxygen, hydrogen, and small amounts of nitrogen, phosphorus, potassium, sulfur, magnesium, and calcium, which are vital components for the management of soil fertility. In the past few decades, attention has been paid toward increasing atmospheric CO_2 concentration, which is one of the major environmental abiotic factors responsible for global environment change (Wayman 1991). Understanding the effects of human-induced environmental change, the impact of increasing CO_2 concentration on terrestrial ecosystems is an area of main concern. Rising CO_2 concentration in the atmosphere has been found to accelerate plant photosynthate and belowground allocation of carbon, providing a potential buffering mechanism against elevated carbon dioxide (King 2011).

AMF are the integral component of most terrestrial ecosystems and are gradually being found to be an important component of the soil organic carbon (SOC) pool. A number of research studies have shown their participation in soil C sequestration, thus predicting AMF contribution in mitigating effect of labile CO_2 on climate change (Treseder 2016; Verbruggen et al. 2016). They play a critical role in the global C cycling as they consume approximately 20% of net plant photosynthate and deposit slow-cycling organic compounds like chitin and glomalin in soils (Smith and Read 2008; Wilson et al. 2009). The SOC is considered as a crucial regulator of C fluxes between both, the biosphere and the atmosphere. The death of

AMF and their hyphae conquer deposition of organic C in soil. Therefore, the global annual flux of carbon into AMF may be significant. The organic carbon added by the AMF hyphae contains recalcitrant substances like glomalin, which may remain intact for a long period and therefore help in sequestering organic C in soil (Wilson et al. 2009; Treseder 2016). Treseder and Turner (2007) found that glomalin deposited in the soils represents tens to hundreds of grams of carbon/m^2/year (a large flux of SOC). The mechanism influencing SOC storage in soil depends mainly on the net primary production and allocation of the plant photosynthate in the above- and belowground structures. AMF largely improve net primary productivity by facilitating plant acquisition of P and N; in exchange, they acquire C from their host plant, and therefore, a portion of net primary productivity is dispensed belowground to AMF. In an ecosystem dominated by mycorrhizal fungi, AMF can receive up to 47% of belowground net primary productivity (Treseder 2016). Therefore, net primary productivity is a major factor in the sequestration of soil organic carbon.

AMF have been considered to be a major carbon source, but their comprehensible role in C sequestration to combat climate change is still undecided. Long-term C deposition in a forest ecosystem depends on the balance between C gains from net primary productivity and C losses by the decomposition of soil organic matter. However, it is uncertain which factor(s) controls C retention and which controls C loss in the forest ecosystems. Leifheit et al. (2015) observed reduced decomposition of woody plant litter in the presence of AMF. Further, the accumulation of carbon in a soil is controlled by the size and activities of the microbial biomass present in soil. In a study, using $^{13}CO_2$-labeling isotope, researchers observed a rapid C flux in the form of glomalin from AMF hyphae to the soil (Johnson et al. 2002; Clemmensen 2013; Treseder and Turner 2007), which is extremely difficult to break down into simple compounds, thereby preventing emission of atmospheric carbon dioxide, suggesting a major role of AMF in global climate change. The sticky nature of glomalin protein enables it to protect the carbonaceous material from rapid degradation in soil and stabilize SOM by promoting soil aggregation (Rillig 2004a; Wu et al. 2014). Wilson et al. (2009) observed a reduction in soil C and N content with AMF suppression that leads to decreased fungal hyphae biomass and consequently decreased accumulation of glomalin. The decreased accumulation of AMF hyphae and glomalin protein could lead to the loss of C and N protected in microaggregates by reducing aggregate stabilization. Thus, AMF largely contribute for the removal of CO_2 from the atmosphere by colonizing plant roots and by the deposition of substantial amount of C in soil.

The activity of soil microbes is generally limited by the availability of labile C in soil. Rising concentration of CO_2 in the atmosphere causes increased microbial activities that improve the decomposition of complex organic materials into simple form eventually releasing nutrients in soil (Cheng and Kuzyakov 2005). Nutrient availability has been found to be one of the very vital factors limiting soil C sequestration. There are two schools of thought; according to one, AMF largely contribute to soil C storage, indirectly by utilizing more atmospheric CO_2 to increase plant growth and biomass production, and a copious amount of plant photosynthate is utilized by the AMF for its own growth and for the growth of extraradical hyphae, which proliferate in the soil vigorously and release organic C as and when senesced and decomposed. However, another school of thoughts advocated a paradigm shift in the role of AMF in climate change. According to them, in the presence of AMF at elevated atmospheric carbon dioxide (CO_2), the extra soil carbon is respired back to the atmosphere, because AMF exert priming effect on saprophytic soil microbes which consequently stimulate additional decomposition of soil organic carbon (Cheng et al. 2012). This work raised a question on the C sequestering capacity of the AMF and their role in mitigating climate change. Further evidences particularly long-term experiments are required to authorize this fact.

To better understand the contribution of AMF in soil carbon sequestration, Clemmensen (2013) investigated a portion of boreal forest islands using a mathematical model to partition soil carbon storage either derived from aboveground parts of plant litter or from underground part—the roots and the microbial biomass like AMF associated with plant roots. They investigated that about 70% of the organic carbon stored in the soil is derived from roots and root-associated microorganisms. They also reported that AMF and roots associated with microbes other than fungi contribute largely deeper in soils where root density is higher, while decomposers are more prevalent in shallow soils. These findings showed that AMF play a significant role in terrestrial ecosystems by sequestering soil carbon. The study suggested that on the death of AMF hyphae, the carbon present in AMF hyphal tissues may remain intact for so many years or even decades. Ecological studies suggested that the longer deposition of carbon in the soil leads to greater soil carbon sequestration. These findings are in accordance with Prescott (2010) who also reported that carbon derived from the soil microbes (like AMF) is not easily decomposed, while it remains unbroken for a longer period than the carbon obtained from plant residues. The longer stay of microbe-driven carbon in the soil may be due to its origin from cell walls, chitin, glucan, and peptidoglycans or polysaccharides like substances, which make these compounds more difficult to be decomposed by decomposers. In this respect, the abundance of AMF in a soil could be directly related to soil carbon storage.

In a disturbed ecosystem or a burned forest, the role of AMF to soil carbon sequestration could be lesser as compared to an intact or unburned forest ecosystem, as the burning makes AM fungal tissues degradable faster than natural decay of fungal tissues. On the other hand, in a disturbed ecosystem, the growth of AM hyphal is less as compared to an intact ecosystem, which is directly proportional to organic carbon accumulation in soil. The role of AMF could be critical in northern forests where instances of wildfire are common. Kasischke and Turetsky (2006) suggested that if the current trend of global warming continues, the contribution of AMF to reduce atmospheric CO_2 may drop, because under such circumstances the cases of wildfire in these areas are likely to increase, which can enhance degradation of AMF-driven carbon in soil that would eventually decrease the soil C sequestration in these regions. Treseder and Holden (2013) suggested that in an ecosystem, the decomposition of mycorrhizal residues including its hyphae and augmentation of plant growth may determine the contribution of AMF in soil carbon sequestration. The role of AMF in improving plant growth determines the extent of the deposition of soil carbon from plant litter. They also highlighted that modern agricultural practices like application of pesticides, excessive use of chemical fertilizers, intensive cultivation, compaction, organic matter loss, and soil erosion adversely affect abundance and efficacy and infectivity of AMF. Instead, majority of agricultural lands lack adequate population of AMF propagules which also affect plant–fungal relationship.

15.12 Future Perspectives and Conclusions

Terrestrial carbon pool in agriculture and forest ecosystems are at risk to any further increase in CO_2 and temperature. It has already been predicted that high temperature and long drought period may lead to wildfires, which could badly influence the soil C storage (Maracchi et al. 2005; Loboda 2012). Under such a scenario, countries across the globe would experience impact of abrupt climate change in terms of hotter days and heat waves even with aggressive easing strategies (Caesar and Lowe 2012). Several regions of sub-Saharan Africa have been projected to become even more prone to severe drought (Rojas et al. 2011). The countries like Russia and the USA have previously experienced a long-term drought due to changes in climate in 2010 and 2012, respectively. Therefore, to combat such problems, it is imperative to minimize emission of greenhouse gases in the atmosphere.

The land use conversion and agricultural activities produce about 30% of total anthropogenic emissions both directly and indirectly (Lal 2012). Therefore, conversion to a restorative land use and adoption of best management practices may be integral to any strategy for mitigating impact of global climate change. The strategy which can be applied to minimize C emission and create a positive ecosystem C budget by enhancing the C pools whether in the soil or biomass could be explored (Lal 2004).

Indeed, soil C sequestration is a feasible strategy for a significant storage of atmospheric CO_2 concentration (Hansen et al. 2008). In terrestrial ecosystems, the response of the soil microorganisms to climate change, including rising temperature and CO_2 levels, is less clear. Arbuscular mycorrhizal fungi are likely to play a critical role in the global carbon cycle. They are globally abundant and play vital roles in C cycling. Their extensive hyphal network in the soil provides an important pathway for the flow of C from roots to bulk soil. The allocation of plant photosynthate to AMF often increases under elevated atmospheric CO_2 level, which stimulates the growth of AMF (Drigo et al. 2010; Cheng et al. 2013). Therefore, it is pertinent to state that global soils could sequester more C through AMF symbioses under eCO_2 condition (Drigo et al. 2008; Orwin et al. 2011). AMF produce copious amount of glomalin, which facilitates the soil particle aggregation that remain stimulates the storage of soil carbon. However, the production of glomalin by

AMF varies among individual AMF taxa, growth conditions, land disturbance, and plant species composition (Treseder and Turner 2007). On the other hand, experiments carried out by Cheng et al. (2012) receive attention toward the priming effect of AMF on saprobes/decomposers. Though it was a short-term study, the possible role of AMF for C sequestration may change under such state of affairs. Therefore, to better understand the role of AMF under ambient (aCO_2) and elevated (eCO_2) carbon dioxide concentrations on the soil C gain or loss, more precise research with long-term experiments are needed to critically evaluate the role of AMF in terrestrial ecosystems.

Treseder (2016) pointed out that current Earth system models do not show accuracy in predicting soil C stocks; therefore, our ability to project future climate change remains tapered. She advocated that the accuracy of ecosystem-scale models can be increased by explicit incorporation of microbial mechanisms and proposed to include the dynamics of individual AMF taxa and glomalin in ecosystem models that can be linked to Earth system models to precisely predict about climate change.

Acknowledgments Authors duly acknowledge the technical support provided by Mr Harsh Sharma, computer analyst, Institute of Life Long Learning, University of Delhi, Delhi, India.

References

Akiyama K, Matsuzaki K, Hayashi H (2005) Plant sesquiterpenes induce hyphal branching in arbuscular mycorrhizal fungi. Nature 435:824–827

Al-Karaki GN, Clark RB (1999) Varied rates of mycorrhizal inoculum on growth and nutrient acquisition by barley grown with drought stress. J Plant Nutr 22:1775–1784

Antibus RK, Lauber C, Sinsabaugh RL, Zak DR (2006) Responses of Bradford-reactive soil protein to experimental nitrogen addition in three forest communities in northern lower Michigan. Plant Soil 288:173–187

Arrigo KR (2005) Marine microorganisms and global nutrient cycles. Nature 437:349–355

Augé RM (2001) Water relations, drought and vesicular-arbuscular mycorrhizal symbiosis. Mycorrhiza 11:3–42

Bago B, Pfeffer PE, Shachar-Hill Y (2000) Carbon metabolism and transport in arbuscular mycorrhizas. Plant Physiol 124:949–958

Bai C, He X, Tang H, Shan B, Zhao L (2009) Spatial distribution of arbuscular mycorrhizal fungi, glomalin and soil enzymes under the canopy of *Astragalus adsurgens* Pall. in the Mu US sandland, China. Soil Biol Biochem 41:941–947

Bardgett RD, Freeman C, Ostle NJ (2008) Microbial contribution to climate change through carbon cycle feedbacks. ISME J 2:805–814

Bonfante P, Anca I (2010) Plants, mycorrhizal fungi and bacteria: a network of interaction. Annu Rev Microbiol 63:363–383

Bonfante P, Genre A (2010) Mechanisms underlying beneficial plant-fungus interactions in mycorrhizal symbiosis. Nat Commun 1:48. doi:10.1038/ncomms1046

Bonfante P, Requena N (2011) Dating in the dark: how roots respond to fungal signals to establish arbuscular mycorrhizal symbiosis. Curr Opin Plant Biol 14:1–7

Caesar J, Lowe JA (2012) Comaparing the impacts of mitigation verses non-intervention scenarios on future temperature and precipitation extremes in the HadGE2 climate model. J Geophys Res 117:1–14

Chabaud M, Genre A, Sieberer BJ, Faccio A, Fournier J, Novero M, Barker DG, Bonfante P (2011) Arbuscular mycorrhizal hyphopodia and germinated spore exudates trigger Ca^{2+} spiking in the legume and non legume root epidermis. New Phytol 189:347–355

Cheng W, Kuzyakov Y (2005) Root effects on soil organic matter decomposition. Agronomy 48:119–140

Cheng L, Booker FL, Tu C, Burkey KO, Zhou L, Shew HD, Rufty TW, Hu S (2012) Arbuscular mycorrhizal fungi increase organic carbon decomposition under elevated CO_2. Science 337:1084–1087

Cheng W, Parton WJ, Gonzalez-meler MA, Phillips R, Asao S, Mcnickle GG, Brzostek E, Jastrow JD (2013) Synthesis and modeling perspectives of rhizosphere priming. New Phytol 201:31–44

Clemmensen KE (2013) Roots and associated fungi drive long-term carbon sequestration in Boreal Forest. Science 339:1615–1618

Del Giorgio PA, Duarte CM (2002) Respiration in the open ocean. Nature 420:379–384

Drigo B, Kowalchuk G, van Veen J (2008) Climate change goes underground: effects of elevated atmospheric CO2 on microbial community structure and activities in the rhizosphere. Biol Fertil Soils 44:667–679

Drigo B, Pijl AS, Duyts H, Kielak AM, Gamper HA, Houtekamer MJ (2010) Shifting carbon flow from roots into associated microbial communities in response to elevated atmospheric CO_2. Proc Natl Acad Sci USA 107:10939–10942

Driver JD, Holben WE, Rillig MC (2005) Characterization of glomalin as a hyphal wall component of arbuscular mycorrhizal fungi. Soil Biol Biochem 37:101–106

Elliott ET, Coleman DC (1988) Let the soil work for us. Ecol Bull 39:23–32

Fellbaum CR, Gachomo EW, Beesetty Y, Choudhari S, Strahan GD, Pfeffer PE, Toby Kiers E, Bücking H (2012) Carbon availability triggers fungal nitrogen uptake and transport in arbuscular mycorrhizal symbiosis. Proc Natl Acad Sci USA 109:72666–72671

Frank AB (1885) Über die auf Würzelsymbiose beruhende Ehrnährung gewisser Bäum durch unterirdische Pilze. Berichte der Deutschen Botanischen Gesellschaft (in German) 3:128–145

Gerdemann JW, Trappe JM (1974) Endogonaceae in the Pacific North West. Mycol Mem 5:1–76

Gutjahr C, Banba M, Croset V, An K, Miyao A, An G, Hirochika H, Imaizumi-Anraku H, Paszkowski U (2008) Arbuscular mycorrhiza-specific signaling in rice transcends the common symbiosis signaling pathway. Plant Cell 20:2989–3005

Hansen J, Sato M, Kharecha P, Beerling D, Berner R, Masson-Delmotte V, Pagani M, Raymo M, Royer DL, Zaches JC (2008) Target atmospheric CO_2: where should humanity aim? Open Atmos Sci J 2:217–231

Hartge KH, Stewart BA (1995) Soil structure. Its development and function. Advances in soil science. CRC Lewis, Boca Raton, FL, pp 424. isbn: 1-56670-173-2

Hodge A, Fitter AH (2010) Substantial nitrogen acquisition by arbuscular mycorrhizal fungi from organic material has implications for N cycling. Proc Natl Acad Sci USA 107:13754–13759

Hooker JE, Black KE (1995) Arbuscular mycorrhizal fungi as components of sustainable soil-plant systems. Crit Rev Biotechnol 15:210–212

Jakobsen I, Rosendahl L (1990) Carbon flow into soil and external hyphae from roots of mycorrhizal cucumber plants. New Phytol 115:77–83

Jastrow JD, Miller RM (1997) Soil aggregate stabilization and carbon sequestration: feedbacks through organomineral associations. In: Lal R, Kimble JM, Follet RF, Stewart BA (eds) Soil processes and the carbon cycle. CRC Press, Boca Raton, FL, pp 207–223

Jastrow JD, Miller RM, Lussenhop J (1998) Contributions of interacting biological mechanisms to soil aggregate stabilization in restored prairie. Soil Biol Biochem 7:905–916

Johnson D, Leake JR, Ostle N, Ineson P, Read DJ (2002) *In situ* (CO_2)-C-13 pulse-labelling of upland grassland demonstrates a rapid pathway of carbon flux from arbuscular mycorrhizal mycelia to the soil. New Phytol 153:327–334

Kasischke ES, Turetsky MR (2006) Recent changes in the fire regime across the North American boreal region—spatial and temporal patterns of burning across Canada and Alaska. Geophys Res Lett 33:1–5

King GM (2011) Enhancing soil carbon storage for carbon remediation: potential contributions and constraints by microbes. Trends Microbiol 19:75–84

Kosuta S, Chabaud M, Lougnon G, Gough C, Denarie J, Barker DG, Becard G (2003) A diffusible factor from arbuscular mycorrhizal fungi induces symbiosis-specific MtENOD11 expression in roots of *Medicago truncatula*. Plant Physiol 131:952–962

Lal R (2004) Soil carbon sequestration impacts on global climate change and food security. Science 304:1623–1627

Lal R (2012) Climate change mitigation by managing the terrestrial biosphere. In: Lal R, Klaus L, Reinhard FH, Bernd US, von Joachim B (eds) Recarbonization of the biosphere. Springer, Dordrecht, pp 17–39

Leifheit EF, Veresoglou SD, Lehmann A, Morris EK, Rillig MC (2014) Multiple factors influence the role of arbuscular mycorrhizal fungi in soil aggregation—a meta-analysis. Plant Soil 374:523–537

Leifheit EF, Verbruggen E, Rillig MC (2015) Arbuscular mycorrhizal fungi reduce decomposition of woody plant litter while increasing soil aggregation. Soil Biol Biochem 81:323–328

Leigh J, Hodge A, Fitter AH (2009) Arbuscular mycorrhizal fungi can transfer substantial amounts of nitrogen to their host plant from organic material. New Phytol 181:199–207

Loboda T (2012) Understanding origins and impacts of drought. Eos 93:417. doi:10.1029/2012EO420007

Lovelock CE, Wright SF, Nichols KA (2004a) Using glomalin as an indicator for arbuscular mycorrhizal hyphal growth: an example from a tropical rainforest soil. Soil Biol Biochem 36:1009–1012

Lovelock CE, Wright SF, Clark DA, Ruess RW (2004b) Soil stocks of glomalin produced by arbuscular mycorrhizal fungi across a tropical rain forest landscape. J Ecol 92:278–287

Maracchi G, Sirotenko O, Bindi M (2005) Impacts of present and future climate variability on agriculture and forestry in the temperate region Europe. Clim Chang 70:117–135

Miller RM, Jastrow JD (2000) Mycorrhizal fungi influence soil structure. In: Kapulnik Y, Douds DD (eds) Arbuscular mycorrhizas: molecular biology and physiology. Kluwer Academic, Dordrecht, pp 3–18

Miller RM, Reinhardt DR, Jastrow JD (1995) External hyphal production of vesicular-arbuscular mycorrhizal fungi in pasture and tallgrass prairie communities. Oecologia 103:17–23

Morton JB, Benny GL (1990) Revised classification of arbuscular mycorrhizal fungi (Zygomycetes): a new order, Glomales, two new suborders, Glomineae and Gigasporineae, and two new families, Acaulosporaceae and Gigasporaceae, with an emendation of Glomaceae. Mycotaxon 37:471–491

Navazio L, Moscatiello R, Genre A, Novero M, Baldan B, Bonfante P, Mariani P (2007) A diffusible signal from arbuscular mycorrhizal fungi elicits a transient cytosolic calcium elevation in host plant cells. Plant Physiol 144:673–681

Nichols KA, Wright SF (2004) Contributions of soil fungi to organic matter in agricultural soils. In: Magdoff F, Weil R (eds) Functions and management of soil organic matter in agroecosystems. CRC Press, Boca Raton, FL, pp 179–198

Nichols K, Wright SF, Dzantor EK (2002) Glomalin: hiding place for a third of the world's stored soil carbon. Agri Res Mag 50:4–7. www.ars.usda.gov/is/AR/archive/sep02/soil0902.htm

Oades JM (1984) Soil organic matter and structural stability: mechanisms and implications for management. Plant Soil 76:319–337

Oehl F, de Souza FA, Sieverding E (2008) Revision of Scutellospora and description of five new genera and three new families in the arbuscular-forming Glomeromycetes. Mycotaxon 106:311–360

Oehl F, da Silva GA, Sánchez-Castro I, Goto BT, Maia LC, Evangelista Vieira HEE, Barea JM, Sieverding E, Palenzuela J (2011) Revision of Glomeromycetes with entrophosporoid and glomoid spore formation with three new genera. Mycotaxon 117:297–316

Orwin KH, Kirschbaum MUF, St John MG, Dickie IA (2011) Organic nutrient uptake by mycorrhizal fungi enhances ecosystem carbon storage: a model-based assessment. Ecol Lett 14:493–502

Pachauri RK, Reisinger A (2007) Climate change 2007: synthesis report. Contribution of working groups I, II and III to the fourth assessment report of the Intergovernmental Panel on Climate Change. IPCC. Cambridge University Press, Cambridge

Powell JR, Parrent JL, Hart MM, Klironomos JN, Rillig MC, Maherali H (2009) Phylogenetic trait conservatism and the evolution of functional tradeoffs in arbuscular mycorrhizal fungi. Proc R Soc Lond B Biol Sci 276:4237–4245

Prescott CE (2010) Litter decomposition: what controls it and how can we alter it to sequester more carbon in forest soils? Biogeochemistry 101:133–149

Read DJ (1991) Mycorrhizas in ecosystems. Experientia 47:376–396

Redecker D, Raab P (2006) Phylogeny of the Glomeromycota (arbuscular mycorrhizal fungi): recent developments and new gene markers. Mycologia 98:885–895

Redecker D, Schüssler A, Stockinger H, Stürmer SL, Morton JB, Walker C (2013) An evidence-based consensus for the classification of arbuscular mycorrhizal fungi (Glomeromycota). Mycorrhiza 23:515–531

Rillig MC (2004a) Arbuscular mycorrhizae and terrestrial ecosystem processes. Ecol lett 7:740–754

Rillig MC (2004b) Arbuscular mycorrhizae, glomalin and soil quality. Can J Soil Sci 84:355–363

Rillig MC, Mummey DL (2006) Mycorrhizas and soil structure. New Phytol 171:41–53

Rillig MC, Steinberg PD (2002) Glomalin production by an arbuscular mycorrhizal fungus: a mechanism of habitat modification? Soil Biol Biochem 34:1371–1374

Rillig MC, Wright SF, Kimball BA, Pinter PJ, Wall GW, Ottman MJ, Leavitt SW (2001a) Elevated carbon dioxide and irrigation effects on water stable aggregates in a Sorghum field: a possible role for arbuscular mycorrhizal fungi. Glob Chang Biol 7:333–337

Rillig MC, Wright SF, Nichols KA, Schmidt WF, Torn MS (2001b) Large contribution of arbuscular mycorrhizal fungi to soil carbon pools in tropical forest soils. Plant Soil 233:167–177

Rillig MC, Wright SF, Eviner VT (2002) The role of arbuscular mycorrhizal fungi and glomalin in soil aggregation: comparing effects of five plant species. Plant Soil 238:325–333

Rillig MC, Ramsey PW, Morris S, Paul EA (2003) Glomalin, an arbuscular-mycorrhizal fungal soil protein, responds to land-use change. Plant Soil 253:293–299

Rojas O, Vrieling A, Rembold F (2011) Assessing drought probability for agricultural areas in Africa with coarse resolution remote sensing imagery. Remote Sens Environ 115:343–352

Rosendahl S (2008) Communities, populations and individuals of arbuscular mycorrhizal fungi. New Phytol 178:253–266

Rosewarne GM, Barker SJ, Smith SE (1997) Production of near-synchronous fungal colonization in tomato for development and molecular analyses of mycorrhiza. Mycol Res 101:966–970

Rosier CL, Hoye AT, Rillig MC (2006) Glomalin-related soil protein: assessment of current detection and quantification tools. Soil Biol Biochem 38:2205–2211

Ruiz-Lozano JM, Azcon R (1995) Hyphal contribution to water uptake in mycorrhizal plants as affected by the fungal species and water status. Physiol Plant 95:472–478

Schimel JP, Bennett J (2004) Nitrogen mineralization: challenges of a changing paradigm. Ecology 85:591–602

Schopf JW, Packer BM (1987) Early Archean (3.3-billion to 3.5-billionyearold) microfossils from Warrawoona Group, Australia. Science 237:70–73

Schüßler A, Walker C (2010) The Glomeromycota. A species list with new families and new Genera (Libraries at the Royal Botanic Garden Edinburgh, Edinburgh; The Royal Botanic

Garden Kew, Kew; Botanische Staatssammlung Munich, Munich; and Oregon State University, Corvallis, OR). pp 1–56

Schüßler A, Walker C (2011) Evolution of the 'plant-symbiotic' fungal phylum, Glomeromycota. In: Poggeler S, Wostemeyer J (eds) Evolution of fungi and fungal-like organisms. Springer, Berlin, pp 163–185

Schüßler A, Mollenhauer D, Schnepf E, Kluge M (2001) A new fungal phylum, the Glomeromycota: phylogeny and evolution. Mycol Res 105:1413–1421

Schüßler A, Martin H, Cohen D, Fitz M, Wipf D (2006) Characterization of a carbohydrate transporter from symbiotic glomeromycotan fungi. Nature 444:933–936

Schwarzott D, Walker C, Schüßler A (2001) Glomus, the largest genus of the arbuscular mycorrhizal fungi (Glomales), is non monophyletic. Mol Phylogenet Evol 21:190–197

Selsted MB, van der Linden L, Ibrom A, Michelsen A, Larsen KS, Pedersen JK, Mikkelsen TN, Pilegaard K, Beier C, Ambus P (2012) Soil respiration is stimulated by elevated CO2 and reduced by summer drought: three years of measurements in a multifactor ecosystem manipulation experiment in a temperate heathland (CLIMATE). Glob Chang Biol 18:1216–1230

Singh BK, Bardgett RD, Smith P, Reay DS (2010) Microorganisms and climate change: terrestrial feedbacks and mitigation options. Nat Rev Microbiol 8:779–790

Six J, Bossuyt H, Degryze S, Denef K (2004) A history of research on the link between (micro) aggregates, soil biota, and soil organic matter dynamics. Soil Tillage Res 79:7–31

Smith SE, Read DJ (1997) Mycorrhizal symbiosis, 2nd edn. Academic Press, San Diego, CA

Smith SE, Read D (2008) Mycorrhizal symbiosis, 2nd edn. Academic Press, San Diego, CA

Smith FA, Smith SE (1997) Structural diversity in vesicular-arbuscular mycorrhizal symbioses. New Phytol 137:373–388

Srivastava D, Kapoor R, Srivastava SK, Mukerji KG (1996) Vesicular arbuscular mycorrhiza-an overview. In: Mukerji KG (ed) Concepts in mycorrhizal research. Kluwer Academic Press, Dordrecht, pp 1–39

Talbot JM, Treseder KK (2011) Dishing the dirt on carbon cycling. Nat Clim Chang 1:144–146

Tinker PB, Nye PH (2000) Solute movement in the rhizosphere. Oxford University Press, Oxford

Tisdall JM, Oades JM (1982) Organic matter and water stable aggregates in soils. J Soil Sci 33:141–163

Torres-Cortes G, Ghignone S, Bonfante P, Schüßler A (2015) Mosaic genome of endobacteria in arbuscular mycorrhizal fungi: transkingdom gene transfer in an ancient mycoplasma-fungus association. Proc Natl Acad Sci USA 112:7785–7790

Treseder KK (2004) A meta-analysis of mycorrhizal responses to nitrogen, phosphorus, and atmospheric CO_2 in field studies. New Phytol 164:347–355

Treseder KK (2016) Model behavior of arbuscular mycorrhizal fungi: predicting soil carbon dynamics under climate change. Botany 94:417–423

Treseder KK, Allen MF (2000) Mycorrhizal fungi have a potential role in soil carbon storage under elevated CO_2 and nitrogen deposition. New Phytol 147:189–200

Treseder KK, Holden SR (2013) Fungal carbon sequestration. Science 339:1528–1529

Treseder KK, Turner KM (2007) Glomalin in ecosystems. Soil Sci Soc Am J 71:1257–1266

van der Heijden MGA, Streitwolf-Engel R, Riedl R, Siegrist S, Neudecker A, Ineichen K, Boller T, Wiemken A, Sanders IR (2006) The mycorrhizal contribution to plant productivity, plant nutrition and soil structure in experimental grassland. New Phytol 172:739–752

Van der Heijden MGA, Bardgett RD, van Straalen NM (2008) The unseen majority: soil microbes as drivers of plant diversity and productivity in terrestrial ecosystems. Ecol Lett 11:296–310

Verbruggen E, Veresoglou SD, Anderson IC, Caruso T, Hammer EC, Kohler J (2013) Arbuscular mycorrhizal fungi—short-term liability but long-term benefits for soil carbon storage? New Phytol 197:366–368

Verbruggen E, Jansa J, Hammer EC, Rillig MC (2016) Do arbuscular mycorrhizal fungi stabilize litter derived carbon in soil? J Ecol 104:261–269

Walker C (1992) Systematics and taxonomy of the arbuscular endomycorrhizal fungi (Glomales)—a possible way forward. Agronomie 12:887–897

Wayman RL (1991) Global climate change and life on earth. Chapman and Hall, New York, p 282
Wilson GWT, Rice CW, Rillig MC, Springer A, Hartnett DC (2009) Soil aggregation and carbon sequestration are tightly correlated with the abundance of arbuscular mycorrhizal fungi: results from long-term field experiments. Ecol Lett 12:452–461
Wright SF (2000) A fluorescent antibody assay for hyphae and glomalin from arbuscular mycorrhizal fungi. Plant Soil 226:171–177
Wright SF, Anderson RL (2000) Aggregate stability and glomalin in alternative crop rotations for the central Great Plains. Biol Fertil Soils 31:249–253
Wright SF, Upadhyaya A (1996) Extraction of an abundant and unusual protein from soil and comparison with hyphal protein of arbuscular mycorrhizal fungi. Soil Sci 161:575–585
Wright SF, Upadhyaya A (1998) A survey of soils for aggregate stability and glomalin, a glycoproteins produced by hyphae of arbuscular mycorrhizal fungi. Plant Soil 198:97–107
Wright SF, Upadhyaya A (1999) Quantification of arbuscular mycorrhizal activity by the glomalin concentration on hyphae. Mycorrhiza 8:283–285
Wright SF, Franke-Snyder M, Morton JB, Upadhyaya A (1996) Time-course study and partial characterization of a protein on hyphae of arbuscular mycorrhizal fungi during active colonization of roots. Plant Soil 181:193–203
Wright SF, Starr JL, Paltineanu IC (1999) Changes in aggregate stability and concentration of glomalin during tillage management transition. Soil Sci Soc Am J 63:1825–1829
Wright SF, Nichols KA, Schmidt WF (2006) Comparison of efficacy of three extractants to solubilize glomalin on hyphae and in soil. Chemosphere 64:1219–1224
Wright SF, Green VS, Cavigelli MA (2007) Glomalin in aggregate size classes from three different farming systems. Soil Tillage Res 94:546–549
Wu QS, Cao MQ, ZouYN HXH (2014) Direct and indirect effects of glomalin, mycorrhizal hyphae, and roots on aggregate stability in rhizosphere of trifoliate orange. Sci Rep 4:5823. doi:10.1038/srep05823
Zhang B, Li S, Chen S, Ren T, Yang Z, Zhao H, Liang Y, Han X (2016) Arbuscular mycorrhizal fungi regulate soil respiration and its response to precipitation change in a semiarid steppe. Sci Rep 6:19990. doi:10.1038/srep19990

Chapter 16
Arbuscular Mycorrhizal Fungi in Hypoxic Environments

Irena Maček

Abstract Hypoxia and even anoxia in plant rhizosphere are common phenomena that can be the consequence of flooding, submergence, soil compaction, or are a specific characteristic of some extreme ecosystems (e.g. due to geological CO_2 release in natural CO_2 springs or mofettes). The frequency and severity of flooding events will dramatically increase in the future, as projected by climate change models. Therefore, understanding the response of different organisms to soil hypoxia, including crop plants, and their interaction with symbiotic and ubiquitous arbuscular mycorrhizal (AM) fungi is becoming increasingly important in order to enhance plant yield and to promote sustainable agriculture in the future. Plants and soil fungi are known to be obligate aerobes and are sensitive to O_2 deficiency since they need a sufficient amount of this gas to support their aerobic metabolism. However, some specific morphological and metabolic adaptations also enable plants to survive in habitats where O_2 availability is severely limited. Moreover, recent reports show that diverse plant root endophytic fungal communities exist in these ecosystems with some specific (new) taxa being reported to even thrive there. This includes obligate biotrophic AM fungi that fully depend on the plant-derived carbon source. A new aspect in the biology of these organisms originating from the research into hypoxic environments is that in addition to carbon, they can also use a plant-derived O_2 source delivered into the submerged organs via plant's root aeration systems (e.g. aerenchyma). Moreover, in the field of community ecology, extreme hypoxic environments (e.g. mofettes) have been shown to represent a powerful tool for the study of slower ecological and evolutionary processes in still largely unexplored soil microbial communities. They can be used to gain insight into the adaptation of native communities to a specific permanent stress (e.g. soil hypoxia) as long-term natural experimental systems. In this chapter a review of the literature investigating AM fungi and their communities in hypoxic environments is presented. Considering this aspect will be essential for our capacity to adequately manage ecosystems and predict ecological and evolutionary

I. Maček (✉)
Biotechnical Faculty, University of Ljubljana, Jamnikarjeva 101, 1000 Ljubljana, Slovenia

Faculty of Mathematics, Natural Science and Information Technologies (FAMNIT), University of Primorska, Glagoljaška 8, Koper, Slovenia
e-mail: irena.macek@bf.uni-lj.si

A. Varma et al. (eds.), *Mycorrhiza - Function, Diversity, State of the Art*,
DOI 10.1007/978-3-319-53064-2_16

responses to global change, with flooding and soil hypoxia being a consistent part of terrestrial ecosystems in the future.

16.1 Introduction

The vast majority of all land plants receive inorganic nutrients via an indirect uptake through the symbiotic arbuscular mycorrhizal (AM) fungi (Fig. 16.1) and not by a direct uptake from the soil (Smith and Read 2008; Hodge et al. 2010). Therefore, with the tendency towards more sustainable food production, arbuscular mycorrhiza may play a large role in sustainable agricultural practices in the future (e.g. Gianinazzi et al. 2010). AM fungi from the phylum *Glomeromycota* form diverse communities in natural habitats. A number of different AM fungal taxa simultaneously interact with the roots of a single plant and the whole community in practically almost any terrestrial ecosystem (e.g. Fitter 2005). Those fungal taxa, however, can provide many different benefits to the plant that extend beyond the typically most exposed P uptake, like improving N uptake and water relations, protecting plants from pathogens and other stress factors. A fungus good in one of these functions is unlikely to be the best in another, which also leads to the diversification of different taxa based on function (Fitter 2005).

Different AM fungal taxa also respond differently to their environment, both biotic (e.g. plant internal environments) and abiotic. Studies on the environmental

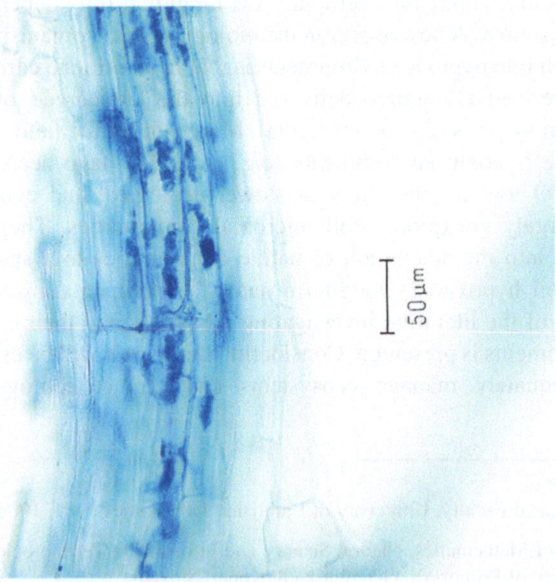

Fig. 16.1 AM fungi-root colonisation with abundant arbuscules in the root cortex of a C4 grass *Setaria pumila*

determinants of the composition of AM fungal communities in natural ecosystems show that AM fungal communities are largely determined by their environmental niche (e.g. soil physicochemical properties and other environmental variables). In one of the relatively rare reports where temporal dynamics has been studied, a combination of both niche and neutral (stochastic) processes in determining AM fungal community composition has also been suggested (Dumbrell et al. 2011). One of the critical abiotic factors affecting cellular metabolism in all eukaryotes, including AM fungi and their host plants, is the O_2 concentration in their environment. The existing reports clearly show that soil aeration and availability of O_2 in the soil atmosphere can act as an important selective pressure on AM fungi and can severely impact AM fungal community composition (e.g. Maček et al. 2011, 2016).

Typically, soil air is not limited in O_2 and contains similar concentrations of this gas as found in the atmosphere (just below 21%). However, in some cases soil O_2 availability can become a limiting factor for survival of aerobic organisms. There can be different causes of soil hypoxia (low O_2 concentration compared to atmospheric conditions) or even anoxia (devoid of O_2), and it can appear locally or in a larger volume of soil. In the case of a local small-scale hypoxia, larger organisms, plant roots and also possibly some filamentous fungi can avoid it by moving or growing into a different soil compartment where O_2 is more abundant. However, when hypoxia is present in a larger volume of soil, and is in addition long-term (e.g. permanently submerged environments, during flooding or in some extreme ecosystems like natural CO_2 springs or mofettes; see the description in the Sect. 16.3.3), this is not possible, and other mechanisms of O_2 supply must take place in order to support aerobic metabolism. Long-term soil hypoxia in mofette areas has been shown to result in unique and temporally relatively stable composition of soil microbial communities (Maček et al. 2011; Šibanc et al. 2014), including AM fungal communities (e.g. Maček et al. 2011, 2016) that are better able to tolerate those conditions. Interestingly, diverse AM fungal communities also exist in environments that are hypoxic. However, the exact physiological and evolutionary mechanisms behind that are yet to be explored.

Hypoxia and its impacts on different aspects of AM fungal ecology and physiology have been chosen to be presented in this chapter for several reasons: (1) it is a common but understudied stress regarding its impacts on AM fungal biology (Maček et al. 2011); (2) it is present in many natural ecosystems worldwide (Perata et al. 2011), with climate change models projections to become even more frequent and severe due to increased areas of flooding in the future (Hirabayashi et al. 2013); and (3) the new insights originating from research into a specific, extreme hypoxic ecosystem—natural CO_2 springs or mofettes—with evidence accumulating and supporting the idea that locally extreme hypoxic environments can serve as natural, long-term experiments in ecology (Maček et al. 2016). This specific ecosystem is a unique example of a plant and root fungal communities subject to well characterised (Vodnik et al. 2006, 2009), localised long-term selection pressure in the form of soil hypoxia (Maček et al. 2011, 2016) and thus enables a more controlled study of this common abiotic stress or on soil microbial communities (e.g. Maček et al. 2011, 2016; Šibanc et al. 2014). In this chapter a range of hypoxic environments with their specificities, in addition to lessons learned regarding AM fungal response, and in particular patterns of their community assembly, are presented.

16.2 Arbuscular Mycorrhiza

In terrestrial ecosystems, symbiotic associations between plant roots and mycorrhizal fungi are near ubiquitous, with 90% of all plant species forming mycorrhiza (Smith and Read 2008). The symbiosis is ancient, over 400 million years old, and was significant in enabling the colonisation of land by plants (Redecker et al. 2002). In exchange for mineral nutrients, the plants supply up to 20% of photosynthates as the only energy source of the fungus (ca. five billion tonnes of C per year) (Bago et al. 2000). The nutrient exchange within plant root cells mainly takes place at the fungus-plant symbiotic interface formed around the finely branched fungal arbuscules (Parniske 2008). Because the fungal hyphae are finer than roots by at least an order of magnitude, the costs to a plant of acquiring nutrients symbiotically will always be much lower than those of doing so by new root growth (Fitter 1991)—which also may be true for some hypoxic environments (e.g. Møller et al. 2013). As with any other mycelial fungus, AM fungi acquire resources in numerous spatially dispersed locations by exploring the soil and moving them within the mycelium to fund growth at favoured locations (Bago et al. 2002; Hughes et al. 2008). Phosphate, for example, is relatively immobile in soil; therefore, localised areas of depletion exist around roots. Fine hyphae covering a greater spatial extent can access more areas and overcome problems of local depletion (Smith and Read 2008). Nutrients are moved in a packaged form between the extra-radical and the intra-radical fungal mycelium (Bago et al. 2002; Hughes et al. 2008; Parniske 2008). The extensive hyphal network of AM fungi also influences the physico-chemical properties of the soil—e.g. stabilisation of structural aggregates (Rillig 2004; Rillig and Mummey 2006)—and directly or indirectly contributes to the release of phosphate from inorganic complexes of low solubility (Finlay 2008). Phosphate is actively imported by fungal transporters (e.g. Harrison and van Buuren 1995; Maldonado-Mendoza et al. 2001) that are present in extra-radical hyphae. Polyphosphate granules are used as transport vehicles to move phosphate—and possibly arginine (N source) and trace elements—to the host root (Parniske 2008). In addition to enhanced phosphorous supply, AM fungi also affect plant's nitrogen assimilation by accelerating decomposition and acquiring N from organic material (Govindarajulu et al. 2005; Hodge et al. 2001). N is taken up by ammonium, nitrate or amino acid transporters in extra-radical hyphae (Fig. 16.2).

This ancient mechanism of plants acquiring mineral nutrients is of particular importance for the developing field of sustainable agriculture. The current data show that within the next 50–100 years, global rock phosphate reserves will deplete (http://www.phosphorusfutures.net), and therefore the price of the commercial fertilisers will increase, becoming even more inaccessible to the poorest and the

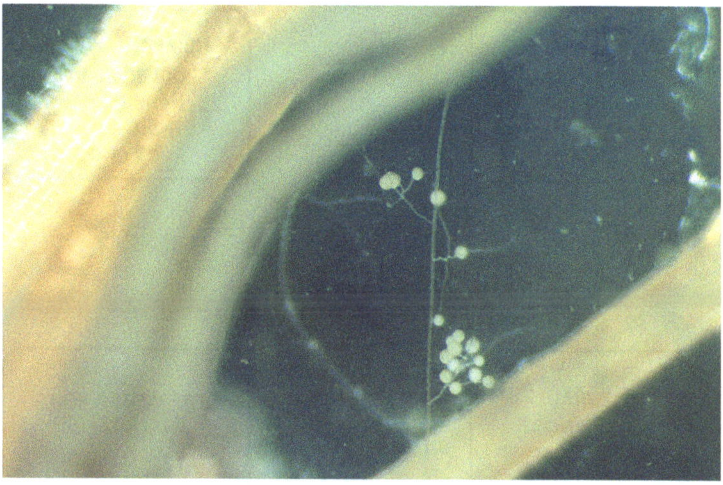

Fig. 16.2 Spores and extra-radical hyphae of AM fungus *Rhizophagus irregularis* around transformed *Medicago truncatula* roots in in vitro culture

hungriest of the world's population. Rock phosphate is the primary, but limited and non-renewable source for large-scale fertilisation in agriculture. How to solve the problems with phosphate supply is a matter of debate, but it becomes increasingly evident that better understanding of the arbuscular mycorrhiza will play a large role in sustainable agricultural practices since the arbuscular mycorrhiza is a key symbiosis for efficient P uptake by plants in natural habitats (e.g. Hodge et al. 2010).

Yet despite its ecological and agricultural importance, astonishingly little is known about ecological and physiological responses of AM fungi to soil hypoxia (Maček et al. 2011, 2012), a common condition in soil (e.g. Perata et al. 2011; Pucciariello et al. 2014). Soil hypoxia primarily affects respiration in aerobes and by this involves both aerobic groups of organisms in the symbiosis, plants and mycorrhizal fungi (Maček et al. 2005, 2011, 2016).

16.3 AM Fungal Diversity in Hypoxic Environments

Hypoxia, or even anoxia, is a common transition property of the soil that appears often in waterlogged and flooded soils or is due to soil compaction (Fig. 16.3). In some ecosystems hypoxia or/and anoxia can also be permanent or long-term. Such ecosystems include submerged environments or also soils at natural CO_2 springs (mofettes) with constant geological gas exhalations consisting mainly of CO_2 originating from the underground reservoirs (e.g. Vodnik et al. 2006, 2009; Maček et al. 2011, 2016).

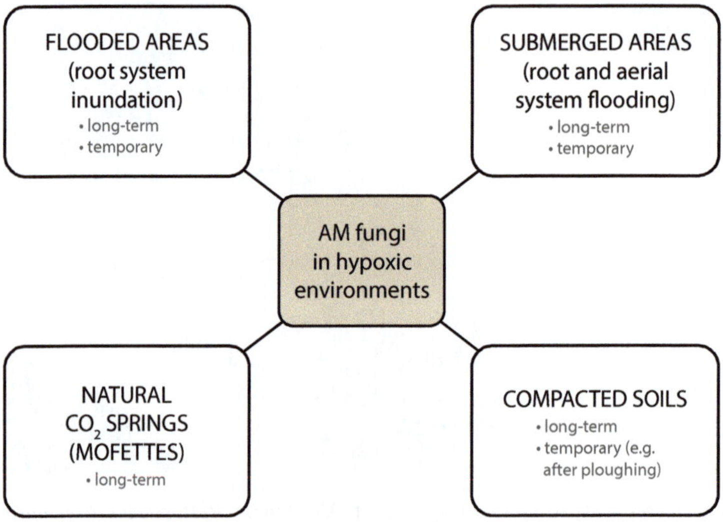

Fig. 16.3 Specific hypoxic environments with confirmed presence of functional and diverse AM fungal communities

In general, hypoxia has been widely studied in plants with currently over 200 records listed on the *Web of Science* portal when using the keywords: (*plant and hypoxia*). There have been several special issues in scientific press on plant anaerobiosis (e.g. *New Phytologist* vol. 190, issue 2, April 2011; *Frontiers in Plant Science*, vol. 5, May 2014). In addition, thematic meetings of the *International Society for Plant Anaerobiosis (ISPA)* are held once every 3 years, and a *New Phytologist Workshop* on flooding biology was held in 2015 in the Netherlands (Voesenek et al. 2016). However, in the context of soil hypoxia caused by flooding and its impact on plant biology, AM fungi are still largely not considered, or as an integrated and important functional part of the plant rhizosphere continuum. When the keywords (*mycorr* and hypoxia*), or (*arbuscul* and hypoxia*), are used, only 10 and 8 records, respectively, have been found in the *Web of Science* (date of the survey, 23 May 2016). This is a disturbing result, which indicates that mycorrhizal symbiosis is largely ignored or neglected when hypoxia is studied in respect to different aspects of plant biology. Most of the studies where hypoxia is at least indirectly studied in relation to mycorrhiza involve flooded and submerged ecosystems. Indeed when the topics (*arbuscul* and flood**) or (*arbuscul* and submerge**) are used, a higher number (however still relatively low) of published studies can be found, with the number of records 123 and 34, respectively (*Web of Science*, 23 May 2016). Among the few studies where hypoxia is directly reported and investigated as a stress factor acting on AM fungal communities, there are reports involving research at a specific extreme environment—natural CO_2 springs or mofettes—with geogenic CO_2 displacing O_2 from soil air, resulting in long-term hypoxia (Maček et al. 2011, 2012, 2016, Table 16.1, see Sects. 16.3.3 and 16.4.2.)

Table 16.1 An overview of molecular studies on diversity and community ecology of AM fungi in a range of hypoxic environments, including some studies from mofettes (natural CO_2 springs), submerged and flooded ecosystems

Environmental factor	AM fungal/plant community	Ecosystem/location	Reference
Soil hypoxia (mofettes or natural CO_2 springs)	Native grassland community	Mofettes (worldwide)	Maček et al. (2016), *Advances in Ecological Research*, special issue *Large Scale Ecology*, vol. 55
	Native grassland community	Mofettes in Stavešinci (Slovenia), Bossoleto (Italy), Cheb basin (Czech)	Šibanc et al. (unpubl.)
	Native grassland community	Mofettes in Stavešinci (Slovenia)	Maček et al. (2011), *Applied Environmental Microbiology*, vol. 77(14)
Submergence (plant root and aerial system flooding)	Native community in submerged isoetid plants; first description of a new AM fungal species *Rhizoglomus melanum*	Oligotrophic lakes (Norway)	Sudová et al. (2015), *Mycological Progress*, vol. 14(3)
	Native community in isoetid vegetation	Oligotrophic lakes (Norway)	Kohout et al. (2012) FEMS *Microbiology Ecology*, vol. 80(1)
	Native community in aquatic macrophytes	Oligotrophic and ultra-oligotrophic lakes (the Netherlands and Norway)	Baar et al. (2011), *Aquatic Botany*, vol. 94 (2)
Flooding gradient (plant root system inundation)	Native community in wetlands	Wetlands—lake (China)	Wang et al. (2016), *Plant and Soil*, vol. 403
	Mangrove communities	Qi-Ao mangrove nature reserve (China)	Wang et al. (2015), *Ann Microbiol*, vol. 65
	Mangrove communities	Qi-Ao mangrove nature reserve (China)	Wang et al. (2011), *PLoS One*, vol. 6(9)
	Rice fields	Rice fields (Italy)	Lumini et al. (2011), *Ecological Applications*, vol. 21(5)

16.3.1 Flooded Soils

The frequency and severity of flooding worldwide is going to increase in the future, with the biggest increase projected by the climate change models for the tropics and Western Europe (Hirabayashi et al. 2013). Therefore, understanding the response of different crop plants to interaction with AM fungi in flooded conditions is becoming increasingly important in order to enhance plant yield and to promote sustainable agriculture. In the special issues on plant anaerobiosis (e.g. *New Phytologist* vol. 190, issue 2, April 2011; *Frontiers in Plant Science*, vol. 5, May 2014), several mechanisms, involved in plant response to flooding stress, the effects floods may have on patterns of plant distribution and biodiversity and the devastating impact on crop growth, are described (e.g. Perata et al. 2011; Pucciariello et al. 2014). However, no studies on any aspect of the plant symbiosis with AM fungi are presented in these volumes, which shows a poor integration of knowledge on arbuscular mycorrhiza into some parts of plant biology, including plant anaerobiosis. Nevertheless, some studies on arbuscular mycorrhiza and AM fungal communities have also been performed in flooded soils. Flooded rice fields and rice plants, in general, are one such system, where substantial progress in interaction between plants and AM fungi and their common response to flooding has been performed (e.g. Vallino et al. 2014) (see Sect. 16.4.1).

In flooded areas different spatial and temporal gradients are a very common feature. The fluctuation between dry and wet (flooded) periods can also result in recolonisation of roots of plants after the water retreats. Such soils have been shown to contain sufficient AM fungal propagules for a new round of colonisation (e.g. Lumini et al. 2011; Wirsel 2004). As a result of living in a dynamic environment, also the composition of AM fungal communities may be more dynamic when compared to 'normal' soil, and plant root colonisation may follow the fluctuations in the water regime. This can also represent a certain stress factor to AM fungi. In a study of the impact of different farming and water regimes on Italian rice fields, *Rhizophagus irregularis* has been shown to be flexible in this system, and its colonisation of rice roots was reversible following a short period of dry conditions (Vallino et al. 2014). However, all the AM fungal taxa may not possess such a high level of plasticity, and more research is needed to shed more light on the AM fungal diversity in these systems, their community composition and its temporal dynamics over a longer time period.

AM fungi are also common in mangrove forests (Radhika and Rodrigues 2007) and were reported to be present in 16 mangrove species at riverine and fringe habitats in Goa, West India (D'Souza and Rodrigues 2013). No molecular analyses of AM fungal communities in the roots of mangrove trees were performed in the study from Goa. However, AM fungal taxa were determined by molecular approaches in a second study that was done in four semi-mangrove plant communities from the Qi'Ao Mangrove Forest Reserve in South China that are subject only to spring and storm high tides (e.g. Wang et al. 2011, 2015). AM fungal spores were extracted from rhizosphere soil and identified using SSU rRNA amplification and Sanger sequencing. In addition, AM fungi in roots were identified. The authors report of six molecular operational taxonomic units (MOTUs) from the *Glomeraceae* family that could not be identified to the genus level and could represent potential new taxa, which is a common feature in many hypoxic environments (e.g. Sudová et al. 2015, see also Sect. 16.3.2).

16.3.2 Submerged Environments

There are only few reports of the AM fungi in submerged environments (in the context of permanent plant root and aerial system flooding). The first report of aquatic plants (*Littorella uniflora* L. Ascherson and *Lobelia dortmanna* L.) having mycorrhizal fungal colonisation was by Søndergaard and Laegaard (1977) in roots from oligotrophic softwater lakes with low P concentrations in Denmark. Further records on the occurrence of AM fungi in macrophytes were published in the 1980s (e.g. Farmer 1985; Tanner and Clayton 1985). In the next two decades, this trend slowly continued, and a few further reports have been published on AM fungal root colonisation of macrophyte vegetation, both from lakes and streams (e.g. Beck-Nielsen and Madsen 2001). However, only very recently and empowered by the newly developed molecular tools, researchers have looked further into AM fungal community composition in these specific ecosystems and have determined the environmental factors that may be of major importance for the development of the AM fungal communities in different aquatic habitats.

The first molecular analysis of AM fungal diversity in aquatic macrophytes from oligotrophic and ultra-oligotrophic lakes in the Netherlands and Norway was performed by Baar et al. (2011). The study reports on diverse AM fungal communities in the roots of *L. uniflora*, with several AM fungal taxa present, including the taxa from the genera *Glomus*, *Acaulospora* and *Archaeospora*. AM fungi occurred more abundantly with low phosphate and high redox values in the lakes than with high phosphate and low redox values (Baar et al. 2011). Interestingly, another report on the AM fungal diversity in oligotrophic lakes shows that organic enrichments of sediments reduce AM fungal colonisation of submerged plant roots, indicating that more organic matter could lead to anaerobic decomposition and less available O_2 in this environment, thus impacting arbuscular mycorrhiza and its functioning in this ecosystem (Møller et al. 2013). Indeed, also older studies show that high levels of AM fungal colonisation in submerged plant roots is correlated to high redox potentials (Wigand et al. 1998; Beck-Nielsen and Madsen 2001), lower depth of habitats with coarse sediment and higher availability of light for photosynthesis (e.g. Wigand et al. 1998), that is, also an internal source of O_2 for plants and potentially also for endophytic fungi.

Surprisingly, however, there are only a few further reports on the AM fungal diversity of submerged aquatic plants (e.g. Kohout et al. 2012; Wang et al. 2011). Recently, a new AM fungal species *Rhizoglomus melanum* was taxonomically described that till now has only been found in lake sediments and was isolated from the rhizosphere of the two aquatic macrophytes, *L. uniflora* and *Isoëtes lacustris*, from the freshwater lake Avsjøen (Norway) (Sudová et al. 2015). Both plant host species, *L. uniflora* and *I. lacustris*, form small, submerged rosettes with a well-developed root system (Beck-Nielsen and Madsen 2001; Farmer 1985; Søndergaard and Laegaard 1977) and are characterised by a continuous lacunal system allowing rapid O_2 diffusion from the shoots to the roots, as well as by high radial O_2 losses from their roots into the sediment (Smolders et al. 2002). Indeed, among aquatic plants, isoetids seem to have the highest degree of root colonisation with AM fungi (Beck-Nielsen and Madsen 2001). The AM fungi appear to be dependent on the high O_2 concentrations in the roots and surrounding root zones of the isoetid plants (Wigand et al. 1998), and this appears to be a consistent and important component of AM fungal habitats in hypoxic soils.

Freshwater research has recently stretched to marine ecosystems; however, the only published study on cultivable root mycobionts of the seagrass *Posidonia oceanica* reports on the colonisation of this species roots with new, undescribed taxa of dark septate endophyte (DSE) fungi (Vohník et al. 2016). To our knowledge, however, there are no published data on the diversity of AM fungi in submerged marine environments. However, observation from seagrass and mangrove forests, along with the new AM fungal species described from oligotrophic lakes, indicates that submerged environments still represent a rich potential source of new fungal taxa that are yet to be discovered.

16.3.3 Mofettes or Natural CO_2 Springs

Natural CO_2 springs, also known as mofettes, are areas with CO_2 gas vents occurring in tectonically or volcanically active sites, where ambient temperature geological CO_2 reaches the surface (Vodnik et al. 2006, 2009; Maček 2013; Maček et al. 2016). In these extreme habitats, severe and relatively constant changes in soil gases take place (Vodnik et al. 2006). In the process of raising the soil CO_2 concentration (up to 99.9%), hypoxia is induced by reducing the partial pressure of O_2, often to very low values (Maček et al. 2012; Maček et al. 2016). Since the early 1990s, mofettes have been used for research of the elevated CO_2 impact on ecosystems, with the main focus on plant responses to atmospheric CO_2 (e.g. Raschi et al. 1997). Only in the recent years, the focus has shifted to researching the mofettophilic communities of a range of different organisms (Maček 2013; Maček et al. 2016), including flora (e.g. Pfanz 2008), soil fauna (e.g. Hohberg et al. 2015; Russell et al. 2011) and soil microorganisms (e.g. Beulig et al. 2015, 2016; Maček et al. 2011; Šibanc et al. 2014); see Maček et al. (2016) for a detailed overview on the recent knowledge on biodiversity and community ecology of different groups of organisms in mofette ecosystems.

The soil gas regime at mofettes, and soil hypoxia in particular, can have strong impact on the communities of obligatory aerobic eukaryotic organisms like plants, soil fauna (e.g. Hohberg et al. 2015) and of AM fungi (Maček et al. 2011, 2012; Maček 2013; Maček et al. 2016) (Fig. 16.4). Thus far, only the research by Maček et al. (2011), studying the impact of elevated CO_2 and soil hypoxia on diversity of AM fungal communities, has focused on the diversity of AM fungi from these habitats. Maček et al. (2011) report on significant levels of AM fungal community turnover (beta diversity) between soil types and the numerical dominance of specific AM fungal taxa when exposed to soil hypoxia. This work strongly suggests that direct environmental selection acting on AM fungi is a major factor regulating

Fig. 16.4 Mofette soil with geological CO_2 exhalations causing long-term soil hypoxia (Stavešinci mofette, Slovenia)

AM fungal communities. However, as noted in the paper, only more intensive sampling, using, for example, a high-resolution amplicon sequencing approach (next-generation sequencing—NGS), can provide further insight into the temporal dynamics of the present AM fungal communities and give more detailed information on the community composition and the differences between sites, in particular the currently lacking information on the presence and the abundance of the rare taxa in these environments (Maček et al. 2011). Therefore, experiments involving mofettes could serve as a good working space for a detailed study of the mechanisms that regulate the evolutionary processes in soil fungi and their functioning in hypoxic environments (Maček et al. 2011, 2016). They could also give us new insight into different pathways of mineral assimilation in hypoxic environments and the role of the fungal partner in those processes (see Sects. 16.4.1 and 16.4.2).

16.3.4 Compacted Soil

Apart from waterlogged and mofette soils, hypoxic or even anoxic environments can also be found in other soil habitats, like underground borrows, microenvironments in soil aggregates and in compacted soils. Here, low O_2 levels can be permanent or, more frequently, temporally present (Hourdez 2012). Soil compaction is common in agroecosystems due to use of heavy field equipment and field traffic. Soils are in particular susceptible to soil compaction when wet. In addition to directly affecting soil structure, soil compaction can also affect roots (growth and activity) and microbial functions in soil. Soil porosity is an important factor affecting soil aeration. Soil compaction changes pore space size, distribution and soil strength and therefore reduces soil gas exchange. Compacted soils have few large pores since when soil particles are pressed together, the pore space between them is reduced (DeJong-Hughes et al. 2001). As the pore space is decreased within a soil, the bulk density is increased. This can lead to locally hypoxic or anoxic conditions.

In intensively managed fields, compacted soils are often maintained by deep tillage. This in addition to compaction (especially the subsoil layers below the ploughing depth) directly negatively impacts microbial communities, where filamentous fungi (also AM fungi) are in particular vulnerable due to mechanical damage (e.g. Helgason et al. 1998; Schnoor et al. 2011). Indeed, some AM fungal taxa disappear in intensively managed soils, and one of the factors affecting this could also be mechanical damage and changes in soil structure and compaction (e.g. Säle et al. 2015). However, it is difficult to test what the effect of the compaction is, *per se*, on soil microbial communities since soil is a complex environment and the communities of soil organisms are usually affected by many factors acting simultaneously.

16.4 AM Fungal Biology in Hypoxic Environments: Where Could the Future Research Go?

The vast majority of the multicellular organisms are aerobes, with some very rare exceptions (Danovaro et al. 2010); thus, their respiration and growth can be severely affected in hypoxic conditions, since nearly all multicellular organisms require O_2 at least during part of their life cycle. The majority of fungi are known to be aerobes, and it is widely accepted that many fungal groups are sensitive to the lack of O_2 in their environments. Since microbes, due to their extremely small body size, cannot avoid hypoxic conditions—in, for example, submerged soils or mofettes—and are too small to flee, they are often forced to evolve and adapt in order to survive, function and grow in those environments. Soil-borne organisms have evolved to tolerate low or rapidly changing O_2 levels and are exposed to occasional hypoxia, where the hypoxic response starts also in fungi (similar as in multicellular eukaryotes) at an O_2 level of about 6% (Simon and Keith 2008).

16.4.1 Aerobic Metabolism and Hypoxia

Respiration is an aerobic process that requires O_2; thus, soil hypoxia will affect respiration in plant roots and fungal hyphae. Respiration is also the main metabolic energy source for the active mineral nutrient uptake into plant roots. Most plant species form mycorrhizas, yet these are to a large extent still neglected by plant physiologists (see Sect. 16.3). One consequence of this neglect is reduced ability to predict plant respiration in any ecosystem (Hughes et al. 2008). The energy requirement for increased rates of plant ion uptake is likely to cause an increase in root respiration in AM symbiosis (Hughes et al. 2008); thus, O_2 availability is crucial for the normal function of the symbiosis. Mitochondria concentrate around the arbuscules (the sites of nutrient transfer) in root cortex cells of colonised *Medicago truncatula* (Lohse et al. 2005), potentially reflecting the provision of adenosine triphosphate (ATP) necessary for nutrient import. The nutrient uptake by an AM plant can be seen as a four-stage process consisting of ion uptake by the external hyphae, ion transport within the fungus, ion export by the internal hyphae and ion uptake by plant root cells, each stage representing a separate ATP demand (Hughes et al. 2008). In the opinion article (Hughes et al. 2008), the authors conclude that there is some evidence that respiration in the globally dominant AM symbiosis can be considerably higher than respiration in non-mycorrhizal plant roots, as also shown to some extent by Atkin et al. (2009). The exact mechanistic understanding of the respiration in mycorrhizal roots or external hyphae has, however, not yet been revealed.

But when exposed to soil hypoxia, respiration of both roots and symbiotic fungi is affected. Plants that have adaptive traits like the ability for aerenchyma formation, or other systems for O_2 transfer to roots, are more likely to survive in these environments and also more likely to support and be able to form symbiotic relationships with other aerobic organisms in their rhizosphere. In plants, the formation of aerenchyma can allow a relatively high level of O_2 in the rhizosphere and sustain aerobic respiration of the roots even in the context of very high soil CO_2 concentrations or submerged soils. AM fungi are likely to be supported by plant-driven O_2 transport into root tissues that can enable aerobic microorganisms to survive in such environments. Aerenchyma are typical adaptive traits of wetland plant species, but can arise also as response to flooding or mineral deficiency (Marschner 2012). However, the authors of a recent study on uninoculated and inoculated rice with *R. irregularis*, grown in dry and flooded conditions, report on the reduced AM fungal colonisation with *R. irregularis* three weeks after inoculation and continuous exposure to flooding (Vallino et al. 2014). This is coincident with the time when rice plants intensively start to form aerenchyma in their root cortex (Vallino et al. 2014). An interesting observation is that despite better aeration of the rhizosphere by aerenchyma, significant reduction of the available space in cortex also reduces root AM fungal colonisation levels. This may be plant and fungus taxa specific; however, it shows that development of AM fungi in the plants with substantial amount of aerenchyma may be limited due to the lack of available space. The results of this study also show that under flooding conditions, AM fungal nutrient transporters are regularly expressed; however, the functional markers of the AM symbiosis reveal a significant decrease in the expression of plant and fungal nutrient transporters during progressive flooding (Vallino et al. 2014).

Increasing plant nutrient uptake by AM symbiosis is also relevant for some submerged plants, especially in conditions of poor nutrient availability in their environment (e.g. submerged vegetation of oligotrophic lakes in Norway, Møller et al. 2013). In the latter study, extensive extra-radical hyphal networks were found in sediments of the submerged isoetid plants (*L. dortmanna* and *L. uniflora*) with a high mean hyphal density (6 and 15 m cm^{-3}, respectively, for each plant species). This is comparable with the density typically found in terrestrial soils (Møller et al. 2013). The hyphal surface area exceeded root surface area by 1.7–3.2 times with the highest density in the main root zone (Møller et al. 2013).

Therefore, research into hypoxic ecosystems has exposed some new aspects of the interaction between plants and mycorrhizal fungi, one of these is also acquiring O_2 from plants by using a plant transport system (e.g. aerenchyma or lacunal system). This is a relatively new concept in AM fungal symbiosis, but is also relevant for other aerobic microbes, either endophytic, living in roots, or rhizosphere (Maček et al. 2011, 2016). Therefore, plant-AM fungal interactions could, in addition to nutrient (trophic) interaction, be expanded to the additional benefit of a positive effect of one species on another by reducing physical or biotic stress in existing habitats and by creating new habitats for AM fungi. In this context this means that some species modify conditions sufficiently to make life more hospitable for others that otherwise would not be able to survive in this environment. The concept is well known in plant literature under the term facilitation and is used to beneficial (non-trophic) interactions that occur between physiologically independent plants and are mediated through changes in the abiotic environment (Brooker and Callaway 2009). The cross-trophic and cross-system level interactions in this concept of facilitation are still a matter of debate and a challenge to a current 'working' definition of facilitation as being limited to the plant-plant interactions, mostly in terrestrial environments (Brooker and Callaway 2009).

16.4.2 AM Fungal Diversity and Community Ecology

A central aim in ecology is to quantify the mechanisms that regulate the diversity of natural communities. AM fungi are one such functionally important group with still insufficiently understood community ecology (Helgason and Fitter 2009; Rosendahl 2008). This is even more prevalent in environmental extremes, including hypoxic environments (Maček et al. 2011, 2016). AM fungi form an extensive mycelial network in soil and so will be subject to strong selection pressures from the abiotic soil environment (e.g. Dumbrell et al. 2010; Maček et al. 2011; Lenoir et al. 2016), and soil hypoxia is one such important abiotic factor (Maček et al. 2011) (Fig. 16.5).

Nevertheless, most of the existing studies on community composition of AM fungi in hypoxic environments were single time-point studies, or sampling has not been done systematically enough to include the temporal component in a consistent manner (Maček et al. 2016). This makes it difficult to predict with a higher certainty the assembly of the typical hypoxia-tolerant AM fungal community throughout different time-points, indicating its long-term temporal stability. However, our preliminary results on AM fungal communities from mofette areas suggest that under permanent (long-term) selective pressure community composition is more

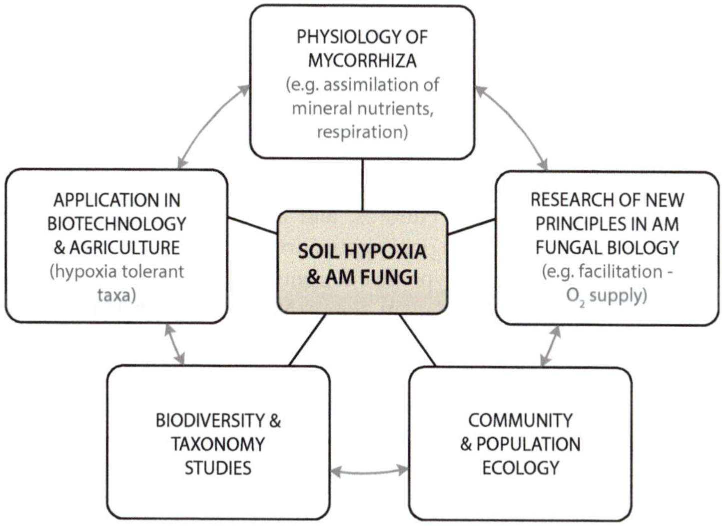

Fig. 16.5 Potential areas and interconnection of current and future research of AM fungal biology and arbuscular mycorrhiza in hypoxic environments

constant compared to the one in a control environment (Maček et al. 2011, 2016). In the latter, stochastic processes and other environmental factors play a much bigger role in structuring the communities among different time-points (e.g. between two vegetation seasons) (Maček et al. 2016; Šibanc et al., unpubl.). The major shifts in obligatory biotrophic AM fungal community composition within and between consecutive years happen each spring, when the winter community supported by low photosynthetic carbon flux into roots is shifted to the summer community (with high photosynthetic carbon flux into roots) and the pattern of how the new community assemblies each year is largely stochastic (Dumbrell et al. 2011). This pattern, however, is much less prominent in areas of high geogenic CO_2 concentrations and hypoxic soil in mofettes, where permanent long-term abiotic selective pressure acts on soil microbial communities (Maček et al. 2016). Therefore, it has been suggested that extreme, persistent and directed abiotic pressure results in a more stable system with highly specific microbial communities, dominated by the adapted and tolerant taxa that are consistently present in high abundance in the soil that is under long-term soil hypoxia (Maček et al. 2011, 2016; Šibanc et al. 2014). The case of mofette AM fungal community composition response shows the potential of some specific (possibly extreme or stressed) environments (e.g. mofettes, submerged environments, flooded soil) to serve as model ecosystems to study some unresolved principles in community ecology (Maček et al. 2016).

16.5 Conclusions

Diversity and community ecology of fungi in hypoxic environments, including mofettes (natural CO_2 springs), submerged and flooded continental or marine environments, remains largely unexplored (Maček et al. 2011, 2016). In a Norwegian oligotrophic lake, a new species of AM fungus *Rhizoglomus melanum* has been recently isolated, grown in pure cultures, and taxonomically described (Sudová et al. 2015). As many biotechnological applications require the capacity to grow in high CO_2 low O_2 environments, hypoxic environments are likely ideal locations for bioprospecting for industrially relevant fungi (Maček et al. 2016). However, the biotechnological potential of these specific ecosystems and their adapted biota is yet to be discovered.

Moreover, climate change models predict an increased frequency and duration of flooding events in the future (Hirabayashi et al. 2013). Especially in the field of sustainable agriculture, understanding the response of crops and their rhizosphere to flooding and soil hypoxia is becoming increasingly important in order to enhance yield and promote sustainable agriculture (e.g. Vallino et al. 2014). A weak overlap of the knowledge on the biology and ecology of AM fungi and the fast-growing research field of plant aerobiosis and responses to flooding (e.g. Perata et al. 2011; Pucciariello et al. 2014) has been noted. Thus, better communication between plant biologists and mycorrhiza researchers is needed in order to address problems related to food production and global changes in the future. Mycorrhiza should be considered as an integral part of plant biology, and hypoxic environments are only one such system where this integration is still largely missing.

Last but not least, many hypoxic environments can be considered as locally extreme environments and could serve as long-term natural experiments in ecology and evolution (Maček et al. 2016). Natural CO_2 springs (mofettes) are an example where the scientific power of this extreme ecosystem is starting to be harnessed for better insights into long-term ecological processes (Maček et al. 2011, 2016; Šibanc et al. 2014). The knowledge on long-term (permanent) stresses will be essential for sustainable future and a better prediction capacity of ecosystem responses to global change, as a permanent and long-term environmental perturbation on itself.

Acknowledgements This work was supported by the Slovenian Research Agency projects J4-5526 and J4-7052, Slovenian Research Agency core funding No. P4-0085, and a Swiss National Science Foundation project (SCOPES). All of the given support is gratefully acknowledged.

References

Atkin OK, Sherlock D, Fitter AH et al (2009) Temperature dependence of respiration in roots colonized by arbuscular mycorrhizal fungi. New Phytol 182:188–199

Baar J, Paradi I, Lucassen ECHET et al (2011) Molecular analysis of AMF diversity in aquatic macrophytes: a comparison of oligotrophic and utra-oligotrophic lakes. Aquat Bot 94:53–61
Bago B, Pfeffer PE, Shachar-Hill Y (2000) Carbon metabolism and transport in arbuscular mycorrhizas. Plant Physiol 124:949–957
Bago B, Pfeffer PE, Zipfel W et al (2002) Tracking metabolism and imaging transport in arbuscular mycorrhizal metabolism and transport in AM fungi. Plant Soil 244:189–197
Beck-Nielsen D, Madsen TV (2001) Occurrence of vesicular-arbuscular mycorrhiza in aquatic macrophytes from lakes and streams. Aquat Bot 71:141–148
Beulig F, Heuer VB, Akob DM et al (2015) Carbon flow from volcanic CO_2 into soil microbial communities of a wetland mofette. ISME J 9:746–759
Beulig F, Urich T, Nowak M et al (2016) Altered carbon turnover processes and microbiomes in soils under long-term extremely high CO_2 exposure. Nat Microbiol 1:1–9
Brooker RW, Callaway RM (2009) Facilitation in the conceptual melting pot. J Ecol 97:1117–1120
Danovaro R, Dell'Anno A, Pusceddu A et al (2010) The first metazoa living in permanently anoxic conditions. BMC Biol 8:30. doi:10.1186/1741-7007-8-30
DeJong-Hughes J, Moncrief JF, Voorhees WB et al. (2001) Soil compaction: causes, effects and control. On line paper on http://www.extension.umn.edu/agriculture/tillage/soil-compaction/
D'Souza J, Rodrigues BF (2013) Biodiversity of arbuscular mycorrhizal (AM) fungi in mangroves of Goa in West India. J For Res 24:515–523
Dumbrell AJ, Nelson M, Helgason T et al (2010) Relative roles of niche and neutral processes in structuring a soil microbial community. ISME J 4:337–345
Dumbrell AJ, Ashton PD, Aziz N et al (2011) Distinct seasonal assemblages of arbuscular mycorrhizal fungi revealed by massively parallel pyrosequencing. New Phytol 190:794–804
Farmer AM (1985) The occurrence of vesicular-arbuscular mycorrhiza in isoetid-type submerged aquatic macrophytes under naturally varying conditions. Aquat Bot 21:245–249
Finlay RD (2008) Ecological aspects of mycorrhizal symbiosis: with special emphasis on the functional diversity of interactions involving the extraradical mycelium. J Exp Bot 59:1115–1126
Fitter AH (1991) Costs and benefits of mycorrhizas: implications for functioning under natural conditions. Experientia 47:350–355
Fitter AH (2005) Darkness visible: reflections on underground ecology. J Ecol 93:231–243
Gianinazzi S, Gollotte A, Binet M-N et al (2010) Agroecology: the key role of arbuscular mycorrhizas in ecosystem services. Rev Mycorrhiza 20:519–530
Govindarajulu M, Pfeffer PE, Jin H et al (2005) Nitrogen transfer in the arbuscular mycorrhizal symbiosis. Nature 435:819–823
Harrison MJ, van Buuren ML (1995) A phosphate transporter from the mycorrhizal fungus *Glomus versiforme*. Nature 378:626–629
Helgason T, Daniell TJ, Husband R et al (1998) Ploughing up the wood-wide web? Nature 394:431. doi:10.1038/28764
Helgason T, Fitter AH (2009) Natural selection and the evolutionary ecology of the arbuscular mycorrhizal fungi (Phylum Glomeromycota). J Exp Bot 60:2465–2480
Hirabayashi Y, Mahendran R, Koirala S et al (2013) Global flood risk under climate change. Nat Clim Chang 3:816–821
Hodge A, Campbell CD, Fitter AH (2001) An arbuscular mycorrhizal fungus accelerates decomposition and acquires nitrogen directly from organic material. Nature 413:297–299
Hodge A, Helgason T, Fitter AH (2010) Nutritional ecology or arbuscular mycorrhizal fungi. Fungal Ecol 3:267–273
Hohberg K, Schulz H-J, Balkenhol B et al (2015) Soil faunal communities from mofette fields: effects of high geogenic carbon dioxide concentration. Soil Biol Biochem 88:420–429
Hourdez S (2012) Hypoxic environments. In: Bell EM (ed) Life at extremes: environments, organisms and strategies for survival. Gutenberg Press, Malta, pp 438–453

Hughes JK, Hodge A, Fitter AH et al (2008) Mycorrhizal respiration: implications for global scaling relationships. Trends Plant Sci 13:583–588
Kohout P, Sýkorová Z, Ctvrtlíková M et al (2012) Surprising spectra of root-associated fungi in submerged aquatic plants. FEMS Microbiol Ecol 80:216–235
Lenoir I, Fontaine J, Sahraoui AL (2016) Arbuscular mycorrhizal fungal responses to abiotic stresses: a review. Phytochemistry 123:4–15
Lohse S, Schliemann W, Ammer C et al (2005) Organisation and metabolism of plastids and mitochondria in arbuscular mycorrhizal roots of *Medicago truncatula*. Plant Physiol 139:329–340
Lumini E, Vallino M, Alguacil MM et al (2011) Different farming and water regimes in Italian rice fields affect arbuscular mycorrhizal fungal soil communities. Ecol Appl 21:1696–1707
Maldonado-Mendoza IE, Dewbre GR, Harrison MJ (2001) A phosphate transporter gene from the extra-radical mycelium of an arbuscular mycorrhizal fungus *Glomus intraradices* is regulated in response to phosphate in the environment. Mol Plant Microbe Interact 14:1140–1148
Maček I (2013) A decade of research in mofette areas has given us new insights into adaptation of soil microorganisms to abiotic stress. Acta Agric Slov 101:209–217
Maček I, Dumbrell AJ, Nelson M et al (2011) Local adaptation to soil hypoxia determines the structure of an arbuscular mycorrhizal fungal community in roots from natural CO_2 springs. Appl Environ Microbiol 77:4770–4777
Maček I, Kastelec D, Vodnik D (2012) Root colonization with arbuscular mycorrhizal fungi and glomalin-related soil protein (GRSP) concentration in hypoxic soils from natural CO_2 springs. Agric Food Sci 21:62–71
Maček I, Pfanz H, Francetič V et al (2005) Root respiration response to high CO_2 concentrations in plants from natural CO_2 springs. Environ Exp Bot 54:90–99
Maček I, Vodnik D, Pfanz H et al (2016) Locally extreme environments as natural long-term experiments in ecology. In: Dumbrell AJ, Kordas RL, Woodward G (eds) Advances in ecological research, vol 55. Academic Press, Oxford, pp 283–323
Marschner P (2012) Marschner's mineral nutrition of higher plants, 3rd edn. Academic Press, San Diego. ISBN-10: 0123849055
Møller CL, Kjøller R, Sand-Jensen K (2013) Organic enrichment of sediments reduces arbuscular mycorrhizal fungi in oligotrophic lake plants. Freshwa Biol 58:769–779
Parniske M (2008) Arbuscular mycorrhiza: the mother of plant root endosymbiosis. Nat Rev Microbiol 6:763–775
Perata P, Armstrong W, Voesenek LACJ (2011) Plants and flooding stress. New Phytol 190:269–273
Pfanz H (2008) Mofetten—kalter atem schlafender Vulkane. RVDL-Verlag, Köln
Pucciariello C, Voesenek LACJ, Perata P, Sasidharan R (2014) Plant responses to flooding. Front Plant Sci 5:4–5
Radhika KP, Rodrigues BF (2007) Arbuscular mycorrhizae in association with aquatic and marshy plant species in Goa, India. Aquat Bot 86:291–294
Raschi A, Miglietta F, Tognetti R et al (1997) Plant responses to elevated CO_2—evidence from natural springs. Cambridge University Press, Cambridge, p 272
Redecker D, Kodner R, Graham LE (2002) *Palaeoglonius grayi* from the Ordovician. Mycotaxon 84:33–37
Rillig MC (2004) Arbuscular mycorrhizae, glomalin, and soil aggregation. Can J Soil Sci 84:355–363
Rillig MC, Mummey DL (2006) Mycorrhizas and soil structure. New Phytol 171:41–53
Rosendahl S (2008) The first glance into the *Glomus* genome: an ancient asexual scandal with meiosis? New Physician 193:546–548
Russell DJ, Schulz H-J, Hohberg K et al (2011) Occurrence of collembolan fauna in mofette fields (natural carbon-dioxide springs) of the Czech Republic. Soil Org 83:489–505
Säle V, Aguilera P, Laczko E et al (2015) Impact of conservation tillage and organic farming on the diversity of arbuscular mycorrhizal fungi. Soil Biol Biochem 84:38–52

Schnoor TK, Lekberg Y, Rosendahl S et al (2011) Mechanical soil disturbance as a determinant of arbuscular mycorrhizal fungal communities in semi-natural grassland. Mycorrhiza 21:211–220
Simon MC, Keith B (2008) The role of oxygen availability in embryonic development and stem cell function. Nat Rev Mol Cell Biol 4:285–296
Smith, Read (2008) Mycorrhizal symbiosis, 3rd edn. London, Academic Press
Smolders AJP, Lucassen E, Roelofs JGM (2002) The isoetid environment: biogeochemistry and threats. Aquat Bot 73:325–350
Søndergaard M, Laegaard S (1977) Vesicular-arbuscular mycorrhiza in some aquatic vascular plants. Nature 268:232–233
Sudová R, Sýkorová Z, Rydlová J et al (2015) *Rhizoglomus melanum*, a new arbuscular mycorrhizal fungal species associated with submerged plants in freshwater lake Avsjøen in Norway. Mycol Prog 14:1–8
Šibanc N, Dumbrell AJ, Mandić-Mulec I, Maček I (2014) Impacts of naturally elevated soil CO_2 concentrations on communities of soil archaea and bacteria. Soil Biol Biochem 68:348–356
Tanner CC, Clayton JS (1985) Effects of vesicular-arbuscular mycorrhizas on growth and nutrition of a submerged aquatic plant. Aquat Bot 22:377–386
Vallino M, Fiorilli V, Bonfante P (2014) Rice flooding negatively impacts root branching and arbuscular mycorrhizal colonization, but not fungal viability. Plant Cell Environ 37:557–572
Voesenek LACJ, Sasidharan R, Visser EJW, Bailey-Serres J (2016) Flooding stress signalling through perturbations in oxygen, ethylene, nitric oxide and light. New Phytol 209:39–43
Vodnik D, Kastelec D, Pfanz H et al (2006) Small-scale spatial variation in soil CO_2 concentration in a natural carbon dioxide spring and some related plant responses. Geoderma 133:309–319
Vodnik D, Videmšek U, Pintar M et al (2009) The characteristics of soil CO_2 fluxes at a site with natural CO_2 enrichment. Geoderma 150:32–37
Vohník M, Borovec O, Kolařík M (2016) Communities of cultivable root mycobionts of the seagrass *Posidonia oceanica* in the northwest Mediterranean sea are dominated by a hitherto undescribed pleosporalean dark septate endophyte. Microb Ecol 71:442–451
Wang Y, Huang Y, Qiu Q et al (2011) Flooding greatly affects the diversity of arbuscular mycorrhizal fungi communities in the roots of wetland plants. PLoS One 6:e24512
Wang Y, Li T, Li Y et al (2015) Distribution of arbuscular mycorrhizal fungi in four semi-mangrove plant communities. Ann Microbiol 65:603–610
Wang Y, Li Y, Bao X (2016) Response differences of arbuscular mycorrhizal fungi communities in the roots of an aquatic and a semiaquatic species to various flooding regimes. Plant Soil 403:361–373
Wirsel SGR (2004) Homogenous stands of a wetland grass harbour diverse consortia of arbuscular mycorrhizal fungi. FEMS Microbiol Ecol 48:129–138
Wigand C, Andersen FO, Christensen KK, Holmer M, Jensen HS (1998) Endomycorrhizae of isoetids along a biogeochemical gradient. Limnol Oceanogr 43:508–515

Chapter 17
Piriformospora indica (*Serendipita indica*): The Novel Symbiont

Uma Singhal, Ram Prasad, and Ajit Varma

Abstract *Piriformospora indica* (*Serendipita indica*) (*Hymenomycetes*, *Basidiomycota*) is a cultivable endophyte that colonizes roots and has been extensively studied. *P. indica* has multifunctional activities like plant growth promoter, bio-fertilizer, immune modulator, bioherbicide, phytoremediator, etc. Growth promotional characteristics of *P. indica* have been studied in enormous number of plants (about 150 plants), and majority of them have shown highly significant outcomes. Certain secondary metabolites from the fungus are reasons behind such promising outputs. Promising outputs of laboratory experiments and small field trials indicated the need for its mass cultivation and usage. For field trials, a formulation "Rootonic" is prepared by mixing *P. indica* biomass in magnesium sulfite (raw talcum powder). The quantity of formulation (Rootonic) to be used per hectare of land for maximum productivity has also been standardized for about 150 plants. *P. indica* has proved to be highly beneficial endophyte with high efficacy in the field. In this chapter, a general view of the journey of *P. indica* from laboratory to field and finally toward industrialization is described.

17.1 Introduction

Soil is one of the organic and inorganic materials on the surface of the earth that provide the medium for plant growth, and rhizosphere is the most active portion of that frontier in which biogeochemical processes influence a host of landscape and global scale processes (McNear 2013). A better understanding of these processes is critical for maintaining the health of the planet and feeding the organisms that live on it (Morrissey et al. 2004). On uprooting a plant, a bulk amount of soil (mud) remains adhered to the rooting system (Fig. 17.1ai). They contain diverse microorganisms like bacteria, fungi, and actinomycetes. On washing the mud, a clean beautiful root architecture is seen (Fig. 17.1aii). The root system of all the land plants are colonized by a special group of fungi named as mycorrhiza. Mycorrhizal

U. Singhal • R. Prasad • A. Varma (✉)
Amity Institute of Microbial Technology (AIMT), Amity University Uttar Pradesh, Block 'E-3', 4th Floor, Sector-125, Noida, Gautam Buddha Nagar, Uttar Pradesh 201313, India
e-mail: ajitvarma@amity.edu

A. Varma et al. (eds.), *Mycorrhiza - Function, Diversity, State of the Art*,
DOI 10.1007/978-3-319-53064-2_17

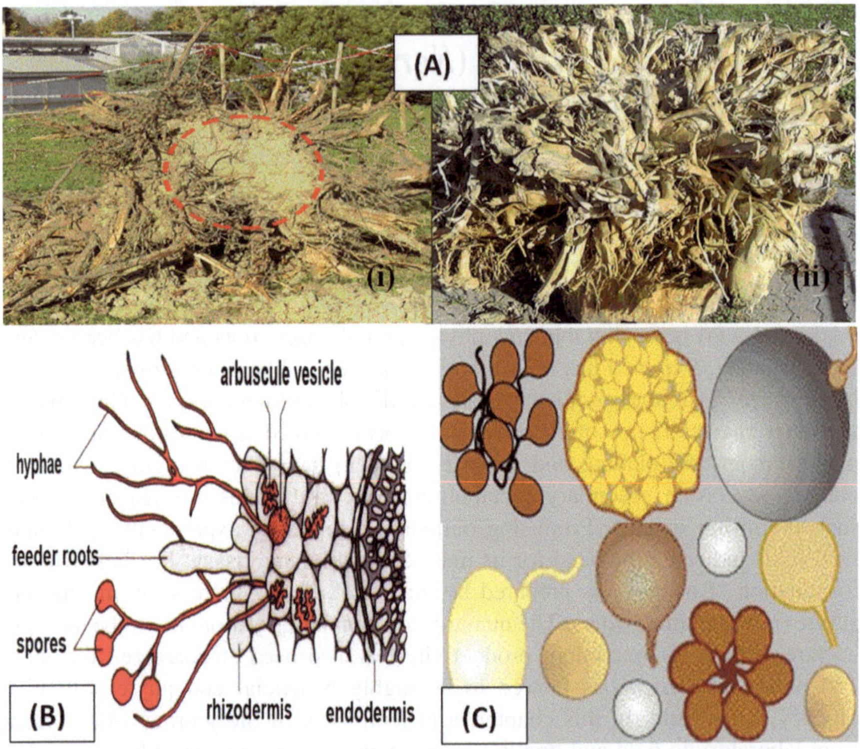

Fig. 17.1 Rhizosphere and spore structure: (**a**) Rhizosphere of plant system (i) pre-washed and (ii) washed; (**b**) anatomy of the root indicating the arbuscule, vesicle, and spores. Note the extramatrical hyphae in the soil and the root cortex; (**c**) spore structure of arbuscular mycorrhiza

fungi are characterized by and named after arbuscules (from the Latin arbusculum, small tree). These structures are formed in the inner root cortex by repeated branching of an intracellular hypha and are considered the site of nutrient exchange (Bonfante and Genre 2010) (Fig. 17.1b). They produce spores of diverse shape, size, and color, either singly or in aggregates (Fig. 17.1c). Spores form as swellings on one or more subtending hypha in the soil or in roots. These structures contain lipids, cytoplasm, and many nuclei. Spores usually develop thick walls with more than one layer and can function as propagules. They promote plant growth, enhance the active ingredient, increase seed production, and protect plants against disease. One of the most characteristic features is the production of arbuscules into the living cortical cell. Although this group of fungus (arbuscular mycorrhiza) was discovered by a German scientist A.B. Frank way back 1885, their biotechnological applications could not be exploited to the level they deserve because they cannot be cultured in the absence of a living root system. Spores/fungal hyphae do not multiply on detaching from the living root system:

Overcoming the above said disadvantage, an endophyte *Piriformospora indica* (*Serendipita indica*) was discovered by Prof Dr. Ajit Varma and his colleagues from Thar Desert of Western India in 1992 from the root system of several xerophytic plants (Verma et al. 1998, 1999). It belongs to a largest class of fungi *Hymenomycetes* within the phylum *Basidiomycota*. A new family *Sebacinaceae* and new order *Sebacinales* was created for this fungus due to its unique features (Weiß et al. 2004; Qiang et al. 2011). This symbiotic fungus not only promotes plant growth but also has other multifunctional activities such as plant growth promoter, bioprotectant, bio-pesticide, helps in enhancing flowering and fruiting, etc. (Gill et al. 2016) (Fig. 17.2). Properties have been patented in Germany (European Patent Office, Muenchen, Germany, Patent No. 97121440.8-2105, Nov.1998) dating back to 1997. *P. indica* is deposited at the Deutsche Sammlung fu¨r Mikroorganismen und Zellkulturen, Braunschweig, Germany (DSM 11827). See also https://www.ncbi.nlm.nih.gov/nuccore/AF014929 Colonization by *P. indica* increases nutrient uptake, allows plants to survive underwater, temperature and salt-stresses, confers (systemic) resistance to toxins, heavy metal ions and pathogenic organisms and stimulates growth and seed production (Harman 2011; Varma et al. 2012). The valuable secondary metabolites excreted by *P. indica* influence early seed germination, better plant productivity, early flowering, etc. (Varma et al. 2012). Use of *P. indica* to increase desiccation tolerance in higher plants has been studied by Varma and his colleagues (2012) and significant increase tolerance was achieved. Genome wide study revealed that its genome is assembled into 1884 scaffolds containing 2359 contigs with an average read coverage of 22 and a genome size of 24.97 Mb. The estimated DNA content of *P. indica* nuclei ranges from 15.3 to 21.3 Mb. To assess the genome completeness of *P. indica* a blast search was performed with highly conserved core genes present in higher eukaryotes (Zuccaro et al. 2009). *P. indica* can be stably transformed by random genomic integration of foreign DNA and that it possesses a relatively small genome as compared to other members of the *Basidiomycota* (Zuccaro et al. 2011). Extensive research on this organism has brought it to an appreciable state and made its field trials and marketing possible. In this chapter a general view of journey of *P. indica* from laboratory to field and finally toward industrialization is described.

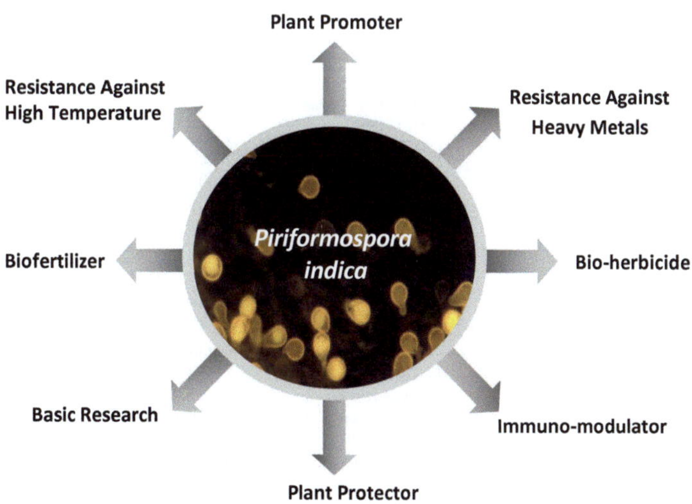

Fig. 17.2 Functional characteristics of fungus

17.2 Recent Taxonomic Position

P. indica, a *Basidiomycete*, resembles in many aspects to arbuscular mycorrhizal fungi (AMF) which, however, belongs to a new family *Sebacinaceae* and new order *Sebacinales*. *Sebacinales* was established to harmonize taxonomic ranking in the major groups of early-diverging *Agaricomycetes* (Weiß et al. 2004). *Sebacinales* is divided into two subgroups: group A or *Sebacinaceae* and group B for which it was proposed as a new family *Serendipitaceae* (Weiß et al. 2016) (Fig. 17.3).

Serendipitaceae also includes *S. indica* (formerly called *P. indica*), the *Sebacinales* species most frequently used in experimental research. This strain, named by reference to its pear-shaped asexual spores, was isolated from the Indian desert soil (Verma et al. 1998) and has been associated experimentally with diverse host plants (Varma et al. 2012). As the related species *Serendipita williamsii* (*Piriformospora williamsii*; Basiewicz et al. 2012), it produces chlamydospores (asexual-resting spores) and hyphae resembling a string of beads, which are also observed in *Serendipita herbamans* and other *Serendipitaceae* (Warcup and Talbot 1988; Riess et al. 2013).

In contrast to AMF, *P. indica* can grow axenically and promotes plant growth, increases resistance of colonized plants against fungal pathogens and tolerance to abiotic stress, and shows further beneficial to plants (Gill et al. 2016). It also alters the secondary metabolites of many plants of economic importance and promotes overall growth and seed production of many plants (Bagde et al. 2010a, b). In contrast to AMF, *P. indica* colonizes *A. thaliana*, a model plant for which a multitude of well-characterized mutants is available. *P. indica* has white to almost hyaline hyphae. The hyphae are thin walled and have a diametric range of

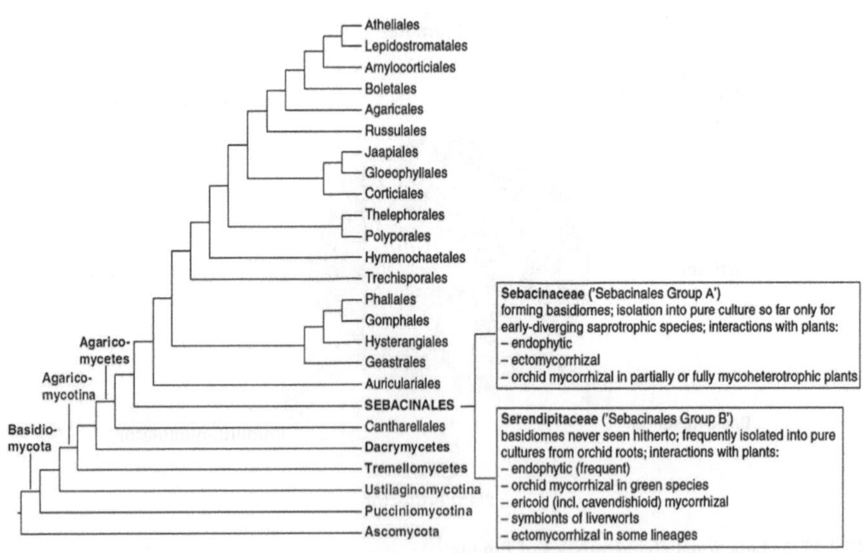

Fig. 17.3 *Sebacinales*: phylogenetic position within *Basidiomycota* and interactions of the two families (*Sebacinaceae* and *Serendipitaceae*) with plants (Weiß et al. 2016)

0.7–3.5 μm. The hyphae are irregularly septate and often exhibit anastomosis. The highly interwoven hyphae appear as intermingled cords and branch irregularly. External deposits, polysaccharides, or hydrophobic proteins can be noticed on hyphal walls at regular intervals. The irregular septation of hyphae accounts for the presence of more than one nuclei in a single compartment. The distinct chlamydospores appear singly or in clusters. Initially the chlamydospores are thin walled and hyaline while they become thick walled and autofluorescent toward maturity. Further, no sexual structures or clamp connections were observed (Varma et al. 2001). The mycelium has a subsurfaced and concentric growth on agar medium. When grown on a solid culture media, very few aerial hyphae were formed. Occasionally the mycelium fabricates periodic rings on agar medium, whereas the structure of the mycelium was homogenous. The morphological characters of the mycelium greatly differ with variations in conditions of cultivation or nutrient compositions of the culture medium. To assess the genome completeness of *P. indica*, a blast search was performed with highly conserved core genes present in higher eukaryotes. A genetic transformation system is established using a fragment of the TEF promoter region for construction of vectors carrying the selectable marker hygromycin B phosphotransferase. It is already shown that *P. indica* can be stably transformed by random genomic integration of foreign DNA and that it possesses a relative small genome as compared to other members of the *Basidiomycota*.

Through 18S rRNA molecular methods and electron microscopy, it was elucidated that this fungus belongs to *Hymenomycetes*. With the help of electron microscopy, the presence of dolipores with non-perforated parenthosomes was elucidated. This further implied that *P. indica* belongs to *Hymenomycetes* (*Basidiomycota*). Sequence comparison showed the close relation of *P. indica* with *Rhizoctonia* group (Varma et al. 2001). *P. indica* is placed as a member of the *Basidiomycetes* order *Sebacinales* by the molecular phylogenetic analysis (Hibbett et al. 2007; Qiang et al. 2012; Weiß et al. 2004). The anamorphic *P. indica* is associated with group B (Basiewicz et al. 2012) and inferred significant changes in physiological and molecular parameter within similar strains of *Piriformospora* (Basiewicz et al. 2012).

17.3 Cultivation and Morphological Characteristics

The fungus has got very simple morphology containing hyphae and pear-shaped large spores. The fungus can be cultivated on simple defined medium both in solid medium and broth. Optimum conditions for growth are temperature of 25 °C ± 2, pH of 6.8, carbon energy of 1.5% glucose. Incubation is done on the rotatory shaker (120 rpm). The best growth is obtained after 7 days of incubation where colonies can be large or small Fig. 17.4.

P. indica has simple septum with dolipores and continuous, straight parenthosomes. It promotes the growth of plants and improves their productivity, increases the drought tolerance of host plants, delays the wilting of leaves, prolongs the aging of callus tissue, and protects the plants from the attack of pathogens

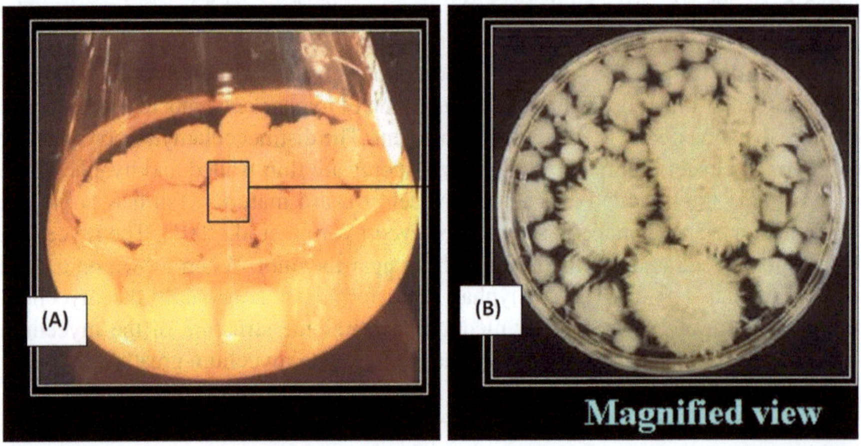

Fig. 17.4 The fungal colony after incubation at 28 °C (**a**) in broth and (**b**) in agar plate

(Kumari et al. 2005; Waller et al. 2008; Zuccaro et al. 2009). It enhances the phosphate uptake by the plants (Yadav et al. 2010). The biological hardening of tissue culture-raised plants with *P. indica* protects them from transplantation shock and increases their survival rate to 90–100%. The stress conditions caused due to acidity, desiccation, and heavy metal toxicity are relieved by *P. indica*. The fungus also induces systemic disease resistance by enhancing the concentration of antioxidants, ascorbate, and glutathione in the plant body to cope up with the oxidative stress caused by pathogens (Waller et al. 2005; Vadassery et al. 2009; Gill et al. 2016). Thus, *P. indica* shows tremendous potential to be used as a biological agent for plant growth promotion and control of plant root disease and a tool for biological hardening of micropropagated plants. Another advantageous feature of the fungus is that it produces a large number of thick-walled, autofluorescent, pear-shaped spores called chlamydospores having longer shelf life. These spores, rather than fungal mycelia, are to be used as bioinoculant for agricultural crops. The spores can be produced easily and can survive unfavorable conditions and germinate on the onset of favorable conditions, which lead again to vegetative growth. All these qualities make spores a good candidate from application point of view.

17.3.1 Growth Conditions of P. Indica

P. indica was cultured and grown on different synthetic media and best in modified Hill and Kaefer medium (Pham et al. 2004; Kumari et al. 2005). Circular agar disks (4 mm in diameter) infested with chlamydospores and actively growing hyphae of *P. indica* are placed onto petri dishes containing solidified Hill and Kaefer medium. Incubation is carried out at 25 °C in the dark for 7–10 days. Broth Erlenmeyer flask is constantly shaken at 80 rpm. After 7–10 days, the petri plate is completely filled up with biomass. In broth Erlenmeyer flask, small and large colonies appear which consist of hyphae and chlamydospores. Spores are extrametrical and intercellular.

17.3.2 Morphogenesis of P. Indica

The morphogenesis of the fungus was monitored for about 7 days in batch cultures. The cylindrical hyphae after 2 days started enlargement at places, and after 5 days, many of the hyphae turned into aggregated spores. At the end of 7 days, typical pear-shaped spores were produced in abundance (Das et al. 2013). Another experiment after 4 days of incubation at 28 °C was transferred in cold at 4 °C, and massive sporulation was found within 24 h. Vice versa, the cultures after 25 h were incubated at 37 °C; the massive sporulation was recorded in Fig. 17.5.

Interestingly, Barman and Prasad group observed that the autofluorescence selectively "labels" the spores in comparison to the hyphal structures, thus offering spatial localization of the spores in the intact culture. This can be the differential concentration of the chromophores in the spores and in the mycelium, with NAD(P)H amount observed to be ca. 40% higher in regions of high metabolic activity. The greater accumulation of such molecules in the spores likely results in enhanced autofluorescence (Siddhanta et al. 2017).

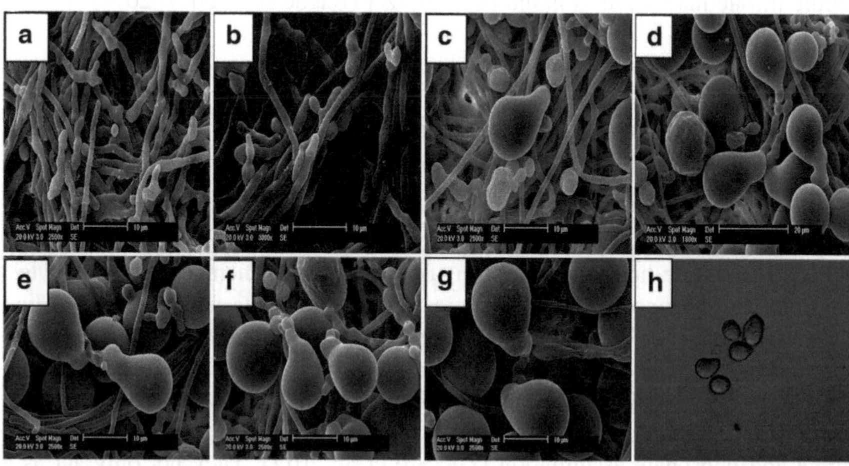

Fig. 17.5 Morphogenesis of *P. indica* (**a, b**) The cylindrical hyphae after 2 days started enlargement at places, and (**c**) after 5 days, many of the hyphae turned into aggregated spores. (**d**) At the end of 7 days, typical pear-shaped spores were produced in abundance. (**e, f**) After 4 days of incubation at 28 °C, they were transferred in cold at 4 °C, and massive sporulation was found within 24 h. (**g, h**) *Vice versa*, the cultures after 25 h were incubated at 37 °C; the massive sporulation was recorded

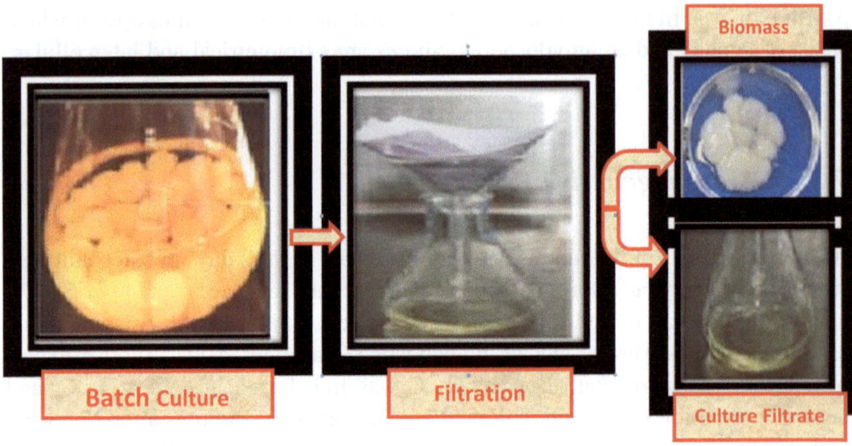

Fig. 17.6 Separation of *P. indica* biomass and culture filtrate

17.3.3 Separation of P. Indica Biomass and Culture Filtrate

One of the unique features of the fungus is that the culture filtrate also acts as an excellent source for plant promotion. Culture filtrate was separated from fungal biomass using a simple filtration procedure (Fig. 17.6). In an independent experiment, the fungus was grown in broth. After 10 days, the biomass was removed. The culture filtrate caused early seed germination and flowering. In the long run, the culture filtrate may serve as liquid bio-fertilizer (Bagde et al. 2011, 2014).

17.4 Applications of *Serendipita indica*

17.4.1 Plant Growth Promotion

P. indica is a wide-host symbiotic fungus and colonizes members of bryophytes, pteridophytes, gymnosperms, and angiosperms—monocots and dicots including orchids and members of the *Brassicaceae* (e.g., *Arabidopsis thaliana*) (Peskan-Berghofer et al. 2004). Plants colonized by *P. indica* display a wide range of beneficial effects including enhanced host growth and resistance to biotic and abiotic stresses, promotion of adventitious root formation in cuttings, and enhanced nitrate and phosphate assimilation (Zuccaro et al. 2011). They not only act as a plant promoter but also as bioprotectant against pathogens (Waller et al. 2005; Deshmukh et al. 2006). Baltruschat et al. in 2008 studied biochemical mechanisms underlying *P. indica*-mediated salt tolerance in barley with special focus on anti-oxidants. *P. indica*-colonized barley roots in salt stress conditions had increased plant growth, elevated the amount of ascorbic acid, and increased activities of antioxidant. These findings have suggested that antioxidants might play a role in both inherited and endophyte-mediated plant tolerance to salinity as reported in *Brassica napus* L. (Chen et al. 2012; Varma et al. 2012). The fungus-treated *Brassica* plants showed significant increase in the size and numbers of their leaves and the weights of their fresh roots, dry roots, and shoots; early flowering; and increased seed yield and oil content. Nutritional analysis revealed that fungus-treated plants had reduced erucic acid and glucosinolate contents and increased accumulation of N, P, K, S, and Zn. Also, RT-PCR results showed that the expression of Bn-FAE1 and BnECR genes, encoding enzymes responsible for regulating erucic acid biosynthesis, was downregulated at mid- and late-life stages during seed development in colonized plants (Binggen Lou-personal communication). Thus, the results confirmed that *P. indica* plays an important role in enhancing growth, seed yield, and seed quality of *B. napus*.

A large number of medicinal plants like *Spilanthes calva*, *Withania somnifera*, *Bacopa monnieri*, *Coleus forskohlii*, and others were inoculated with the *P. indica* in pots as well as in fields to study its influence on the host plants (Das et al. 2013). The effect of *P. indica* on growth of *B. monnieri*, *A. vasica*, and *S. alva* is given in Fig. 17.7a.

17.4.2 Protection Against Pathogens and Insects

In addition to plant promotion, this fungus also protects the plant against pathogens and insects. Some of the plants' growth promotional factors involved in *Arabidopsis* protection are glucosinolates, ethylene, etc. *P. indica* co-inoculation protected bottle gourd and *Nicotiana tabacum* from other fungal and viral infections leading to healthy growth of plant (Fig. 17.7b).

17.4.2.1 Glucosinolates

A second class of plant glycosides, called the glucosinolates or mustard oil glycosides, break down to release defensive substances. Found principally in the *Brassicaceae* and related plant families, glucosinolates break down to produce the compounds responsible for the smell and taste of vegetables such as cabbage, broccoli, and radishes. Glucosinolate breakdown is catalyzed by a hydrolytic enzyme called a thioglucosidase or myrosinase that cleaves glucose from its bond with the sulfur atom. These defensive products function as toxins and herbivore repellents (Wittstock et al. 2004). Like cyanogenic glycosides, glucosinolates are stored in the intact plant separately from the enzymes that hydrolyze them, and they are brought into contact with these enzymes only when the plant is crushed. Several studies have reported that glucosinolates exhibit growth inhibition or feeding deterrence to a wide range of general herbivores such as birds, slugs, and generalist insects (Giamoustaris and Mithen 1995, 1996). It was also found that plants respond to herbivore or insect damage by systematically accumulating higher levels of

Fig. 17.7 Application of *P. indica* in plant growth promotion and protection of plants against pathogens. (**A**) Effect of *P. indica* on in vitro grown *Bacopa monnieri* and *A. vasica*, pronounced growth response and flowering in *S. alva* after inoculation with *P. indica*. (**a**) Control (without *P. indica*), (**b**) inoculated with *P. indica*. (**B**) *P. indica* protects against plant pathogens like fungi and viruses. (**a**) Bottle gourd infested with insects and virus in the field; (**b**) bottle gourd plants treated by *P. indica* are healthy; (**c**) *Alternaria longipes* infection status of untreated *Nicotiana tabacum* and (**d**) *P. indica*-colonized plants

glucosinolates and thus presumably increasing their resistance (Martin and Müller 2006). Usually, it is the indole glucosinolates which become induced.

17.4.2.2 Ethylene

The plant response to damage by insect herbivores involves both a wound response and the recognition of certain insect-derived compounds referred to as elicitors. Although repeated mechanical wounding can induce responses similar to those caused by insect herbivorous in some plants, certain molecules in insect saliva can serve as enhancers of this stimulus. In addition, such insect-derived elicitors can trigger signaling pathways systemically, thereby initiating defensive responses in distant regions of the plant in anticipation of further damage. After being regurgitated by an insect, elicitors become part of its saliva and are thus applied to the feeding site during herbivory. Plants then recognize these elicitors and activate a complex signal transduction pathway that induces their defenses. Ethylene is one of the signaling compounds induced by insect herbivory. In many cases, the concerted action of ethylene is necessary for the full activation of induced defenses (Arimura et al. 2008).

17.4.3 Stress Tolerance

Abiotic stresses are often interrelated, either individually or in combination; they cause morphological, physiological, biochemical, and molecular changes that adversely affect plant growth and productivity and ultimately yield. Heat, drought, cold, and salinity are the major abiotic stresses that induce severe cellular damage in plant species, including crop plants. *P. indica* also shows tolerance toward stresses of extremes of climate that are very high and very low temperature and also salt stress (Gill et al. 2016). It also shows synergism with other plant growth bacteria like *Azospirillum brasilense*.

17.5 The Secret of Plant Promotion, Value Addition, and Early Flowering

Fungus interactions are characterized by a more efficient nutrient uptake from the soil due to a better hyphal penetration into the soil compared to the penetration of the thicker root hairs. The plant delivers phosphorus assimilates to the fungus. In mycorrhizal associations, plants acquire phosphate from the extensive network of fine extra-radical hyphae of fungus, which extend beyond root depletion zones to mine new regions of the soil. Several studies have indicated that the interaction alters the pathway for nitrogen metabolism whereby transferring more nitrogen nutrients to the plants. Preliminary studies indicate that *P. indica* influences the sulfate reduction leading to formation of sulfur proteins and glutathione contents. This, in turn, influences the resistance against water deficiency and drought. The supply of the fungus with carbon sources and the faster growth of colonized plants require the breakdown of starch which is deposited in the root amyloplasts. Thus, it is not surprising that one of the major starch-degrading enzymes, the glucan-water dikinase, is activated by the fungus.

The fungus regulates the uptake and transportation of important macronutrients like iron, zinc, manganese, copper, etc. Interaction of plants results in synthesis of important phytohormones. The cumulative effect of macro-micronutrients and phytohormones influences the plant metabolism—the value addition, early flowering, and plant growth. Massive proliferation of useful rhizospheric microorganisms sustains soil fertility (organic farming).

17.6 Step Forward Toward Commercialization

To enhance the usage so that the benefits of the fungus are used by a common farmer, the fungus was formulated with magnesium sulfite which acts as a carrier. For this, 2% (w/w) of formulation served as effective and stable carrier. On an average, the CFU count was maintained as 10^9 and moisture 20%. The protocol for formulation is given in Fig. 17.8.

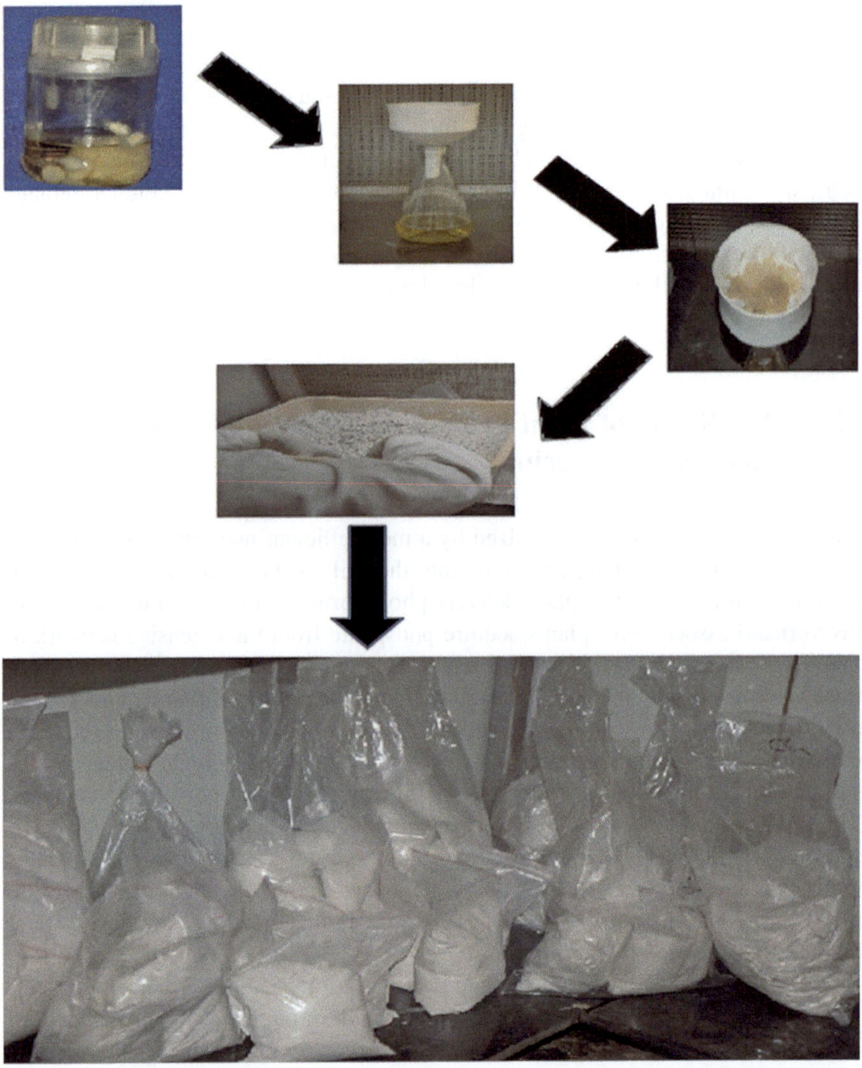

Fig. 17.8 Steps for the preparation of formulation

17.6.1 *Formulation of Rootonic for Field Application*

The Rootonic bio-fertilizer was formulated to enhance its handling, storage, propagation, and overall convenience of use by a common farmer. To enhance the usage so that the benefits of the fungus are used by a common farmer, it was formulated with magnesium silicate which acts as a carrier, and quantity for field trial was optimized. For this, 2% (w/w) of formulation served as effective and stable carrier. On an average, the colony forming unit (CFU) count was maintained as 108 and moisture 20%. Protocol for seed treatment for field trial has been given as Fig. 17.9 and the quantity of formulation required for the seed treatment.

Fig. 17.9 Protocol for seed treatment

17.7 Conclusion

P. indica is a rewarding organism with its huge and distinguished properties. Colonization by *P. indica* increases nutrient uptake and allows plants to survive in drought, salt stress, and temperature stress. Excellent plant growth promotion, resistant against extremes of climate and bio-protecting capability of the organisms, has paved way for its varied field applications. Large field trials at various locations in India showed beneficial effects of *P. indica* on plant growth and development. Promising outputs of field trials showed that it should be used at a large scale so that common farmers are benefited and finally countries' economy is at profit. Increase in productivity of certain crop upon interaction with *P. indica* will increase total land usage. Field trials of the same are done by formulating biomass with powder and inoculating the mixture into the root of plants. The formulation is termed "Rootonic." The journey from *P. indica* to Rootonic is exciting and very fulfilling. Large-scale production and application of the product are still under process, and we are looking forward to its commercialization soon.

Acknowledgment Ajit Varma is thankful to Department of Science and Technology and Department of Biotechnology for partial financial funding and to DST-FIST for providing confocal microscope facility.

References

Arimura G, Garms S, Maffei M et al (2008) Herbivore-induced terpenoid emission in *Medicago truncatula*: concerted action of jasmonate, ethylene and calcium signaling. Planta 227:453. doi:10.1007/s00425-007-0631-y

Bagde US, Prasad R, Varma A (2010a) Interaction of *Piriformospora indica* with medicinal plants and of economic importance. Afr J Biotechnol 9:9214–9226

Bagde US, Prasad R, Varma A (2010b) Characterizations of culture filtrate of *Piriformospora indica* by HPTLC analysis. Asian J Microbiol Biotechnol Environ Sci 12:805–809

Bagde US, Prasad R, Varma A (2011) Influence of culture filtrate of *Piriformospora indica* on growth and yield of seed oil in *Helianthus annus*. Symbiosis 53:83–88

Bagde US, Prasad R, Varma A (2014) Impact of culture filtrate of *Piriformospora indica* on biomass and biosynthesis of active ingredient aristolochic acid in *Aristolochia elegans* Mart. Int J Biol 1:29–37

Baltruschat H, Fodor J, Harrach BD, Niemczyk E, Barna B, Gullner G, Janeczko A, Kogel KH, Schaefer P, Schwarczinger I, Zuccaro A, Skoczowski A (2008) Salt tolerance of barley induced by the root endophyte *Piriformospora indica* is associated with a strong increase in antioxidants. New Phytol 180:501–510

Basiewicz M, Weiss M, Kogel KH, Langen G, Zorn H, Zuccaro A (2012) Molecular and phenotypic characterization of *Sebacina vermifera* strains associated with orchids, and the description of *Piriformospora williamsii* sp. nov. Fungal Biol 116:204–213

Bonfante P, Genre A (2010) Mechanisms underlying beneficial plant–fungus interactions in mycorrhizal symbiosis. Nat Commun 1:48. doi:10.1038/ncomms1046

Chen HJ, Wu SD, Huang GJ, Shen CY, Afiyanti M, Li WJ, Lin YH (2012) Expression of a cloned sweet potato catalase SPCAT1 alleviates ethephon-mediated leaf senescence and H_2O_2 elevation. J Plant Physiol 169:86–97

Das A, Prasad R, Srivastava RB, Deshmukh S, Rai MK, Varma A (2013) Co-cultivation of plants with medicinal plants: case studies. In: Varma A, Kost G, Oelmuller R (eds) *Piriformospora indica*: sebacinales and their biotechnological applications. Springer, Berlin, Heidelberg, pp 149–171

Deshmukh S, Hückelhoven R, Schäfer P, Imani J, Sharma M et al (2006) The root endophytic fungus *Piriformospora indica* requires host cell death for proliferation during mutualistic symbiosis with barley. Proc Natl Acad Sci USA 103:18450–18457

Giamoustaris A, Mithen R (1995) The effect of modifying the glucosinolate content of leaves of oilseed rape (*Brassica napus ssp. oleifera*) on its interaction with specialist and generalist pests. Ann Appl Biol 126:347–363

Giamoustaris A, Mithen R (1996) The effect of flower colour and glucosinolates on the interaction between oilseed rape and pollen beetles. Entomol Exp Appl 80:206–208

Gill SS, Gill R, Trivedi DK, Anjum NA, Sharma KK, Ansari MW, Johri AK, Prasad R, Pereira E, Varma A, Tuteja N (2016) *Piriformospora indica*: potential and significance in plant stress tolerance. Front Microbiol. doi:10.3389/fmicb.2016.00332

Harman GE (2011) Multifunctional fungal plant symbiont: new tools to enhance plant growth and productivity. New Phytol 189:647–649

Hibbett DS, Binder M, Bischoff JF, Blackwell M, Cannon PF, Eriksson OE, Huhndorf S, James T, Kirk PM, Lücking R, Lumbsch HT, Lutzoni F, Matheny PB, McLaughlin DJ, Powell MJ, Redhead S, Schoch CL, Spatafora JW, Stalpers JA, Vilgalys R, Aime MC, Aptroot A, Bauer R, Begerow D, Benny GL, Castlebury LA, Crous PW, Dai Y-C, Gams W, Geiser DM, Griffith GW, Gueidan C, Hawksworth DL, Hestmark G, Hosaka K, Humber RA, Hyde KD, Ironside JE, Kõljalg U, Kurtzman CP, Larsson K-H, Lichtwardt R, Longcore J, Miadlikowska J, Miller A, Moncalvo J-M, Mozley-Standridge S, Oberwinkler F, Parmasto E, Reeb V, Rogers JD, Roux C, Ryvarden L, Sampaio JP, Schüßler A, Sugiyama J, Thorn RG, Tibell L, Untereiner WA, Walker C, Wang Z, Weir A, Weiss M, White MM, Winka K, Yao Y-J, Zhang N (2007) A higher-level phylogenetic classification of the fungi. Mycol Res 111:509–547

Kumari R, Pham GH, Prasad R, Sachdev M, Srivastava A, Yadav V, Verma PK, Sharma S, Malla R, Singh A, Maurya AK, Prakash S, Pareek A, Rexer KH, Kost G, Garg AP, Oelmueller R, Sharma MC, Varma A (2005) *Piriformospora indica*: fungus of the millennium. In: Podila GK, Varma A (eds) Basic research and applications of mycorrhizae, vol 1. IK International, India, pp 259–295

Martin N, Müller C (2006) Induction of plant responses by a sequestering insect: relationship of glucosinolate concentration and myrosinase activity. Basic Appl Ecol 8:13–25

McNear DH Jr (2013) The rhizosphere—roots, soil and everything in Between. Nat Educ Knowl 4:1

Morrissey J, Dow J, Mark G, O'Gara F (2004) Are microbes at the root of a solution to world food production? EMBO Rep 5:922–926

Peskan-Berghofer T, Shahollari B, Giong PH, Hehl S, Markert C, Blanke V, Kost G, Varma A, Oelmuller R (2004) Association of *Piriformospora indica* with *Arabidopsis thaliana* roots represents a novel system to study beneficial plant–microbe interactions and involves early plant protein modifications in the endoplasmic reticulum and at the plasma membrane. Physiol Plant 122:465–447

Pham GH, Kumari R, Singh A, Sachdev M, Prasad R, Kaldorf M et al (2004) Axenic cultures of *Piriformosporaindica*. In: Varma A, Abbott K, Werner D, Hampp R (eds) Plant surface microbiology. Springer, Berlin, pp 593–616

Qiang X, Weiss M, Kogel KH, Schaefer P (2011). *Piriformospora indica*—a mutualistic basidiomycete with an exceptionally large plant host range. Mol Plant Pathol 13:508–518

Qiang X, Weiss M, Kogel KH, Schäfer P (2012) *Piriformospora indica*—a mutualistic basidiomycete with an exceptionally large plant host range. Mol Plant Pathol. doi:10.1111/J.1364-3703.2011.00764

Riess K, Oberwinkler F, Bauer R, Garnica S (2013) High genetic diversity at the regional scale and possible speciation in *Sebacina epigaea* and *S. incrustans*. BMC Evol Biol 13:102

Siddhanta S, Paidi SK, Bushley K, Prasad R, Barman I (2017) Exploring morphological and biochemical linkages in fungal growth with label-free light sheet microscopy and Raman spectroscopy. ChemPhysChem 18(1):72–78

Vadassery J, Tripathi S, Prasad R, Varma A, Oelmüller R (2009) Monodehydroascorbate reductase 2 and dehydroascorbate reductase 5 are crucial for a mutualistic interaction between *Piriformospora indica* and *Arabidopsis*. J Plant Physiol 165:60–70

Varma A, Verma S, Sudha, Sahay N, Britta B, Franken P (1999) *Piriformospora indica*—a cultivable plant growth promoting root endophyte with similarities to arbuscular mycorrhizal fungi. Appl Environ Microbiol 65:2741–2744

Varma A, Singh A, Sudha, Sahay NS, Sharma J, Roy A, Kumari M, Rana D, Thakran S, Deka D, Bharati K, Franken P, Hurek T, Blechert O, Rexer KH, Kost HA, Hock B, Maier W, Walter M, Strack D, Kranner I (2001) *Piriformospora indica*: a cultivable mycorrhiza-like endosymbiotic fungus. In: Hock B (ed) Mycota IX series. Springer, Berlin, pp 123–150

Varma A, Tripathi S, Prasad R, Das A, Sharma M, Bakshi M, Arora M, Rastogi K, Agrawal A, Kharkwal AC, Tsimilli-Michael M, Strasser RJ, Bagde US, Bisaria VS, Upadhyaya CP, Malla R, Kost G, Joy, Sherameti I, Chen Y, Ma J, Lou B, Oelmuller R (2012) The symbiotic fungus *Piriformospora indica*: review. In: Hock B (ed) The mycota XXL. Springer, Berlin, Heidelberg, New York

Verma S, Varma A, Rexer KH, Hassel A, Kost G, Sarbhoy A, Bisen P, Butehorn B, Franken P (1998) *Piriformospora indica*, gen. et sp. nov., a new root-colonizing fungus. Mycologia 90:896–902

Waller F, Achatz B, Baltruschat H, Fodor J, Becker K, Fischer M, Heier T, Hückelhoven R, Neumann C, von Wettstein D, Franken P, Kogel K-H (2005) The endophytic fungus *Piriformospora indica* reprograms barley to salt-stress tolerance, disease resistance, and higher yield. Proc Natl Acad Sci USA 102:13386–13391

Waller F, Mukherjee K, Deshmukh SD, Achatz B, Sharma M, Schäfer P, Kogel KH (2008) Systemic and local modulation of plant responses by *Piriformospora indica* and related Sebacinales species. J Plant Physiol 165:60–70

Warcup JH, Talbot (1988) Mycorrhizal associations of isolates of *Sebacina vermifera*. New Phytol 110:227–231

Weiß M, Selosse M-A, Rexer K-H, Urban A, Oberwinkler F (2004) Sebacinales: a hitherto overlooked cosm of heterobasidiomycetes with a broad mycorrhizal potential. Mycol Res 108:1003–1010

Weiß M, Waller F, Zuccaro A, Selosse M-A (2016) Sebacinales—one thousand and one interactions with land plants. New Phytol 211:20–40. doi:10.1111/nph.13977

Wittstock U, Agerbirk N, Stauber EJ, Olsen CE, Hippler M, Mitchell-Olds T, Gershenzon J, Vogel H (2004) Successful herbivore attack due to metabolic diversion of a plant chemical defense. Proc Natl Acad Sci USA 101:4859–4864

Yadav V, Kumar M, Deep DK, Kumar H, Sharma R, Tripathi T, Tuteja N, Saxena AK, Johri AK (2010) A phosphate transporter from the root endophytic fungus *Piriformospora indica* plays a role in phosphate transport to the host plant. J Biol Chem 285:26532–26544

Zuccaro A, Basiewicz M, Zurawska M, Biedenkopf D, Kogel K-H (2009) Karyotype analysis, genome organization, and stable genetic transformation of the root colonizing fungus *Piriformospora indica*. Fungal Genet Biol 46:542–550

Zuccaro A, Lahrmann U, Güldener U, Langen G, Pfiffi S, Biedenkopf D et al (2011) Endophytic life strategies decoded by genome and transcriptome analyses of the mutualistic root symbiont *Piriformospora indica*. PLoS Pathogen 7:e1002290. doi:10.1371/journal.ppat.1002290

Chapter 18
Mass Cultivation of Mycorrhiza-Like Fungus *Piriformospora indica* (*Serendipita indica*) by Batch in Bioreactor

Uma Singhal, Manpreet Kaur Attri, and Ajit Varma

Abstract *Piriformospora indica (Serendipita indica)* is an axenically cultivable root endophytic fungus which exerts plant growth promoting effects on its host plants. To enable commercial production of its chlamydospores, *Serendipita indica* was cultivated in a 10 l batch bioreactor condition in such that they result in maximum biomass during growth phase and in maximum chlamydospore yield during subsequent sporulation phase. An enhancement of 100% in overall biomass productivity (0.18 g l^{-1} h^{-1}) and reduction of about 70% in the time (60 h) required to achieve the maximum spore yield (9.25×10^9 spores ml^{-1}) was achieved in comparison to the original batch culture grown in Hill and Kaefer medium. The high chlamydospore yield obtained promises to be economical for commercial production of *P. indica*.

18.1 Introduction

Piriformospora indica (*Serendipita indica*), a root endophytic fungus, was isolated from the rhizosphere of the woody shrubs *Prosopis juliflora* and *Zizyphus nummularia* growing in desert interior of Rajasthan, India. It exhibits most of the beneficial characteristics of arbuscular mycorrhizal fungi (AM fungi). Like AM fungi, it has a broad and diverse host spectrum and exerts plant growth-promoting effects on its host plants (Singh et al. 2000). But the most important advantage of *P. indica* over AM fungi is that it is a facultative symbiont and can be easily cultivated axenically on a variety of synthetic media (Varma et al. 2001). This is the reason that among the microbes capable of exerting plant growth-promoting effects, which can be used as biological agents against plant pathogens, spore formers are receiving increasing attention in agriculture as potential alternative to chemicals in the form of biofertilizers and biopesticides (Casula and Cutting 2002).

U. Singhal • M.K. Attri • A. Varma (✉)
Amity Institute of Microbial Technology (AIMT), Amity University Uttar Pradesh, Block 'E-3', 4th Floor, Sector-125, Noida, Gautam Buddha Nagar, Uttar Pradesh 201313, India
e-mail: ajitvarma@amity.edu

A. Varma et al. (eds.), *Mycorrhiza - Function, Diversity, State of the Art*,
DOI 10.1007/978-3-319-53064-2_18

The fermentation medium influences the nutritional and physiochemical environment and therefore directly affects productivity and process economics (Zhang and Greasham 1999). A suitable medium thus supports vegetative growth and also the subsequent production of spores. Media optimization is therefore an important consideration in the development of bioprocesses that can produce affordable agricultural bioinoculants. Hence, this chapter describes the mass cultivation of *P. indica* by batch culture and bioreactor for enhancement of its growth for large-scale production of spores.

18.2 Symbiont Cultured on Solid or in Liquid Media

The fermentation unit in industrial microbiology is analogous to a chemical plant in chemical industry. A fermentation process is a biological process and, therefore, has requirements of sterility and use of cellular enzymatic reactions instead of chemical reactions aided by inanimate catalysts, sometimes operating at elevated temperature and pressure. Industrial fermentation processes may be divided into two main types, with various combinations and modifications. These are batch fermentations and continuous fermentations (bioreactor).

18.2.1 Methods

18.2.1.1 Microorganism, Culture Maintenance, and Inoculum Preparation

Usually, the stock culture is maintained on slants containing Hill and Kaefer medium (Prasad et al. 2005) supplemented with 15 g/l agar. Inoculate the slants, incubated at 30 °C for 10 d, and then store at 4 °C. For the preparation of inoculum, grow *P. indica* on Hill and Kaefer medium (Kaefer 1977) in a petridish.

18.2.2 Media Composition

A large variety of synthetic and complex media are employed for the activation of fungal strains.

(a) MMN 1/10 (Herrmann et al. 1998)

Composition (g/l)	
$CaCl_2 \cdot 2H_2O$	0.07
$MgSO_4 \cdot 7H_2O$	0.15
NaCl	0.03

(continued)

$(NH_4)_2HPO_4$	0.03
KH_2PO_4	0.05
Trace elements (mg/l)	
$(NH_4)_6Mo_7O_{24}·4H_2O$	0.09
H_3BO_4	1.55
$CuSO_4·5H_2O$	0.13
KCl	3.73
$MnSO_4·H_2O$	0.84
$ZnSO_4·7H_2O$	0.58
Fe-EDTA (mg/l)	
FeSO4	8.50
EDTA	1.50
Agar	20.0 g

(b) Modified aspergillus medium (Varma et al. 2001)

The media composition was the same; except yeast extract, peptone, and casamino acid were reduced to 1/10 in quantity.

(c) M4 N (Mukerji et al. 1998)

Composition (g/l)	
D-Glucose	10.0
$(NH_4)2HPO_4$	0.25
KH_2PO_4	0.50
$MgSO_4·7H_2O$	0.15
$CaCl_2·2H_2O$	0.05
Ferric citrate (2% ferric citrate, 2% citric acid v/v)	7.0 ml
NaCl	0.025
Thiamine HCl	100.0 mg
MES	2.5
Malt extract	1.5
Yeast extract	1.5
Agar	15.0
pH	5.6

(d) MMNC (Kottke et al. 1987)

Composition (g/l)	
Glucose	10.0
$CaCl_2·2H_2O$	0.07
$MgSO_4·7H_2O$	0.15
NaCl	0.03
$(NH_4)2HPO_4$	0.25
KH_2PO_4	0.5
Casein-hydrolysate	1.0
Malt extract	5.0

(continued)

Trace elements (mg/l)	
$(NH_4)_6Mo_7O_{24}\cdot 4H_2O$	0.02
H_3BO_4	1.55
$CuSO_4\cdot 5H_2O$	0.13
KCl	3.73
$MnSO_4\cdot H_2O$	0.85
$ZnSO_4\cdot 7H_2O$	0.58
Fe-EDTA (mg/l)	
$FeSO_4$	8.5
EDTA	1.5
Vitamins (mg/l)	
Thiamine	0.1
Riboflavin	0.1
pH	5.6
Agar	20.0 g

(e) MS (Murashige and Skoog 1962)

Chemicals (mg/l)	
Macronutrients	
NH_4NO	30.5
KNO_3	1650.0
$CaCl_2\cdot 2H_2O$	900.0
$MgSO_4\cdot 7H_2O$	440.0
KH_2PO_4	370.0
Micronutrients	
KI	170.0
H_3BO_3	0.83
$MnSO_4\cdot H_2O$	6.20
$ZnSO_4\cdot 7H_2O$	15.60
$NaMoO_4\cdot 2H_2O$	8.60
$CuSO_4\cdot 5H_2O$	0.25
$CoCl_2\cdot H_2O$	0.025
Iron source	
$Na_2\cdot EDTA$	0.025
$FeSO_4\cdot 7H_2O$	37.30
Vitamins	
Nicotinic acid	27.8
Pyridoxine HCl	0.5
Thiamine HCl	0.1
Glycine	2.0
Myo-inositol	100.0
Agar	0.7% (w/v)
Sucrose	3.0% (w/v)
PH	5.6–5.7

Each chemical was dissolved in bidistilled water individually. pH of the medium was adjusted using 1 N NaOH/HCl before autoclaving at 121 °C, 15 lbs. for 20 min. Stock solutions were stored at 4 °C except organic supplements, which were stored at 20 °C.

(f) WPM ("Woody Plant Medium" for Populus) Ahuja (1986)

Composition (g/l)	
Sucrose	20.0
K_2SO_4	1.00
Ca $(NO_3)_2 \cdot 4H_2O$	0.73
NH_4NO_3	0.40
$MgSO_4 \cdot 7H_2O$	0.37
Myo-inositol	0.10
Agar	7.00

Add 700 ml H_2O and adjust pH to 5.8 using 3.7% HCl (ca. 9.5 ml)

Add after autoclaving sterile phosphate solution (0.17 g KH2PO4 dissolved in 270 ml H_2O) + 15 ml NaOH (saturated)

10 ml of trace element stock solution (see below)

10 ml Fe-EDTA (see below)

10 ml Glycine stock solution (100×: solve 20 mg in 100 ml)

1 ml Thiamine stock solution (1000×: solve 10 mg in 100 ml)

1 ml Nicotinic acid stock solution (1000×: solve 50 mg in 100 ml)

1 ml $CaCl_2$ stock solution (1000×: solve 3.6 g in 50 ml)

250 ml Pyridoxine stock solution (4000×: solve 40 mg in 100 ml)

100 ml $CuSO_4$ stock solution (10,000×: solve 25 mg in 100 ml)

Sterilize by filtration before adding.

100× trace element stock solution (g/l, autoclave, store at 4 °C):

$MnSO_4 \cdot H_2O$	2.23
$ZnSO_4 \cdot 7H_2O$	0.86
H_3BO_4	0.62
Ammonium molybdate	0.10
KI	0.09

100× Fe-EDTA stock solution

Dissolve 0.128 g $FeSO_4$ and 0.172 g EDTA at 60 °C in 100 ml H_2O store at 4 °C, 20 g.

$CaCl_2 \cdot 2H_2O$	0.07
$MgSO_4 \cdot 7H_2O$	0.15
NaCl	0.03
$(NH_4)_2HPO_4$	0.03
KH_2PO_4	0.05
Trace elements (mg/l)	
$(NH_4)_6Mo_7O_{24} \cdot 4H_2O$	0.018
H_3BO_4	

(g) MMN (Modified Melin-Norkrans) (Johnson et al. 1957)

Composition (g/l)	
NaCl	0.025
KH_2PO_4	0.5
$(NH_4)_2HPO_4$	0.25
$CaCl_2$	0.05
$MgSO_4$	0.15
$FeCl_3$	0.001
Thiamine hydrochloride	83.0 ml
Trypticase peptone	0.1% (w/v)
Glucose monohydrate	1.0% (w/v)
Malt extract	5.0% (w/v)
Trace elements from stock	10.0 ml/l
Trace elements (stock) (g/l)	
KCl	3.73
H_3BO_3	1.55
$MnSO_4{\cdot}H_2O$	0.85
$ZnSO_4$	0.56
$CuSO_4$	0.13

pH was adjusted to 5.8 with 1 N HCl/NaOH.

All the stocks were stored at 4 °C except thiamine hydrochloride which was stored at 20 °C.

(h) Malt Extract (Galloway and Burgess 1962)

Composition (g/l)	
Malt extract	30.0
Mycological peptone	5.0
Agar	15.0
pH	5.4

(i) Potato Dextrose Agar (PDA) (Martin 1950)

Composition (g/l)	
Potato peel	200.0
Dextrose	20.0
Agar	15.0
Distilled water	1.0 l

Skin of potatoes was peeled off, cut into small pieces, and boiled (200 g) in 500 ml of water till they were easily penetrated by a glass rod. After filtration through cheese cloth, dextrose was added to the filtrate. Agar was dissolved and the required volume (1 l) was made up by the addition of water. The medium was autoclaved at 15-lb. pressure for 20 ml.

(j) Aspergillus Medium (Kaefer 1977)

Composition (g/l)	
Glucose	20.0
Peptone	02.0
Yeast extract	01.0
Casamino acid	01.0
Vitamin stock solution	01.0 ml
Macroelements from stock	50.0 ml
Microelements from stock	02.5 ml
Agar	10.0
$CaCl_2$ 0.1 M	1.0 ml
$FeCl_3$ 0.1 M	1.0 ml
pH	6.5
Macroelements (major elements Stock) (g/l)	
$NaNO_3$	120.0
KCl	10.4
$MgSO_4 \cdot 7H_2O$	10.4
KH_2PO_4	30.4
Microelements (trace elements) Stock (g/l)	
$ZnSO_4 \cdot 7H_2O$	22.0
H_3BO_3	11.0
$MnCl_2 \cdot 4H_2O$	5.0
$FeSO_4 \cdot 7H_2O$	5.0
$CoCl_2 \cdot 6H_2O$	1.6
$CuSO_4 \cdot 5H_2O$	1.6
$(NH_4)_6Mo_7O_{27} \cdot 4H_2O$	1.1
Na_2EDTA	50.0
Vitamins (%)	
Biotin	0.05
Nicotinamide	0.5
Pyridoxal phosphate	0.1
Amino benzoic acid	0.1
Riboflavin	0.25

pH was adjusted to 6.5 with 1 N HCl. All the stocks were stored at 4 C except vitamin which was stored at 20 °C.

(k) Jaggery (Sugar Cane Juice concentrate)

Composition (w/v)	
Grounded jaggery	4%
pH	6.5

The grounded jiggery was dissolved in distilled water.

Chemical composition (in %) of Jaggery is:

Sucrose	60–85
Glucose and fructose	5–15
Protein	0.4
Fat	0.05
Minerals	0.6–1.0
Calcium	0.4
Magnesium and phosphorus	0.045
Iron	11

This formulation is patented (Patent number: 944/DEL/2012 dt: 27.03.12).

18.3 Growth in Batch Culture on Shaker

Many fungi effectively sporulate in submerged cultures containing suitable medium with optimized nutrients. Thus, fungal cultures can be produced aseptically on large scale using shake flasks. The sporulation can be enhanced by transferring the jar in cold for 24 h.

18.3.1 Preparation of Media

Various media compositions are employed for the production of fungal cultures in shake flaks. Hill and Kaefer medium along with other media such as YPG media, Potato Dextrose Broth (PDB), and Malt extract are effectively employed for the submerged cultivation of fungal cultures. The media are prepared following the standardized media compositions in distilled water. The desired pH of the composed media is adjusted using acid or base and quantified using a pH meter (Bagde et al. 2010).

18.3.2 Sterilization of Medium

Prior to inoculation of media for the production of cultures, the media should be sterilized. This avoids growth of undesired microbes along with the cultures. Sterilization of medium is done under high pressure saturated stream at 121 °C for 15–20 min in autoclaves.

18.3.3 Inoculum Preparation

The stock cultures of fungus are maintained on slants containing complex medium (Hill and Kaefer 2001) augmented with 15 g/l agar (Prasad et al. 2005). The slants after being inoculated were incubated at 30 °C for 10 days and later on stored at 4 °C. For the preparation of inoculum, the fungal cultures were initially grown in a petri dish containing Kaefer medium (Kaefer 1977). At the time of inoculation, agar discs of approximately 8 mm were then punched out using sterilized cork borer. These discs were then used for inoculation of seed cultures.

18.3.4 Growth in Flasks/Jars on Shaker

The submerged cultures were raised in 500 ml of Erlenmeyer flasks containing 100 ml of complex media, most probably Hill and Kaefer medium. The flasks were inoculated with 5 ml of freshly prepared inoculum at 30 °C under constant shaking at 200 rpm on a rotary shaker. Flasks may be replaced by 200 ml jars.

18.3.5 Harvesting of Biomass

After maximum growth is attained, the biomass is harvested. The culture is filtered and the biomass is separated. The culture can also be centrifuged at 3000 rpm for separation of filtrate and biomass. The filtrate thus obtained can be employed for the study of bioactives produced by the fungal cultures. The biomass separated is used as a bioinoculant by mixing the 2% (w/v) biomass with sterilized magnesium sulphite or vermiculite.

18.4 Production of Fungal Cultures in Fermenters

Fermenters provide optimized environmental and nutritional conditions for the large-scale production of microbial cultures. The constant administration of conditions at variable stages in fermenters enables a more efficient scale-up of microbial cultures. The submerged conditions enhance the uptake of nutrients resulting in stimulation of the biochemical processes. Fermentation of the microbes can be accomplished through the following three processes: batch, continuous, and fed-batch.

Batch culture comprises of a closed system which encompasses an initial restricted availability of nutrient. The batch fermentation is employed for the production of biomass as well as primary and secondary metabolites. Further in fed-batch systems, the exponential growth phase can be prolonged by the continuous addition of fresh culture to the system. This addition results in the continuous culture system. The continuous culture systems require media which is designed for substrate limited growth. Thus, these systems effectively maintain microbial population in exponential growth where cultures grow at a constant rate and biomass concentration for extended periods.

There is yet another system called the fed-batch system. The batch cultures constantly being fed with medium without the culture fluid being removed corresponds to the fed-batch culture (Yoshida et al. 1973). It is initially established in batch mode and further fed accordingly depending upon the conditions required by the culture. The fed-batch cultures control the organism's growth rate which is related to the specific rate of oxygen uptake. Thus, fed-batch culture systems are readily used in fermentation technology.

18.4.1 Medium for Optimal Growth

The media used for fermentation greatly influences the nutritional requirements as well as physiochemical environment and thus directly effects productivity and process economics (Zhang and Greasham 1999). Therefore, a suitable media should invariably support vegetative growth and production of spores. The optimum growth conditions are observed in a modified Kaefer media with peptone, 3.0; yeast extract, 3.0; KH_2PO_4, 1.83; and $MgSO_4.7H_2O$, 0.65 g/l. The concentration of other components was the same as in the original Kaefer medium without $NaNO_3$ and KCl, while the glucose concentration was 20 g/l (Kumar et al. 2011a, b).

18.4.2 Sterilization of the Fermenter

Prior to the initiation of the production process, the fermenter needs to be sterilized. The fermentation media and the fermenter can be sterilized together or separately. The fermenter is sterilized by channeling steam into the vessel *via* all entries and releasing the steam slowly through air outlet. The jackets or coils of the fermenter are sterilized by heating them with stream. Also the steam pressure is maintained at 15 psi inside vessel for 20 min approximately for thorough sterilization.

18.4.3 Cultivation in Fermenter

For all the fermentation processes, an active 2% inoculum raised in an optimized medium is used. The initial pH is calibrated at 6.5. As the biomass production is initiated, there is uptake of glucose which decreases the pH to between 5.5 and 6.0

in late log phase. Since the optimum pH for sustainable growth of *P. indica* is 5.8, there is no requirement for pH control in fermenter systems where the fungal cultures are grown on media containing complex nitrogen sources. The temperature range is in between 20 and 35 °C. However, for optimized growth the fungal cultures are grown at a temperature of 30 °C. The fungus grows best at lower agitation and low oxygen concentrations (Varma et al. 2001). Thus, the cultures are grown at 200 rpm and 60% working volume.

18.4.4 Recovery of Biomass Produced

After the desired biomass is obtained, the production process is terminated. The biomass produced in the fermenter vessel is removed. The produced biomass is then filtered, separating the filtrate from the biomass. After separation, the biomass obtained is then formulated by mixing with sterilized magnesium sulphite, talcum powder, or vermiculite (described in Chap. 17).

18.5 Cultivation in Bioreactor

Growth may be defined as an irreversible increase in the volume of an organism, usually accompanied by an increase in biomass. Mycelial fungi exhibit extension growth of hyphae, accompanied by an increase in biomass. Unicellular fungi (e.g., Yeasts) may exhibit an increase in individual cell volume, accompanied by an increase in biomass. But collectively, the number of yeast cells within a culture (i.e., cell concentration) may also increase, resulting in an increase in biomass of the culture as a whole.

18.5.1 Batch Fermentations

The fermenter vessel is filled with the prepared media up to its 60% capacity. The operational parameters such as pH, temperature, dissolved O_2 (DO), agitation and duration of the process, etc., are fed into the fermentor's central processing unit (CPU) at the start of the experiment for microbial fermentation. The media is steam sterilized in a pure culture process. The inoculum of a pure culture is added to the fermenter, from a separate pure culture vessel. Fermentation proceeds, and after the proper time the contents of the fermenter are taken out for further processing. The fermenter is cleaned and the process is repeated. Thus, each fermentation is a discontinuous process divided into batches.

18.5.2 Continuous Fermentation

Growth of microorganisms during batch fermentation conforms to the characteristic growth curve, with a lag phase followed by a logarithmic phase. This, in turn, is terminated by progressive decrements in the rate of growth until the stationary phase is reached. This is because of limitation of one or more of the essential nutrients. In continuous fermentation, the substrate is added to the fermenter continuously at a fixed rate. This maintains the organisms in the logarithmic growth phase. The fermentation products are taken out continuously. The design and arrangements for continuous fermentation are somewhat complex.

18.5.3 Aerobic Fermentations

A number of industrial processes, although called "fermentations'," are carried on by microorganisms under aerobic conditions. In older aerobic processes, it was necessary to furnish a large surface area by exposing fermentation media to air. In modern fermentation processes, aerobic conditions are maintained in a closed fermenter with *submerged cultures.* The contents of the fermenter are agitated with au impeller and aerated by forcing sterilized air (Fig. 18.1).

18.5.4 Growth Kinetics in Liquid Media: Batch Culture

For estimating the growth of a mycelial fungus growing in a liquid medium, we might first have to filter off the liquid medium and then determine the dry mass of the mycelium.

To estimate growth of a yeast species in liquid culture, we could:

- Either filter the culture and determine the dry mass of all the yeast cells together (i.e., the biomass of the culture).
- Or we could estimate the concentration of cells in the culture, using either a haemocytometer or optical density readings.

If we plotted biomass or cell concentration against time, we might obtain the following characteristic S-shaped growth curve as shown in Fig. 18.2.

1. Characteristic S-shaped Growth Curve:

- During an initial Lag phase, the rate of growth or cell division is very slow.
- Growth or cell division then starts to accelerate into the exponential phase—for example, with a unicellular organism (e.g., Yeast species) any 1 cell produces 2 in a given period of time, those 2 produce 4, the 4 produce 8, 8 produce 16, and so on. This exponential phase represents the period when the fungus is growing or multiplying most rapidly. This phase will continue until one or more nutrients

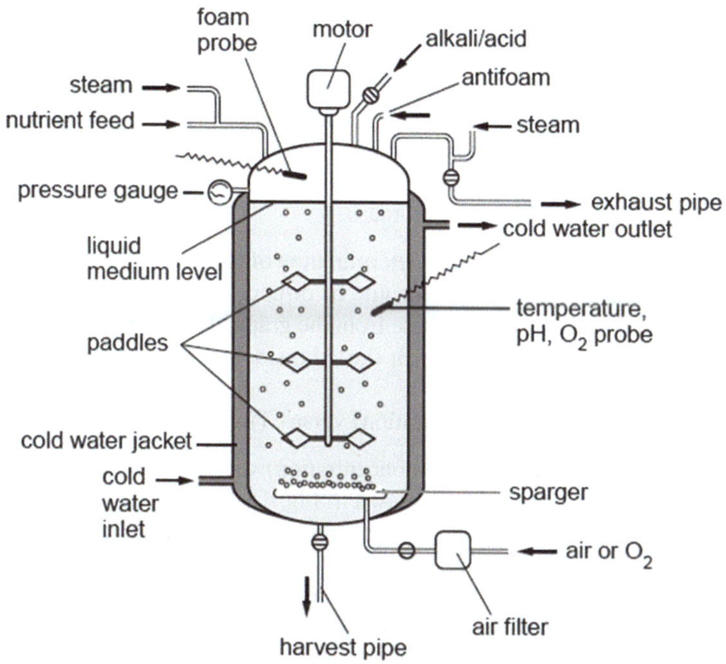

Fig. 18.1 Aerobic fermenter

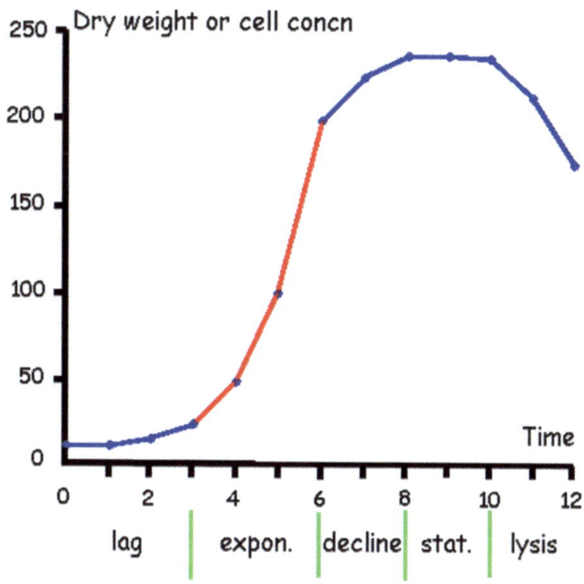

Fig. 18.2 Characteristic S-shaped growth curve

become limiting, oxygen becomes depleted, and/or metabolic by-products accumulate to toxic levels, when.

- Growth will start to decelerate (Decline).
- This may be followed by a stationary phase, during which there is no discernible change in cell concentration or biomass.
- Finally, one may observe a phase of cell death and lysis—which results in a decrease in cell number and/or biomass.

Often the most interesting part is determination of the rate of growth taking place during the exponential phase. But it would be difficult to determine the overall rate of growth during the exponential phase from the graph above, i.e., the red section of this graph, because the rate of growth (i.e., the slope of this region of the graph) changes with time.

2. Log_e (biomass or cell concentration) versus Time:

- If one plot $\log_e$ (dry mass or cell concentration) versus Time, we obtain a graph with this characteristic shape as shown in Fig. 18.3.
- The exponential phase of growth is now represented by the linear (straight line) red region.
- The slope of this red region is now constant and represents the specific growth rate (or relative growth rate) of the fungus $= \mu$.
- μ is a measure of the rate of change in biomass or cell concentration relative to the biomass or cell concentration already present.

Fig. 18.3 Log_e (biomass or cell concentration) vs. Time

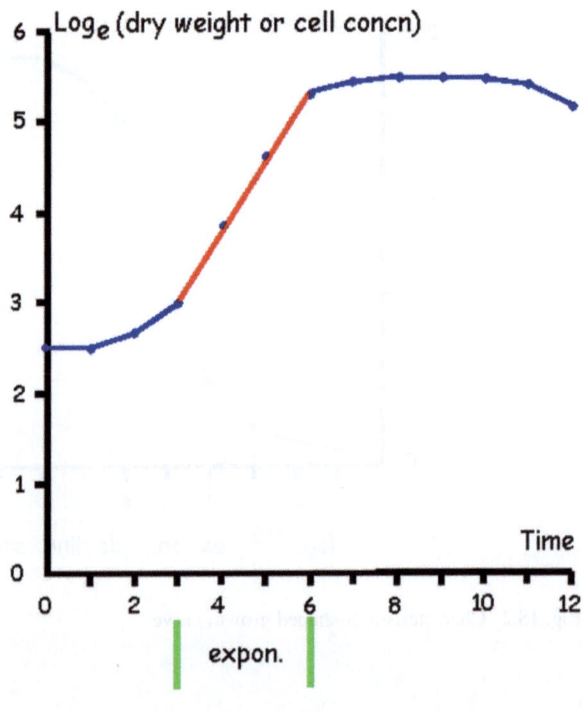

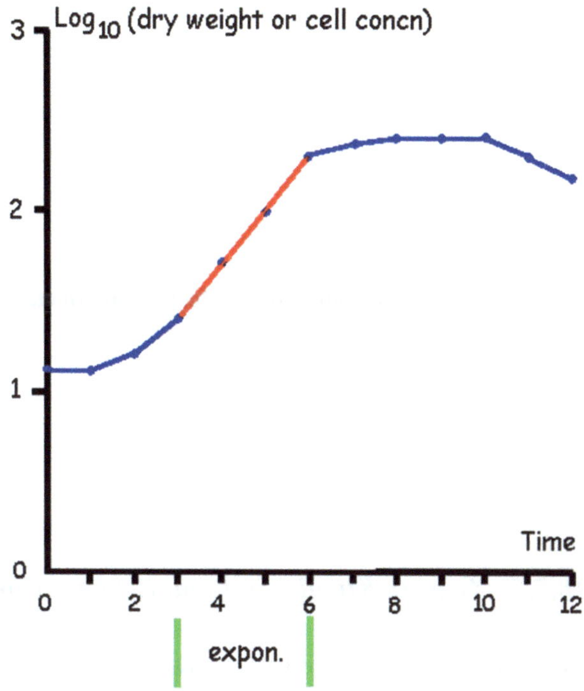

Fig. 18.4 Log_{10} (biomass or cell concentration) vs. Time

- With rate of change in biomass or cell concentration (i.e., dn/dt, change in biomass divided by change in time) calculation, rate of change relative to the biomass or cell concentration already present (i.e., $(dn/dt)/n = \mu$) is also being calculated.
- If all the conditions are optimal for growth of the fungus, then the maximum specific growth rate (μ_{max}) is obtained—this is characteristic for any particular organism.

3. Log_{10} (biomass or cell concentration) versus Time:

- One could plot log_{10} (dry mass or cell concentration) versus Time as shown in Fig. 18.4.
- This provides a graph with a shape similar to that above (Fig. 18.3)—but the values on the y-axis will be different.
- So the logarithmic values in our calculation of μ will have to be converted to log_e, by multiplying them by 2.303—because these organisms are exhibiting exponential growth.

4. Doubling Time (or Generation Time) T_d:

- The specific growth rate of a unicellular organism (e.g., Yeast or bacterium) is also related to the doubling time or generation time (T_d) of the organism.

- This is the time it takes for all the cells present in the culture to double in number.
- $T_d = (\log_e 2)/\mu$
- So, $\mu = (\log_e 2)/T_d$
- The doubling time in graph 3 above is 1 h (see red region), so $\mu = 0.693\ \mathrm{h}^{-1}$.

The graphs illustrated above are characteristic of batch cultures:

- No additional nutrients are added to the culture vessel once it has been inoculated and incubation has commenced, and the only environmental factor controlled is the temperature of incubation—nutrients, oxygen levels, and pH will change as incubation proceeds and the culture grows.
- Therefore, this is essentially a "closed" system.
- Batch cultures are used in some industries because valuable microbial products (e.g., antibiotics, ethanol, organic acids) accumulate in the medium during the stationary phase of growth.

18.5.5 Growth Kinetics in Liquid Media: Continuous Culture

An alternative to the liquid batch culture system is continuous culture in a liquid medium:

- This involves the continuous addition of fresh culture medium to the vessel and the withdrawal (by means of an overflow devise) of a corresponding volume of old, spent medium, which will contain some of the microbial cells.
- The apparatus used is called a chemostat as shown in Fig. 18.5.

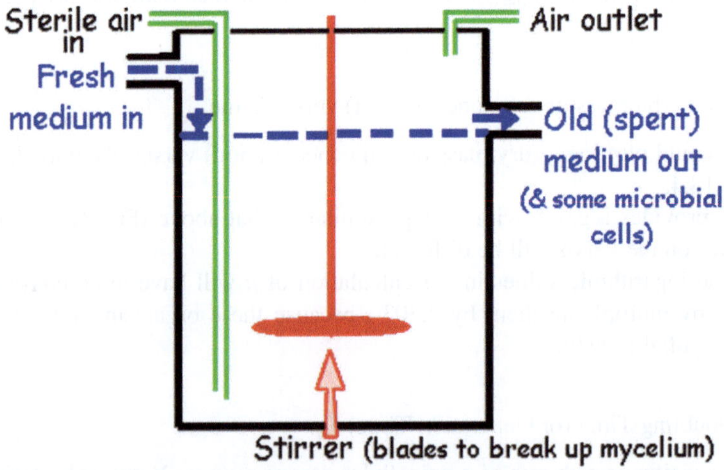

Fig. 18.5 Growth kinetics in liquid media: continuous culture

Chemostat

- Many environmental factors (e.g., pH, oxygen levels, nutrients, and temperature) can be controlled very precisely throughout the incubation period.
- The culture is stirred continuously—ensuring nutrients and oxygen reach the cells and metabolic products are distributed away from them.
- Chemostats may be used to culture yeasts or mycelial fungi.
- When culturing a mycelial fungus, the stirrer consists of blades which prevent large mycelial pellets from forming. In this system, the fungus continues growing exponentially because it's continuously supplied with fresh nutrients and oxygen, and the pH is controlled.
- But the actual rate of that exponential growth will depend upon the rate of flow of the culture medium through the culture vessel, i.e., dilution rate.
- The rate of exponential growth = specific growth rate μ = dilution rate D.
- $\mu = D = f/v$, where f = flow rate (ml h^{-1}) and v = volume of the culture (ml).
- If all environmental factors are optimal for growth, then the maximum specific growth rate (μ_{max}) will be achieved.

18.6 Mass Cultivation of *P. indica* in 10 l Fermenter

P. indica, which mimics AMF, represents a good model system to understand the molecular basis of photo- and mycobiont interaction. Its application in horticulture or agriculture as a potent biofertilizer and biocontrol agent is economically and practically feasible through the easy propagation of a fungal inoculum using liquid or axenic cultures.

It is shown that the fungus can be grown axenically on different synthetic media. Among the tested media, the best growth reported to be on Hill and Kaefer medium (2001) which is reported from different authors (Varma et al. 1999, 2001; Pham et al. 2004). However, significant quantitative and morphological changes are detected when the fungus is grown on different nutrient composition with no apparent negative effect on plants (Kumar et al. 2011a, b).

A 3 or 7 or 10 l bioreactor was used to grow *P. indica* on optimized Hill and Kaefer medium as shown in Fig. 18.6 to establish the best conditions for a maximal biomass and spore production for scale-up studies.

When *P. indica* was grown n 7-l bioreactor on optimized Hill and Kaefer medium containing (20.0 g/l glucose, 1.0 g/l peptone, 1.0 g/l yeast extract, 1.0 g/l Casein acid hydrolysate, 50.0 ml/l macroelement, 2.5 ml/L microelement stock solution, 1.0 ml/l vitamin stock solution, 1 ml/l $CaCl_2$, 0.1 M, and 1.0 ml/l $FeCl_3$), a maximum dry cell weight of 7.36 g/l was obtained after 42 h of growth. The value of biomass yield and the specific daily growth rate were 0.79 and 1.15, respectively. The fungus initiated the sporulation after 48 h, and a spore yield of 9.25×10^9 spores/ml was achieved after 60 h of growth. The early sporulation in this case may be due to rapid consumption of glucose. Due to more efficient mixing and

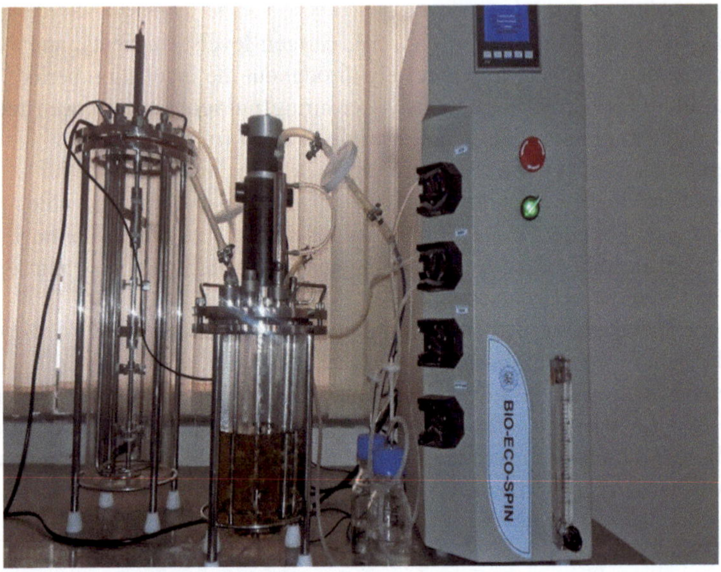

Fig. 18.6 Production of biomass in bench Fermenter (7 l)

homogenized fungal suspension, the growth of fungus was faster in the bioreactor and resulted in early depletion of the carbon source and thereby early sporulation compared to a shake flask. A complete growth profile of *P. indica* on modified Hill and Kaefer medium has been depicted. The pattern of pH profile was quite similar in all these experiments where complex nitrogen sources were present in the growth medium. The uptake of glucose caused a decrease in pH of fermentation broth which might be due to the generation of acidic metabolites. The growth of fungus remained unaffected as long as the pH during the log phase was not reduced below 4.5. Besides this, it was found that the optimal mass cultivation of *P. indica* is achieved on soil extract-enriched media and jaggery (extracted from *Saccharum officinarum*) that contains 60–85 g/l sucrose, 5–15 g/l glucose and fructose, 0.4 g/l protein, 0.05 g/l fat, 0.6–1.0 g/l minerals (0.4 g/l calcium, 0.045 g/l magnesium, and phosphorus), and 11% iron. The soil extract enriched with some nutrients is suitable for mass production of the endophyte, up to 14 days. Hill and Kaefer (2001) medium looks to give better results in a longer run. Soil extract and jaggery which are economically feasible need to be optimized for getting a higher biomass of *P. indica* at fermentor scale.

18.6.1 Measurement of Cell Growth, Growth Yield, and Specific Growth Rate

The growth of *P. indica* was expressed in terms of dry cell weight (DCW) per liter of culture broth, which was determined by filtering a known volume of culture broth through Whatman No. 1 filter paper, drying to a constant weight in vacuum oven at 60 °C for about 48 h, and weighing the dry mass. Growth yield (YX/S) was calculated as grams of biomass produced per gram of substrate consumed. The specific growth rate (μ) was calculated from the Equation $=1/X \times dx/dt$, where X is the biomass concentration (g/l) at time t.

18.6.2 Measurement of Chlamydospores

P. indica produced pear-shaped chlamydospores, which were attached to the mycelium.

The spores were dislodged by adding 1 ml of Tween 80–100 ml of culture broth, vortexing, grinding in a mixer grinder, and sonicating for 5 min each. After their detachment, the spores were counted with a hemocytometer.

18.7 Conclusions

Significant increases in biomass productivity and reductions in time to achieve maximum spore yield were obtained for *P. indica* by using batch bioreactor for its mass cultivation. The medium appears useful for economical mass production of spore-rich *P. indica* biomass for agricultural and horticultural applications.

Acknowledgment Ajit Varma is thankful to Department of Science and Technology and Department of Biotechnology for partial financial funding and to DST-FIST for providing confocal microscope facility.

References

Ahuja MR (1986) Aspen. In: Evans DA, Sharp WR, Ammirato PJ (eds) Handbook of plant cell culture 4, techniques and applications. Macmillan, New York, pp 626–651

Bagde US, Prasad R, Varma A (2010) Mass cultivation of Piriformospora indica in new Brunswick fermenter and its formulation as biofertilizer. Asian J Microbiol Biotechnol Environ Sci 12:911–916

Casula G, Cutting SM (2002) *Bacillu sprobiotics*: spore germination in the gastrointestinal tract. Appl Environ Microbiol 68:2344–2352

Gallowey LD, Burgess R (1962) Applied mycology and bacteriology, 3rd edn. Leonard Hill, London, pp 54–57

Herrmann S, Munch JC, Buscot F (1998) A gnotobiotic culture system with oak microcuttings to study specific effects of mycobionts on plant morphology before, and in the early phase of, ectomycorrhiza formation by *Paxillusin volutus* and *Piloderma croceum*. New Phytol 138:203–212

Hill TW, Kaefer E (2001) Improved protocols for aspergillus medium: trace elements and minimum medium salt stock solutions. Fungal Genet News-Lett 48:20–21

Johnson CN, Stout PR, Broyer RC, Carlton AB (1957) Comparative chlorine requirements of different plant species. Plant Soil 8:337–353

Kaefer E (1977) Meiotic and mitotic recombination in *Aspergillus* and its chromosomal aberrations. Adv Genet 19:33–131

Kottke I, Guttenberger M, Hampp R, Oberwinkler F (1987) An in vitro method for establishing mycorrhizae on coniferous tree seedlings. Trees 1:191–194

Kumar M, Yadav V, Kumar H, Sharma R, Singh A, Tuteja N, Johri AK (2011a) *Piriformospora indica* enhances plant growth by transferring phosphate. Plant Signal Behav 6:723–725

Kumar V, Sahai V, Bisaria VS (2011b) High-density spore production of *Piriformospora indica*, a plant growth promoting endophyte, by optimization of nutritional and cultural parameters. Bioresour Technol 102:3169–3175

Martin JP (1950) Use of acid, rose bengal and streptomycin in the plate method for estimating soil fungi. Soil Sci 69:215–232

Mukerji KG, Mandeep, Varma A (1998) Mycorrhizosphere microorganisms: screening and evaluation. In: Varma A (ed) Mycorrhiza manual. Springer, Berlin, pp 85–98

Murashige T, Skoog F (1962) A revised medium for rapid growth and bioassay with tobacco tissue cultures. Physiol Plant 15:431–487

Pham GH, Kumari R, Singh A, Malla A, Prasad R, Sachdev M, Kaldorf M, Buscot F, Oelmüller R, Hampp R, Saxena AK, K-H Rexer, Kost G, Varma A (2004) Axenic culture of symbiotic fungus *Piriformospora indica*. Plant surface microbiology, pp 593–613

Prasad R, Pham HG, Kumari R, Singh A, Yadav V, Sachdev M, Garg AP, Peskan T, Hehl S, Sherameti I, Oelmüller R, Varma A (2005) Sebacinaceae: culturable Mycorrhiza-like endosymbiotic fungi and their interaction with nontransformed and transformed roots. In: Declerck S, Strullu D-G, Fortin A (eds) In vitro culture of mycorrhizas. Soil biology, vol 4. Springer, Berlin

Singh A, Sharma J, Rexer KH, Varma A (2000) Plant productivity determinants beyond minerals, water and light. *Piriformospora indica* – a revolutionary plant growth-promoting fungus. Curr Sci 79:101–106

Varma A, Verma S, Sahay NS, Bütehorn B, Franken P (1999) *Piriformospora indica*, a cultivable plant-growth-promoting root endophyte. Appl Environ Microbiol 65:2741–2744

Varma A, Singh A, Sahay NS, Sharma J, Roy A, Kumari M, Rana D, Thakran S, Deka D, Bharti K, Franken P, Hurek T, Blechert O, Rexer KH, Kost G, Hahn A, Hock B, Maier W, Walter M, Strack D, Kranner I (2001) *Piriformospora indica*: a cultivable mycorrhiza-like endosymbiotic fungus. In: Hock B (ed) Mycota IX. Springer, Berlin, pp 123–150

Yoshida F, Yamane T, Nakamoto K (1973) Fed-batch hydrocarbon fermentations with colloidal emulsion feed. Biotechnol Bioeng 15:257–270

Zhang J, Greasham R (1999) Chemically defined media for commercial fermentations. Appl Microbiol Biotechnol 51:407–421

Index

A
ABC-type transporter, 136
Abiotic, 330
  stress, 51, 281, 359
Absorption, 110
Acaulospora, 5, 240, 337
  A. longula, 14
Acaulosporaceae, 310
Achromobacter, 65
Actinomycetes, 51, 349
Adhesins, 48
Aerenchyma, 294, 296, 341
Aerobacter aerogenes, 66
Aerobic fermentations, 376
Aerobiosis, 344
Aeromonas, 66
AFLP. *See* Amplified fragment length polymorphism (AFLP)
Agroecosystems, 283, 339
Alcaligenes, 43
Alkaloids, 138
Allelopathic compounds, 236
Alpine forests, 207
AM. *See* Arbuscular mycorrhiza(s) (AM)
Amanita, 112, 113, 115
  A. muscaria, 63, 68
Amaranthaceae, 11
Ambispora gerdemannii, 80
American truffles, 244–245
AMF. *See* Arbuscular mycorrhizal fungi (AMF)
Amphibious species, 295–297
Amplified fragment length polymorphism (AFLP), 118
Anaerobiosis, 334
Anastomosis, 257, 353
Angiosperms, 10, 40, 356
Anoxic, 292
Anthropogenic, 188
Antibiotics, 51
Antioxidants, 136, 243, 354
AON. *See* Autoregulation of nodulation (AON)
Apigenin, 135
Apoplasmic, 20
Apoplastic, 13
  interface, 310
Appressoria formation, 158
Aquatic macrophytes, 335
Arabidopsis, 357
  A. thaliana, 356
Arbuscular mycorrhiza(s) (AM), 1, 4, 39, 100, 102, 134, 178, 307, 332–333
Arbuscular mycorrhizal fungi (AMF), 75, 306
  colonization, 283
  diversity, 283
  spore, 283
Arbuscules, 3, 12, 13, 76, 102, 178, 264, 283, 294, 332, 350
Arbutoid mycorrhiza, 308
Archaebacteria, 63
Archaeospora, 240, 337
Archaeosporales, 47
Arctostaphylos uva-ursi, 207
Armillaria melea, 280
Arthrobacter chlorophenolicus, 45
Arum type, 3
Ascogonia, 233
Ascogonial fusion, 233
Ascomata, 243
Ascomycete fungi, 212

A. Varma et al. (eds.), *Mycorrhiza - Function, Diversity, State of the Art*,
DOI 10.1007/978-3-319-53064-2

Ascomycetes, 196, 201, 232, 277, 308
Ascomycetous fungi, 109
Ascomycota, 4, 100, 310
Ascomycotina, 40, 61
Ascorbate, 354
Ascorbic acid, 356
Assimilated carbon, 112
Associated bacteria, 63–65
Asymbiotic stage, 150
Asymptomatic endophytic fungi, 85
Asymptomatic mycoviruses, 85
Autochthonous microorganisms, 52
Autofluorescent, 354
Autoregulation of mycorrhization, 157–161
Autoregulation of nodulation (AON), 147
Auxin, 144
  transport, 145
  transport regulators, 136
Azospirillum, 65, 67
  A. brasilense, 359
Azotobacter, 67
  A. chroococcum, 45
Azygosporous, 11

B
Bacillus, 43, 63, 65
  B. thuringiensis, 52
Bacopa monnieri, 357
Bacteria-like organisms (BLOs), 46
Bacterial infection, 144
Bacteroids, 148
Basidiomycetes, 179, 201, 232, 308
Basidiomycota, 4, 100, 310, 351, 353
Basidiomycotina, 40, 61
Basidiospores, 117
Batch culture, 376–380
Batch fermentations, 375
Betula
  B. ermanii, 198
  B. papyrifera, 206
Betulaceae, 11, 188
Bioactive compounds, 309
Bioavailability, 137
Biochanin A, 160
Biodegradation, 137
Biofertilizers, 10, 351, 365
Biofilm formation, 44
Biofilms, 43
Biogeochemical cycle(ing), 50, 188, 213, 314
Bio-herbicide, 351
Biological diversity, 112
Biomass, 292
Biopesticides, 365
Bioprotectant, 351
Bioprotector, 10
Bioreactor, 366, 375–381
Bioremediation, 182–183
Biotic, 330
  stress, 51, 281
Biotrophic, 19, 134
Biotrophs, 178
Biotrophy, 64
BLOs. *See* Bacteria-like organisms (BLOs)
Boletus, 112, 113, 115
Boreal forest islands, 320
Bradyrhizobial symbiosis, 161
Bradyrhizobium, 48, 161
Branching factor, 151
Brassicaceae, 4, 11, 356, 357
Brassica napus, 357
Bryophytes, 10, 356
Burkholderia, 43, 44, 46, 48, 63, 65

C
Calcium-calmodulin-dependent protein kinase, 153
Calcium oscillations, 141
Calmodulin-dependent kinase, 143
Candidatus, 46
Cantharellus, 115
  C. cibarius, 122
Carbohydrate transporters, 264
Carbon cycling, 318–321
Carbon Flow, 181–182
Carbon pool, 321
Carbon sequestration, 320
Carbon transport, 265
Carex elata, 297
Casein acid hydrolysate, 381
Ca spiking, 310
Castanea sativa, 78
Catabolic plasmid, 52
Cellulase, 19
Cenococcum geophilum, 80, 110
Chalcone synthase-GusA (CHS–GusA), 145
Chemoattractants, 86
Chemostat, 381
Chemo tactic, 42
Chenopodiaceae, 11
Chinese truffles, 242–244
Chitin, 320
Chitinolytic enzymes, 51
Chlamydospores, 352, 353, 383
Chlorinated phenols, 182

Chlorogenic acids, 36
Chromatograms, 30
Chronosequence, 194
CHS–GusA. *See* Chalcone synthase-GusA (CHS-GusA)
Cistus ladanifer, 207
Citrate, 34
Claroideoglomus etunicatum, 282
Climate change, 305, 306, 320
CO_2 concentration, 321
^{13}CO2-labeling isotope, 319
^{13}CO2 pulse-labelling, 233
Coenocytic hyphae, 150
Coleus forskohlii, 357
Colonization, 258, 332
Community ecology, 342–343
Compacted soil, 339–340
Continuous culture, 380–381
Continuous fermentation, 376
Coralloid, 107
Cortinariaceae, 200
Corynebacterium, 43
Coumestrol, 140
Crassulaceae, 4
Cross facilitation, 48–49
C-selection, 191
C sequestration, 307, 319
Cyanobacteria, 306
Cyanogenic glycosides, 357
Cylindrocarpon macrodidymum, 280
Cyperaceae, 11
Cytokinin, 144

D

Dark septate endophytes (DSEs), 275, 277–278, 293, 338
Debris-degrading enzymes, 18
Decomposers, 320
Desert forests, 313
Devonian era, 308
Devonian period, 40
Diazotrophs, 67
Dichotomous, 107
Dichotomy, 107
Dipterocarpaceae, 113, 188
Disease incidence, 30, 31
Diversispora, 240
  D. epigaea, 276
Diversisporaceae, 310
Diversisporales, 47
DNA sequencing, 80
Dolipores, 353
Drought, 254, 321, 362
Drought tolerance, 86
DSEs. *See* Dark septate endophytes (DSEs)
Dynamics, 117–118

E

"Early-stage" fungi, 118
Early-successional system, 197
ECM. *See* Ectomycorrhizal (ECM)
ECMRs. *See* Ectomycorrhizas (ECMRs)
Ecological Niche, 42–47
Ecotypes, 238
Ectendomycorrhizas, 100, 308
Ectomycorrhiza (EM), 4, 39, 102–104, 188, 198, 307
Ectomycorrhizae, 232, 234, 235
Ectomycorrhizal (ECM), 100, 188, 254
  associations, 11
  communities, 204
  Mushrooms, 110–111
Ectomycorrhizal fungi, 11, 12, 181, 190
Ectomycorrhizas (ECMRs), 61, 206
Ectoparasitic nematode, 280
Ectotrophic, 10, 101, 103
Elicitors, 358
EM. *See* Ectomycorrhiza (EM)
Endocellular mode, 49
Endogenous enzymes, 43
Endomycorrhizas (ENMRs), 61
Endophytes, 28
Endosymbionts, 46
Endosymbiotic intracellular, 42
Energy currency, 181
ENMRs. *See* Endomycorrhizas (ENMRs)
Enterobacter, 65, 66
Entomoplasmatales, 47
Environmental stresses, 306
Enzymes, 109
Epidermal cells, 108
Epigeous, 103
Epigeous sporocarps, 188
EPS. *See* Extracellular polysaccharides (EPS)
Ericoid mycorrhizas, 4, 39, 100, 179, 307
Escherichia coli, 66
Ethylene, 358
Eucalyptus camaldulensis, 64
Eumycota, 100
Eutrophication, 134
Exodermal cells, 108
Exopolysaccharide, 43
External hyphae, 341
Extracellular enzymes, 210, 317

Extracellular lipopolysaccharides, 144
Extracellular penetration, 12
Extracellular polysaccharides (EPS), 144
Extracellular proteins, 138
Extramatrical hyphae, 191
Extramatrical mycelium, 17
Extraradical hyphae, 40, 313, 320
Extra-radical hyphal, 341
Extraradical mycelia, 45
Extraradical mycelium, 14, 40, 47, 103, 107, 109–110
Extraradical network, 45

F
Facultative saprotrophytic fungi, 12
Fagaceae, 188
Fagaceae forests, 76
Fagus sylvatica, 78
Fed-batch system, 374
Fermentations, 376
Fertility islands, 202
Fertilizers, 134
Fire-Affected Sites, 205–207
Flavonoid biosynthesis, 88
Flavonoids, 135–138, 143, 145
Flavonolaglycones, 136
Flavonones, 139
Flooded soils, 335–336
Fluorescent pseudomonads, 51
Forest ecosystems, 204, 319
Formononetin, 158
454 pyrosequencing, 234
Frankia, 67
Free-living soil, 65
Free riders, 40
Fructose, 34
Functional diversity, 263
Fungal colonisation, 292–295, 299
Fungal community(ies), 212, 331
Fungal diversity, 342–343
Fungal endophytes, 278–281, 292
Fungal gene expression, 317
Fungal-host compatibility, 264–265
Fungal identity, 262–263
Fungal richness, 196
Fungal vitality, 64
Fungi, 28
Fungi association, 12
Fungicides Carbendazim, 161
Fungus–Root Interface, 20–21
Funneliformis mosseae, 282
Fusariosis, 28
Fusarium oxysporum, 28, 32
Fusarium wilt, 35

G
Gaeumannomyces, 277
Gametophytes, 40
Gasses, 306
Geaster, 115
Generation time, 379–380
Genetic analysis, 238
Genetic diversity, 238, 257
Genome sequencing, 124
Genome size, 351
Geosiphon pyriforme, 63
GFP-tagged bacteria, 64
Gigaspora
  G. margarita, 14, 42, 43, 47, 49
  G. rosea, 68
Gigasporaceae, 262, 310
Glacier Forefronts, 195–197
Global carbon cycle, 306
Global change, 111–112
Glomalean, 40
Glomales, 29
Glomalin, 181, 307, 313–316
Glomerales, 47, 277
Glomerales, 310
Glomeromycota, 4, 39, 100, 232, 253, 255, 262, 308, 311, 330
Glomeromycotina, 40, 61
Glomus, 42, 337
  G. aggregatum, 282
  G. etunicatum, 154
  G. fistulosum, 49
  G. intraradices, 14, 28, 32, 52, 154, 317
  G. macrocarpum, 154
  G. mosseae, 51, 52
Glucan, 320
Glucose, 34
Glucosinolates, 357–358
Glutathione, 354, 359
Glycogen, 108
Glycol-proteins, 181
Glycosides, 136, 357
Gramineae, 156
Grapevines, 276–277
Greenhouse effect, 134
Greenhouse gas(es), 306, 321
Growth kinetics, 380–381
Gymnosperms, 40, 356

H

Hair roots, 100
Hartig net, 12, 13, 102, 103, 107–109
Heavy metals, 254
Hebeloma
  H. crustuliniforme, 211
  H. cylindrosporum, 16
  H. sacchariolens, 110
Helophytes, 297–298
Helper bacteria, 68–70
Hemicellulase, 19
Hemocytometer, 383
Herbicides, 240
Hertig's net, 308
Heterokaryon incompatibility, 239
Heterorhizic, 105
Heterothallic, 235
Heterotrophic, 306
Heterotrophs, 101
Hexose transporter, 63
Hill and Kaefer medium, 372
Histidine-rich glycoproteins, 148
Hormonal regulation, 134
Host recognition, 140
Host specificity, 116–117
Humus, 312
Hydraulic conductivity, 87
Hydrolytic, 109
Hydrophilic, 30, 87
Hydrophobic, 87, 137, 313
  interactions, 144
  proteins, 353
Hydrophobins, 16
Hydroxycinnamates, 36
Hygromycin B phosphotransferase, 353
Hymenomycetes, 351, 353
Hyphomycetes, 278
Hyphopodium, 102
Hyphosphere, 41, 42, 50
Hypogeous, 103, 201, 232
Hypogeous fungi, 195
Hypoxia, 331, 340
Hypoxia-tolerant, 342
Hypoxic, 292, 294, 340

I

IAA oxidase, 146
IAA Production, 67
Immobilization, 66
Immune signaling, 142
Immuno-modulator, 351
Incubation period, 124
Indole glucosinolates, 358
Induced systemic resistance, 280
Industrial fermentation, 366
Infection pocket, 144
Inoculation, 203
Interactive space, 41–42
Intercellular penetration, 12
Intermittent flooding, 295
Inter-retrotransposon amplified polymorphism (IRAP), 118
Intracellular hypha, 350
Intraradical mycelia, 40
Intraradical mycelium, 14, 47, 265
Ion uptake, 341
IRAP. *See* Inter-retrotransposon amplified polymorphism (IRAP)
Iron chelation, 66
Isoflavones, 139
Isoflavonoids, 135, 138, 140, 145
  glucosides, 136
ITS barcoding, 80

J

Juncaceae, 11

K

Kaempferol, 135, 146
Klebsiella pneumoniae, 66
K-selection, 191

L

Laccaria bicolor, 89, 110
Lactarius, 108, 112, 113, 115
Larix kaempferi, 198
Lateral roots, 103
Leccinum, 112, 115
Lectin, 64
Leucojum aestivum, 297
Lichens, 99
Ligand–receptor interaction, 48
Lignase, 19
Lignin biosynthesis, 36
Lipochitooligosaccharides, 264, 310
Lipophilic, 30
Littorella uniflora, 337
Lobelia dortmanna, 337
Lotus japonicus, 14, 153, 309
Low porosity, 201
Lupinus polyphyllus, 154
Lycopersicon esculentum, 52

Lycopods, 40
Lysophosphatidylcholine, 157
Lythrum salicaria, 295

M

Macronutrients, 279, 359
Macrophytes, 337
Macroptilium atropurpureum, 141
Malate, 34
MAMP. *See* Microbe-associated molecular pattern (MAMP)
Mangrove communities, 335
Mannitol, 108
Mantle, 12, 76, 102, 107–108
Mating types, 238
Medicago
  M. sativa, 139, 263
  M. truncatula, 62, 341
Medicarpin, 140, 158
Mediterranean
  climate, 77
  forests, 81, 85
Mesophilous, 213
Mesorhizobium loti, 140
Metabolic pathways, 88
Metagenomics, 124
Methylobacterium, 65
MHB. *See* Mycorrhization helper bacteria (MHB)
Microaggre gates, 319
Microarthropod community, 237
Microbe-associated molecular pattern (MAMP), 142
Microbial biofilms, 52
Microbial biomass, 312, 319, 320
Microbial cells, 380
Microbial community(ies), 68, 86, 331, 340
Microbial processes, 306
Microbial resistance, 36
Microenvironments, 42, 196
Microheterotrophs, 181
Micronutrient, 279
Microsclerotia, 293, 297, 300
Microsymbiont, 134, 141–142
Middle lamella, 109
Mineralization, 66, 214
Mine Spoils, 201–204
Mobilization, 110
Mofettes, 331, 335, 338–339
Molecular dialogue, 138
Molinietum caeruleae, 298
Monopodial mycorrhizae, 105
Monotropoid mycorrhizas, 100, 307
Moroccan woodlands, 81
Morphogenesis, 355
Mosses, 40
MRE. *See* Mycoplasma-related endobacteria (MRE)
MRs. *See* Mycorrhiza (MRs)
Mucilaginous, 45
Multi-digitate mode, 109
Mushrooms, 82, 99, 104
Mycelial exudates, 42
Mycelial networks, 206
Myc factors, 152, 264, 310
Mycobacterium, 65
Mycobionts, 40, 381
Mycodiversity, 79–82
Mycoheterotrophic, 10, 182
Mycophagous, 195, 232
Mycoplasma-related endobacteria (MRE), 259
Mycoplasmatales, 47
Mycorrhiza (MRs), 10, 28, 61, 99, 100, 243, 307–308, 349
Mycorrhiza formation, 124
Mycorrhizal associations, 75, 100, 180, 197, 359
Mycorrhizal colonization, 200
Mycorrhizal contribution, 178–179
Mycorrhizal dependency, 279
Mycorrhizal ecology, 192
Mycorrhizal fungi, 17, 179, 293, 308
Mycorrhizal mycelium, 42
Mycorrhizal networks, 204, 209, 281–282
Mycorrhizal pathway, 14
Mycorrhizal relationship, 10
Mycorrhizal successions, 190
Mycorrhization, 62, 68, 85–88, 135, 150, 154, 212, 236
Mycorrhization helper bacteria (MHB), 48, 68
  Streptomyces, 68
Mycorrhizospheres, 41, 42, 52, 64, 69, 70
Mycorrhizospheric soil, 183
Mycotrophic, 151
Mykorrhiza, 40, 178
Myrosinase, 357
Myrtaceae, 188

N

Native grassland community, 335
Necrotrophy, 64
Neoflavonoids, 135
Net primary productivity, 319

Next-generation sequencing (NGS), 47, 80, 339
Niches, 138, 331
Nicotiana tabacum, 357
Nitrogen
  cycling, 317–318
  fixation, 67
  transport, 265
Nitrogenase, 67, 148
  enzyme, 147
Nod factor signal, 142
Nod gene, 135
Nodulation, 135, 143, 148–150
Nodule development, 143
Nodule Organogenesis, 147–148
Nodule primodia (NP), 145, 147
Nonmycorrhizal, 4
Non-renewable, 333
Nostoc, 67
NP. *See* Nodule primodia (NP)
Nutrient absorption, 101
Nutrient acquisition, 188, 198
Nutrient acquisitor, 136
Nutrient circulation, 118–120
Nutrient cycling, 210, 311
  microbial communities, 189
Nutrient mobilization, 43
Nutrient transporters, 341
Nutrient uptake, 14, 296

O
Obligate biotrophophic fungi, 12
Obligate biotrophs, 308, 313
Oligosaccharides, 313
Oligotrophic, 335, 337
Optimal growth, 374
Orchidaceae, 307
Orchid mycorrhizas, 4, 39, 100, 307
Organic matter, 294
Orobanchaceae, 4
Orthologous copies, 293
Osmolytes, 85
Ozone, 112

P
Paenibacillus, 49, 65
  P. brasiliensis, 43
  P. validus, 48, 49, 64
PAM. *See* Periarbuscular membrane (PAM)
Panicum hemitomon, 294
Pantoea, 65
Parenthosomes, 353
Paris type, 3
Pathogenic infection, 142
Paxillus involutus, 64, 110, 182
PDA. *See* Potato Dextrose Agar (PDA)
PDB. *See* Potato Dextrose Broth (PDB)
Pear-shaped, 353
Peptidoglycans, 320
Periarbuscular membrane (PAM), 155
Peridium, 233
Pesticides, 321
Phalaris arundinacea, 297
Phanerochaete velutina, 182
Phenolic compounds, 48
Phosphatases, 50, 66
Phosphate, 332
Phosphate Solubilization, 66
Phosphate-solubilizing bacteria (PSB), 50
Phosphate uptake, 263, 354
Phosphodiesterase, 18
Phosphomonoesterase, 18
Phosphorus, 277, 292, 295, 297
  assimilates, 359
Photoassimilates, 182
Photobionts, 28
Photochemical efficiency, 297
Photosynthesis, 306
Photosynthetic capacity, 85
Photosynthetic rates, 316
Phragmites australis, 298
Phyllosphere, 65
Phylloxera, 283
Phylogenetic analysis, 308
Phylogenetic group, 262
Phytoalexin, 136, 154
Phytobionts, 40
Phytoestrogen, 161
Phytohormones, 48, 359
Phytopathogenic, 28
Phytopathogenic fungi, 35, 280
Phytophthora cinnamomi, 51
Piceirhiza bicolorata, 211
Pinaceae, 11, 188, 196, 245
Pinus banksiana, 16, 206
Pinus sylvestris, 182
Piriformospora
  P. indica, 28, 32, 351, 354–356, 365
  P. williamsii, 352
Pisolithus, 115
  P. tinctorius, 110, 202
Pisonia grandis, 103
Plantago lanceolate, 282
Plant diversity, 112
Plant–fungal interface, 155

Plant pathway, 14
Plant protector, 351
Polyaromatic hydrocarbons, 182
Polygonaceae, 11
Polyphosphates, 108, 265, 332
Polysaccharides, 48, 320
Population, 117–118
Posidonia oceanica, 338
Post-agricultural sites, 211–214
Potato Dextrose Agar (PDA), 29
Potato Dextrose Broth (PDB), 372
Pre-symbiotic phase, 47
Primary successional sites, 194–204
Primordium formation, 147
Productivity, 50
Propagules, 195
Prosopis juliflora, 365
Proteaceae, 4, 11
PSB. *See* Phosphate-solubilizing bacteria (PSB)
Pseudomonas, 43, 63, 65
  P. chlororaphis, 45
  P. fluorescens, 42, 45
  P. putida, 49
Pseudomycorrhizas, 103
Pteridophytes, 10, 40, 356
Pure culture, 124
Putrefaction, 178
Pyranoisoflavones, 151

Q
Quercetin, 146
Quercus
  Q. robur, 78, 210
  Q. suber, 78
Quinic acid, 36

R
Radices, 154
Raffinose, 49, 65
Random amplified polymorphic DNA (RAPD), 118
Recalcitrant, 307
  substances, 319
Reforestation, 238
Renewable, 189
Repeat SSR markers, 118
Resident spores, 206
Resistance
  heavy metals, 351
  high temperature, 351
Respiration, 306
Response to stress, 241
Retama sphaerocarpa, 52
Rhizobiaceae, 67
Rhizobia inoculation, 143
Rhizobial infection, 142
Rhizobial Invasion, 144–147
Rhizobial symbiosis, 150
Rhizobium, 65, 134
  R. leguminosarum, 42, 149
Rhizobium–legume symbiosis, 151
Rhizobiumlupini, 140
Rhizoctonia, 353
  R. solani, 51
Rhizoglomus melanum, 335, 337, 344
Rhizomorphs, 17, 103, 109, 124
Rhizophagus
  R. intraradices, 276, 282
  R. irregularis, 14, 276, 282, 336, 341
Rhizoplane, 65
Rhizopogon mycorrhizas, 180
Rhizosphere (RS), 34, 42, 62, 65, 136–137, 292, 334
Rhodococcus, 49
Root
  architecture, 179–180, 279
  colonization, 200
  exudates, 30, 32, 34, 36, 48, 310
  exudation, 86
  hairs, 298
  tips, 80, 208
Root-fungus association, 103
Root–microbe interactions, 137
Rootonic, 360, 362
RS. *See* Rhizosphere (RS)
R-selection, 191
Russula, 112, 113, 115

S
Salicaceae, 188, 196, 200
Salinity, 254
Salmonella, 66
Salt stress, 362
Sand dune systems, 200–201
Saprotrophic mycelia, 18
Saprotrophs, 178
Schoenus
  S. ferrugineus, 297
  S. nigricans, 297
Scleroderma, 115
Sclerotia, 103, 117, 206
Sebacinaceae, 351, 352
Sebacinales, 351, 352

Secondary metabolism, 36
Secondary metabolites, 41, 280
Secondary succession, 190, 204
Secondary successional, 192
sites, 204–214
Sedges, 292
Septoglomus viscosum, 276
Serendipita
S. herbamans, 352
S. indica, 28, 351, 365
S. williamsii, 352
Serendipitaceae, 352
Setaria pumila, 330
Sheathing mycorrhiza, 12, 103
Shikimate esters, 36
Shorea robusta, 104–106, 113
Siderophores, 19, 51, 66
Simple sequence repeat (SSR), 118
Singhal, Uma, 349–362, 365–383
Single-strand conformational polymorphism (SSCP), 118
Sinorhizobium meliloti, 140
SOC. *See* Soil organic carbon (SOC)
Soilborne diseases, 52
Soil-borne organisms, 340
Soilborne pathogens, 51, 161
Soil-Disturbed Sites, 201–204
Soil formation, 189
Soil glue, 313–314
Soil hypoxia, 341
Soil inoculum, 123
Soil matrix, 317
Soil microbiota, 210
Soil organic carbon (SOC), 318
Soil organic matter, 318
Soil richness, 178
Soil salinity, 294
Solanum lycopersicum, 28
Sorghum tricolor, 29
SoyaSignal, 161
Species diversity, 208
Species richness, 208
Species specificity, 64
Spilanthes calva, 357
Spinacia oleracea, 154
Sporangium-inhibitors, 51
Spore germination, 32
Sporocarps, 123, 195, 196, 199–201, 213
SSCP. *See* Single-strand conformational polymorphism (SSCP)
S-selection, 192
SSR. *See* Simple sequence repeat (SSR)
Starch-degrading enzymes, 359
Sterilization, 374
Steroids, 138
Streptomyces, 49
Stressor, 331
Stress tolerance, 359
Strigolactones, 47, 264, 309, 310
Submerged cultures, 376
Submerged environments, 337–338
Submerged soils, 340
Subsoil, 340
Succinate, 34
Sucrose gradient, 29
Suillus
S. bovinus, 64
S. sibiricus, 112
S. variegatus, 182
Sulfur proteins, 359
Sustainable agriculture, 10
Symbiosis, 178, 332
Symbiosome(s), 145, 147–148
Symbiotic, 196
association, 104
Symplasm, 20
Symplast, 87
Synergism, 359
Systemic acquired resistance, 280

T

T3SS. *See* Type III secretion system (T3SS)
Taraxacum officinale, 282
Taxonomical richness, 208
TEF promoter region, 353
Temperature stress, 362
Terrestrial ecosystems, 320
Thioglucosidase, 357
Thiram, 161
Tophthora diseases, 83
Topography, 207, 282
Translocation, 40, 110
Transplantation shock, 354
Trehalose, 44, 108
Trichoderma, 35
Tricholoma, 213
Trifolium repens, 139
2,4,6-trinitrotoluene, 182
Truffles, 232
Trypan blue, 31
Tsuga heterophylla, 209
Tuber aestivum, 236–238
Tuber borchii, 235–236
Tuberculate, 107
Tuber magnatum, 234–235
Tuber melanosporum, 123, 238–242
Type III secretion system (T3SS), 144
Typha latifolia, 294

V
Variovorax, 65
Vesicles, 12, 13, 283
Vibrio anguillarum, 66
Vibrio cholerae, 66
Vineyard management, 282
Vitamins, 138
Volatile organic compounds, 62
Volatiles, 48
Volcanic deserts, 197–199

W
Water-holding capacity, 201
Water regime, 292
Weed control, 283
Wetland, 292
Wetland plant, 297
Wilcoxina micolae, 211
Withania somnifera, 357

Y
Yersinia, 66
YPG media, 372

Z
Zizyphus nummularia, 365
Zygomycota, 232, 253

MIX
Papier aus verantwortungsvollen Quellen
Paper from responsible sources
FSC® C105338

If you have any concerns about our products,
you can contact us on
ProductSafety@springernature.com

In case Publisher is established outside the EU,
the EU authorized representative is:
Springer Nature Customer Service Center GmbH
Europaplatz 3, 69115 Heidelberg, Germany

Printed by Libri Plureos GmbH
in Hamburg, Germany